American Institute of Chemical Engineers

# 4th Topical Conference on Natural Gas Utilization 2004



Held at the 2004 AIChE Spring National Meeting

April 25-29, 2004
New Orleans, Louisiana, USA

**Printed from e-media with permission by:**

Curran Associates, Inc.
57 Morehouse Lane
Red Hook, NY 12571
www.proceedings.com

**ISBN: 978-1-60423-477-0**

**Some format issues inherent in the e-media version may also appear in this print version.**

American Institute of Chemical Engineers

# 4th Topical Conference on Natural Gas Utilization 2004

Held at the 2004 AIChE Spring National Meeting

AIChE

April 25-29, 2004
New Orleans, Louisiana, USA

American Institute of Chemical Engineers

4th Topical Conference on Natural Gas Utilization
2004

# Table of Contents

# T9 - Natural Gas Utilization

**The Mariscal Sucre LNG Project – a downstream perspective**

*Jesús Del Vecchio (Shell Venezuela), Emile de Jong (Shell Global Solutions)*

Between 1978 and 1982, substantial quantities of natural gas were discovered in the Caribbean Sea, north of the Paria peninsula, Venezuela. The resources are contained in 4 fields: Rio Caribe, Mejillones, Patao and Dragon, which collectively trace a linear footprint sequence east to west over a distance of some 95 km.

Following earlier initiatives to develop these fields, the Mariscal Sucre LNG (MSLNG) project was established in June 2002 with the signing of a Framing Agreement between the Sponsors and MEM, followed by a Preliminary Development Agreement (PDA) in November 2002. Respective holdings are PDVSA[1] (60%), Shell (30%), Mitsubishi (8%) and 2% reserved for Venezuelan entities. The project targets an initial LNG export level of 4.7 Mtpa into the US markets by 2009 via a newly built LNG plant. Further, the project provides for future supply of domestic gas of up to 300 mmscfd to internal Venezuelan markets.

Key challenges for the project were the need for a robust, cost competitive LNG process design, capable of dealing with upstream uncertainties in feedgas quality, volume of gas production, and development planning. An evaluation of LNG processes shows a very good fit between the MSLNG project and Shell's DMR process with large electric motors as drivers.

The LNG GameChanger™ design, based on the air-cooled Dual Mixed Refrigerant (DMR) process, has the flexibility to balance the refrigerant loads and therefore allows optimum use of the available driver power, under varying operational conditions and with changes in feedgas quality. Another feature of the design is that it employs a parallel compressor/driver line-up using more widely available equipment. The latter is enhanced by the choice for electric motors as drivers for the main compressors. The concept as foreseen for MSLNG offers a flexible design capacity up to at least 6 Mtpa, allowing a perfect match of plant size with upstream views on optimum gas production plateau level.

This paper provides background information on the project, current status and an overview of the key technical challenges from a downstream perspective.

[1] Petroleos De Venezuela, SA

## Introduction to the Mariscal Sucre LNG Project in Venezuela

Between 1978 and 1982, substantial quantities of natural gas were discovered in the Caribbean Sea, north of the Paria peninsula, Venezuela. The resources are contained in 4 fields: Rio Caribe, Mejillones, Patao and Dragon, which collectively trace a linear footprint sequence east to west over a distance of some 95 km.

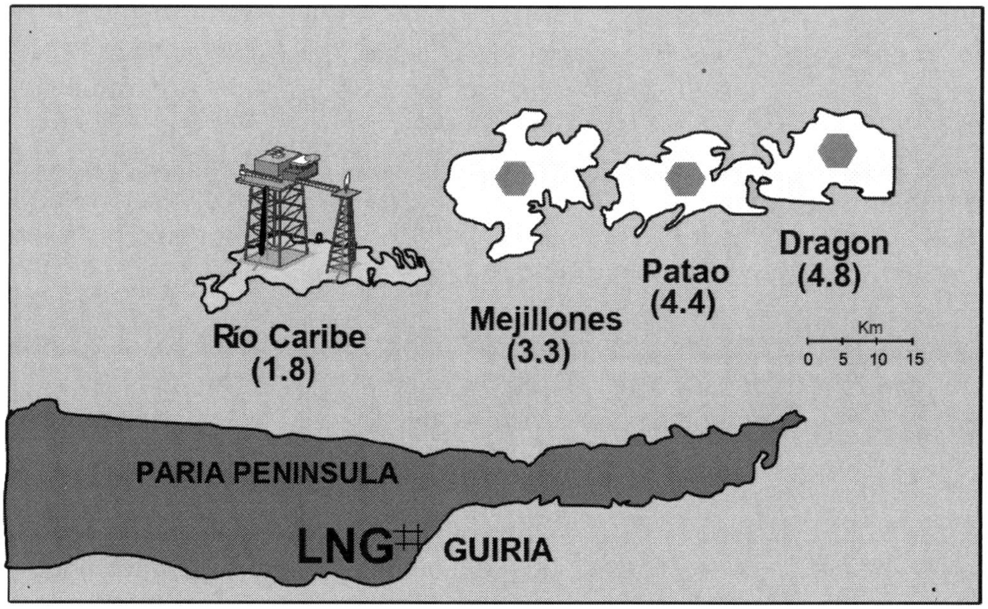


Figure 1. Location of fields

In 1988, Lagoven invited Shell to participate in formulating a development plan for these fields, which resulted in the joint delivery of a feasibility report for an integrated LNG project – Proyecto Cristobal Colon (PCC). Initially, this project aimed for 2 trains of 2.05 Mtpa each, and later 2 trains of 2.85 Mtpa each were considered. However, the PCC project turned out to be not economically viable against the then current economic parameters and was stopped in 1996. A development trail then followed until June 1999, when Shell submitted new proposals for a revised project to PDVSA[2] and MEM (Ministry of Energy and Mines). The basis for this proposal was a single train of 4.7 Mtpa, aircooled, and using the Dual Mixed Refrigerant process. This project, Venezuela LNG, was however cancelled in 2001.

Following a re-structuring of the venture, the Mariscal Sucre LNG (MSLNG) project was established in June 2002 with the signing of a Framing Agreement between the Sponsors and MEM, followed by a Preliminary Development Agreement (PDA) in November 2002. Respective holdings are PDVSA (60%), Shell (30%), Mitsubishi (8%) and 2% reserved for Venezuelan entities. The project targets an initial LNG export level of 4.7 Mtpa, into the US markets by 2009. Further, the project provides for future supply of domestic gas of up to 300 mmscfd to internal Venezuelan markets.

[2] Lagoven is no longer an entity and has been absorbed into the PDVSA organization

From a downstream perspective, key challenges for the project are:

- Plant location
- Uncertainty in feedgas composition and optimum gas production level
- Selection of appropriate technology
- Meeting Venezuelan expectations regarding national content and technology transfer

## Plant location and pipeline routing

In 2001 a site comparison study was carried out for the Venezuela LNG (VLNG) export plant. Starting from a long list of some 21 areas, a short-list was prepared with 8 possible locations, likely to be the most acceptable from an environmental, social, health and technical viewpoint. A further site comparison resulted in a recommendation to select the NE Güiria site as the area for the VLNG project development. Locations on the north side of the paria peninsula were discarded for environmental reasons, and the need for a breakwater. SW Güiria was nominated second best alternative.

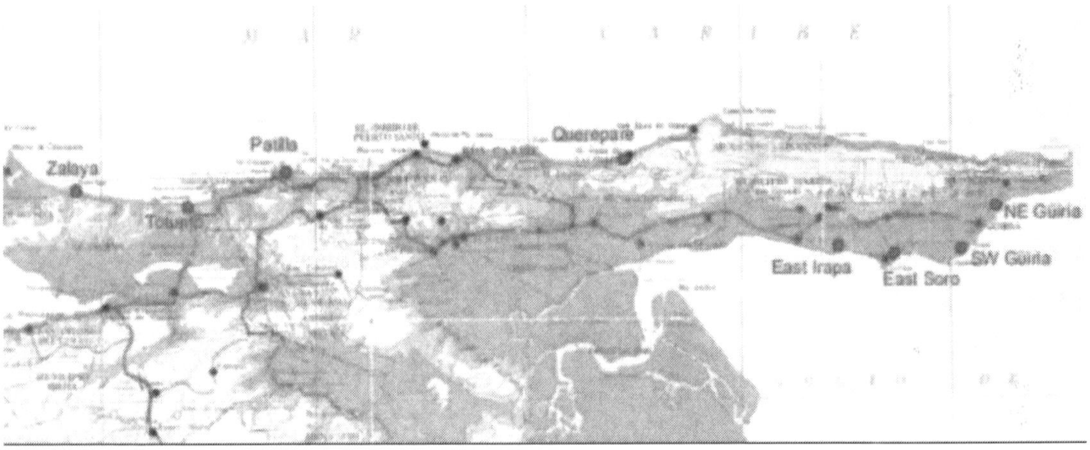


Figure 2. Possible site locations.

The LNG project triggered the idea of developing an industrial complex, which was then further studied by PDVSA and resulted in the Complejo Industrial Gran Mariscal Ayacucho (CIGMA) project. The fact that this required more area, and because of more gas in the area, further supported the choice for the site location in the gulf of Paria. However, the SW Güiria site now proved to be the preferred area because of the potential availability of a large piece of land. The CIGMA area (see Figure 3 below) comprises 6.265 ha of flat woodland with few inhabitants. Also the availability of deep water (>13 m.) near shore is in favour of SW Güiria. Within the complex, a preferred side was selected (refer to Figure 3 below) for the MSLNG project. Figure 4 shows an early impression of the aircooled LNG plant projected on the future site.

As the gas is located north of the Paria Peninsula, several options are considered for the pipeline routing.

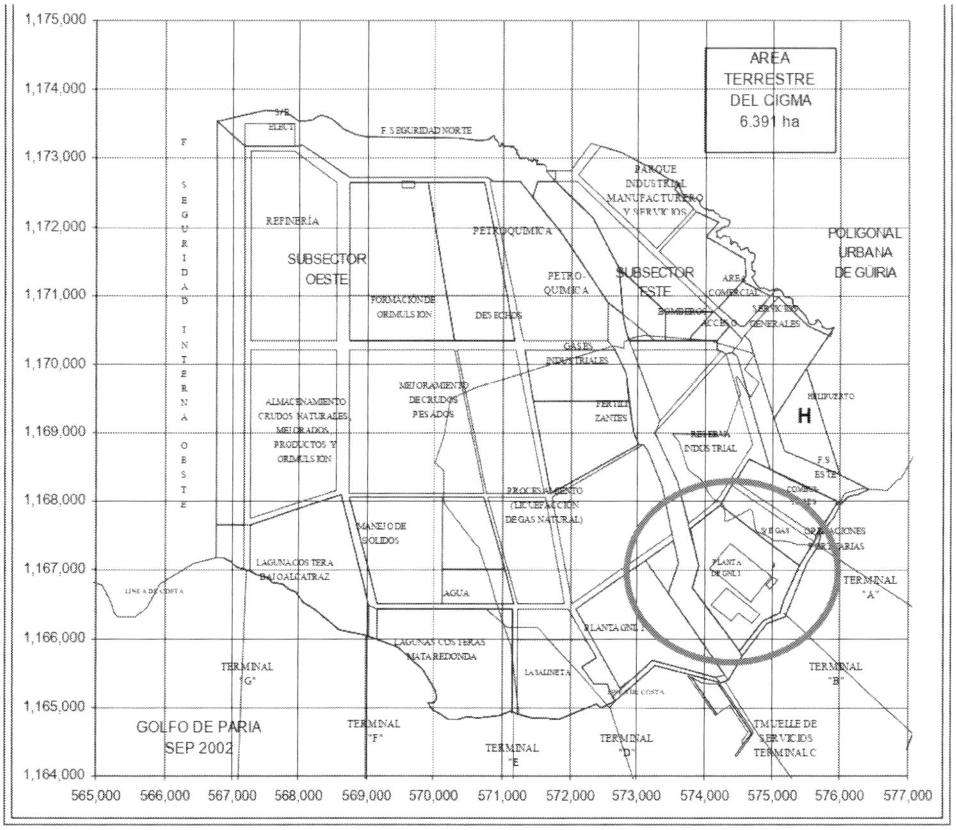


Figure 3. Location of sites within CIGMA area. (The circled area “Planta de GNL 3” is the selected site; “Planta GNL 1” is the alternative site)

Figure 4. MSLNG plant as foreseen near Guiria, Paria Peninsula, Venezuela

**Feedgas composition**

With the exception of the Rio Caribe field, the gas reserves from the other fields are very lean and do not contain significant quantities of propane and heavier. The positive aspect of this is that the LNG produced in the plant facilities is lean and, as such, ideally suited for the US market, which generally sets lower HHV specifications than the traditional markets in the Far East and Europe.

A drawback however is that the depletion of Rio Caribe field needs to be carefully managed. Acceleration of the Rio Caribe reserves, which is desirable to enhance condensate revenues, could leave the LNG plant without refrigerant make-up components at a certain point during the lifetime of the project. Accelerated Rio Caribe production, and hence import of refrigerant components is considered.

**LNG technology options considered**

For the liquefaction technology, the specific project challenges translated in the need for a robust, maximum value LNG process design, capable of dealing with upstream uncertainties in feedgas quality, volume and rate of gas production (influencing plant capacity).

The target capacity range was about 5 Mtpa. The main options considered in detail were the $C_3$/MR and the DMR processes. For both processes, gasturbines (GE F7) and electric motors (up to 60 MW shaft power into process) were considered as drivers.

In current baseload LNG liquefaction plants the $C_3$/MR process is the most widely used process. The most recent LNG train projects have been based on the application of 2 large GE-7 gas turbines for both the propane and MR compression service. The Shell Dual Mixed Refrigerant (DMR) process [1,2,7] was considered as alternative because of its potential. The use of a mixed refrigerant instead of propane in the precooling cycle elegantly overcomes the inherent limitations of the $C_3$MR design; the additional degree of freedom resulting from the use of two mixed refrigerant cycles allows full utilization of power in a design driven by gasturbines or electric motors. The DMR process uses spool-wound exchangers in precooling instead of kettles, reducing equipment count and plant complexity. This process has been selected for the Sakhalin LNG project [7] currently under construction, with an annualized design capacity of 10 Mtpa (two trains).

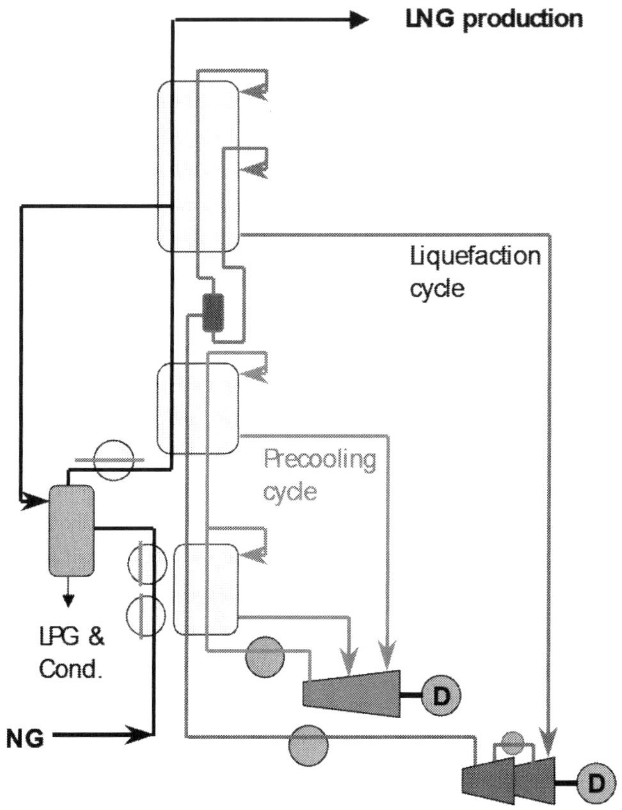

Figure 5. The Shell DMR process line-up.

Recent developments of the DMR process have further improved process efficiency. An example is the application of a three stage pre-cooling cycle. On a like for like basis, the DMR process efficiency calculated for tropical conditions is some 5% better compared to $C_3MR$, clearly confirming the efficiency potential of the DMR process [1,2,3]. As a result, this enables a production of 5.5 Mtpa LNG under tropical conditions, using 2 GE-F7 machines.

## The fit of the DMR electrical technology with the Mariscal Sucre LNG project

Electrically-driven LNG trains can compete with gasturbine-driven trains provided the increase in cost is compensated by the increased availability. It is noted that for a single train project at capacities up to 5 Mtpa there is little difference between a gasturbine driven design using 2 GE F7 machines, and an electrically driven design. However, the electric drive concepts are not bound by a maximum capacity of about 5 Mtpa because motor sizes are variable compared to the discrete power output of gasturbines. Further, there are other benefits of the electric drive like variable speed, less need to reduce refrigerant inventory for start-up (which helps with a lean gas), flatter and more continuous production profile. Furthermore, it is noted that the MSLNG project plans for more than one train. With this in mind, the electric drive route is very attractive. This because multiple trains can share a common powerplant, and economy of scale can be maximized in the powerplant. The choice for electric drivers also allows for maximizing future synergies across the various LNG plants and other industries on the CIGMA complex, supporting the Venezuelan ambitions.

Comparing $C_3$/MR and DMR, the DMR process allows for deeper pre-cooling, allowing full utilization of compressor and driver power in a design. Due to this, the DMR designs are showing both for mechanical and electrical drivers a larger capacity for similar capex (and hence a better project value). Another benefit of the flexibility of the DMR process is the possibility to better handle uncertainties and changes in feedgas composition, as anticipated in the future, and allow maximum plant performance once in operation.

Overall, it was found that a combination of 4 electric drivers with the benefits of the DMR process offers best value to the project. This concept, as developed by Shell [4,5,6], shows a configuration employing 2 parallel compressors in the precooling cycle and 2 parallel compressors in the liquefaction cycle (driven by 4 variable speed motors of equal power) (see Figure 6 below). The parallel line-up ensures an excellent plant availability (349 days per annum), and allows a large train with moderately sized equipment. The motor size is variable and determines the capacity (45-60 MW for 4.5 to 6 Mtpa). It is particularly attractive at capacities from 5 Mtpa upwards, where this design outperforms the traditional designs employing 2 large GE F7 gasturbines. The capacity flexibility up to 6 Mtpa allows for an optimum match between plant size and upstream gas production level for the MSLNG project.

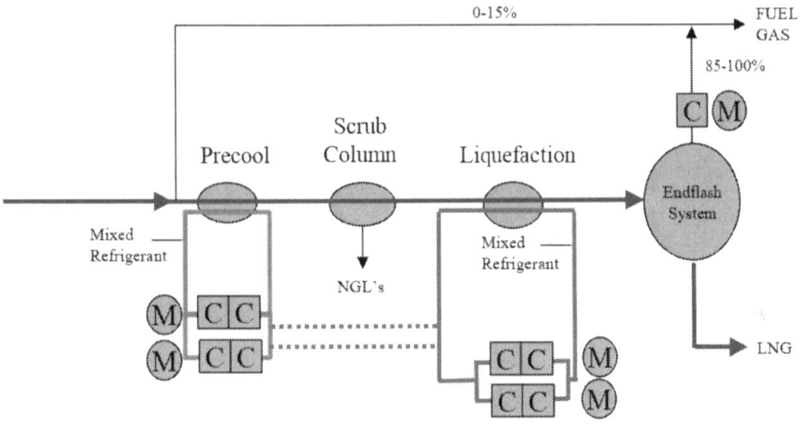


Figure 6. Configuration

## Way forward

Following the completion of the feasibility studies under the Preliminary Development Agreement and subject to a further agreement amongst the partners, a Basis of Design will be developed, followed by Basic Design Packages. It is expected that this will be done in 2004, to allow the start of a Project Specification phase early 2005.

## References

[1] Comparison of baseload LNG processes. Kees Vink, Rob Klein Nagelvoort. LNG12, 1998, Perth, Australia.

[2] Liquefaction process comparison of C3MR and DMR under tropical conditions. Rob Nibbelke, Steve Kauffman, Barend Pek, GPA March 2002, Dallas, Texas.
[3] Comparing five LNG processes. Henri Paradowski and Phil Hagyard, Technip. Hydrocarbon Engineering October 2003.
[4] Changing the LNG Game. Cas Groothuis, Dave Fletcher, Rob Klein Nagelvoort. LNG13, 2001, Seoul, Korea.
[5] The LNG GameChanger™ Technology. Eva Heyman, Alan Bliault, Barend Pek, GASEX 2002.
[6] Persian LNG – A Giant Awakes. Ate Visser, Colin Bowkley. World Gas Conference Tokyo 2003.
[7] Engineering design challenges for the Sakhalin LNG project. Wim Dam, Siew-Mung Ho. GPSA Conference San Antonio, March 2001, Texas.

## Contact details

Jesus Del Vecchio
Shell Venezuela
Torre Shell, Av. Principal de las Mercedes
Piso 12, Urb. El Rosal
Caracas, Miranda Venezuela,
tel. 58-212-955-4156, fax
Email: jesus.delvecchio.shell.com

Emile de Jong
Shell Global Solutions
Carel van Bylandtlaan 30
The Hague, The Netherlands, 30
tel. 31703771541, fax
Email: emile.dejong@shell.com

# BP'S BIG GREEN TRAIN: BENCHMARKING NEXT GENERATION LNG PLANT DESIGNS

*Jeff Sawchuk, Team Leader LNG and Products*
*Richard Jones, Project Director*
*BP*
*Houston, Texas 77253*

*Charles Durr, Technology Vice President, LNG*
*Keenis Davis, Process Manager*
*Kellogg Brown & Root, Inc.*
*Houston, Texas 77002*

*Prepared for Presentation at AIChE Spring National Meeting*
*4$^{th}$ Topical Conference on Natural Gas Utilization*
*New Orleans, LA, April 2004*

## ABSTRACT

Through the Big Green Train Project, BP with KBR as the engineering contractor undertook an extensive program that evaluated key aspects of design and engineering of a Baseload LNG Liquefaction plant. The study results provided a high level of confidence that "Next Generation" LNG plant designs are achievable and that the industry standards will be raised for future LNG plants based on design, capacity, life cycle cost and green house gas emissions. BP set aggressive targets and the project team met these challenges through close collaboration between the owner, contractor and process licensor.

The Big Green Train plant designs were based on LIQUEFIN™ liquefaction technology process that was developed by Axens/IFP, Institut Français du Pétrole. The project team set a program to determine if the other available liquefaction process technologies could deliver similar results. The team developed a benchmarking basis of design that contained sufficient level of detail on the design parameters without penalizing any of the LNG process technologies. Although this was a significant effort, it was required to ensure a fair and unbiased comparison of the technologies. This effort would not have been possible without the tremendous support of the LNG technology Licensors who participated.

This paper shows that any of the participating process Licensors' technologies can be used for "Next Generation" LNG plants. While process technology has some influence on the cost of a LNG plant, success is really achieved by the owner/contractor project team working in conjunction with the process licensor toward common goals.

## INTRODUCTION

Market pressures for lower cost liquefied natural gas (LNG) and BP's need to monetize its ever expanding gas resources were the driving forces behind BP's desire to develop "Next Generation" LNG technology that could set a new industry standard for future LNG plant designs based on plant capacity, capital cost per tonne of annual LNG, energy efficiency and green house gas emissions. In 2002, BP and KBR completed the Select Stage engineering of the Big Green Train (BGT) project which was essentially a Pre-FEED study for an LNG project. During this study, a significant number of engineering studies were performed to evaluate economical and technical drivers of LNG production facility designs. The LNG Licensor technology used for these studies was the Axens/IFP's LIQUEFIN™ process.

As part of the Select Stage work scope, an LNG process technology benchmarking effort was undertaken to compare the LIQUEFIN™ process to the available LNG processes. The following companies participated in the benchmarking effort:

- Black & Veatch Pritchard, Inc.
- Air Products and Chemicals, Inc.
- Linde AG

The objectives of the benchmarking were to gain a full understanding of the benefits and limitations of each process for achieving large capacity, energy efficient and low cost plant design, and to enable a technical and commercial rating of each of the processes with respect to the All Electric LNG plant design that was developed during the BGT project. The All Electric plant design was based on a power island that consisted of three Frame 7FA dry low $NO_x$ (DLN) gas turbine generators. Each is equipped with a two pressure level heat recovery steam generator (HRSG) to utilize the hot exhaust gas. One Frame 7FA and its HRSG were considered a spare to provide an N+1 sparing philosophy. This allowed for higher plant availability than gas turbine driven LNG plants.

## SUMMARY

General conclusions drawn from the benchmarking effort were that Big, Green and Low Cost LNG trains were achievable with all of the LNG process technologies that were evaluated. We also determined that all of the Licensor's LNG plant designs used in the benchmarking had the potential to be further optimized. Finally, $CO_2$ emissions were strongly influenced by the optimization of the fit between the power plant output and load factor, and the LNG plant power requirements.

One of the main activities of the BGT project was to evaluate a large range of refrigeration driver-compressor options. The various LNG plant designs with the different refrigeration driver-compressor configurations were then ranked based on metrics including life cycle costs, capital cost per tonne of LNG, energy efficiency, plant availability and $CO_2$ emissions. The BGT evaluation culminated with the selection of three distinct LNG plant designs that were studied in more detail.

Three key project objectives were established for the Select Stage of the BGT Project.

1. Determine the maximum capacity for the LNG plant based on next generation LNG technology, and identify any current bottlenecks and/or limiting equipment in the design of larger trains.
2. Reduce the current engineering, procurement, and construction (EPC) cost per annual tonne of LNG produced by 25 percent as compared to Atlantic LNG Train 1 (ALNG1).
3. Reduce the current level of $CO_2$ emissions per tonne of LNG capacity by 50 percent (compared to ALNG1).

## BENCHMARKING APPROACH

The approach to the Big Green Train benchmarking effort began with developing a process basis of design, which was based on the All Electric LNG plant design. This basis of design document was carefully designed to provide sufficient structure and constraints to ensure that the plant designs by the various process licensors were comparable without placing design restrictions that could result in a disadvantaged plant design. Achieving this balance was a critical element in this effort.

The design basis document set forth the specific site conditions, process data, fuel gas requirement, product specifications, compressor guidelines, and rundown and storage system parameters. A block flow diagram and benchmark deliverable documents were also prepared.

The project team approached the Licensors in individual discussions to determine their interest in participating in the BGT benchmarking effort. Meetings were held with each of the Licensors to provide background on BGT and provide the rationale for the benchmarking effort.

The process design basis, benchmarking deliverables and process block flow diagram were then provided to the participating LNG Licensors. The project team and Licensors then had regular interaction prior to their submission of the deliverables to help ensure that all the designs were on a consistent basis.

After receiving the submissions by each of the Licensors, follow-up meetings and conversations occurred to review and discuss issues as well as to obtain clarifications. After receiving the final designs from the Licensors, BP and KBR ran process simulations, and reconfigured the Licensor's designs to increase the consistency in the basis of designs.

Upon reaching a final process design, BP and KBR sent the process simulations to each of the Licensors for their comments on our design changes. The purpose of this was to reach consensus with the Licensors on the final design. Consensus was reached on the designs with each of the Licensors; however, each Licensor indicated areas that they would do differently. The changes were generally accepted by the Licensors and not believed to make a material difference.

## PROCESS DESIGN BASIS

The process design basis document set specific conditions and parameters that were to be adhered to in the LNG plant designs. Site specific information provided to the Licensors were the ambient temperature range, barometric pressure range, and relative humidity as provided in the following table:

**Table 1. Site Specific Information**

| Parameter | Minimum | Maximum |
|---|---|---|
| Barometric Pressure | 100.0 kPa | 102.1 kPa |
| Ambient Temperature | 22° C | 32° C |
| Relative Humidity | ~84 % | 100% |

The feed gas was to be delivered at the plant boundary at 4000 kPa, and had a gas composition as shown in the following table:

**Table 2. Feed Gas Composition**

| Component | Mol % |
|---|---|
| $N_2$ | 0.1 |
| $CO_2$ | 0.2 |
| $C_1$ | 93.9 |
| $C_2$ | 3.3 |
| $C_3$ | 1.2 |
| $C_4$ | 0.7 |
| $C_{5+}$ | 0.6 |
| Total | 100 |

Minimum propane, butane and pentanes plus recovery levels were specified as were the natural gas liquids (NGL) and LNG product specifications parameters as provided in the following table:

**Table 3. Product Specifications**

| | |
|---|---|
| LNG product specification | |
| High Heating Value | 1010 Btu/scf to 1075 Btu/scf |
| $H_2S$ | < 4 ppmv |
| Total sulfur | < 35 mg/m$^3$ |
| Nitrogen | < 0.5 mol% |
| $C_{5+}$ | < 0.1 mol% |
| Ethane product specification | |
| $C_1$ | < 1.0 mol% |
| $C_{3+}$ | < 1.5 mol% |
| Propane-Plus product specification | |
| $C_2/C_{3+}$ | < 2.0 mol% |

A total plant fuel gas maximum was provided to help ensure that the designs were comparable. A further restriction on the fuel gas was set so that if end flash was used the volume could not exceed 90% of the total fuel requirement.

## Compressor Guidelines

Compressor guidelines and restrictions were provided to the Licensors to ensure consistency for comparing the plant designs. Electric motors were required as drivers for all compressors. Total compression horsepower was set to 332 MW; however, certain constraints were set to control the allocation of the horsepower, including allocation to the individual refrigeration compressor string. The Licensor did have some discretion in the compressor service options within the constraints that were provided. The Licensors were also required to consider the following key parameters in their compressor design:

Aerodynamics

- Suction flow
- Mach number
- Head per stage
- Flow coefficient
- Impeller tip speeds
- $N/N_{critical}$

Mechanical

- Bearing size

- Bearing span
- Seal size
- Log decrement
- Casing design pressure
- Power/impeller

Efficiencies for compressors, expanders and pumps were also supplied to the Licensors for use in their simulations for consistency purposes.

### Cooling Medium

Air-cooling was chosen as the cooling medium, and the minimum approach temperatures for the following were specified to ensure consistency between Licensor assumptions:

- Sensible cooling
- Condensing services
- Subcooling services

### LNG Rundown and Storage

Providing an end flash system was left to the discretion of the process Licensor. The Licensor was required to calculate the amount of boil-off gas from the LNG storage and loading area based on supplied parameters for flashing, heat leak, pump energy and vapor displacement.

### Emissions

Consistent with BP commitment of no harm to the environment, no venting or flaring was allowed except for emergency situations.

### Block Flow Diagram

The following block flow diagram was provided to the Licensors to provide an indication of the requirements for the LNG facility.

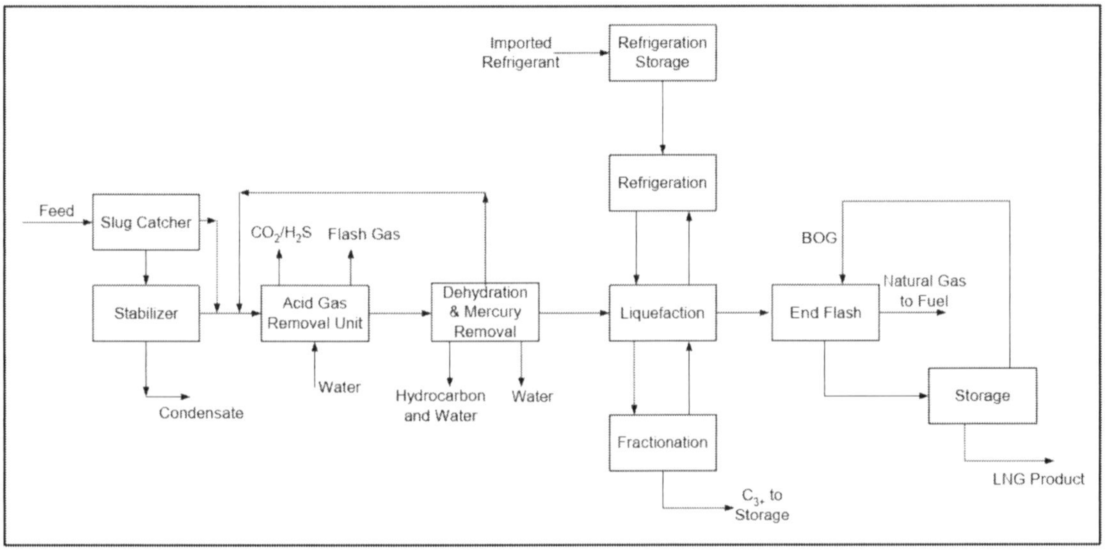


**Figure 1. Block Flow Diagram**

## Option for Alternative Design

An option for the Licensor to provide an alternate design was provided to allow them to submit an additional plant design that could afford benefits that could not be offered with the conforming design. This design would then be considered on its own merits and not used in the benchmarking. The alternative design was also required to be within the specified total power and certain other constraints.

## Evaluation Priorities

The priorities used to evaluate the LNG process technologies included life-cycle economics, specific power, capital cost (Capex) per annual tonne of LNG production, $CO_2$ tonne per tonne of LNG production, modular plant design for ease of expansions, overall plant availability, and plant efficiency (defined as energy content of the total products divided by the energy content of the plant feed stream). By setting the evaluation priorities, the evaluation of the Licensor designs was easily accomplished by the project team.

## Licensor Proposal Content

Each process Licensor was required to provide technical and commercial deliverables to facilitate the evaluation of their designs. This information was considered critical to the benchmarking effort.

The technical deliverables included general discussion of process, capacity, process flowsheet showing primary process control and brief description, heat and material balance, sized equipment list, process power consumption list, process data for the refrigerant compressors, main liquefaction heat exchanger duties, flows, approximate size, weight and any additional information the Licensor felt was pertinent to the evaluation such as experience list, depressurization requirements for short and long term plant stoppages, start-up times and flaring estimates during start-ups.

Commercial information such as pricing for any proprietary equipment, licensing fee, overall schedule for any proprietary equipment and engineering services, process and mechanical warranties, process efficiency and plant availability was also requested.

## PROCESS DESCRIPTIONS

All the Licensors used and accounted for feed gas compression, feed pretreatment, demethanizer or scrub column overhead compression, an end flash system and boil off gas (BOG) compressors.

### IFP/Axens

As previously mentioned, the BGT plant designs were based on the IFP/Axen's LIQUEFIN™ process, which is a dual mixed refrigerant process. A description is provided in the following text.

For the LIQUEFIN™ process, dry natural gas feed from the dehydration unit was cooled down and partly condensed before being sent to the demethanizer system where an expander-recompressor was used to drop the pressure such that component separation could occur. A deethanizer system was utilized to achieve the $C_2$ content and vapor pressure specification of the $C_{3+}$ cut. A suitable $C_2$ stream was produced to be used as make-up for the two mixed refrigerant circuits, MR1 and MR2. The excess $C_2$ was chilled and re-injected in the LNG; however, the system was designed so that the $C_2$ could be exported in the future as a product.

At the outlet of the cryogenic exchanger, the high-pressure liquefied natural gas was sent to an LNG turbine, where the pressure was reduced and electricity was generated. The cold liquid from the turbine was partially vaporized as it passed through a Joule-Thompson (J-T) valve and sent to the end flash system. The liquid phase separated was sent by means of the LNG pumps to the LNG storage tanks.

A heavy mixed refrigerant (MR1) was used in the pre-refrigeration cycle. Electric motors were utilized to drive parallel compressors used for the pre-refrigeration cycle. The compressor discharge was desuperheated, condensed and subcooled using air coolers.

The main heat exchanger was composed of parallel cold boxes, each containing several cores (plate fin heat exchangers) in parallel for the pre-refrigeration exchanger, and the same number of cores in parallel for the cryogenic exchanger. Multiple levels of cooling were provided in the pre-refrigeration exchanger.

A light mixed refrigerant (MR2) was used in the cryogenic cycle. Identical electric motors were used to drive the parallel MR2 compressors. The air-cooled MR2 compressor discharge was then sent to the main exchange line in the pre-refrigeration/cryogenic cold boxes.

At the outlet of the cryogenic exchangers, the subcooled MR2 liquid from the different cold boxes was collected and sent to a liquid turbine to extract energy and generate electricity. A typical LIQUEFIN™ process scheme is illustrated in Figure 2.

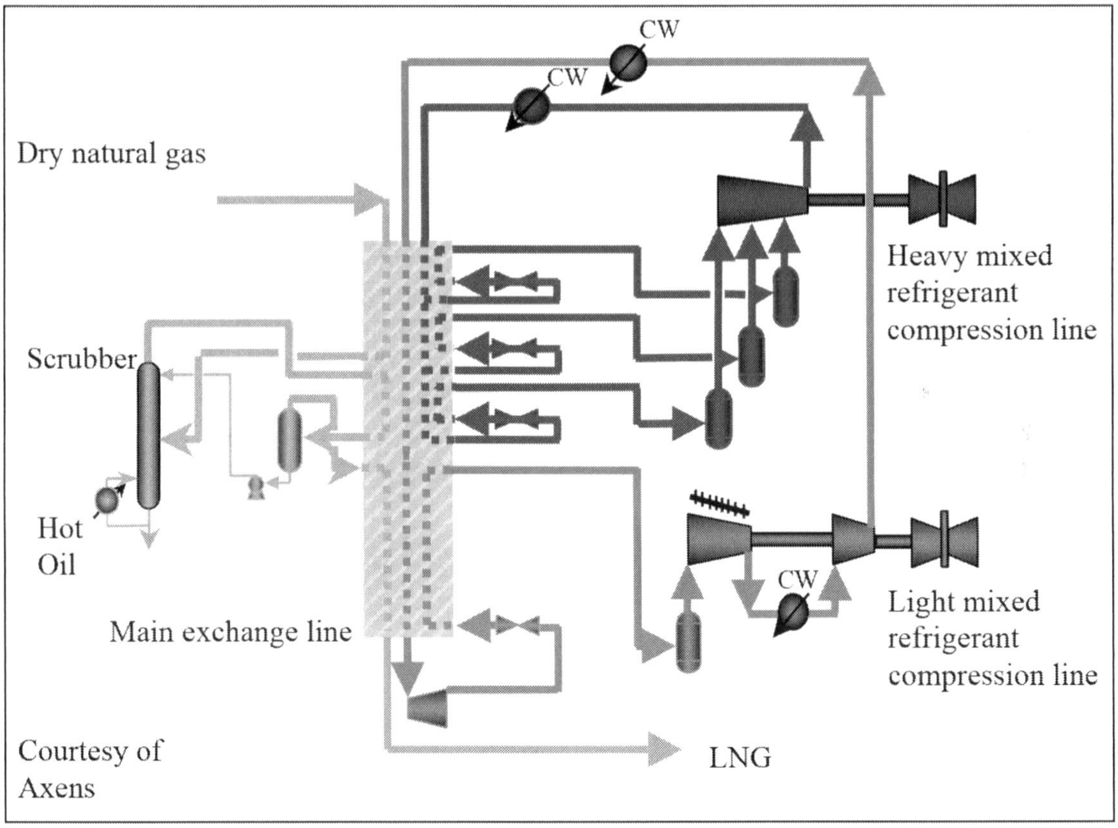


**Figure 2. Axens LIQUEFIN™ Process**

## Black & Veatch

Black & Veatch proposed their proprietary single mixed refrigerant process PRICO® for their BGT plant design for the benchmarking. A description of their liquefaction process is below.

In the PRICO® process, dry natural gas feed from the dehydration unit was cooled down and partly condensed in the refrigeration exchanger before being sent to the demethanizer system where an expander-recompressor was used to drop the pressure such that component separation could occur. A deethanizer system achieved the $C_2$ content and vapor pressure specification of the $C_{3+}$ cut and prepared a suitable $C_2$ stream to be used as make-up for the mixed refrigerant circuit. The excess $C_2$ was chilled and re-injected in the LNG; however, the system was designed so that the $C_2$ could be exported in the future as a product.

The compressed demethanizer overhead vapor was sent to the refrigerant exchanger to be liquefied and subcooled. At the outlet of the refrigerant exchanger, the high-pressure liquefied natural gas was sent to an LNG expander, where the pressure was reduced and energy extracted in the form of electricity. The cold liquid from the turbine was partially vaporized as it passed through a J-T valve and sent to the end flash system. The liquid phase was pumped to LNG storage tanks using the LNG transfer pumps.

Electric motors were used to drive the parallel refrigerant compressors on the mixed refrigeration cycle. Both the liquid and vapor from the refrigerant discharge separators were sent to the refrigerant exchanger in parallel with the dry feed. The refrigerant exchanger was composed of parallel cold boxes, each containing several cores (plate fin heat exchangers). The mixed refrigerant was cooled to the same temperature as the LNG, flashed to a lower pressure with a J-T valve, and re-injected into the refrigerant exchanger to provide refrigeration for LNG production.

Figure 3 is a schematic of a typical PRICO® process.

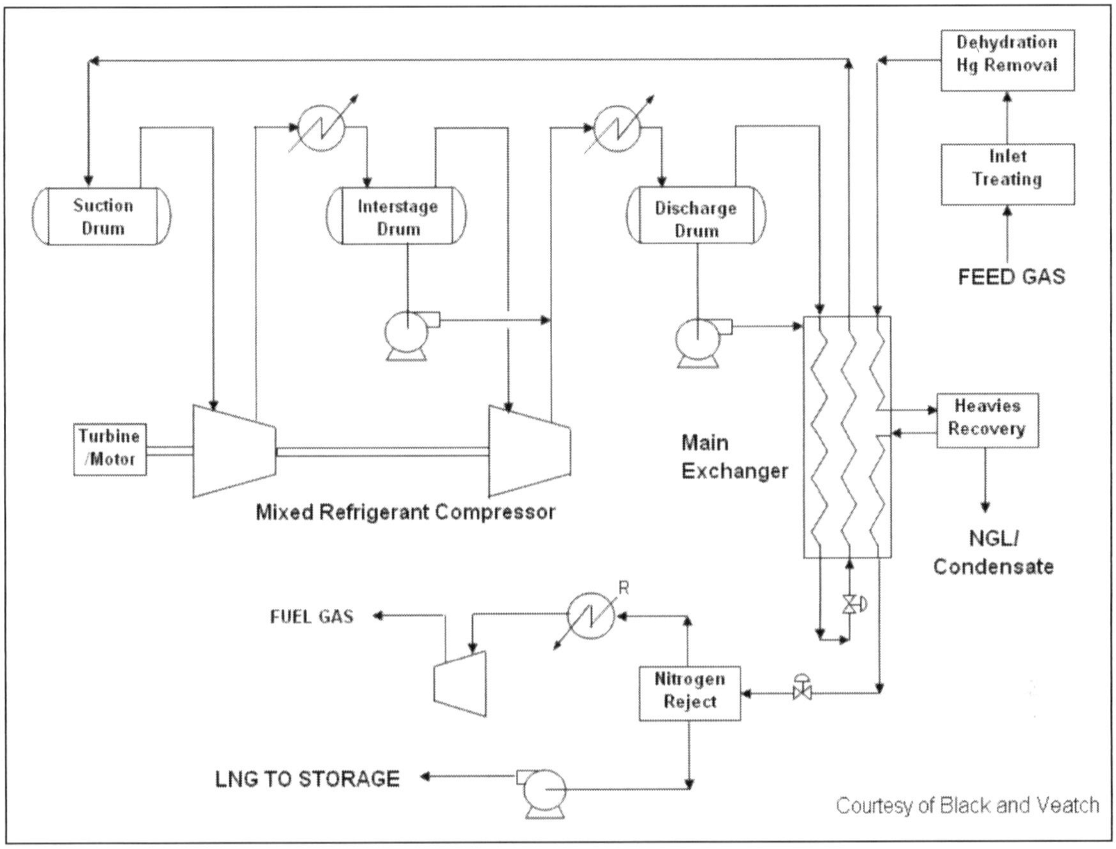


**Figure 3. Black & Veatch PRICO® Process**

## APCI

APCI proposed their AP-X™ process for their BGT plant design. The AP-X™ process is a propane, mixed refrigerant, and nitrogen based LNG liquefaction process. A description is provided in the following discussion.

In the AP-X™ process, dry natural gas feed from the dehydration unit was cooled and partly condensed by propane refrigeration and heat integration. The stream was then sent to the scrub column system where heavier components were separated from the gas to be liquefied. Vapor from the column was fed to the main cryogenic heat exchanger (MCHE) to continue the chilling and liquefaction process.

Air Products used an integrated lean oil scrub system for recovering refrigerant make-up and LPG components. LPG recovery could alternately be achieved using a front-end expander plant, an integrated expander plant, or a conventional scrub column depending upon the desired recovery.

The high-pressure liquefied natural gas from the MCHE was directed to the LNG subcooler to be cooled further with cryogenic nitrogen. The LNG was then sent to the LNG hydraulic turbine, where the pressure was reduced and energy extracted in the form of electricity. The cold liquid from the turbine was partially vaporized as it passed through a J-T valve and was sent to an end flash system. The liquid phase was pumped to LNG storage tanks using the LNG transfer pumps.

Electric motors were used to drive the compressors on the propane refrigeration cycle. The propane refrigeration system used a progression of stages/pressure levels to chill and/or condense the process streams. The compressor discharge was air-cooled to desuperheat, condense, and subcool the propane refrigerant.

Electric motors were used to drive the compressors for the mixed refrigerant (MR) system also. The MR was chilled, condensed and subcooled before flowing through the MR hydraulic turbine, where the pressure was reduced and energy extracted in the form of electricity. The cold liquid from the turbine was partially vaporized as it passed through a J-T valve and was sent to the top of the MCHE to provide refrigeration.

The nitrogen refrigeration system provided the final level of subcooling for the high pressure LNG before it was flashed and sent to storage. Electric motor driven compressors and expander/recompressors or companders are used for this system. The refrigerant remained single phase throughout the cycle.

A typical schematic for AP-X™ process is provided in Figure 4.

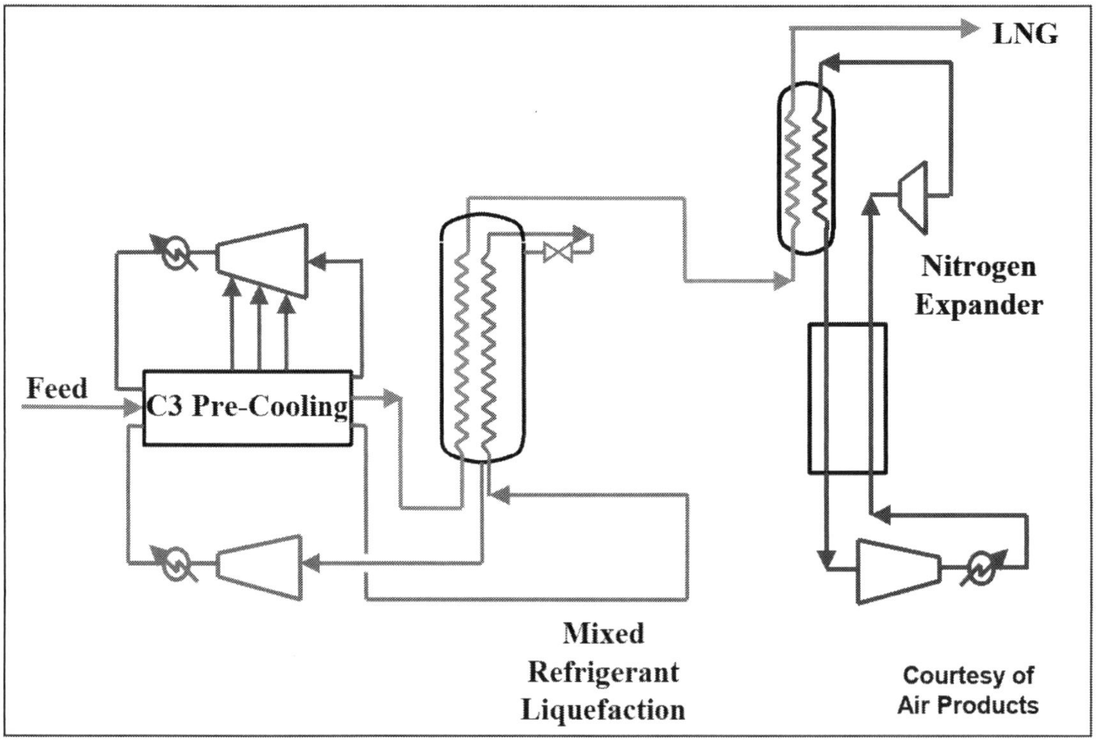


**Figure 4. APCI AP-X™ Process**

**Linde AG**

Linde AG proposed their mixed fluid cascade (MFC®) process, which utilized a triple mixed refrigerant LNG liquefaction process, for their design in the BGT benchmarking effort. A description of the Linde process is provided in the following text.

In the MFC® process, dry natural gas feed from the dehydration unit was cooled down and partly condensed and separated. The vapor flowed through the feed gas expander and on to the recontactor column. The purpose of the recontactor column was to scrub heavier components from the gas to be liquefied. Vapor from the column was compressed and fed forward for liquefaction and subcooling. The bottoms liquid from the column was fed to the deethanizer.

The deethanizer achieved the $C_2$ content and vapor pressure specification of the $C_{3+}$ product. The deethanizer bottoms liquid was sent to fractionation to recover propane, butane and $C_{5+}$ products. Fractionation was required for the MFC® process so that a portion of the butane could be chilled and recycled to the recontactor column for recovering LPGs. Ethane was re-injected into the LNG when a separate product was not required.

The compressed vapor from the recontactor was cooled before being liquefied and subcooled using MR in the spiral wound liquefier and subcooler exchangers. At the outlet of the subcooler, the high-pressure liquefied natural gas was directed to the LNG hydraulic turbine, where the pressure was reduced and energy extracted in the form of electricity. The cold liquid from the turbine was partially vaporized as it passed through a J-T valve and was sent to an end flash system. The liquid phase was pumped to LNG storage tanks using the LNG transfer pumps.

The precooling mixed refrigerant (MR1) was used to handle the higher temperature refrigeration services in the MFC® process. Electric motors powered these refrigerant compressors; and parallel cold boxes each containing several core exchangers (plate fin heat exchangers) were used for the heat transfer. The liquefaction mixed refrigerant (MR2) was used to handle the intermediate temperature refrigeration services in the MFC® process. In particular, this included the liquefaction range for the high-pressure natural gas. Electric motors powered the liquefaction cycle compressor and liquefaction occurred in a spiral wound liquefier. MR3 was used to handle the low temperature refrigeration services in Linde's process, which consisted of subcooling LNG and MR3. Electric motors were also used to drive the subcooling cycle compressors. A spiral wound subcooler was used for the heat transfer.

Figure 5 represents a typical MFC® process scheme.

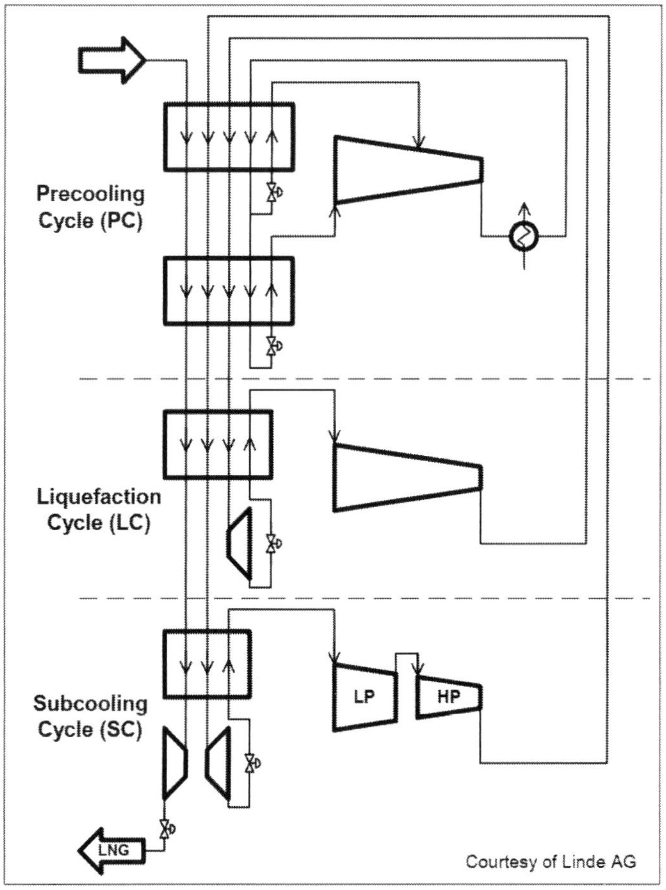


**Figure 5. Linde AG MFC® Process**

## RESULTS

The results of the benchmarking showed that all of the Licensors can provide Big (large capacity) LNG plants, which is illustrated in the following graph. We feel that further optimization can be performed with each of the Licensor's plant designs. The optimization is expected to result in a significant narrowing of the range of plant capacities. Consequently, Big LNG plant designs can be

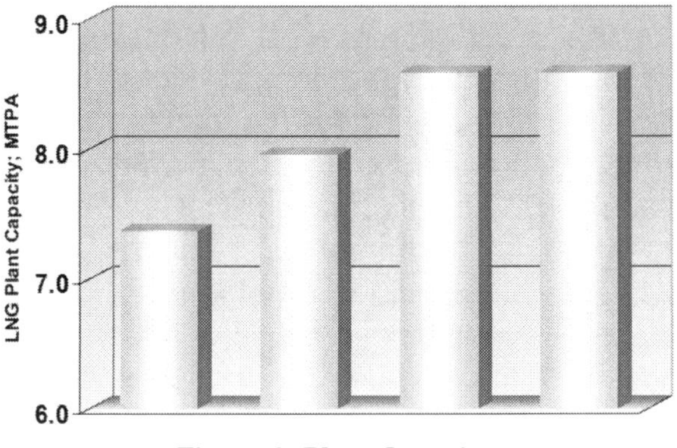


**Figure 6. Plant Capacity**

achieved with all of the LNG process technologies that were evaluated. The typical sweet spot issues that have differentiated the process licensors at other capacity ranges did not come into play for the BGT design.

The results of the $CO_2$ emissions metrics for the BGT benchmarking designs showed that all of the LNG process technologies evaluated generated world-class emission levels that are truly "Next Generation" quality. The net result showed that two of the designs met the 50 percent $CO_2$ reduction target versus Atlantic LNG Train 1, and the other two designs provided well over a 40 percent reduction. With additional optimization, the emission levels of all the LNG process technologies are expected to equalize at or above the 50 percent $CO_2$ reduction.

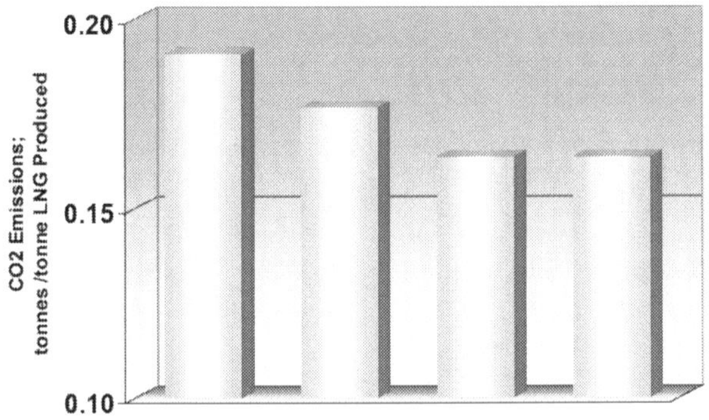


Figure 7. CO2 Emissions

Auto-consumption for the plant designs is a measure of the energy efficiency of the LNG plant designs. The designs developed during the benchmarking effort yielded energy efficiencies that are far superior to the existing LNG plant designs, and reflect what the project sponsors should expect for energy efficiency for their future LNG facilities. By focusing on the design philosophies and expectations regarding energy efficiency early in the development of the LNG project, owners, the engineering contractor and the process Licensor can jointly develop "Next Generation" LNG plant designs that are very energy efficient, and have ultra-low $CO_2$ emission levels.

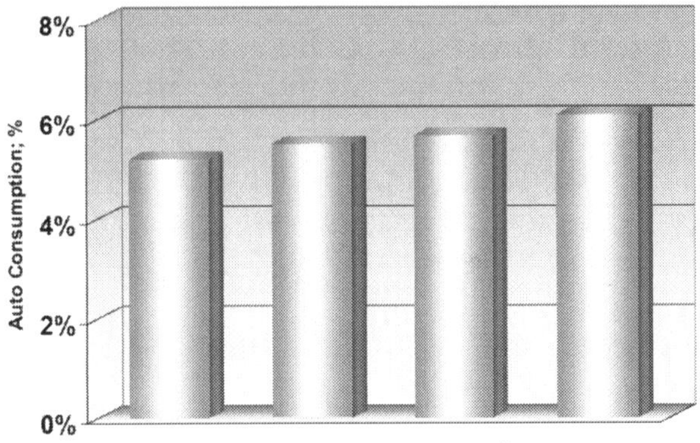


Figure 8. Autoconsumption

The benchmarking results for capital cost reduction generated for the different LNG process technologies show that all of the BGT designs would set new cost benchmarks in the industry. The following chart illustrates the range of cost reductions in EPC cost per annual tonne of LNG for the

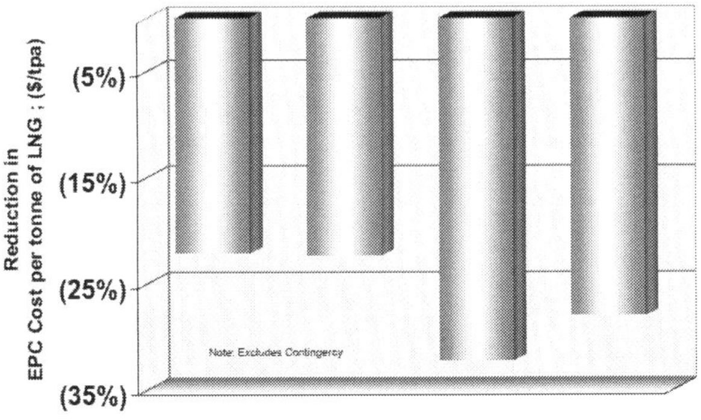


Figure 9. EPC Cost Reduction

Licensors, and demonstrates that nearly all meet the cost reduction metric of twenty-five percent in EPC Cost per annual tonne of LNG. All of the LNG plant designs have room for further reduction in cost through additional optimization. Consequently, all of the plant designs are expected to meet or exceed the twenty-five percent cost reduction versus Atlantic LNG Train 1.
Note that a random order was chosen for depicting the various process Licensors in each of the preceding bar graphs.

## CONCLUSIONS

All of the LNG process Licensor technologies evaluated demonstrated that they could provide Big, Green and Low Cost LNG trains. This is evident since the metrics set forth in the BGT project have been or can be met by all of the process licensors.

BGT LNG plant designs are not LNG process dependent.

The success of the benchmarking efforts was a direct result of the strong cooperation of the owner, contractor and process licensors. Creating an environment of openness and trust early in the project between the project sponsor, the engineering contractor and the LNG process Licensor is a key element required to achieve the technical step changes necessary to achieve "Next Generation" LNG plant designs. Early cooperation can also provide the highest quality plant design and generate the highest project value to all of the participants. Typically the owner and the contractor have a greater understanding of the overall LNG project and equipment issues. Therefore it is important that their experience be drawn upon for the development of the licensor's design basis.

## ACKNOWLEDGEMENTS

The authors wish to express their appreciation to Air Products and Chemicals, Inc.; Axens; Black & Veatch Pritchard, Inc; BP Trinidad and Tobago; General Electric Power Systems; Institut Français du Pétrole; Kellogg, Brown and Root project team; Linde AG; Nuovo Pignone; and Siemens for their tremendous support and efforts associated with the BGT Project.

## BIO for RICHARD JONES

Richard Jones holds a Bachelor of Science in Petroleum Engineering from The University of Tulsa, and has been with BP for 28 years. He has held various positions over his tenure in upstream operations; drilling and completions; reservoir engineering; gas supply optimization; negotiator for gas pipeline, gas processing and gas sales settlements and agreements; international business development; development of power generation business; and project management of GTL and LNG value-chain projects. Major international projects have included LNG regas in South America, Europe and Asia; LNG liquefaction in Africa and the Middle East; Gas-to-Liquids plant in Middle East, upstream field development in Middle East; and power generation plants Asia. Current assignment is Project Manager for developing BP's Next Generation LNG technology and plant designs via Big Green Train Project, and BP's Facilities Representative for the Angola LNG project.

# LOWERING LNG UNIT COSTS THROUGH LARGE AND EFFICIENT LNG LIQUEFACTION TRAINS – WHAT IS THE OPTIMAL TRAIN SIZE?

*Anthony Eaton, Rick Hernandez, and Allyn Risley*
*ConocoPhillips Company*
*600 N. Dairy Ashford*
*Houston, TX 77077*

*Phil Hunter, Amos Avidan, and John Duty*
*Bechtel Corporation*
*3000 Post Oak Blvd.*
*Houston, TX 77056*

*Prepared for Presentation at AIChE Spring National Meeting*
*4th Topical Conference on Natural Gas Utilization*
*New Orleans, LA, April 2004*

## Abstract

The international LNG trade is undergoing an unprecedented period of growth, and LNG plant owners keep searching for ways to lower costs by benefiting from economies of scale. Bechtel has recently designed and is constructing the world's largest LNG train, with a nominal capacity of over 5 million tonnes per annum (mtpa), utilizing the Phillips Optimized Cascade LNG Process. Bechtel and ConocoPhillips have also engaged in a detailed study of larger LNG trains through our LNG Product Development Center, and conclude that LNG trains as large as 8 mtpa are feasible, and could be cost-effective.

Although such "mega" LNG trains are feasible they may not be suitable for everyone. We see LNG plant designs in the coming decade falling broadly into three groups, with nominal capacities around 4 mtpa, 5 mtpa, and 8 mtpa. These different train designs are likely to co-exist in the coming years, as different owners find it advantageous to choose one vs. another. The basic design of 4 mtpa trains (until recently considered very large trains) will continue to suit new grassroots projects with limited gas supply or committed gas sales, or in cases where copying an existing train design is the most cost-effective option. Approximate 5 mtpa trains could be suited for expansion or grass roots projects that require lower LNG unit costs, and have the available gas supply and sales. An 8 mtpa "mega" train would be most suitable for an expansion train at a very large existing LNG complex targeting a large, distant market, with virtually unlimited gas supply.

One question facing designers of large-capacity trains is: "How does a single 8 mtpa train compare in cost and operational flexibility vs. two 4-mtpa trains?" Our study to answer this question was based on the Phillips Optimized Cascade LNG Process with three refrigerants (propane, ethylene, and open-loop methane). The study explored innovative configurations based on Frame 7 and Frame 9 gas turbines, liquid and gas expanders, utilization of heat recovery systems and electric drives. Plant capacities studied ranged from 3.3 mtpa to 8 mtpa.

We have concluded that large trains can be feasible and cost-effective. The compressors utilized in our studies represent some of the world's largest but they meet acceptable guidelines for major design parameters. Other design parameters investigated included: feasible line and valve sizes, size of heat exchangers, cold boxes and other key equipment.

One major concern with "mega" LNG trains is significant reduction in LNG production when one of the refrigerant compressor trains goes out of service. This problem can be mitigated with the Phillips Optimized Cascade LNG Process "two-train-in-one" reliability concept which allows the LNG train to operate at up to 75% capacity even when one compressor string trips. LNG trains utilizing this design have demonstrated on-stream availability greater than 95%, the highest in the LNG Industry. The outstanding safety, reliability, and high operating factors of the LNG industry keep building up confidence in larger trains, and the trend to design larger trains will keep accelerating.

## Introduction

The international LNG trade continues expanding rapidly. The trend toward even larger capacity plants is a drive in the industry. LNG plants in the future will have nominal capacities up to 8.0 MTPA. This rising capacity requires major focus on train configuration, compressor selection, and economies of scale. This can be seen in Figure 1.

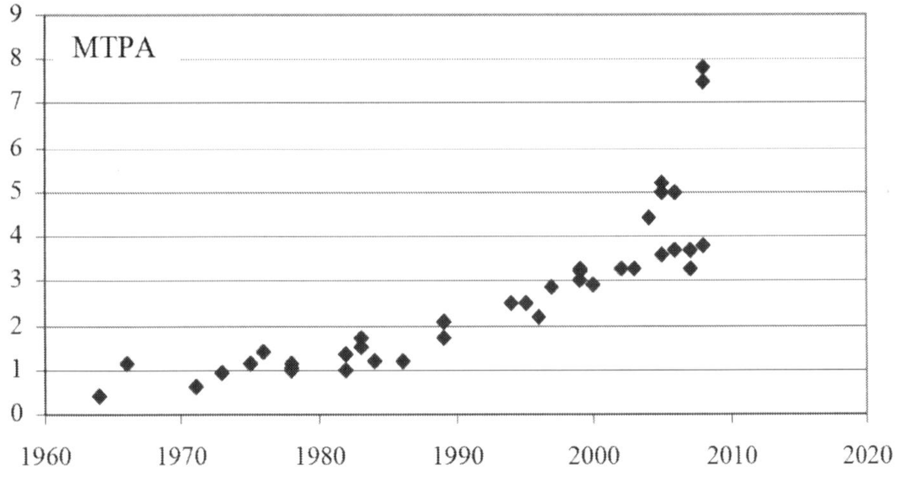


**Figure 1.** Trend in LNG train size (MTPA) in the future years

As the LNG trade increases and becomes more than a small niche-market, owners continue to look for ways to lower costs by benefiting from economies of scale. The effort has concentrated, as plant capacity grows, on building larger single LNG train plants.

We have concluded that large trains are feasible and cost-effective. To achieve the desired results, it requires innovation, "out of the box" thinking, technological rigor, and a "can-do" attitude. For example, the compressors utilized in our studies represent some of the world's largest, yet they meet all design parameter guidelines.

Our study focused on the scale-up of the Phillips Optimized Cascade LNG Process (POCLP) to achieve the higher plant capacity at approximately 8.0 MTPA. Phillips Petroleum Company developed the original Optimized Cascade LNG process in the 1960's. The objective was to develop a liquefaction technology that permitted easy start-up and smooth operation for a wide range of feed gas conditions. This process was first used in 1969 at Phillips Petroleum Company's Kenai, Alaska LNG facility. The facility was constructed by Bechtel and was the first to ship LNG to Japan. It is the first LNG project in the world to achieve 34 years of uninterrupted supply to its Japanese customers. Figure 2 provides an overall schematic of a typical POCLP.

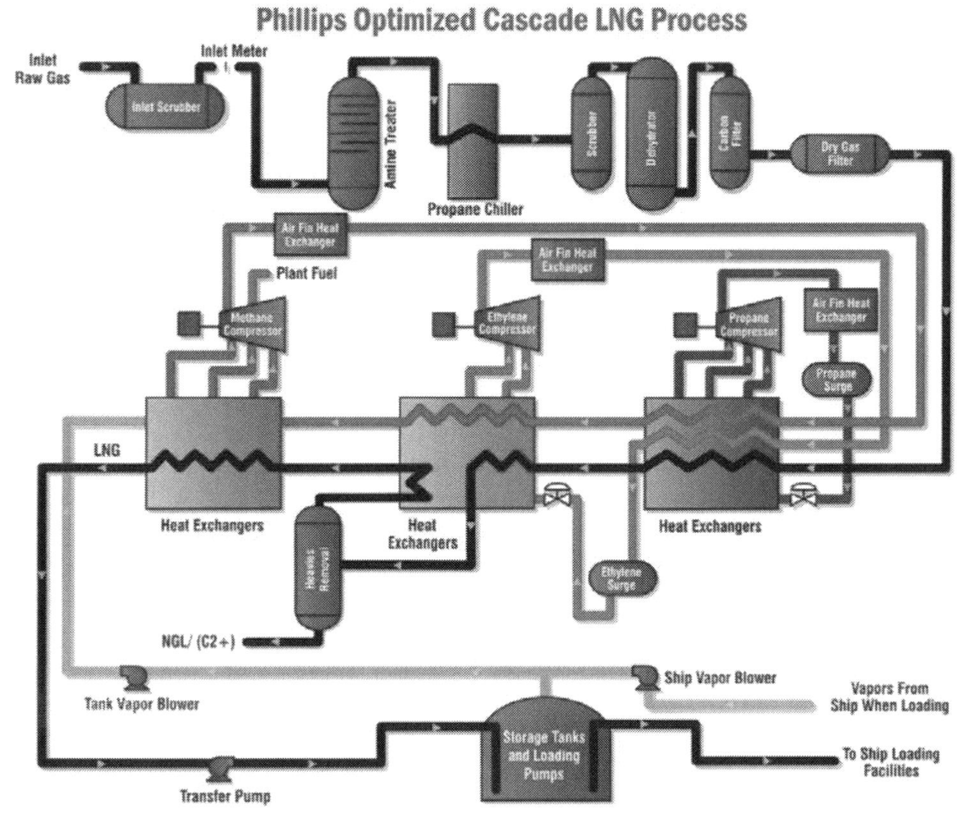


**Figure 2.** Phillips Optimized Cascade LNG Process – Train Size 1.0 to 8.0 MTPA

The POCLP is able to provide designs with high thermal efficiency and achieve a design that is optimized for project economics. The process utilized proven technology and equipment and has a wide range of operations flexibility. The operational flexibility can be seen in Table 1. Turndown rates to 10% are achievable for long-term operation. Due to the pure component systems, the plant provides an ease of startup and operation. The plants boast a low utility requirement and reduced flaring requirements since the refrigerants are not flared on typical upset conditions. This leads to reduced requirements for maintenance and operations staffing.

| Operational Flexibility – In Percent of PFD Design Production | | |
|---|---|---|
| | Low | High |
| Overall | 10 | 105 |
| Full Rate | 90 | 105 |
| One Turbine/Compressor Down | 60 | 80 |
| Half Rate | 30 | 60 |
| Idle | 10 | 30 |

**Table 1.** Operations Flexibility in the Phillips Optimized Cascade LNG Process

The POCLP also provides for a "2-Trains-in-1" reliability option to offer high availability. Recent reliability and availability analysis shows overall plant availability of greater than ninety five percent. This has been demonstrated in operation at Kenai for over 34 years, and in Atlantic LNG for over 5 years. Figure 3 provides an overall layout of the "2-Train-in-1" concept.

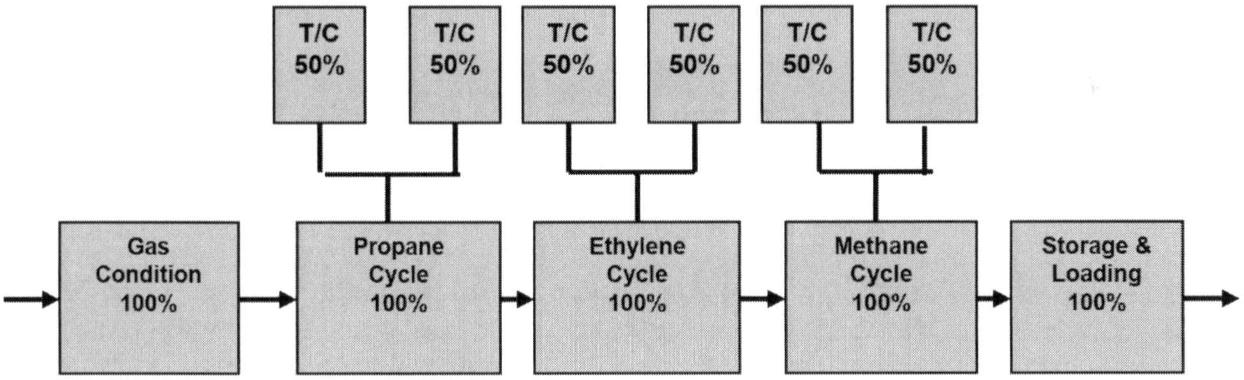


**Figure 3.** "Two-Trains-in-One" Operational Availability Concept in the POCLP

## Study Basis

Table 2 provides an outlook at the various configurations that have been investigated for the Phillips Optimized Cascade LNG Process.

| DESCRIPTION | UNITS | FRAME 5 | FRAME 7EA | FRAME 7 C2/C3 | FRAME 9E C2/C3 | FRAME 5 2 TRAINS |
|---|---|---|---|---|---|---|
| LNG Production | MTPA | 3. ++ | 5. ++ | 5. ++ | 7. ++ | 6++ |
| Comparative Total Direct Cost per TPA | | 100.0 | 77.7 | 84.2 | 72.0 | 80.4 |

**Table 2.** Driver/Compressor Comparison Summary for the Phillips Optimized Cascade LNG Process

The innovative configuration shown in Figure 4 is designed to use large, single shaft GE Frame gas turbines in order to achieve a plant nominal capacity of 8.0 MTPA using the Phillips Optimized Cascade LNG Process. The plant configuration and layout of turbo-compressors are driven by the requirements to maximize plant reliability, maintain plant flexibility, meet environmental targets, minimize capital expenditure, and reduce operating expenditures.

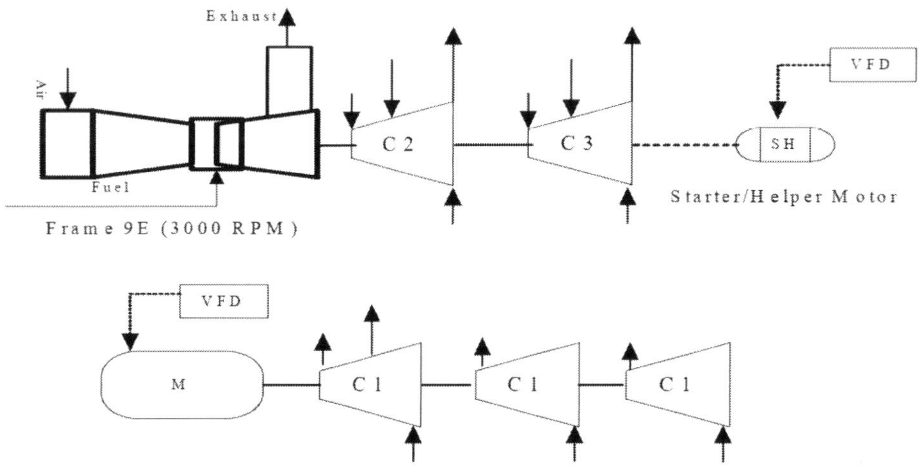


**Figure 4.** Large Train Concept in the Phillips Optimized Cascade LNG Process

The "two-trains-in-one" configuration achieves higher reliability and maintains flexibility advantages since the train can operate at capacities up to 75% with one of the two lines down. The 8.0 MTPA line-up illustrated produces approx. 110 MW per gas turbine string and 60 MW per electric motor string. Once the compressor string is on-line, the starting motor is utilized as a helper to achieve the desired LNG production. Propane and ethylene are on the some shaft. Both strings are redundant in a 2x 50% configuration.

As part of the study, the Cold Boxes have been redesigned and are available for the 8.0 MTPA facilities. Overall requirements for weight and dimensions to allow shallow draft delivery have been maintained. Equipment, in terms of vessels, pipe, valves, and fittings, is available for the train size considered.

To further improve overall facility capacity, capability, reliability, and efficiency, engineered options are available as follows:

- Partial/Full Waste Heat Recovery
- Liquid Expander / Generator
- Expander / Compressor
- Inlet Air Chilling for GT drivers
- Electric motor drives

The technology for enhanced removal of NGL and LPG in the Phillips Optimized Cascade LNG Process has been improved for the most recent trains designed. Based upon technology improvements and observances of the operation of the previous designed trains, a more efficient process was devised for enhanced NGL/LPG recovery. The process was improved to add reflux and heat exchange systems to the NGL and LPG removal system. The alterations provide improved operability and flexibility of the product specifications. It also provides the capability of deep LPG recovery when desired. The design can accommodate for increased ethane recovery from the natural gas feed stream. This allows the designed facility to alter operations to maximize the facility product economic drivers. The process can easily shift from a NGL recovery to LPG recovery to ethane recovery configuration depending upon the economic drivers.

**Design Basis**

- 2 Frame 9's, 2 60 MW electric motors, 2-in-1, NP compressors study, Cold box design and shallow draft delivery, full NGL recovery and integration, other critical equipment such as valves
- Option to the design basis: WHR, liquid expanders, inlet air chilling, all electric drive
- Utilizing 2 – 160,000 m3 tanks and 3 – 200,000 dedicated ships

Table 3 provides typical adjustments to base for the options in the Phillips Optimized Cascade LNG Process.

| Option | Availability | Production Efficiency | Production | Thermal Efficiency |
|---|---|---|---|---|
| | | | | |
| Base Case | --- | --- | --- | --- |
| Inlet Air Chilling | + 0.0 | +2.81 | +6.01 | -0.60 |
| Waste Heat Recovery | +0.75 | +0.17 | -0.98 | +1.19 |
| Liquid Expander | +0.0 | +1.49 | +4.04 | +0.45 |
| All Electric Motors | +2.57 | +1.07 | +0.0 | -0.871 |

**Table 3.** Options Summary for the Phillips Optimized Cascade LNG Process

Assumptions for the above table are:

1) Inlet Air Chilling from 79F to 60F and will be further supported by future studies
2) Waste Heat Recovery has 12 in of outlet duct losses resulting on loss of available power and will be further supported by future studies
3) Assumes 75% isentropic efficiency for the liquid expander and is supported by current studies.

The electric motor case is designed to equivalent capacity and supported by current studies.

**Schedule**

- Pre-FEED studies complete, ready to start FEED/EPC immediately
- Milestone schedule
    - Process targets – 28 to 32 months EPC
    - Tanks target - 32- 34 months EPC

## Startup

Start-up of the facility at larger sizes is essentially the same as the smaller trains. The single component refrigerants provide the ease of establishing and managing the refrigeration systems. The basic exchanger equipment utilized allows for improved and faster cool down rates for equipment as well as ease of control of cool down of equipment. The "two-train-in-one" approach in refrigeration units gives better flexibility during machine trips/shut downs, which corresponds to reduced LNG production versus total loss of production. The flaring requirements are minimal since there are no refrigerants flared during trips/shut downs. The design provides for a simple and easy to operate utility support systems. The control system has been simplified to facilitate easier start up or shut down. The process lends itself to quick recovery time after trips/shut downs and LNG production is back to full rates for a cold restart within 2-4 hours, warm restart within 4-8 hours. The factors lead to low staffing requirements for commissioning and start up of facility. Also, reduced numbers of cryogenic equipment and piping with no specialty items such as spiral wound heat exchangers.

## Operability

The operability of the larger designs is expected to change with the use of differing plant turbine and compressor combinations. However, many of the aspects of a Phillips Optimized Cascade LNG Process plant are expected to stay the same. The plant turndown capability can range from 0-100% whilst maintaining good efficiencies. Single component refrigerants are easier to maintain versus other refrigerant types since there are no compositional adjustments to maintain efficiencies. There is no flaring of refrigerants and minimal flaring of process gases/liquids during trips/shut downs. There is a simplified control scheme for DCS and for equipment for ease of operation. The design allows for quick evacuation of compressor casings for quick restart when necessary. The process will maintain the same quick recovery time after trips/shut downs. Therefore, LNG production will be back to full rates for a cold restart within 2-4 hours, warm restart within 4-8 hours. The major maintenance programs can be executed while the plant is still on reduced production. And finally, there are low staffing requirements for plant operation and maintenance.

## Conclusions

We envision LNG plant designs in the coming decade falling broadly into three groups, with nominal capacities around 4 mtpa, 5 mtpa, and 8 mtpa. These different train designs are likely to co-exist in the coming years, as different owners find it advantageous to choose one vs. another. The basic design of 4 mtpa trains (until recently considered very large trains) will continue to suit new projects with limited gas supply or committed gas sales, or in cases where copying an existing train design is the most cost-effective option. Approximate 5 mtpa trains could be suited for expansion or grass roots projects that need to lower LNG unit costs, and have the available gas supply and sales. An 8 mtpa "mega" train would be most suitable for an expansion train at a very large existing LNG complex targeting a large, distant market, with virtually unlimited gas supply.

We have concluded that large trains can be feasible and cost-effective. The compressors utilized in our studies represent some of the world's largest but they meet acceptable guidelines for major design parameters. Other design parameters investigated included feasible line and valve sizes, size of heat exchangers, cold boxes and other key equipment.

One major concern with "mega" LNG trains is significant reduction in LNG production when one of the refrigerant compressor trains goes out of service. This problem can be mitigated with the Phillips Optimized Cascade LNG Process "two-train-in-one" reliability concept which allows the LNG train to operate at up to 75% capacity even when one compressor string trips. LNG trains utilizing this design have demonstrated on-stream availability of greater than 95%, the highest in the LNG Industry. The outstanding safety, reliability, and high operating factors of the LNG industry keep building up confidence in larger trains, and the trend to design larger trains will keep accelerating.

## MFC®s3
## the Formula for High Capacity LNG Base Load Plants

*Heinz C. Bauer, Hubert E. Franke, Marc T. Schier*
*heinz.bauer@linde-le.com, hubert.franke@linde-le.com, marc.schier@linde-le.com*
*Linde AG, Linde Engineering Division*
*Dr.-Carl-von-Linde-Str. 6-14*
*82049 Hoellriegelskreuth / Germany*

*Geir A. Owren*
*gaow@statoil.com*
*Statoil ASA*
*Arkitekt Ebbells Vei 10, Rotvoll*
*7005 Trondheim / Norway*

*Prepared for Presentation at AIChE Spring National Meeting*
*4th Topical Conference on Natural Gas Utilization*
*New Orleans, LA, April 2004*



## ABSTRACT

Many LNG base load projects, which are under discussion or already on the drawing boards, are designed for train capacities between 5 and 8 mtpa LNG. Conventional liquefaction processes using only two refrigerant cycles are limited to about 5 mtpa LNG, unless parallel strings within a single train are considered. The Statoil-Linde Mixed Fluid Cascade (**MFC®**) process overcomes this limitation with a third cycle, which shifts the limit of a real single train concept with proven compressor drivers to above 8 mtpa LNG. A symmetric (**s**) compressor concept equally loads three (**3**) identical drivers by load balancing between the refrigeration cycles. Further, dedicated spiral wound heat exchangers in the liquefaction and the subcooling cycles are fit for the full plant capacity in the mentioned range.

## INTRODUCTION

Main challenges in the design of base load LNG plants are high plant capacity (5 to 8 mtpa) and low gross calorific value (1050 BTU/SCF) especially for the US market. To demonstrate the recent achievements of the MFC® technology, which is jointly owned by Statoil ASA and Linde AG, a detailed process evaluation study has been performed. As basis of this study two typical gas compositions (Table 1) have been selected, which represent **lean** and rich natural gas streams as they are found in large gas reservoirs (for easy identification all results related to the **lean** gas are shown by bold lines or characters).

**Table 1**

| mol% | **Lean Gas** | Rich Gas |
|---|---|---|
| $N_2$ | **5.00** | 5.00 |
| $CH_4$ | **88.93** | 84.07 |
| $C_2H_6$ | **3.96** | 5.58 |
| $C_3H_8$ | **1.37** | 2.73 |
| $C_4H_{10}$ | **0.48** | 1.34 |
| $C_5H_{12}$ | **0.17** | 0.65 |
| $C_6H_{14}$ | **0.06** | 0.32 |
| $C_7H_{16}$ | **0.02** | 0.16 |
| $C_8H_{18}$ | **0.01** | 0.08 |
| Benzene | **0.01** | 0.08 |
| | **100.00** | 100.00 |

**Figure 1**

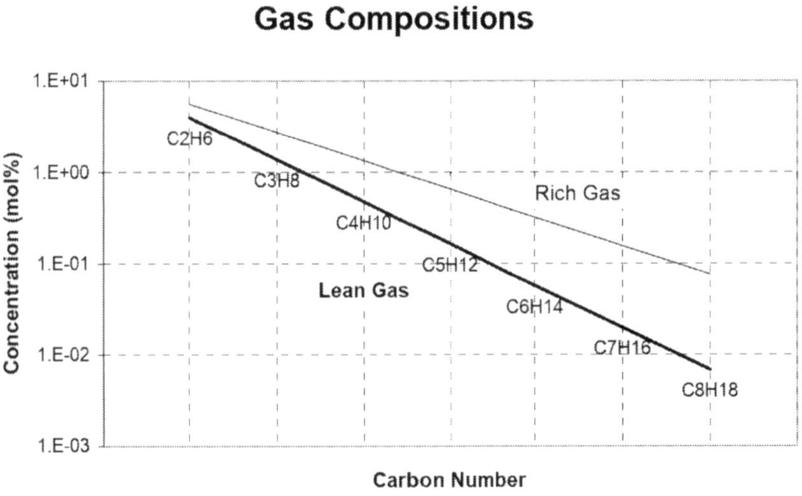


To allow for interpolation between the results for the rich and **lean** gas a smoothened concentration profile has been selected, in which the C2 to C8 hydrocarbons show an exponential decline (see Figure 1).

## LOAD BALANCING (MFC®s3)

The Mixed Fluid Cascade (MFC®) Process as known e.g. from the US patent 6,253,574 uses three independent refrigeration cycles. Obviously, the best economy of scale can be achieved when all three cycle compressors are driven by the same size and type of driver. In this paper all figures refer to the use of three General Electric MS 7121 EA (Frame 7) unless otherwise stated. If highest availability is of the essence, the three refrigeration cycles can be designed with two times 50% gas turbine/compressor trains. In this case six GE MS 6581 B (Frame 6) would replace the three Frame 7s.

**Figure 2**

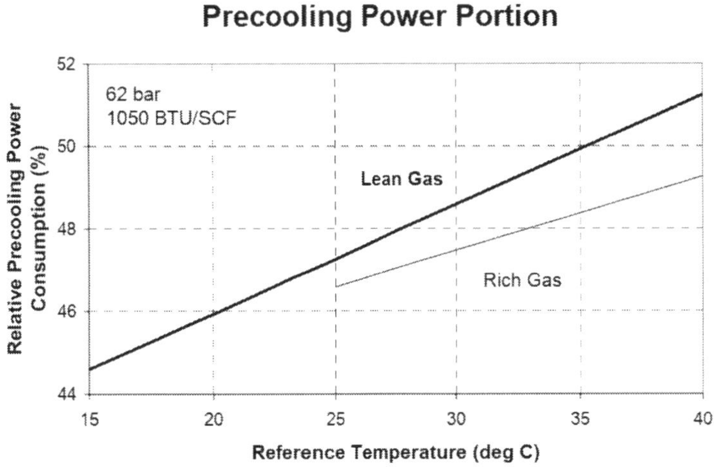

All LNG plants require an extraction at least of those hydrocarbons, which would freeze in the LNG under storage conditions (e.g. aromatics and $C_5$plus). In an LNG plant usually precooling is considered as first cooling step between ambient temperature and extraction of the mentioned hydrocarbons.

**Figure 3**

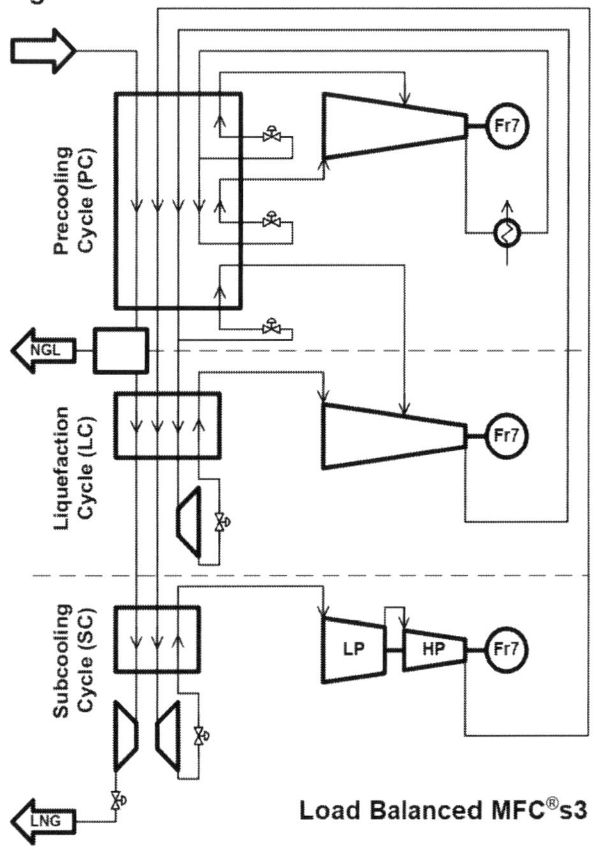


**Load Balanced MFC®s3**

The precooling portion of the overall power demand of all refrigeration compressors for the two gases defined in Table 1 is shown in Figure 2 as a function of a reference temperature. This is the temperature, under which all main process streams (natural gas, refrigerant fluids) enter into the cryogenic heat exchangers. The lower the reference temperature and the richer the gas gets the smaller the required compressor power for precooling becomes. This situation can be addressed reasonably well by designers of dual flow liquefaction processes[1,2,3], if the power mismatch between precooling and liquefaction plus subcooling is compensated by helpers for the gas turbines.

A process with three refrigeration cycles as the MFC® process offers a much wider field for even load distribution between the cycles. If part of the refrigerant of the liquefaction cycle (LC) is vaporized under elevated pressure in the precooling section and is fed to the LC compressor as a side stream, a perfect load balancing between all three refrigeration cycles can be achieved (Figure 3) for all real situations. This feature is a major aspect of a cost effective design for large production capacities. As all three (**3**) cycles are **s**ymmetrically driven this arrangement is referred to as MFC®s3.

## ADVANCED CALORIFIC VALUE CONTROL (ACVC)

Traditionally, Far East LNG importers like Japan and South Korea accepted LNG with high gross calorific value (GCV). No special consideration had to be made to reduce the GCV. The rising demand for LNG in the US and Europe, however, comes along with much more severe requirements for GCV reduction, as existing pipeline specifications have to be met. A low GCV value coincides with a low critical pressure of LNG (Figure 4), which is in

**Figure 4**

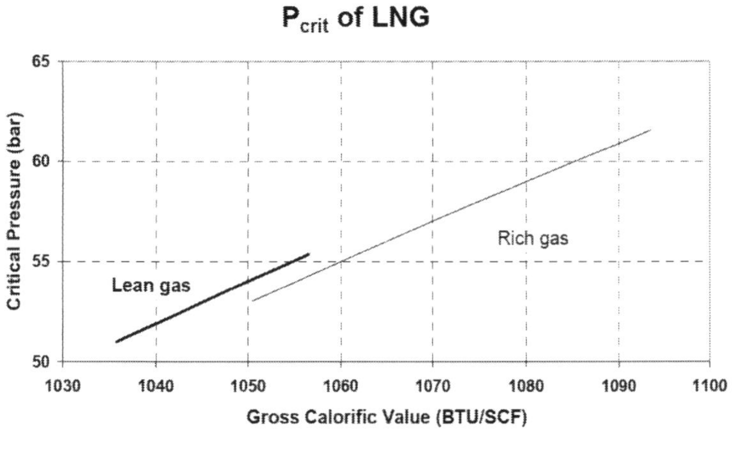

many cases lower than the natural gas pressure. To maintain the operability of the NGL extraction, its operating pressure has to be reduced below the critical pressure of the LNG product. Conventional systems, which use a scrub column for NGL extraction are less suitable for deep NGL recovery as the specific energy consumption (SEC) increases significantly. Linde/Statoil have developed the proprietary Advanced Calorific Value Control (ACVC) system, which overcomes this problem. Figure 5 shows a comparison between a scrub column design and the ACVC design based on the rich gas case (62 bar feed gas pressure, 35 deg C reference temperature). If a GCV of 1050 BTU/SCF has to be reached more than 10% energy can be saved with the ACVC design. It should be mentioned that the increase of the SEC at lower GCV is as least partly caused by its definition as the NGL by-product is not taken into account.

**Figure 5**

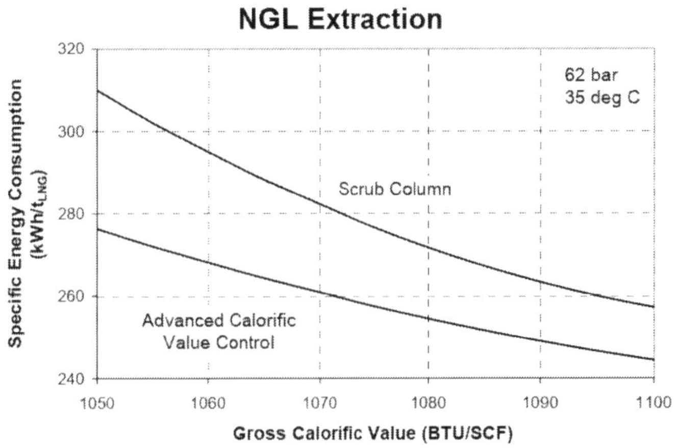


**Figure 6**

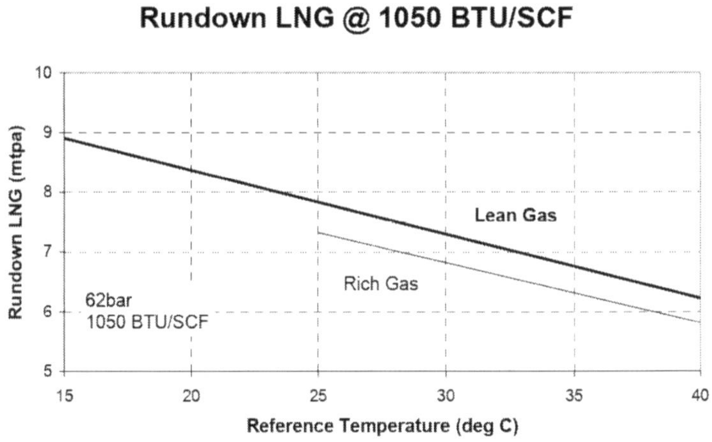


**Figure 7**

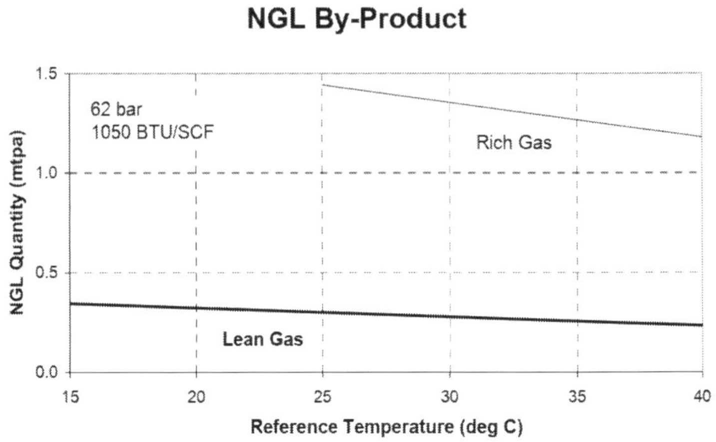


Figure 6 shows the rundown LNG quantity of a plant based on three Frame 7 gas turbines as drivers for the refrigeration compressors. In this case the starters are not used as supplementary power source.

The significant amount of extra NGL by-product as shown in Figure 7 contributes a lot to the operating revenues of the LNG plant. Irrespective of GCV requirements, NGL extraction is an elegant method to improve the overall economics of the plant, especially when the natural gas composition is on the rich side.

## SPECIFIC ENERGY CONSUMPTION

Absolute figures for the specific energy consumption (SEC) of a base load LNG process are reported frequently[4]. NGL extraction, though required in reality, is mostly not considered.

In order to generate information with immediate use in the conceptual phase of an LNG plant, a detailed parameter study has been performed on the basis of a MFC®s3 process with NGL extraction down to a GCV of 1050 BTU/SCF by means of the ACVC concept.

**Figure 8**

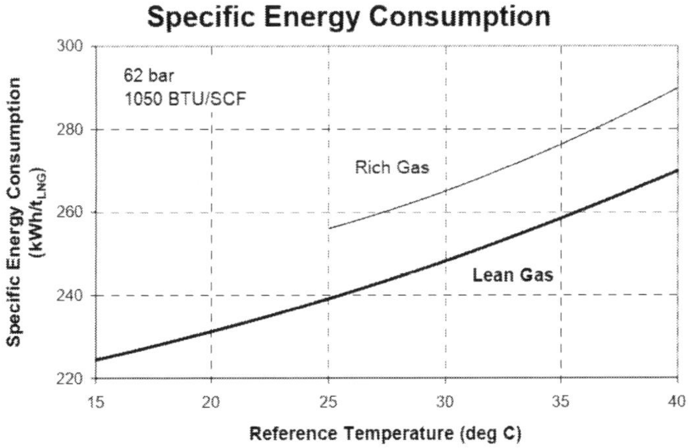


**Figure 9**

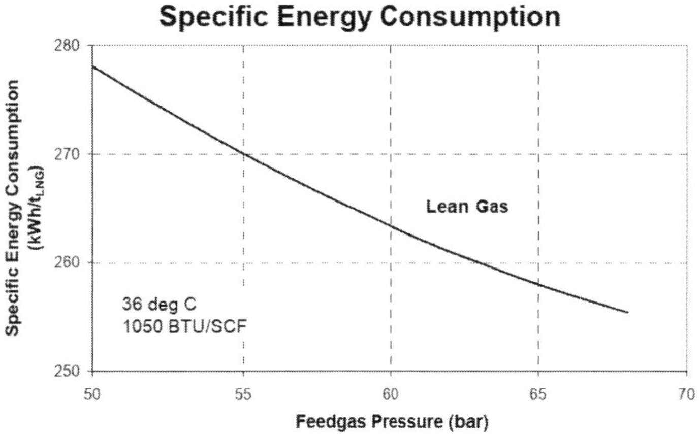


Figure 8 shows results for the **lean** and the rich gas case based on a feed gas pressure of 62 bar at inlet to subcooling for various reference temperatures. Figure 9 shows similar results for the **lean** gas with variable feed gas pressure at 36 deg C reference temperature. Both Figures would show lower SEC values, if the total amount of liquid products (LNG+NGL) would be taken into account.

## PROCESS DESIGN EXAMPLE

In order to provide actual figures for a realistic process design a large LNG train has been studied. On the basis of the **lean** gas composition with 62 bar and 35 deg C at the inlet to precooling a conceptual process design was made. The refrigeration compressors are driven by Frame 7's with additional 20 MW on each shaft, which have been recruited from the starter/helpers. The resulting LNG rundown amounts to 8.5 mtpa at 333 stream days, which is accompanied by an additional quantity of 0.4 mtpa NGL ($C_3$plus hydrocarbons). The specific energy consumption of the refrigeration compressors is 259 kWh/$t_{LNG}$. A pre-

**Figure 10**

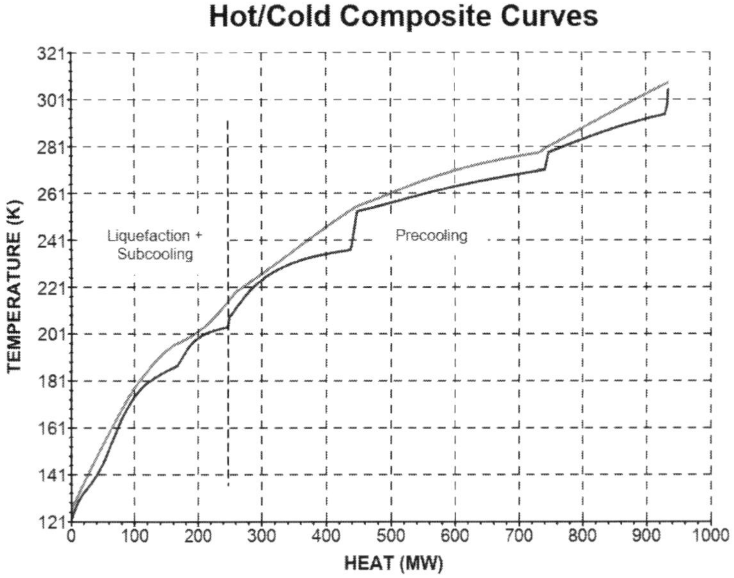

**Table 2**

| Compressor | GE Model | Polytropic Efficiency |
|---|---|---|
| Precooling | 3MCL1403 | 80 |
| Liquefaction | 2MCL1406 | 82 |
| Subcooling LP | AN250 | 88 |
| Subcooling HP | BCL1007 | 81 |

**Table 3**

| Heat Exchanger | Heating Surface | Shipping Weight |
|---|---|---|
| Liquefaction | 18,000 $m^2$ | 120 t |
| Subcooling | 35,000 $m^2$ | 220 t |

liminary selection of the cycle compressors is shown in Table 2.

Figure 10 shows the composite curves for all hot and cold streams in the cryogenic heat exchangers. About three quarters of the heating surface are required for precooling, only one quarter for liquefaction and subcooling. Precooling uses cost effective manifolded aluminum plate-fin heat exchangers, while liquefaction and subcooling is implemented in robust spiral wound heat exchangers[5] with a single coil for each service (see Table 3).

## CONCLUSIONS

With the second generation of MFC® based LNG plants large capacities can be achieved without technological risk, as well proven gas turbines, compressors and heat exchangers are used. A low gross calorific value of the LNG as mandated in the US and Europe provides extra income as a result of efficient and cost effective NGL recovery. SRI Consulting's[6] analysis indicates that the low specific power consumption, combined with the large single-train capacity, results in very competitive economics for the MFC® process.

## ACKNOWLEDGEMENT

The authors appreciate the contribution of GE Power Systems, Oil & Gas (Nuovo Pignone S.p.A.) to the reality check of the refrigeration cycle compressors.

## REFERENCES

1 J. C. Bronfenbrenner, The Air Products Propane Precooled/Mixed Refrigerant LNG Process, LNG Journal, Nov./Dec. 1996, page 25-27

2 K. J. Vink, R. Klein Nagelvoort, Comparison of Baseload Liquefaction Processes, LNG 12 Conference, Perth, Australia, May 1998

3 Béatrice Fischer, A new process to reduce LNG cost, AIChE Spring National Meeting 2002, Second Topical Conference on Natural Gas Utilization, Conference Proceedings, page 191-197

4 Henri Paradowski, Phil Hagyard, Comparing five LNG processes, Hydrocarbon Engineering, Vol. 8, Number 10, October 2003, page 32

5 Helmut Reithmeier, Manfred Steinbauer, Rudolf Stockmann, Arne O. Fredheim, Oddvar Jørstad, Pentti Paurola, Industrial Scale Testing of a Spiral Wound Heat Exchanger, LNG 14 Conference, Doha, Qatar, March 2004

6 Marcos A. Cesar, Liquefied Natural Gas by the Mixed Fluid Cascade Process, PEP Review 2002-7, November 2003

# THE AP-X™ LNG PROCESS: A FLEXIBLE APPROACH FOR REDUCING LNG UNIT COST

*Mark J. Roberts*
*Dr. Yu-Nan Liu*
*James C. Bronfenbrenner*
*Joseph M. Petrowski*
*Air Products and Chemicals, Inc.*
*Allentown, PA, USA*
*www.apci.com*

## ABSTRACT

The Air Products propane pre-cooled, mixed refrigerant LNG process (C3MR) has been the workhorse of the LNG industry for more than 30 years. It has been applied in LNG plants producing from 1 to 5 Mta of LNG per train, using steam and gas turbine drivers, sea water and air cooling, rich and lean feeds containing a broad range of nitrogen, with and without LPG extraction. The process has proven to be efficient, flexible, reliable, and cost competitive.

The AP-X™ LNG Process will increase single train capacities to 7-10 Mta. This capacity expansion has the potential to bring down the capital cost of LNG significantly. By producing warmer LNG at pressure, today's C3MR process capacity can be increased up to 80% with the same proven equipment in use today. The warmer LNG is then subcooled to storage temperature by an all vapor, nitrogen refrigerant in a compression/expansion cycle that has been used for years to liquefy oxygen, nitrogen, and natural gas. The efficiency of this process is as good as what has been achieved in the industry by any other process, while the capital cost of the liquefier is reduced significantly.

This paper will present applications of the AP-X™ LNG Process to achieve maximum capacity in a single train with a variety of compressor drivers. This range of applications demonstrates that the AP-X™ LNG Process brings significant economies of scale to the industry, reducing the capital cost of LNG while maintaining the efficiency, flexibility, and reliability of the C3MR process.

---

™ AP-X is a trademark of Air Products and Chemicals Inc.

## INTRODUCTION

Air Products is the leader in supplying liquefaction process technology and MCR® Main Cryogenic Heat Exchangers (MCHE's) to the LNG industry. As the industry has grown, the solutions offered by Air Products have evolved to meet the increasing demand to lower cost. Air Products' involvement in the baseload LNG industry began with the first all mixed refrigerant (single-MR) liquefier for Esso, Libya. That first plant was commissioned in 1970 and has capacity of 0.6 Mta per train. Since that time Air Products' leadership in the industry has continued from the successful introduction of the propane precooled, mixed refrigerant process (C3MR) in 1972 at Brunei, to trains currently under construction utilizing the Split-MR™ machinery configuration [1] with capacities approaching 5 Mta. This paper will describe a new liquefaction process that Air Products believes is key to meeting the challenge of increased demand and lower unit cost for LNG production.

Economies of scale continue to favor increasing train size for baseload LNG plants in order to drive down the unit cost of LNG. Figure 1 illustrates this historical trend. To meet this demand for lower cost, LNG liquefaction technology has and will continue to evolve.

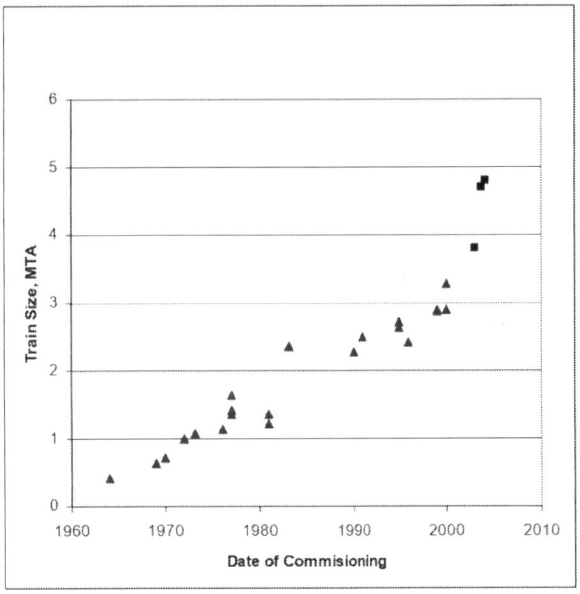


The AP-X™ LNG Process achieves not only high capacity in a single train, but also can incorporate high LPG recovery, lower LNG heating value for new markets, and maximum efficiency. This range of applications demonstrates that the AP-X™ LNG Process brings significant economies of scale to the industry, reducing the capital cost of LNG while maintaining the efficiency, flexibility, and reliability of the C3MR process.

## AP-X™ LNG PROCESS

While the C3MR process remains the preferred option in many cases, there exists substantial developing demand for larger train sizes. For example, trains using multiple GE Frame 7 or Frame 9 gas turbine drivers or large electric motors

® MCR is a registered trademark of Air Products and Chemicals Inc.
™ Split MR is a trademark of Air Products and Chemicals Inc.

can be configured. While it remains feasible to further increase train capacity with a C3MR process, new designs must be developed for several major equipment items at capacities exceeding 5.0 Mta. For example, the propane and centrifugal MR compressors are approaching single casing flow limits at current world scale LNG plant production levels.

In response to continuing customer demand for increased LNG train capacity and lower unit cost, Air Products has developed and patented [2] the AP-X™ LNG Process. The AP-X™ process cycle is an improvement to the C3MR process in that the LNG is subcooled using a simple, efficient nitrogen expander loop instead of mixed refrigerant. Other embodiments include a dual MR version where another MR refrigeration loop is used for pre-cooling and nitrogen is likewise used for subcooling.

In addition to improving the efficiency, the use of the nitrogen expander loop makes greatly increased capacity feasible. It does this by reducing the flow of *both* propane and mixed refrigerant. Volumetric flow of mixed refrigerant at the low-pressure compressor suction is about 60% of that required by the C3MR process for the same production. Mass flow of propane is about 80% of that required by the C3MR process.

With the new AP-X™ process, train capacities in excess of 8 Mta are feasible in tropical climates, in existing compressor frame sizes, without duplicate/parallel compression equipment, and using a single spool-wound MCHE of a size currently being manufactured.

The nitrogen expander loop is a simplified version of the cycle employed by Air Products in hundreds of air separation plants and nitrogen liquefiers worldwide. Experience has shown these plants to be simple to operate and very reliable. Many of these plants are remotely operated, including shutdowns and restarts. The nitrogen cycle has also been employed by Air Products with similar success in small, stand-alone LNG peak-shaving plants.

The AP-X™ process cycle is depicted below in figure 2. As is the case with $C_3$-MR process, propane is used to provide cooling to a temperature of about -30 °C. The feed is then cooled and liquefied by mixed refrigerant, exiting the MCHE at a temperature of about -120 °C. Final subcooling of the LNG is done using cold gaseous nitrogen from the nitrogen expander. Figure 3 shows the equipment layout for the liquefaction and subcooling sections of an AP-X™ train. Coil-wound heat exchangers are used to liquefy and subcool the LNG, while the nitrogen economizer uses brazed aluminum plate-fin heat exchangers.

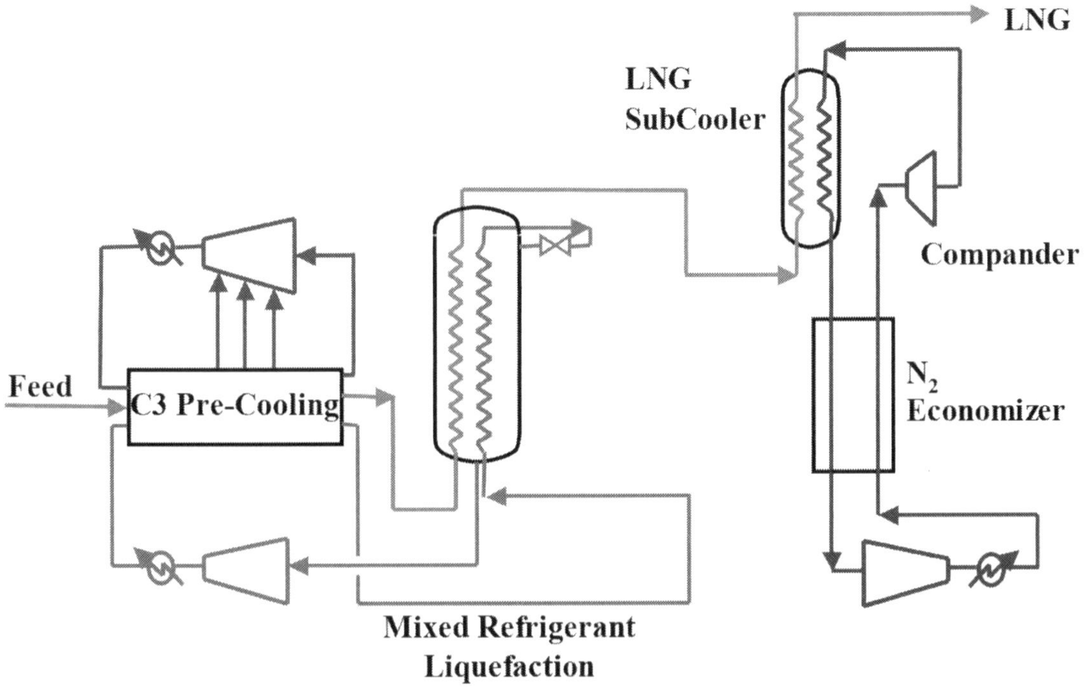


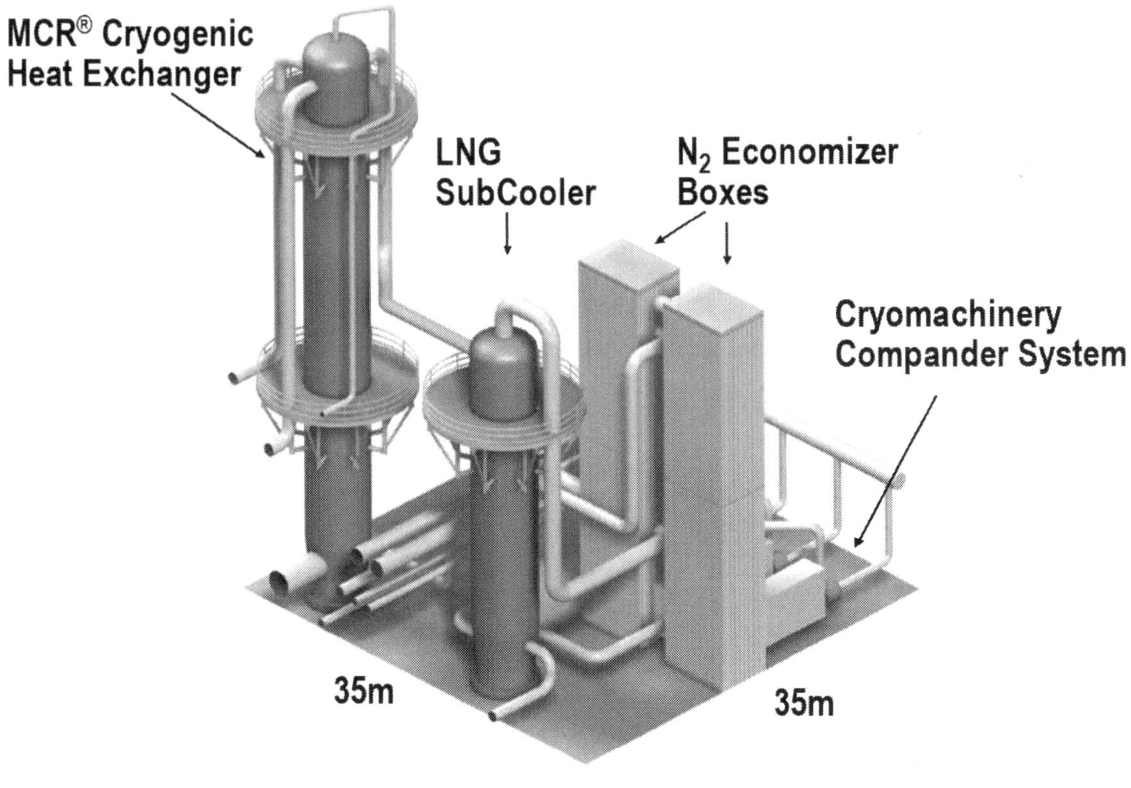


Figure 3: AP-X™ LNG process layout

### Expandable and flexible

It is possible to operate an AP-X™ train at a reduced production rate of about 65% without the nitrogen expander loop by adjusting the composition of the mixed refrigerant inventory.

The ability to operate an AP-X™ process plant in C3MR mode can also be exploited as an expandable plant. For example, a producer may choose to invest in a nominal 5 Mta C3MR train with plans to expand production later to a level of up to 8 Mta by adding the nitrogen expander cycle as the market develops.

Retrofitting an existing plant with a nitrogen expander cycle to subcool the LNG is also possible, although the production increase will be more modest due to bottlenecks with existing equipment.

## POSSIBLE DRIVER CONFIGURATIONS

The power split between $C_3$, MR, and N2 is flexible, and can be manipulated by changing the temperature range of the three refrigerant loops. This feature, and the use of the Split-MR™ machinery configuration, allows considerable flexibility in matching compressor driver sets, whether Frame 7's, Frame 9's, electric motors or combinations of the above.

### Three Frame 7EA drivers

A nominal production of 7.5 Mta in a tropical environment is possible using the AP-X™ process with three Frame 7 drivers. The configuration of drivers is shown in figure 4. In this case, the Split-MR™ machinery configuration is used with 20 MW starter/helper motors on two of the drivers.

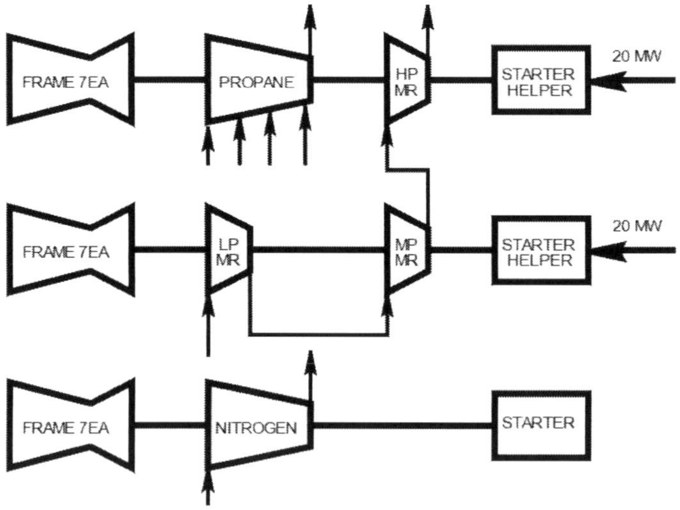

### Two Frame 9E drivers

Frame 9 drivers are very attractive for large trains. They provide 50% more power than a Frame 7 driver at a cost increase of less than 20%. It is possible to achieve the same production from two Frame 9 drivers as from three Frame 7 drivers. Frame 9 drivers also have an advantage in that their lower operating speed (3000 RPM) allows larger capacities in a single propane compressor casing. Figure 5 shows a configuration producing a nominal 7.5 Mta using two Frame 9 drivers and 20 MW helper motors. This configuration uses a modification of the Split-MR™ machinery configuration in which propane and LP MR compressors are on one driver with nitrogen and MP/HP MR compressors on the other driver.

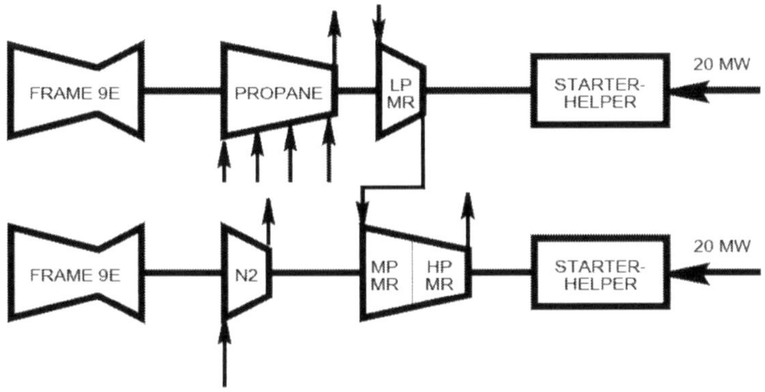


### Three Frame 9E drivers

With the AP-X™ process, three Frame 9 drivers can be used to achieve a nominal production in the range of 8-10 Mta in a tropical environment. It is also possible to use a combination of Frame 9 and Frame 7 drivers at the low end of this range, depending on the specific project requirements.

At the low end of the 8-10 Mta production range, a significant amount of power can be generated from starter/generator motors using three Frame 9 drivers thus unloading the power requirement of the power generation facility. Each compression service (propane, mixed refrigerant and nitrogen) can be driven separately allowing maximum operational simplicity.

To achieve a production rate at the high end of the 8-10 Mta range, the Split-MR™ machinery configuration can be used to balance the driver power requirement of the three Frame 9 drivers and to reduce the size of the starter/helper and starter/generator motors. For a production rate approaching

10 Mta two propane compressor casings will be required to stay within the limits of proven compressor designs.

Figures 6,7 and 8 show configurations utilizing three Frame 9 drivers for liquefaction capacities of 8,9 and 10 Mta respectively. At the low end of the range (figure 6) a train will export about 60 MW of electricity, while at the high end of the range (figure 8) a train will import about 10 MW of refrigeration power from the grid.

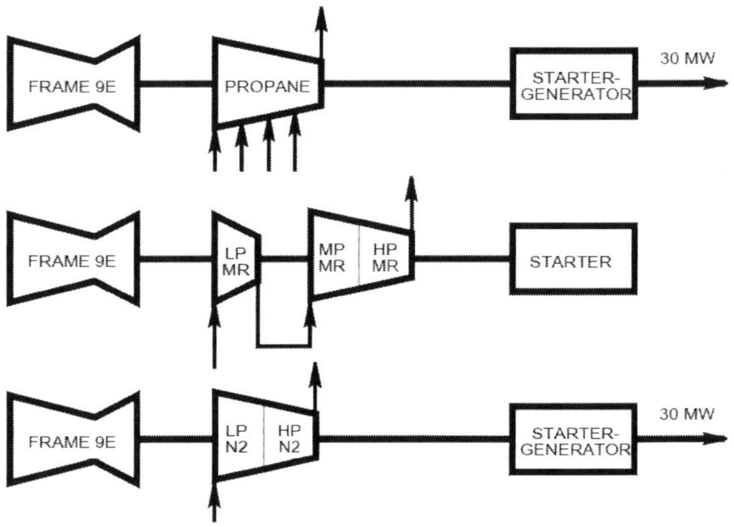


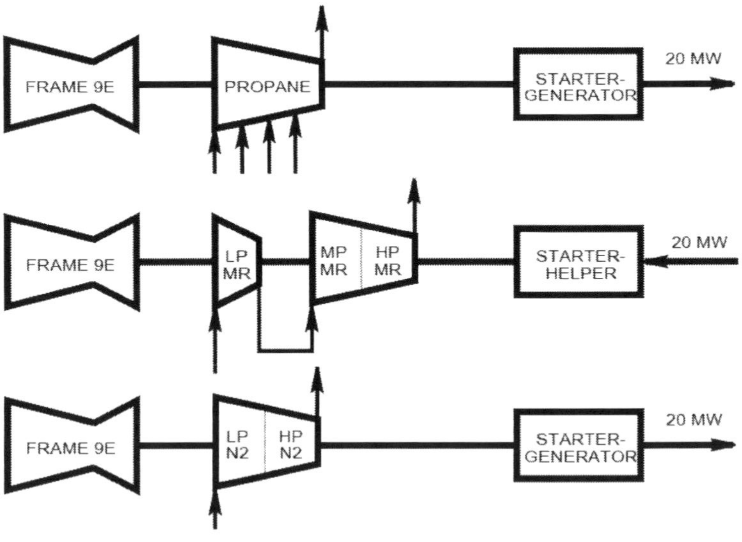

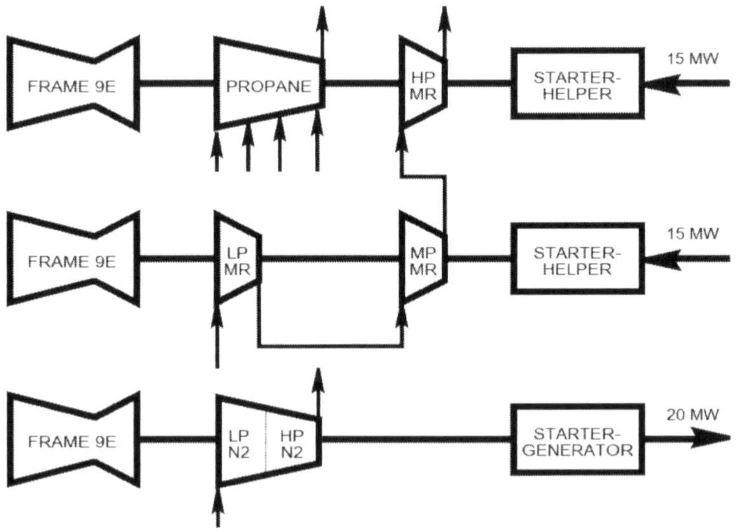


### Electric motor drive

The use of electric motors as compressor drivers is gaining increasing interest in the LNG industry. Electricity can be generated in a combined cycle power plant to greatly reduce the $CO_2$ emitted per unit LNG produced. Electric motors are also a very flexible drive system for the AP-X™ process. Liquefaction capacities from 7-10 Mta can be easily achieved, with the specific arrangement depending upon the desired capacity and the maximum motor size considered proven by the LNG owner. Systems may also be considered which use combinations of electric motor and gas turbine drive for the refrigeration compressors. With the use of a Variable Frequency Devices, the speed of each compressor can be optimized not only to achieve maximum compressor aerodynamic efficiency at a design point but also at the warm and cold ambient temperature extremes.

Depending on the desired capacity and the maximum electric motor size, many different configurations are possible. Figure 9 shows one example; for a train producing 8 Mta using electric motors up to 65 MW in size.

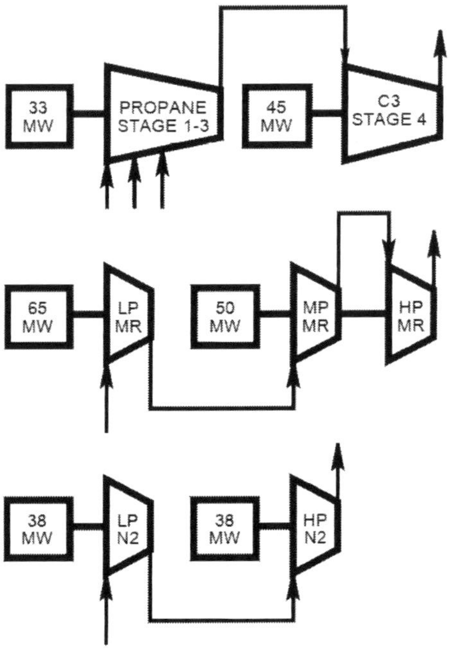


## CAPITAL COST

The AP-X™ meets the industry demand for the economies of scale that can result from larger single train capacities. The unit cost of LNG production is lowered significantly with the AP-X™ process. Merlin Associates have recently completed an independent study comparing the cost of a single 8 Mta AP-X™ train to that of two 4 Mta C3MR trains. The results are summarized in Table 1 Overall plant facility costs are reduced by about 11% with a total cost saving of about 140 MM$. For the C3MR option, some process units such as acid gas removal and the endflash system could potentially be combined depending on owner preference. The resulting savings are only about 10 MM$[3] however.

**Table 1: Capital cost comparison**

| | | |
|---|---|---|
| **Estimate Basis** [1] | | |
| Process | C3MR | AP-X™ |
| Plant Capacity, mtpa | 8 | 8 |
| No. Trains | 2 | 1 |
| Train Capacity, mtpa | 4 | 8 |
| **Capital cost -millions US$** [2] | | |
| Plant Facilities | 1,272 | 1,131 |
| Marine Facilities | 24 | 24 |
| Temporary Infrastructure | 52 | 52 |
| Total | 1,348 | 1,206 |
| US$/tpa | 168 | 151 |

Notes:
(1) Air cooled plant; Warm climate
(2) Merlin Associates estimate in constant 7/1/03 US$

## SUMMARY

The AP-X™ LNG process is a hybrid of two proven refrigeration process, a C3MR process for pre-cooling and liquefaction followed by a gas expander cycle for LNG subcooling. The process is very flexible and can be implemented using Frame 7's, Frame 9's, or electric motors as main drivers for the refrigeration compressors to achieve capacities from 7-10 Mta. The Capital savings for one AP-X™ LNG process train as compared to two C3MR trains is greater than 10%

## REFERENCES CITED:

1. Liu, Y.N., et. al. "Reducing LNG Costs by Better Capital Utilization" LNG 13 International Conference 2001, Seoul, Korea
2. Roberts, et. al. "Hybrid Cycle for the Production of Liquefied Natural Gas" US Patent 6,308,521
3. Telephone conversations, APCI and Merlin Associates, October 2003

**Session T9008** **Paper No. 68a**

# U.S. DOE NATURAL GAS TO HYDROGEN PROGRAM PLAN

Venkat Venkataraman and Dexter Sutterfield, U.S. Department of Energy,
National Energy Technology Laboratory
and David Olsen, Solutions Through Innovative Technologies, Inc.

Spring 2004 National Meeting of the American Institute for Chemical Engineers,
April 25, 2004, New Orleans, LA, for the session on Conversion of Natural Gas to Hydrogen

Unpublished



**Key words:** hydrogen, natural gas, methane, steam-methane reforming, ion transport membrane, synthesis gas

**Disclaimer**



**Abstract**

The United States Department of Energy is planning a multi-year research program with industry to develop new, more energy- and cost- effective technologies to convert natural gas to hydrogen in both central and distributed plants. This program will be conducted in parallel with the coal program's hydrogen generation research that has a much more challenging synthesis gas mixture with significant contaminants that have to be removed. The goal of the natural gas to hydrogen program is to accelerate development of technologies that will help the U.S. progress toward a hydrogen economy. The planned DOE Natural Gas to Hydrogen Program Plan will be presented along with results from discussions with industry on research needs and priorities.

The United States Department of Energy (DOE), in response to president George W. Bush's National Energy Policy[1] and initiatives for a National Hydrogen Economy[2], has developed programs within the DOE offices of Energy Efficiency and Renewable Energy (EERE), Nuclear Energy, Science and Technology, and Fossil Energy (DOE FE). The series of publications (Figure 1) to plan and initiate research, development and demonstration (RD&D) along the path to a "Hydrogen Economy" have been outlined in a National Hydrogen Vision[3], DOE Roadmap[4], DOE FE Hydrogen Posture Plan and the Coal and Natural Gas to Hydrogen Program plan.[5] DOE FE developed plans based on both policy and input from stakeholders for RD&D for hydrogen production from both coal and natural gas. This paper outlines DOE's hydrogen from natural gas RD&D plans. This is a multi-year research program with industry to develop new, more energy and cost effective technologies to convert natural gas to hydrogen in both central and larger distributed plants. DOE's EERE office has a similar RD&D natural gas to hydrogen program but focuses on small (< 1,000 Kg/day hydrogen) distributed generation.

On earth, hydrogen ($H_2$) is not naturally occurring. Hydrogen, like electricity, is an energy carrier - not an energy source. Thus $H_2$ has to be made from a primary energy source: fossil energy (coal, natural gas, petroleum and kerogen) or a renewable source such as biomass, or a hydrogen containing material such as water. Nuclear energy, solar or hydro/wind power can be used to generate $H_2$ from $H_2O$ either directly or through electrolysis of water. In the long-term, nuclear energy is anticipated to produce most $H_2$ with a small percentage of total demand from renewables. In the U.S., natural gas and coal are the only affordable fossil energy resources that can produce the large volumes of $H_2$, in the near to mid-term, for the U.S. to begin the transition to a hydrogen economy. In the near-term, natural gas can provide startup $H_2$ as it is the fuel of choice today in the commercial production of $H_2$ and has a widespread high-volume infrastructure for delivery of natural gas to potential central and distributed $H_2$ production facilities.

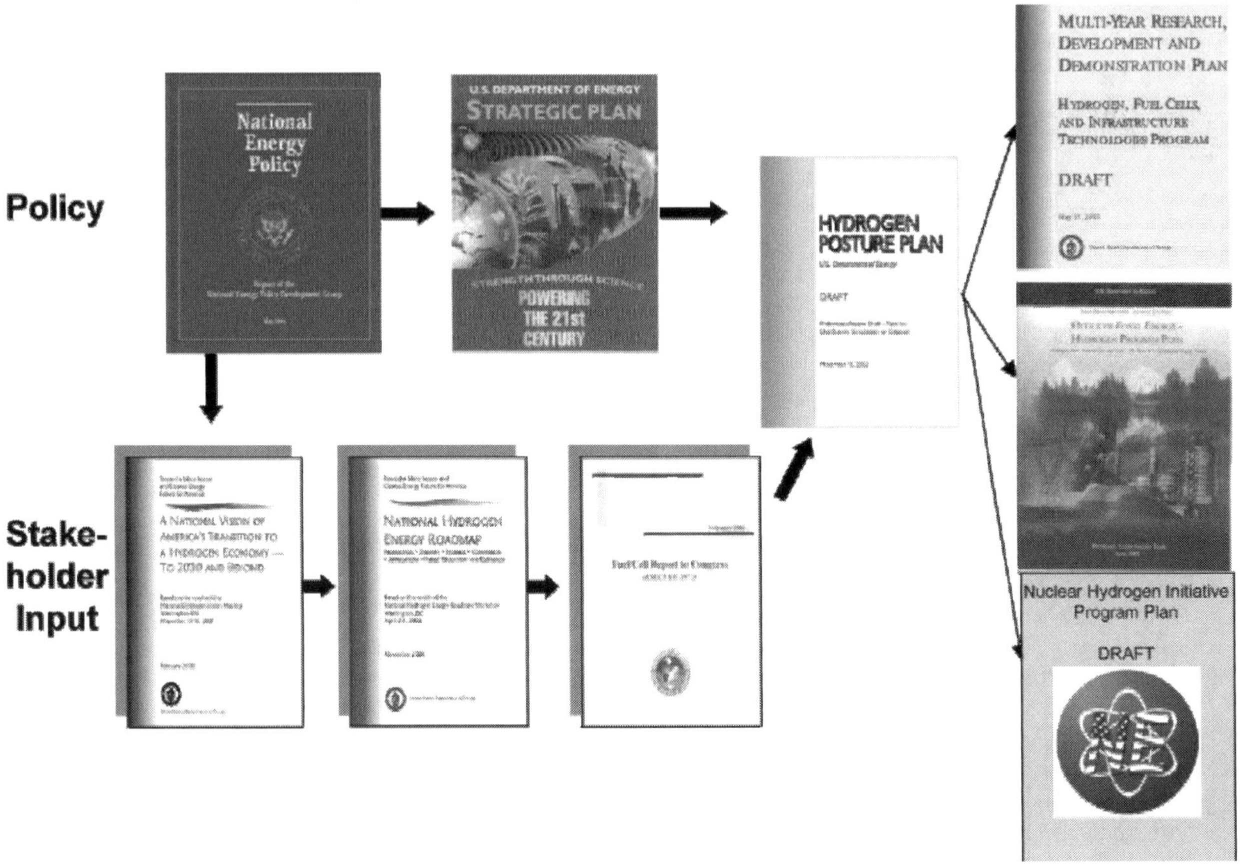


**Figure 1. Planning documents that outline future research on hydrogen.**

Why chose natural gas as a starting feedstock for $H_2$ production? Historically, man has been moving from energy sources with low hydrogen to carbon ratio (H/C ratio), predominantly biomass (including wood and peat), toward the use of energy sources with a higher H/C ratio as shown in Figure 2 by Ausubel, American Scientist, March-April 1996.[6]

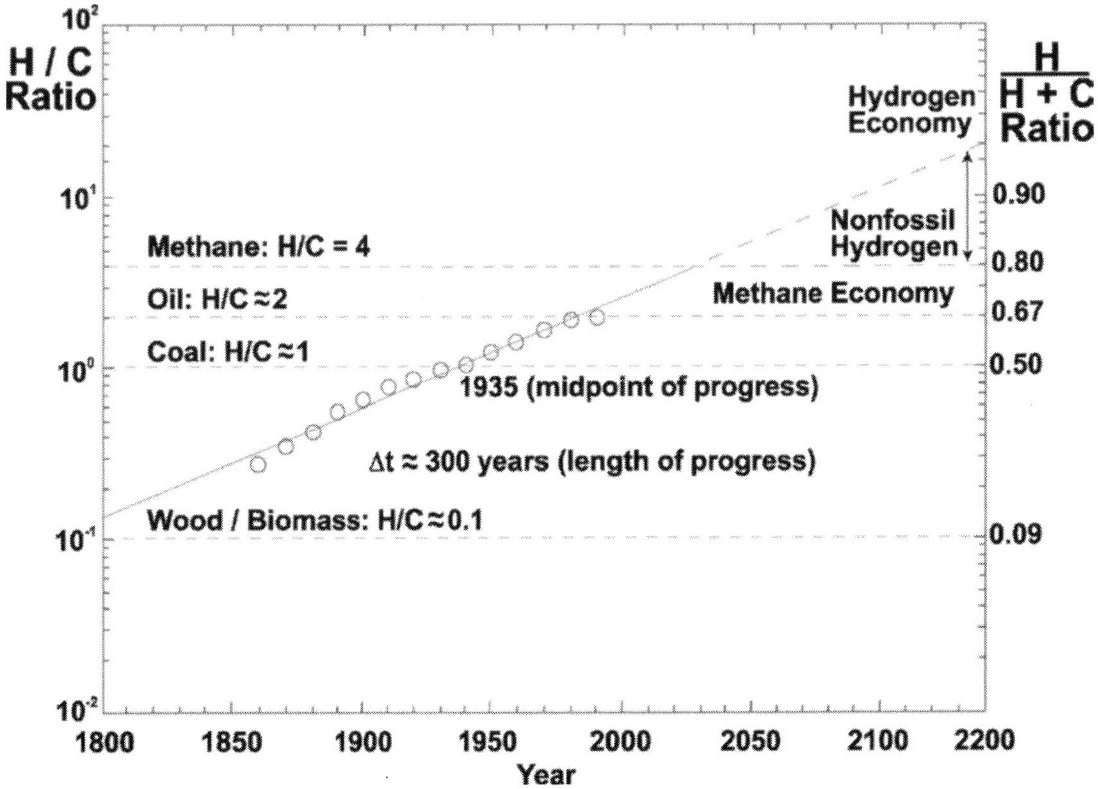


**Figure 2. Hydrogen / carbon ratios in fuels and our evolution toward a hydrogen economy.**[6]

Hydrogen has to be generated from a hydrogen-containing material. Natural gas (methane, $CH_4$) has the highest hydrogen to carbon ratio of the fossil fuels (4:1). By contrast: coal with a ratio of 1:1 has 4 times more carbon per hydrogen. Biomass with a H/C ratio of 0.1 to 1 has 40 times more carbon per molecule than a molecule of methane and thus 40 times more carbon if carbon dioxide ($CO_2$) has to be captured and sequestered. Thus, molecular ratios of feedstocks for conversion to $H_2$ greatly influence both the amount of $H_2$ produced and the amount of $CO_2$ generated that needs to be captured and sequestered.

Selection of natural gas as a starting feedstock has a lot of pros and cons:
Pros:

- Natural gas has the most favorable H/C ratio of available feedstocks.
- Natural gas has extensive collection and delivery infrastructure in North America. This allows for convenient delivery of a feedstock to potential $H_2$ plants, small and large scale throughout the U.S.
- Today, steam-methane reforming (SMR) is the predominant method for $H_2$ produced in the U.S. Steam reforming of naphtha, gasification (partial oxidation, POX) is a small percentage.[7, 8] Catalytic reformers and other refinery offgases are the largest source of $H_2$ within a petroleum refinery, merchant $H_2$ from SMR is used to supplement $H_2$ needs.
- SMR is a catalytic reaction at 1,500°F and is the most efficient and most widely used $H_2$ production technology. While SMR is a mature technology, advances can and are being made in DOE FE's Advanced Synthesis Gas Program that will reduce $H_2$ cost.
- The product of SMR, synthesis gas, goes through a water-gas shift reaction that can produce a $CO_2$ stream that is of commercial quality that, with minor cleanup can be used for food processing. The flue gas from natural gas fired heaters associated with the process yield high quality flue gas (low sulfur oxides).
- As a feedstock, natural gas will remain the dominant feedstock for synthesis gas generation now and for the foreseeable future due to ample supply, moderate price (long-term contracts - not spot market), high-quality (low sulfur content), high $H_2/CO$ ratio in the synthesis gas, moderate capital cost (as compared to coal or biomass processing plants) - even without factoring in the added costs of developing a delivery infrastructure and/or larger volume sequestration of $CO_2$ for these lower H/C ratio solid fuels.
- Startup of a hydrogen economy based on natural gas in a distributed system (local plants rather than gasoline station size) would require only one new infrastructure pipeline for $CO_2$ collection/sequestration, as opposed to large coal or

biomass processing central plants where a feed stock delivery system, $H_2$ distribution system and a $CO_2$ collection/sequestration system would be required.

Cons:

- Natural gas prices have been volatile within recent years in response to growing demand caused by construction of new inexpensive gas fueled electrical generation and restricted access to new gas resources to counteract the effects of natural decline on production capacity in gas reservoirs.
- Demand and supply issues on natural gas are highlighted in a recent National Petroleum Council report "Balancing Natural Gas Policy – Fueling the Demands of a Growing Economy".[9] Adequate supply, even with Alaskan North Slope gas and imports of LNG will remain a problem if gas use for electrical generation continues to increase.
- The hydrogen pathway of energy use is higher than direct use of the natural gas as fuel.

SMR is one way to make $H_2$. Partial oxidation to produce synthesis gas (Figure 3) of coal or natural gas is also used to produce $H_2$ and other chemicals. In refineries, $H_2$ is produced as a by-product of naphtha reforming, any supplemental $H_2$ (95%) is produced from SMR.[7] All these generation methods contribute to meeting the current U.S. $H_2$ demand (year 2000) of about 9 million tons per year, of which 1.5 million tons is merchant $H_2$ that is sold to refineries and chemical industry (mostly in the manufacture of ammonia and other nitrogen-based fertilizers).[8] Reforming liquid hydrocarbons (such as near zero sulfur gasoline, alcohols or ethers) can also produce $H_2$ and is being done today in vehicles fitted with onboard reformers and fuel cells.

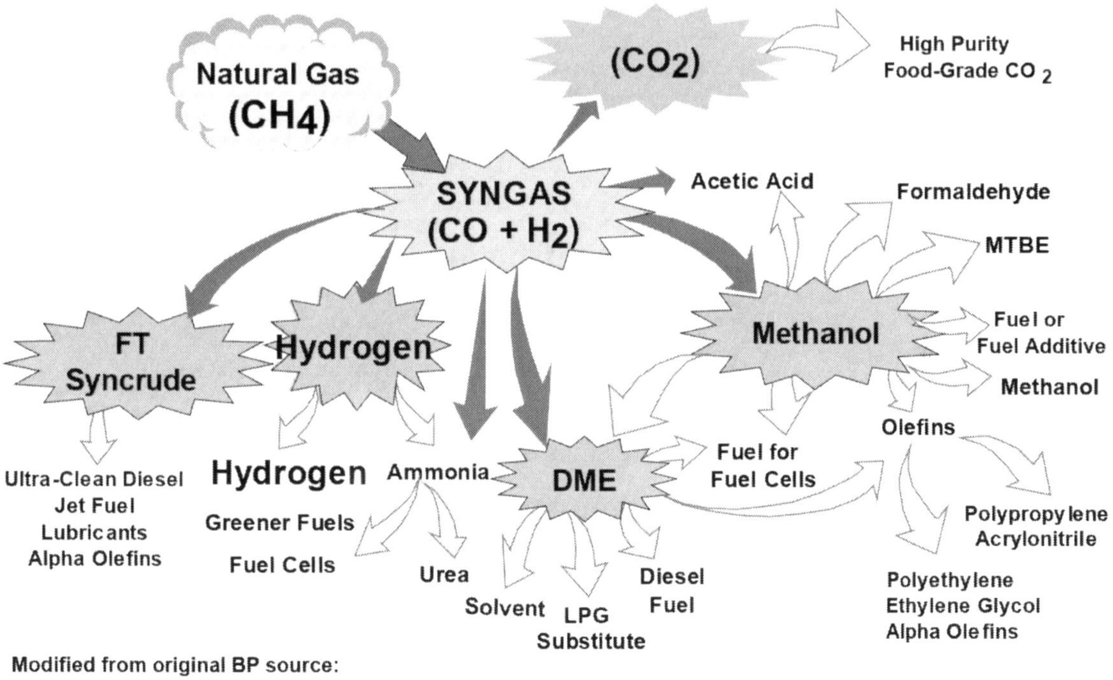


**Figure 3. Relationship between synthesis gas, starting materials and products.**

Much of the current pipeline infrastructure is designed to transport and deliver liquid hydrocarbon fuels: gasoline, diesel, jet fuel, etc. Use of gaseous fuels for transportation (compressed natural gas) is a very small fuel market. Use of less energy dense fuels, such as gaseous $H_2$, poses even more challenges. Because gaseous $H_2$ has a low fuel density and requires large-volume, high-pressure fuel tanks to provide a vehicle with reasonable refueling range, even in very efficient fuel cell vehicles. Thus reforming of clean, near zero or sulfur-free, liquid fuels onboard may prove to be a better route to supplying $H_2$ in the transportation side of a hydrogen economy, but massive volumes of $H_2$ will still be needed to produce these liquid fuels.[10, 11] In order for the Nation to move toward a "hydrogen economy" the source of $H_2$ in the near-term looks to be natural gas. Use of the existing natural gas supply infrastructure to supply fuel for both central and distributed hydrogen plants is one way to bridge the transition between the current liquid fuel infrastructure and a future economy based on $H_2$. Natural gas is a near-term, not a mid-or long-term, solution because natural gas is in limited supply and it's best use is in distributed systems home heating. Coal can extend the bridge of time. Natural gas at the current time and for probably the next 15 years, will be the lowest cost fuel for production of $H_2$, Table 1.

| Resource | Technology | Hydrogen Cost (\$/MMBtu) / (\$/kg)* | Year Technology is Available |
|---|---|---|---|
| Natural Gas** | Steam Methane Reforming, PSA, No Sequestration | 5.54 / 0.75 | Current |
| Natural Gas** | ITM Synthesis Gas Generation, Advanced Membrane Separation, $CO_2$ capture | 4.15 / 0.56 | 2013 |
| Coal | Gasification, Shift, PSA, No Sequestration | 6.83 / 0.92 | Current |
| Coal | Advanced Gasification, Membrane Separation, $CO_2$ Sequestration | 5.89 / 0.79 | 2015+ |
| Coal*** | Advanced Gasification, Membrane Separation, Solid Oxide Fuel Cell for Co-Production of Power, $CO_2$ Sequestration | 2.40 / 0.32 | 2020+ |
| Biomass | Pyrolysis to bio-oil followed by steam reforming (overall efficiency limits use and practicality) | (9 – 16) / (1.21 – 2.16) | 2015+ |
| Nuclear | Sulfur-Iodine Cycle (Thermochemical Process) | 9.70 / 1.31 | 2020+ |
| Electrolysis | Electricity Cost at 4 cents/kWh | (19 – 22) / (2.56 – 2.97) | Current |

*A kg of $H_2$ is roughly equivalent to a gallon of gasoline.
**These two cases are based upon a natural gas price of $3.15/MMBtu. Hydrogen costs will increase or decrease from these values at roughly 1.5 times the change in natural gas price above or below $3.15/MMBtu.
***The hydrogen cost in this case is based upon achievement of the associated Vision 21 Program goals. The value of power produced in the process is assumed to be 53.6 mils/kWh.

**Table 1. Cost comparison of selected hydrogen production technologies.[5]**

## Summary of Research Needs (Conversion of Natural Gas to Hydrogen) from Industry Discussion

On June 4-5, 2003, DOE FE held a Merit Review in Houston, Texas, of 5 DOE/industry cost-shared Advanced Synthesis Gas and Clean Fuels projects. Due to the availability of industry experts, technology developers, energy and industrial gas companies, DOE held a public facilitated open discussion of conversion of natural gas to $H_2$. The objective of the discussion was to get input from experts to identify technology gaps for research in producing and delivering hydrogen from natural gas. Participants were encouraged to define research needs, priorities, conflicts and timing in technologies that have potential to produce hydrogen from natural gas at lower cost and/or higher efficiency. This review has assisted DOE in planning the move toward a hydrogen economy and ideas/suggestions are being incorporated into DOE's Hydrogen from Natural Gas Program Plan currently being developed. The discussion proved insightful for all participants and may stimulate further discussion and insight that could lead to higher-efficiency processes and lower-cost hydrogen. Discussion of research needs (including questions on policy, comments and concerns of participants) in the breakout sessions are categorized and summarized as the following bullets (some ideas/comments are repeated from above because they were voiced by participants and some comments could fall under multiple categories):

**Right Fuel? Right Direction?**

- In light of limited supply and higher prices, is natural gas the best fuel for a transition fuel along the way to $H_2$ or should it be only a home heating/industrial fuel? Use natural gas only to meet current demand for $H_2$
- Are short-term solutions with challenges of supply, cost and $CO_2$ the route to go or should all the hydrogen RD&D effort go to long-term solutions - nuclear? Should DOE omit work on fuels that produce $CO_2$ and work on only permanent solutions/routes for a hydrogen economy?

- Is current technology (SMR and/or partial oxidation) sufficient (volume and economics) to make the volumes of $H_2$ needed to test the viability of a hydrogen economy?
- Should efforts go to making $H_2$ generation on demand (on-board reforming of a liquid) for use in a fuel cell?
    - Use of near zero sulfur liquid reformable high-density, liquid fuels
    - Development of sulfur tolerant catalysts for reformers
    - Development of methods to remove CO from $H_2$, which acts as a poison in fuel cells

**Process Improvements**

- Heat transfer maximized in large and small scale plants
- From POX or SMR, need more robust catalysts

**Separation and Purification**

- Need better, high temperature membranes / processes to separate CO, $H_2$, $CH_4$ and $CO_2$. Need to have high-purity $H_2$ with < 10 ppm CO
- Need clean separation of synthesis gas components (CO from $H_2$) before water gas shift
- Need better natural gas cleanup before use in $H_2$ plants
- Sequestration of $CO_2$ adds to complexity of design and cost, impacting the economics of $H_2$ plants

**New Process Concepts Needed**

- Need for a "one box" process at high pressure and moderate temperature for $H_2$ production and $CO_2$ capture
- Direct conversion of methane to methanol or to other high-density liquid fuel as a hydrogen carrier

**Infrastructure Concerns**

- Reformable higher energy density and lower flammability liquid fuels may be more acceptable than $H_2$
- Plant size – economy of scale problems are immense.
- "Mass production of plants - thousands" as opposed to "one of a kind"
- Can $CO_2$ be captured from a distributed $H_2$ system? Economically? Would higher $CO_2$ emissions be acceptable?
- Gaseous $H_2$ has 730 miles of pipeline in the U.S., but many orders of magnitude larger system is needed for nationwide $H_2$ delivery system. Right-of-way access for new infrastructure is critical because of energy density of $H_2$, pipelines need to be at pressures >3000 psi.
- Is there a need for both a gaseous $H_2$ system and a sequestration $CO_2$ system?

**Chicken and Egg**

- Gaseous $H_2$ versus higher density liquid fuels (ammonia, alcohols, ethers, near zero sulfur gasoline, etc.) as hydrogen carriers?
- Early or late entrance into a hydrogen market? Will a consumer market for $H_2$ develop? When? At what rate? What volume? Will the investment become a stranded investment?
- What are limitations of increased $CO_2$? Beneficial uses of $CO_2$?

**Risk Management**

- Permitting plants with revolutionary new technology is increasingly difficult and time consuming – increasing risk to investors.
- Will the public accept the risks of high-pressure $H_2$ tanks – increasing risk to investors?
- Can we overcome product liability of selling products with known "dangerous properties or dangerous defect" – some technologies have transitioned from the "abnormally dangerous" to "normally dangerous" classification.
- Are industrial gas and energy companies increasing risk to their investors?
- Entire hydrogen system must come together (production, storage, distribution and end use) at a market penetration (>30%) to gain market acceptance. Can this hurdle be overcome?

**Cost versus Return on Investment**

- Will the public and investors accept the risks and cost of a Hydrogen System displacing an existing fuel delivery system?
- Is higher energy use to produce $H_2$ more acceptable then current liquid fuel system or more modest modifications to current fuel system (onboard reformable liquid fuels)?
- Consumers even with Government action (Federal mandates, tax incentives, etc.), end up paying the price for the fuel they use. What is the cost versus benefit of a hydrogen economy? In the near term? In the mid to long-term?
- Is the cost of moving $CO_2$ emissions from millions of sources (including motor vehicles) to local spot sources ($H_2$ production plants with $CO_2$ capture) worth the energy loss and cost when the U.S. has so many other priorities?

## DOE's Natural Gas to Hydrogen Program Plan

DOE FE anticipates a 10-year program (completion about 2015) to develop and demonstrate (with industry, national laboratories and non-government organizations) natural gas to hydrogen conversion technologies that will reduce the cost of $H_2$ production to the range of $4.15 MMBtu $H_2$ based on a natural gas cost of $3.15/MMBtu. Part of these efforts build on developments with industry in DOE's Advanced Synthesis Gas and Clean Fuels programs (where synthesis gas, membrane technology and high-temperature $H_2$ separation have been studied) and on the immense volume of hydrogen RD&D conducted internationally in which DOE participates. Details of a working program Natural Gas to Hydrogen Multi-Year Program Plan are anticipated within the next few months.

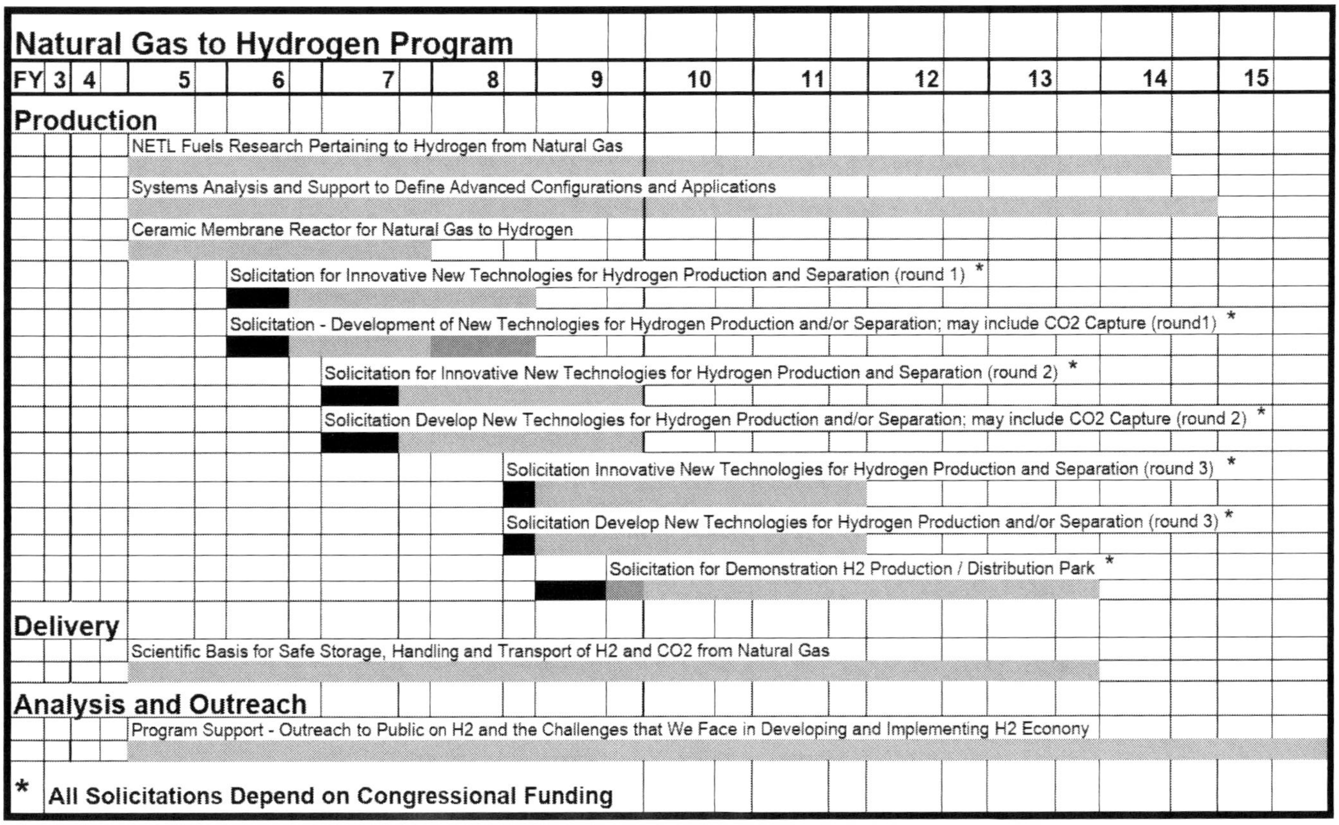


**Figure 4. DOE FE's natural gas to hydrogen program GANTT chart with program completion by 2015.**

### REFERENCES


1. National Energy Policy. Report of the National Energy Policy Development Group, May 2001.
2. President George W. Bush in his 2003 State of the Union address to Congress.
3. U.S. DOE, A National Vision of America's Transition to a Hydrogen Economy – to 2030 and Beyond, 2003.
4. U.S. DOE, National Hydrogen Energy Roadmap, 2003
5. Hydrogen from Natural Gas and Coal: The Road to a Sustainable Energy Future, June 2003 DOE FE Hydrogen Program Plan
6. Ausubel, Jesse H. American Scientist, March-April 1996..
7. SFA Pacific Inc. Hydrogen and Synthesis Gas: Technology, Economics and Market Outlook, January 1998.
8. Katsaros, Arthur. Air Products and Chemicals, Inc. U.S. Industrial Hydrogen Infrastructure Presentation. November 2001.
9. National Petroleum Council, Balancing Natural Gas Policy – Fueling the Demands of a Growing Economy, National Petroleum Council, September 2003
10. National Petroleum Council, 1995. Research, Development, and Demonstration Needs of the Oil and Gas Industry, National Petroleum Council, http://www.npc.org.
11. National Petroleum Council, U.S. Petroleum Refining – Assuring the Adequacy and Affordability of Cleaner Fuels, National Petroleum Council, June 2000.

## Synthesis Gas Production

*Peter Seier Christensen, Kim Aasberg-Petersen, Martin Østberg, and Thomas S. Christensen, Haldor Topsøe A/S, Research & Development Dept., DK-2800 Lyngby, Denmark*

### 1. Introduction

Large-scale processes for conversion of natural gas into liquids such as methanol or synthetic diesel by the Fischer-Tropsch synthesis are attracting increasing attention with several plants under construction or in the design phase. The cost of the synthesis gas production constitutes a large fraction of the plant investment. Autothermal Reforming (ATR) with oxygen is the most economic and preferred technology for production of synthesis gas for these applications.

The use of fuel cells for combined production of power and heat is being intensively pursued as a means of increasing the electrical efficiencies and of reducing the environmental impact of power production. The required conversion of the hydrocarbon feedstock into a gas acceptable for the fuel cell may take place either by steam reforming or by (catalytic) partial oxidation (CPO). In the latter case, air is used as the oxidant.

This paper presents pilot plant experiments for CPO and ATR and discusses the advantages of the technologies in small and large scale.

### 2. Autothermal Reforming

A Gas-to-Liquids (GTL) plant for production of synthetic diesel consists of synthesis gas generation (SGU), Fischer-Tropsch synthesis, and a section for product work-up and purification. The first section is the main contributor to the investment of the total plant. In Fig. 1, the process concept for synthesis gas production by ATR [1] is illustrated. In a GTL plant, Carbon Dioxide is typically added to the ATR feed gas to adjust the $H_2/CO$-ratio in the product gas to around 2.

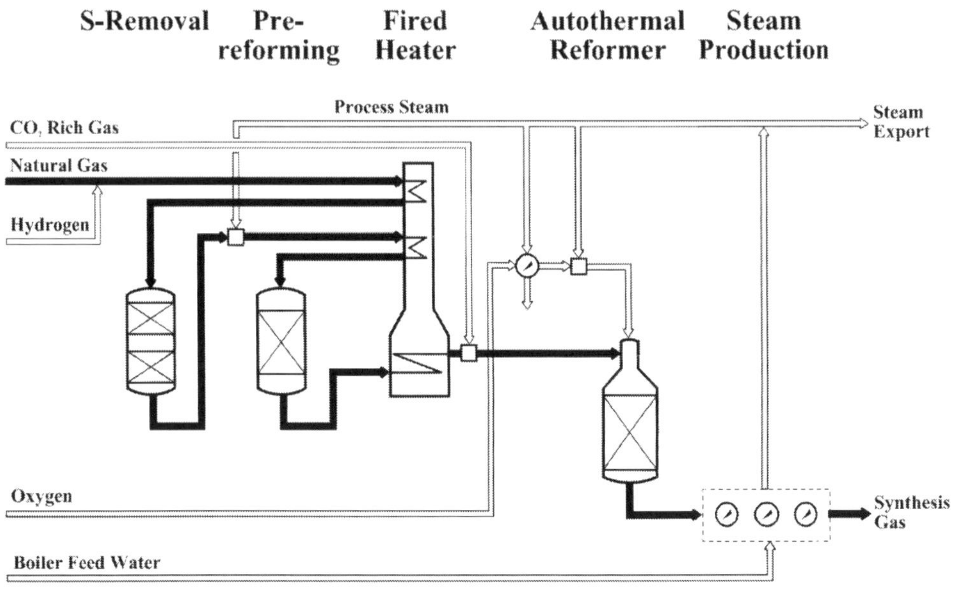


Figure 1 Synthesis gas production by ATR [1]

The ATR reactor principle is illustrated in Fig. 2. The compact ATR reactor contains a burner, a combustion chamber, and a catalytic bed placed in a refractory lined vessel. The prereformed natural gas reacts with oxygen and steam in a substoichiometric flame. In the catalyst bed, the methane steam reforming and shift reactions are equilibrated typically with a nickel catalyst.

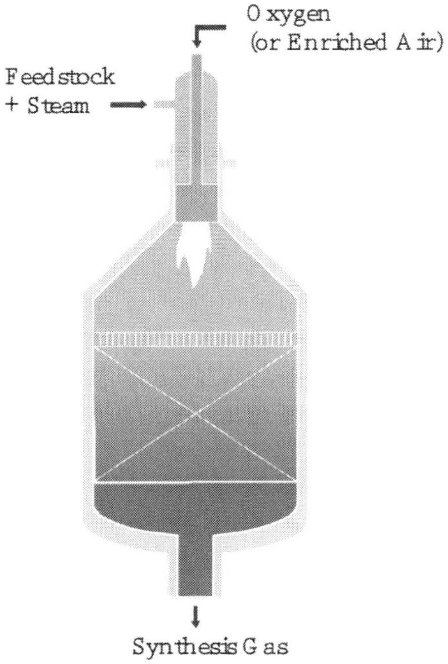


Figure 2: Principle of ATR [2,3]

Haldor Topsøe A/S have commercialised the ATR technology at a steam-to-carbon ratio of 0.6. A European plant has been in operation for around two years and other large-scale plants are in advanced stages of engineering or construction in Nigeria and Qatar.

The cost of the SGU, including the Air Separation Unit (ASU), is the largest fraction of the plant capital. Reduction of the steam-to-carbon (S/C) ratio is one option for further optimising the ATR-based front-end. In Table 1, key parameters are given for an ATR-based front-end showing the advantages of reducing the S/C-ratio. The oxygen consumption, the recycle flow and the total flow through the SGU are lower which results in reduced capital cost per unit of syngas produced. The fraction of hydrogen and carbon monoxide in the synthesis gas is also increased. The influence of the preheating temperature of the feed to the ATR is also shown in Table 1. It is evident that a high preheat temperature is desirable to minimise the oxygen consumption considering the high cost of the ASU.

Table 1 Relative flows and syngas inert concentration in ATR-based front-end for production of equivalent amount of hydrogen and carbon monoxide ($H_2/CO$ = 2.00) in a GTL plant at various conditions. The desired $H_2/CO$-ratio is obtained by addition of pure $CO_2$

| | | | |
|---|---|---|---|
| Hydrocarbon feed temperature (° C) | 650 | 650 | 450 |
| S/C (mole/mole) | 0.6 | 0.4 | 0.6 |
| $O_2$-consumption | 100 | 98 | 109 |
| $CO_2$-addition | 100 | 62 | 92 |
| ATR exit flow | 100 | 93 | 100 |

Operation at lower S/C-ratio reduces the margin to carbon formation in the prereformer and to soot formation in the ATR. Fundamental studies and extensive pilot plant experiments have been performed to both understand and push the limits towards lower S/C-ratios. In Table 2, examples of operating conditions without soot formation are given. Operation for an extended period of time at S/C < 0.40 has been demonstrated in Topsoe's process demonstration unit.

Table 2: ATR pilot plant tests at low S/C-ratios [2,4]

| **Test** | **A** | **B** | **C** | **D** |
|---|---|---|---|---|
| S/C | 0.21 | 0.51 | 0.60 | 0.36 |
| $CO_2/C$[1] | 0.01 | 0.19 | 0.01 | 0.01 |
| $O_2/C$[1,2] | 0.59 | 0.62 | 0.58 | 0.57 |
| Exit temperature (° C) | 1065 | 1025 | 1020 | 1022 |
| Pressure (bar) | 24.5 | 27.5 | 28.5 | 28.5 |
| $H_2/CO$ (mole/mole) | 1.96 | 1.80 | 2.30 | 2.15 |

[1]: Mole/mole of hydrocarbon atoms

[2]: The $O_2$ ratio is about 5% higher than truly adiabatic reactors with same exit temperature

## 3. Catalytic Partial Oxidation

The principle of CPO is illustrated in Fig. 3. The hydrocarbon feed and the oxidant are mixed in an inlet zone upstream the catalyst bed. A thermal shield is often employed between the inlet zone and the catalyst bed to avoid overheating. In the catalytic section, the mixture reacts by partial and total combustion along with the methane steam reforming and shift reactions. The latter two reactions are typically in equilibrium at the reactor outlet and the effluent is free of oxygen. The catalyst is normally based on noble metals and the space velocity is in many cases very high.

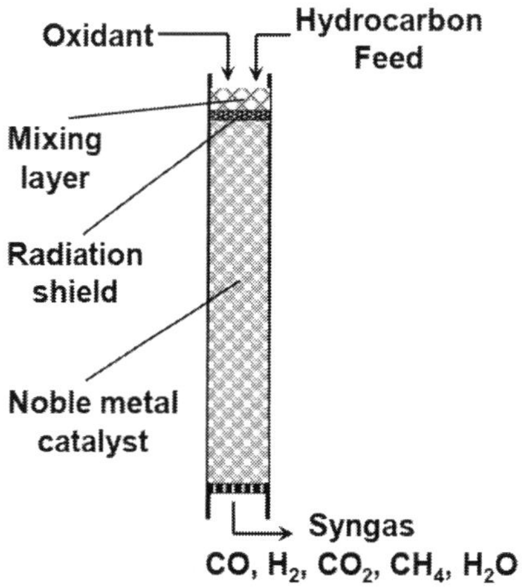


Figure 3 CPO principle

The presence of a flammable mixture in the inlet zone upstream the catalyst may in some cases make use of CPO at high inlet temperatures difficult. This is especially the case at elevated pressures. In Table 3 the auto-ignition temperature is given for natural gas in air as a function of pressure. The auto-ignition temperatures are lower with oxygen as oxidant. For operation below the auto-ignition temperature, the inlet feed temperatures of the hydrocarbon feedstock and oxidant must be kept low. For GTL CPO does not appear economical considering that the ASU is about 40% of the total SGU cost [1,6].

Table 3 Auto-ignition temperatures for natural gas in air [5]

| Pressure (bar) | Auto-ignition temperature (° C) |
|---|---|
| 1 | 465 |
| 4 | 313 |
| 20 | 267 |
| 40 | 259 |

Experiments with CPO have been conducted at the Topsoe pilot plant in Houston, Texas. Selected results at various pressures with and without steam in the feed are presented in Fig. 4. In all cases, the methane conversion is close to equilibrium.

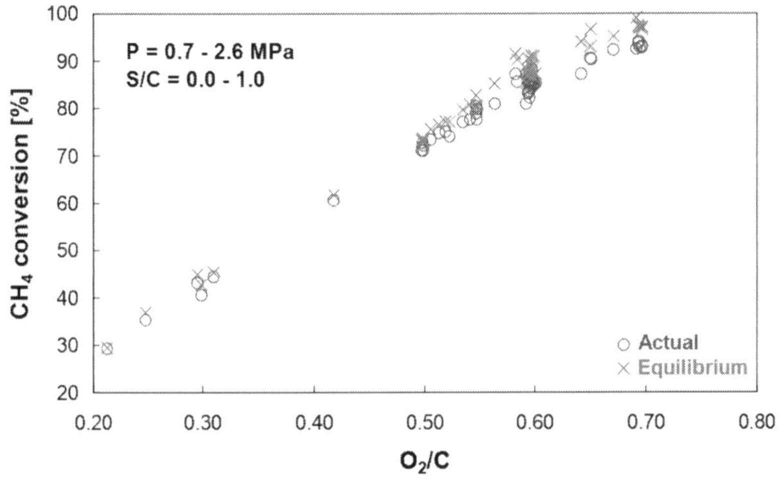


Figure 4 Methane conversion in CPO pilot plant experiments with air as oxidant [7]

The use of CPO with air as oxidant for conversion of hydrocarbon feed stocks ranging from natural gas to diesel in fuel cell plants is attracting increased attention. A main advantage of applying CPO in a fuel cell plant includes its autothermal nature, its quick start-up time and the fact that no steam is required in some cases. The alternative to CPO is the use of steam reforming. Steam reforming has the advantage of a higher electrical efficiency and a higher hydrogen yield at the reformer exit. In addition no noble metal catalyst is needed. The choice of CPO or steam reforming will depend on the actual objectives and site specific issues for a given unit.

## 4. Conclusions

Oxygen-blown Autothermal Reforming is the preferred technology for large scale synthesis gas production for GTL-plants. Reducing the steam-to-carbon ratio will further improve plant economics. Pilot plant experiments have demonstrated operation at steam-to-carbon ratios much lower than practiced in industry today.

The use of CPO at high pressure and elevated feed temperatures may be difficult due to the existence of a flammable mixture in the inlet zone. For fuel cell applications, air-blown CPO may be preferred to steam reforming due to the fast start-up, compact nature, and the lack of steam requirements in some cases. However, the drawback of using CPO for fuel cell applications is a lower overall electrical efficiency.

## 6. References

[1] K. Aasberg-Petersen, T.S. Christensen, C. Stub Nielsen, I. Dybkjær: Recent Developments in Autothermal Reforming and Prereforming for Synthesis Gas Production in GTL Applications, ACS Petroleum Chemistry Div. Preprints: Vol. 47, P. 96-97

[2] T.S. Christensen, M. Østberg, J.-H. Bak Hansen, Process Demonstration of Autothermal Reforming at Low Steam-to-Carbon Ratios for Production of Synthesis Gas, AIChE Annual Meeting, Reno, USA, Nov. 4-9, 2001
[3] J.R. Rostrup-Nielsen, J. Sehested, J.K. Nørskov, Hydrogen and Synthesis Gas by Steam and $CO_2$-Reforming, Advances in Catalysis, Adv.Catal. 47 (2002)
[4] T.S. Christensen, I.I. Primdahl, Improved Syngas Production Using Autothermal Reforming, Hydrocarbon Proc., 73 (1994) 39
[5] Physical Properties of Natural Gases, Groningen: N.V. Nederlandse Gasunie, 1988
[6] I. Dybkjær, T.S. Christensen, Syngas for Large-Scale Conversion of Natural Gas to Liquid Fuels, Stud. Surf. Sci. Catal., 136 (2001) 435
[7] K. Aasberg-Petersen, L. Basini and A. Guarinoni, M. Østberg, Catalytic Partial Oxidation of Natural Gas, AIChE Spring Meeting, New Orleans, USA, March 10 - 14, 2002

# Development of the ITM Syngas Ceramic Membrane Technology

*Joseph M. Abrardo[1], Douglas L. Bennett[1], Michael F. Carolan[1], Christopher M. Chen* [1], William L. Schinski[2], James J. Steppan[3], William E. Waldron[1]*

[1] *Air Products and Chemicals, Inc., 7201 Hamilton Blvd, Allentown, PA, 18195 USA*
[2] *ChevronTexaco Energy Research and Technology Co., 100 Chevron Way, Richmond, CA, 94802 USA*
[3] *Ceramatec, Inc., 2425 South 900 West, Salt Lake City, UT, 84119 USA*



Prepared for Presentation at the AIChE Spring National Meeting
New Orleans, LA, April 26, 2004



## Abstract

The ITM Syngas Team, led by Air Products and including ChevronTexaco, Ceramatec, and other partners, in collaboration with the U.S. Department of Energy, is developing Ion Transport Membrane (ITM) technology for the production of synthesis gas. ITM ceramic membranes are fabricated from non-porous, multi-component metallic oxides and operate at high temperatures with exceptionally high oxygen flux and selectivity. Oxygen from low-pressure air permeates, as oxygen ions, through the ceramic membrane and is consumed through chemical reactions, thus creating a chemical driving force that pulls oxygen ions across the membrane at high rates. The oxygen reacts with natural gas in a partial oxidation process to produce a hydrogen and carbon monoxide mixture (synthesis gas).

The ITM Syngas process is a breakthrough technology that combines air separation and high-temperature synthesis gas generation into a single ceramic membrane reactor, resulting in significant savings in the capital cost of synthesis gas production. Because synthesis gas is a feedstock for a range of different processes, ITM Syngas represents a technology platform that has numerous applications, such as gas-to-liquids; hydrogen; ultra-clean fuels, including liquid transportation fuels; and chemicals such as methanol.

Significant advances have been made in developing and scaling up the ITM Syngas technology. This paper describes the selection of membrane materials, the scale up of ceramic membrane fabrication, and the operation of a 24,000 SCFD Process Development Unit.

* Correspondence should be addressed to C. M. Chen, email: chencm@airproducts.com, tel: +1 (610) 481-3315

## Introduction

The ITM Syngas Team, led by Air Products and including ChevronTexaco, Ceramatec, and other partners, in collaboration with the U.S. Department of Energy, is developing Ion Transport Membrane (ITM) technology for the production of synthesis gas. ITM ceramic membranes are fabricated from non-porous, multi-component metallic oxides and operate at high temperatures with exceptionally high oxygen flux and selectivity. A conceptualization of the ITM Syngas technology is shown in Fig. 1. Oxygen from low-pressure air permeates, as oxygen ions, through the ceramic membrane and is consumed through chemical reactions, thus creating a chemical driving force that pulls oxygen ions across the membrane at high rates. The oxygen reacts with natural gas in a partial oxidation process to produce a hydrogen and carbon monoxide mixture (synthesis gas).

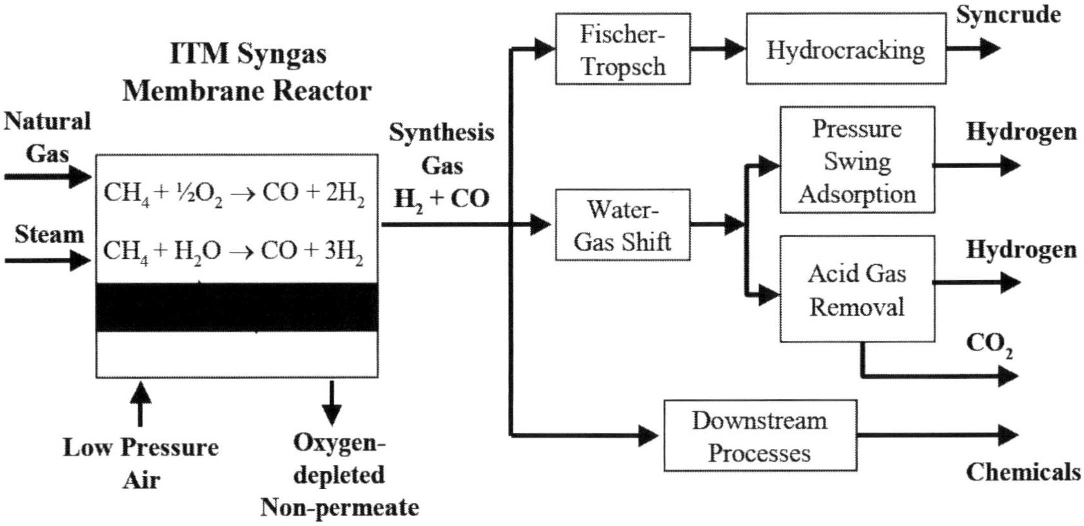


Fig. 1. Conceptual ITM Syngas process with several applications.

## Technology Applications

The ITM Syngas process is a breakthrough technology that combines air separation and high-temperature synthesis gas generation processes into a single ceramic membrane reactor, with significant savings in the capital cost of synthesis gas production.[1-4] Because synthesis gas is a feedstock for a range of different processes, ITM Syngas represents a technology platform that has numerous applications, such as gas-to-liquids; hydrogen; ultra-clean fuels, including liquid transportation fuels; and chemicals such as methanol (Fig. 1).

Process designs and cost estimates were developed for single-train Fischer-Tropsch Gas-to-Liquids plants producing 6,000 to 18,000 barrels/day. The capital cost of the synthesis gas process was projected to be 30% lower for ITM Syngas compared to conventional technology employing oxygen-blown Autothermal Reforming (ATR).

Applications to co-produce hydrogen and carbon dioxide, such as use of the hydrogen as a decarbonized fuel, were also evaluated. ITM Syngas processes producing 150 to 760 MMSCFD (million standard cubic feet per day) of fuel-grade hydrogen were compared to conventional processes based on oxygen-blown ATR with cryogenic air separation to supply oxygen. Economic evaluation of the ITM Syngas process showed the potential for 30% capital cost savings in the syngas generation process area and over 20% capital cost savings for the overall synthesis gas/$H_2$/$CO_2$ production process. The ITM Syngas process also had a higher thermal efficiency of 74% compared to 71% for the oxygen-blown ATR process. The capital cost and efficiency advantages for ITM Syngas are due to the combination of oxygen separation and high temperature synthesis gas production into a single unit operation and the ability to use low-pressure air as the oxidant source. The hydrogen product is a de-carbonized fuel suitable for centralized power generation and for distribution to local stationary or mobile applications, including fuel cells.

**Materials Development**

The ITM Syngas process places severe demands on the membrane material. The membrane must simultaneously meet the criteria of being thermodynamically stable in the high-pressure, reducing, natural gas feed and intermediate synthesis gas; being thermodynamically stable in the low-pressure, oxidizing air feed; having sufficient mixed electronic and oxygen ion conductivity to achieve economically attractive oxygen fluxes; and having the requisite mechanical properties (strength, creep, and expansion) to meet lifetime and reliability criteria. Through the careful tailoring of the membrane composition, a patented family of materials [5] developed by the ITM Syngas team meets these criteria.

The economics of the syngas production look most attractive when a high-pressure natural gas feed and a low-pressure air feed are used to produce a high-pressure synthesis gas product to match downstream process pressures, thus avoiding the expenses associated with compressing the air or the synthesis gas. A large hydrostatic pressure difference will exist across the membrane from the synthesis gas side of the membrane to the air side of the membrane. The requirement of a long service life for the membrane will call for the membrane material to have a sufficiently low creep rate at process temperatures to avoid excessive creep deformation or creep rupture of the membrane.

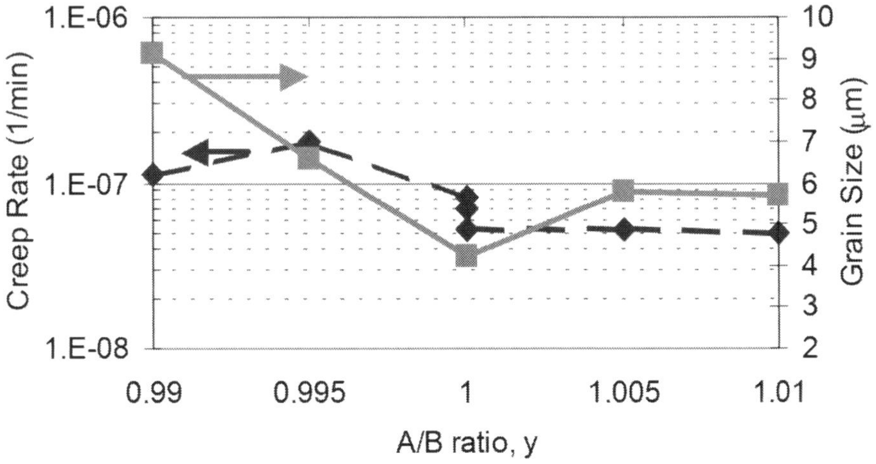


Fig. 2. Creep rate as a function of A/B cation ratio, y

The family of materials which has been carefully designed to meet this condition are perovskites with the formula $(La_{1-x}Ca_x)_yFeO_{3-\delta}$ where $0<x<0.5$; $1.0<y$ and $\delta$ makes the compound charge neutral. The A/B cation ratio, y, has a large effect on the observed creep rates. Dense bars of a specific composition within the family of materials were prepared where y=0.98, 0.99, 1.00, 1.01 and 1.02. Fig. 2 shows the 4-point bending creep rate [6] as a function of the A/B ratio, y, along with the grain size [7] of the samples. The creep rate decreases by factor of 2 as the A/B ratio increases from less than 1.0 to > 1.0. This is even more impressive when one considers that the creep rate can increase significantly with decreasing grain size [8]. The samples (A/B ratio of y > 1) with the lowest creep rates, had a much smaller grain size than the samples (y < 1.0) with the fastest creep rates. Therefore, lower creep rates and longer membrane life can be obtained by using A/B ratios > 1.0.

**Ceramic Membrane Fabrication**

Since the initiation of work to develop planar ITM Syngas membranes, the active area of membrane articles has increased approximately 140 fold. The first planar articles were small disks that featured membrane thicknesses similar to the commercial requirements. Next, rectangular pilot-size wafers were fabricated that had over 30 times the active area of the disks. These hermetic double-sided "wafers" had an active ITM membrane on both primary surfaces of the planar body and a common internal air flow system. The pilot-size wafers included all the critical engineering features found in commercial-scale wafers, and therefore provided an excellent test vehicle for the development of commercial-scale full-size wafer fabrication processes.

Development work began in 2001 to fabricate full-size wafers. An example of a full-size wafer is shown in Fig. 3. A double-sided planar ITM Syngas wafer was designed that will utilize commercially practiced fabrication methods, including tape casting, laser cutting, lamination, and co-sintering. The design features a symmetric planar structure composed of four separate active regions – two on each primary planar surface and flanking the central air manifold as shown in Fig. 3 as the dark regions. Air flows inside the wafer from end to end in a series of long parallel channels built into the substructure of the wafer. Channel dimensions were optimized to minimize flow resistance and maximize membrane support. A porous layer was included on the exterior of the active ITM wafer regions (dark regions) to help support the thin separating layer and control reaction rates.

An ITM Syngas module can be created by stacking wafers vertically. An example ITM Syngas module is shown in Fig. 4, which shows a 3-wafer stack built from pilot-size wafers. The wafers were joined and sealed to ceramic "spacers" that create an open channel for the flow of natural gas and synthesis gas between wafers (Fig. 4). All ceramic parts were joined and sealed using a ceramic sealant suitable for operating temperatures up to 1100°C.

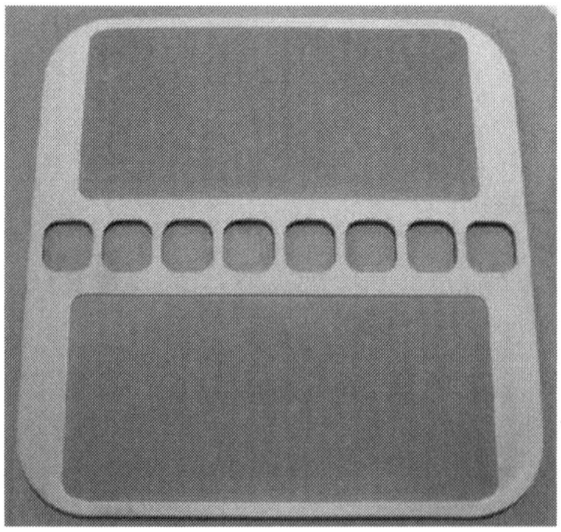

Fig. 3. Commercial-scale full-size wafer showing two active regions (darker area) on either side of the central air manifold.

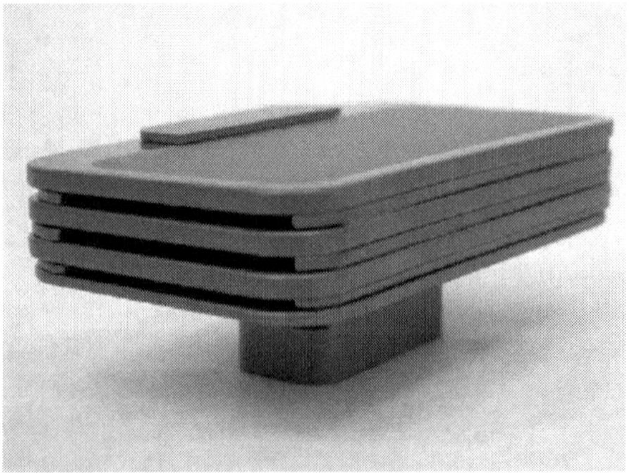

Fig. 4. 3-wafer module built from pilot-size wafers with all-ceramic seals.

**Process Development Unit**

A laboratory Process Development Unit (PDU) was constructed at Air Products and is used to test planar ceramic membranes. The PDU operates at nominally 24 MSCFD (thousand standard cubic feet per day) of synthesis gas capacity, and is used to evaluate membrane performance under full process operating conditions. The PDU, shown in Figure 5, is suitable for testing pilot- and full-sized membranes.

Recent PDU tests with pilot-size membranes included operation under synthesis gas conditions at temperatures between 825 and 900°C and pressures between 215 and 425 psig. Membranes demonstrated the ability to operate through a range of temperature, pressure, and composition conditions.

Figure 5. The ITM Syngas Process Development Unit.

## Conclusions

Significant progress has been made in materials selection, membrane design and fabrication, and catalyst development, which are necessary steps toward commercializing the ITM Syngas technology. The potential economic benefits of the technology continue to be very promising.

## Acknowledgement

The authors gratefully acknowledge the work of ITM Syngas team members at Air Products and Chemicals Inc., Ceramatec Inc., ChevronTexaco Energy Research and Technology Co., Norsk Hydro, Eltron Research Inc., McDermott Technology Inc., Penn State University, University of Pennsylvania, and Pacific Northwest National Lab. The co-funding by the U.S. DOE (DE-FC26-97FT96052) is also gratefully acknowledged.

## References

[1] S. Nataraj, R.B. Moore, S.L. Russek, U.S. Patent 6048472 (2000).
[2] S. Nataraj, S.L. Russek, U.S. Patent 6077323 (2000).
[3] S. Nataraj, S.L. Russek, U.S. Patent 6110979 (2000).
[4] S. Nataraj, P. N. Dyer, S.L. Russek, U.S.Patent 6114400 (2000).
[5] P.N. Dyer, M.F. Carolan, D. Butt, R.H.E. VanDoorn, and R.A. Cutler, Mixed Conducting Membranes for Syngas Production, US Patent No. 6492290 B1 (2002).
[6] D. Cramer and D. Richerson, Mechanical Testing Methodology for Ceramic Design and Reliability, Marcel Decker, Inc., New York, 1998.
[7] American Society of Testing and Materials, ASTM C1211-92.
[8] A. Hynes and R. Doremus, Critical Reviews in Solid State and Materials Sciences, 21(2), 129-187 (1996).

## The application of oxygen transport membranes for syngas or hydrogen production

*E.T. (Skip) Robinson, Principal Investigator, Torix, Inc.*
*Siv Aasland, Staff Engineer, Statoil, ASA*
*Jack Chen, Senior Development Associate, Praxair, Inc.*

Prepared for presentation at Spring AIChE Meeting, April 26, 2004, Session T9008, Natural Gas Utilization, Synthesis Gas Generation, paper 68d. Unpublished, AIChE Shall Not Be Responsible for Statements or Opinions Contained in Papers or Printed in its Publications.

### Introduction

Praxair, Statoil and Foster Wheeler are engaged in the development of oxygen transport membranes for the production of syngas and hydrogen from natural gas. Syngas (hydrogen and carbon monoxide in a 2:1 molar ratio) is growing in importance as the intermediate feedstock for Fischer Tropsch synthesis of ultra-clean liquid fuels or synthetic crude oil. Hydrogen demand by U.S refineries is increasing due to a heavier crude slate and EPA mandated reductions in diesel fuel sulfur content. In addition, hydrogen is being touted as the next generation ultra-clean fuel for transportation and power generation.

Oxygen transport membranes (OTM) have the potential to lower the costs, improve yields and reduce NOx and $CO_2$ emissions in comparison to conventional steam methane reforming for hydrogen production or autothermal reforming for syngas production.

### Oxygen transport membranes

The oxygen transport membranes in this work are ceramic mixed conducting materials that permit both oxygen anions and electrons to move freely through the material. The membranes are inorganic crystalline metal oxides with the general composition ABOx, where A and B are cations and x is a number that satisfies the valences of A and B. The crystalline structure allows oxygen ions to diffuse in the direction of decreasing oxygen partial pressure, provided sufficient electronic conductivity is present to permit the counter migration of electrons. The oxygen separation process, illustrated in Figure 1, involves four steps: 1) molecular oxygen diffuses to the membrane surface on the air side, 2) molecular oxygen dissociates to oxide ions, consuming electrons, 3) oxide ions diffuse across the membrane by moving through oxygen vacancies, electrons move in the opposite direction, and 4) oxygen ions recombine or react with process gas on the fuel side, liberating electrons. The rate of oxygen diffusion (oxygen flux) is proportional to the log of the oxygen partial pressure gradient. For good ionic conducting materials at high temperature, other factors might be rate limiting such as transport of oxygen to the membrane surface or diffusion of oxidation products away from the opposite surface. At lower temperatures, ionic conductivity, an intrinsic property of the ceramic material, is often rate limiting. Most materials require temperatures in excess of 600°C and a partial pressure gradient (log [$pO_2$ air/$pO_2$ fuel]) of four or more to achieve an oxygen flux of commercial interest[1].

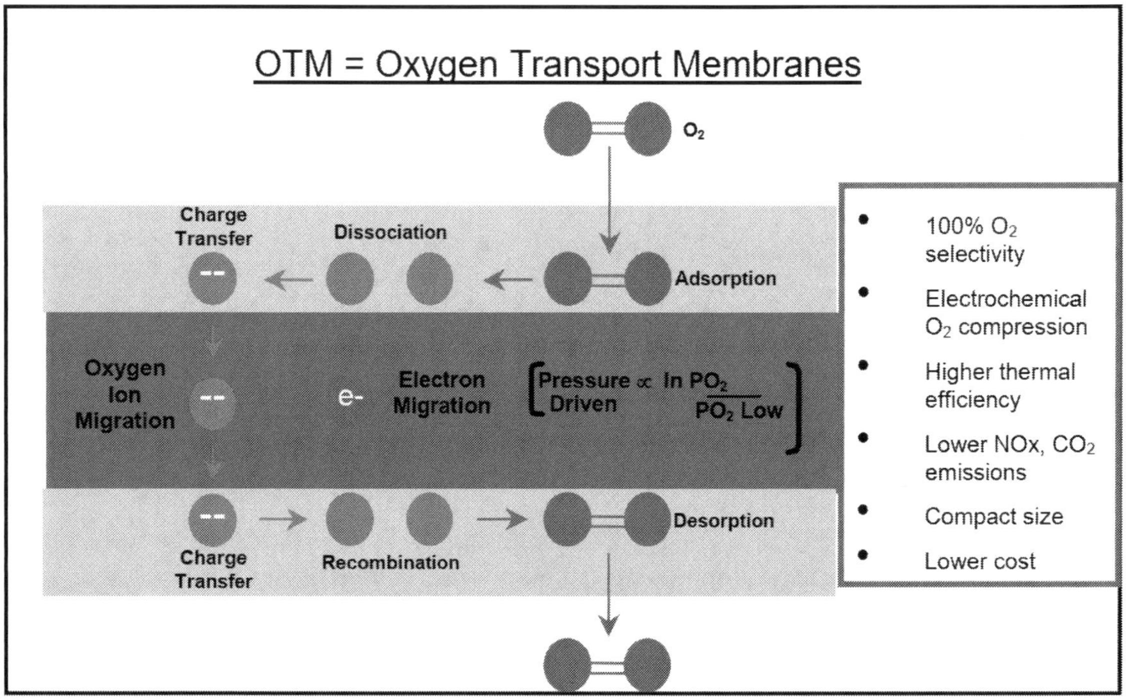


**Figure 1**. Oxygen transport membrane separation mechanism

Optimization of oxygen transport membranes for commercial applications

The critical performance factors for a ceramic oxygen transport membrane are oxygen flux, strength and durability. Oxygen flux, usually reported in units of standard cubic centimeters of oxygen per minute per square centimeter ($sccm/cm^2$), will determine the amount of membrane surface area and reactor volume required for a given application. Commercially attractive fluxes have been achieved and demonstrated at the pilot plant scale. However the major hurdle has been membrane robustness, durability and life.

An OTM must be thin (<1 mm) to achieve the flux target, strong to withstand static pressure differentials of 15 to 30 bar and durable to survive mechanical stresses caused by thermal and chemical cycling, while always remaining leak-tight. Durable materials are also physically and chemically stable over the expected range of environmental conditions (temperature, pressure, gas composition). The program focus for the last two years has been to achieve these goals in a single system. The successful approach has been to develop composite structures consisting of a very robust porous substrate coated with a gas tight, thin film, durable oxygen transport membrane. The film and substrate materials must be chemically compatible and have similar thermal expansion behavior. The substrate must have high strength and low creep rate at the maximum expected temperature of operation. The high temperature strengths (four point bend) of a lead candidate substrate and OTM film materials are shown in Figure 2.

The OTM film must be uniformly dense, thin, strongly bound to the substrate, and durable enough to survive numerous thermal cycles and oxygen partial pressure gradients that can induce

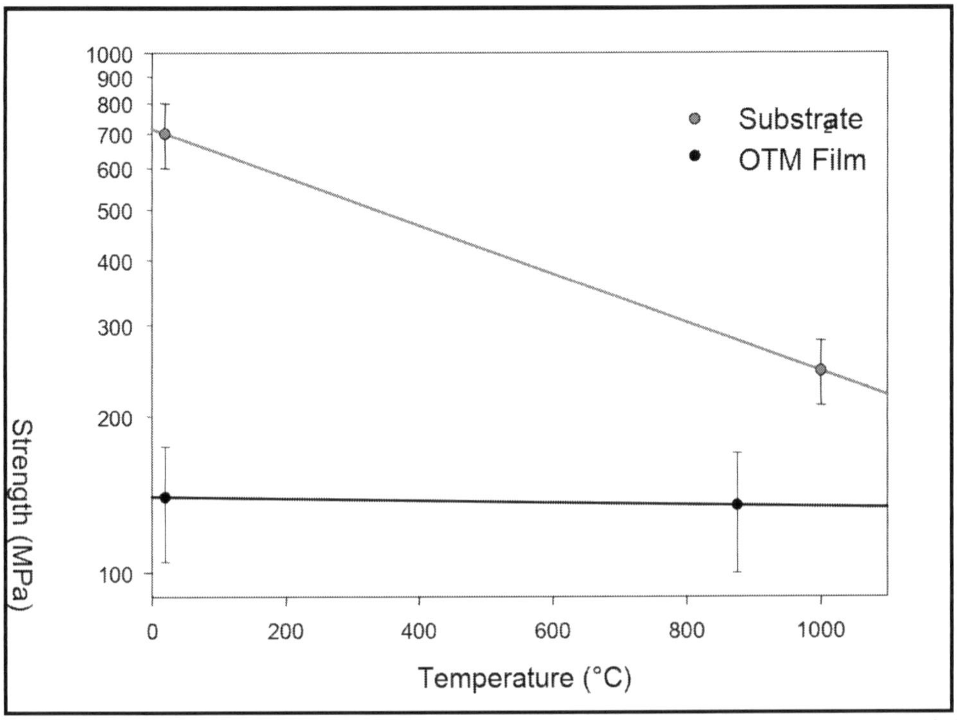


**Figure 2.** High temperature strength of lead substrate and OTM film materials

deleterious stress profiles in the membrane itself. Most highly conductive OTM materials will expand under low oxygen partial pressure. This so called "chemical expansion" will place part of the membrane in tension which can lead to failure. Figure 3 shows the chemical expansion behavior of a number of OTM material candidates as they are subjected to a range of oxygen partial pressures.

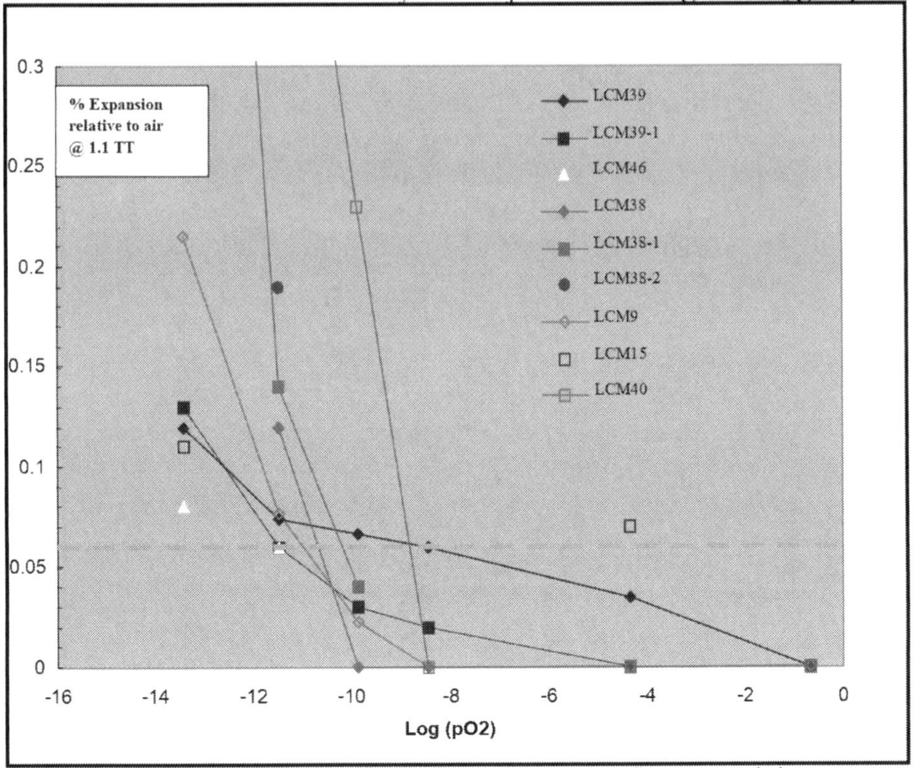


**Figure 3:** Chemical expansion behavior of OTM film materials

Optimization of the film material requires developing compositions with minimal chemical expansion, adequate oxygen flux and sufficient strength while retaining thermal expansion and chemical compatibility with the substrate materials. Figure 4 shows the effect of varying one component of a multiphase material on flux.

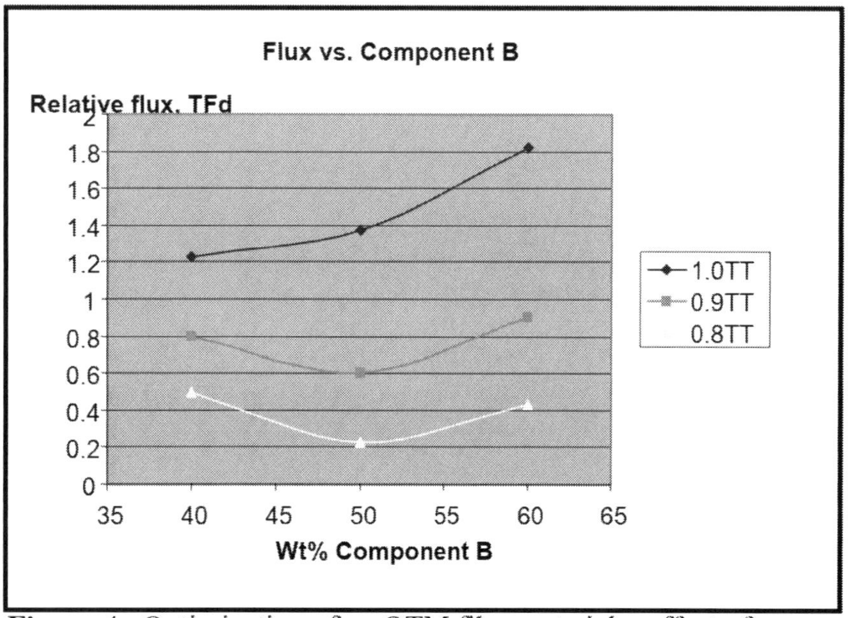


**Figure 4.** Optimization of an OTM film material – effect of composition on flux

Films and substrate candidate materials that meet minimum standards for compatibility, flux and strength are formed into bi-layer coupons and thermally cycled in ovens. Materials that pass the oven cycling test are then fabricated into disks which represent a small cross-section of a full size OTM element. The disks are operated at commercial target temperatures and gas compositions, and cycled to represent start-up and shut-down events. This durability/life test is operated for a minimum of 500 hours. At completion of the run, the disk is removed, leak tested, sectioned and examined by SEM. Results of a recent test are shown in Figures 5. After 500 hours at the target temperature (TT) and 10 thermal and 12 chemical cycles, the flux (TFc) did not decline, nor did the leak rate increase. Post mortem showed the film to be intact, firmly bonded to the substrate and air tight. This system is now being scaled-up to continue testing in larger pilot plants.

Applications of oxygen transport membranes for GTL syngas

An excellent application of an OTM system is the production of syngas gas for Fischer-Tropsch gas-to liquid synthesis (FTGTL). FTGTL is a commercially proven process for producing synthetic crude oil or ultra-clean transportation fuels from natural gas or coal. A typical FTGTL process is illustrated in Figure 6. High purity oxygen is reacted with natural gas and steam at high pressure

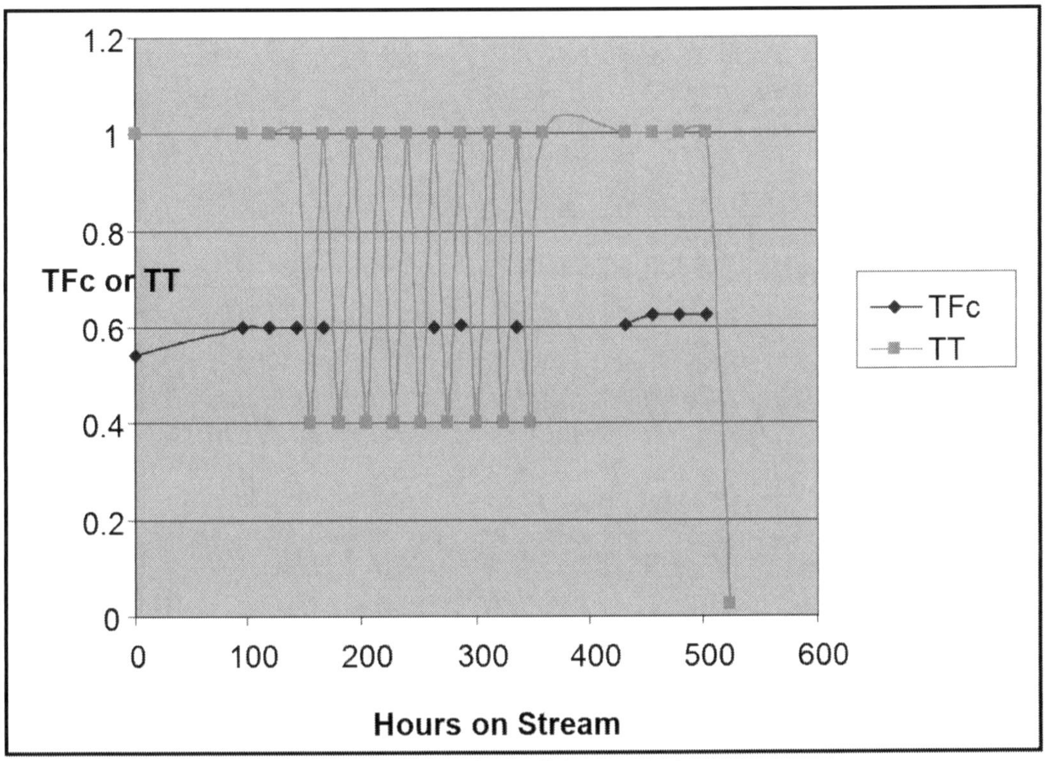


**Figure 5.** Life/durability test of a lead candidate OTM system

and temperature (15 to 30 bar, >950°C) in an auto-thermal reformer (ATR) to generate a synthesis gas with a H2/CO ratio of ~ 2. A catalyst is included in the reaction chamber to ensure the gas composition achieves near equilibrium composition. The major cost of the ATR is the air separation unit which includes air compression to 6 bar, refrigeration and cryogenic distillation, heat exchange and final compression of the purified oxygen stream to 30 bar.

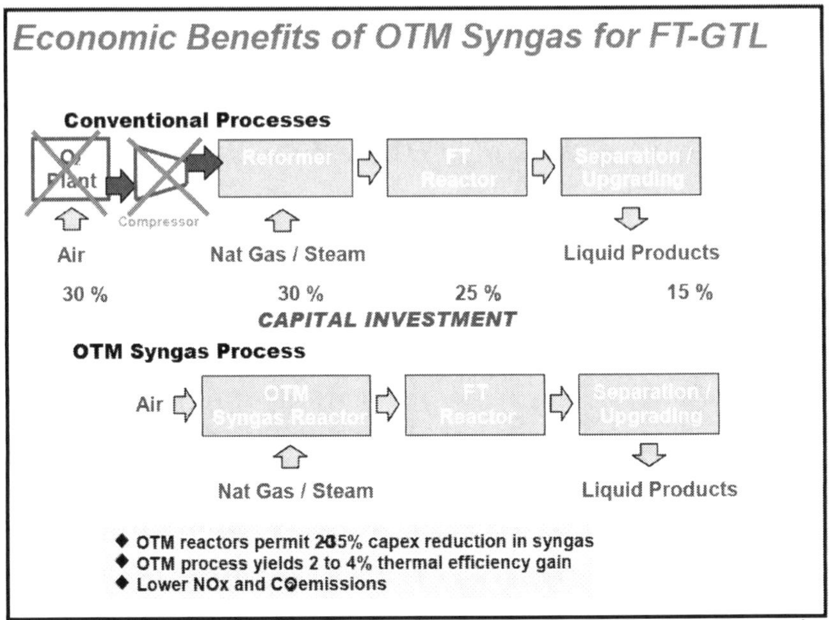


**Figure 6.** Economic benefit for an OTM syngas Fischer Tropsch gas to liquid facility

If an OTM is incorporated into the ATR, the air separation unit and oxygen compressor can be eliminated; and the air compressor can be replaced with a low pressure air blower. Prior studies have shown that the syngas plant capital costs can be reduced by 20 to 35% with OTM auto-thermal reforming. Additional benefits include a much smaller equipment footprint, 2% to 4% higher thermal efficiency, and lower NOx and $CO_2$ emissions.

Application of oxygen transport membranes for hydrogen production

OTM systems can also be applied to the conversion of natural gas to hydrogen. Recent studies by Foster Wheeler have compared OTM assisted hydrogen plants to state-of-the art steam methane reforming. The design basis is a Gulf Coast plant producing 100 million scfd of 99.99% hydrogen. The feedstock is natural gas with 95.5% methane, 1.8% $C_2$ plus, 1.8% $CO_2$ and the balance nitrogen. Although this study is not complete, interim results show that OTM assist can improve hydrogen yield 8.2% and reduce NOx emissions by 33%. An improved configuration, still under study, suggests hydrogen yield improvements approaching 21% are possible. In this latter case NOx emissions are expected to be over 50% less than the SMR base case. In all cases, $CO_2$ emission reductions are commensurate with the improvement in hydrogen yield. In the lowest NOx case, the OTM assisted hydrogen plant will enable complete $CO_2$ capture at significantly lower cost than any SMR installation.

Summary and Conclusions

Significant progress has been made in the development of high temperature ceramic oxygen transport membranes and systems. Recent advances in OTM life and durability will expedite the commercialization of this important technology. OTM assisted syngas or hydrogen production are expected to lower costs, improve thermal efficiency and accrue substantial environmental benefits in comparison to state-of-the-art processes now in use. An OTM based hydrogen plant has the potential to achieve the highest hydrogen yield from natural gas with near zero NOx emissions and low cost $CO_2$ capture and so could play a key role in a hydrogen economy.

Acknowledgements

This work was supported by the National Energy Technology Laboratory, U.S Department of energy under contract DE-FC26-01NT41096.

The authors are grateful for the contributions from the OTM alliance scientists and engineers, and in particular Kjersti Kleveland, Bente Krogh, Marit Dahle and Morten Brustad of Statoil; Jonathon Lane, John Sirman, Hancun Chen, Prasad Apte, Bart van Hassel , and Dick Van Slooten of Praxair; and Ann Hooper, Leo Bonnell and John Hemmings of Foster Wheeler.

---

[1] Robinson, E.T., Oxygen transport membranes for ultra-clean fuel production, ACS Fuel Chemistry Pre-prints, March 23-27, 2003, 47(1)

# Combine Reforming of Methane over $LaFe_{1-x}BxO_3$ (B= Co, Ni) Perovskites as Catalysts Precursors

M. R. Goldwasser[1*], M. L. Lugo, M. E. Rivas[1], E. Pietri[1], M. J. Pérez-Zurita[1], M. L. Cubeiro[1], A. Griboval-Constant[2] and G. Leclercq[2]

1. *Centro de Catálisis, Petróleo y Petroquímica, Escuela de Química, Fac. Ciencias, UCV, Apartado 47102, Los Chaguaramos, Caracas, Venezuela.*
2. *Université des Sciences ET Technologies de Lille, Laboratoire de Catalyse Hétérogene et Homogène, 59655 Villeneuve D'Ascq, Cedex, France.*

*Corresponding author: Mireya R. Goldwasser; E-Mail: mgoldwas@reacciun.ve;
Phone: (58212)7812349, Fax: (58212) 2392162

**Abstract.**

The effect of the presence of iron in a series of $LaBO_3$ (B = Co, Ni) perovskites as catalyst precursors in the combine reforming of methane with $CO_2$ and $O_2$ for the production of syngas was studied. The solids were synthesized by the citrate sol gel method and characterized by IR, BET Surface Area, XRD, TPR and ICP, under the condition of fresh, reduced and used samples. The results showed that through this synthesis it is possible to obtain highly crystalline, homogeneous and pure solids, with well-defined structures. After reduction, structural breakdown of the perovskite occurs and stabilization by the presence of iron of highly dispersed metallic particles from the cation on position B (Ni, Co) over the corresponding oxides was evidenced, decreasing deactivation by avoiding the sinterization of metals by coke formation on the surface of the catalysts.

**Introduction**

Addition of oxygen to both carbon dioxide and steam reforming of methane can reduce the carbon deposition on the catalyst surface and increase methane conversion. This combine reaction emerges as an alternative of particular interest since it allows working at lower temperatures, due to the exothermic character of the partial oxidation reaction, which not only improves the $H_2/CO$ ratio in the product, but traduces in lower energy consumption and a more economical way of methane conversion [1, 2]. However, deactivation of the catalyst by coke formation is one serious problem present in reforming reactions. This serious problem could be inhibited by the use of highly dispersed metals species [3, 4]. The use of precursors such as perovskite like oxides $ABO_3$, in which A-site cation is a rare earth and/or alkaline earth and B-site cation is a transition metal, could be the answer to these problems. Perovskites not only fulfill the stability requirements, but also by further reduction of B-site cations which remain distributed in the structure, they result in the formation of a well dispersed and stable metal particle catalyst. The use of perovskite type oxides as catalyst precursors appears as an interesting alternative due to the possibility of producing highly dispersed metallic particles over the corresponding oxides, diminishing catalyst deactivation by coke.

## 2. Experimental

Perovskite type oxides $LaCoO_3$, $LaCo_{0.6}Fe_{0.4}O_3$, $LaCo_{0.6}Ni_{0.4}O_3$, $LaFe_{0.8}Ni_{0.2}O_3$ ,$LaFe_{0.6}Ni_{0.4}O_3$, $LaFe_{0.4}Ni_{0.6}O_3$, and $LaNiO_3$ were synthesized by the citrate sol gel method. Adequate amounts of solutions of the nitrate salts of A and B-site cations precursors were added to a solution of citric acid (99.5 Riedel-de Haën) with an excess of ethylene glycol (99.5 Riedel-de Haën) as the organic polydentate ligant as previously reported [3,4].

Solids were characterized before and after catalytic tests by means of different techniques. Inductively Coupled Plasma Emission Spectroscopy (ICP) using a Perkin Elmer ICP/ 5500 instrument analyzed the chemical compositions (Co, Ni, Fe and La) of the perovskite. The IR spectra of the final product were recorded in a Thermo Nicolet Nexus FT-IR 470 in the range 1200-400 $cm^{-1}$. BET specific surface areas were determined by nitrogen-argon adsorption at 77K with an $N_2$ /Ar ratio of 30/70 using a Micromeritics model ASAP 2010. X-ray diffraction experiments were conducted using a Siemens D-8 advanced diffractometer with a Cu K$\alpha$ radiation for crystalline phase detection between 20° and 90° (2$\theta$) and under temperature program reduction in $H_2$. JCPDS-ICDD standard spectra software was used to determine the phases and particle size [5].

Activity tests were performed using 200 mg of catalyst in a 20-mm ID quartz reactor at atmospheric pressure operated in a fixed-bed continuous flow system ($CH_4/CO_2/O_2$= 4/1/2, Ar as diluents,750-850 ºC, 24$Lh^{-1}g^{-1}$ hourly space velocities). Before the catalytic tests, the solids were reduced in $H_2$ flow (20 mL/min, T=700°C, 8h). After reduction, the system was swept with Ar for 15 min. and adjusted to reaction temperature. The water produced during reaction, was condensed before passing the reactants and products to the analyzing system, which consisted of an on-line gas chromatograph (Perkin Elmer Auto System XL) equipped with a TCD and provided with a Carbosieve SII 80/100 (12'x1/8' OD SS). The conversions of $CH_4$ and $CO_2$ were defined as the converted $CH_4$ and $CO_2$ per total amount of $CH_4$ and $CO_2$ fed respectively. The selectivity for product "i" was calculated based in carbon balance and defined as $\%S_i = N_i / (NCH_4 (c) + N CO_2 (c))$ 100, where $NCH_4$ (c) and $NCO_2$ (c) are the amount of methane and carbon dioxide converted.

## 3. Results and Discussion

Table 1 lists the ICP chemical analysis of as-synthesized perovskites. A close similarity between analytical and nominal values in each case is observed, confirming the coincidence of real and nominal formula and the advantage of the sol gel synthesis. However, an oxygen deficiency was shown by most perovskites.

**Table 1. ICP Chemical Analysis.**

| Perovskite | % La (n) | % La (exp.) | % Co (n) | % Co (exp.) | % Ni (n) | % Ni (exp) | % Fe (n) | % Fe (exp.) | % Ox (n) | % Ox (exp.) |
|---|---|---|---|---|---|---|---|---|---|---|
| $LaCoO_3$ | 56.50 | 52.49 | 23.97 | 23.79 | | - | - | - | 19.52 | 18.23 |
| $LaCo_{0.6}Ni_{0.4}O_3$ | 56.52 | 52.22 | 14.38 | 14.19 | 9.55 | 9.77 | - | - | 19.53 | 19.76 |
| $LaCo_{0.6}Fe_{0.4}O_3$ | 56.78 | 52.27 | 14.45 | 14.87 | - | - | 9.13 | 8.48 | 19.62 | 19.35 |
| $LaNi_{0.2}Fe_{0.8}O_3$ | 57.08 | 54.29 | - | - | 4.82 | 4.76 | 18.35 | 17.63 | 19.73 | 18.99 |
| La $Ni_{0.4}Fe_{0.6}O_3$ | 56.95 | 54.05 | - | - | 9.62 | 9.20 | 13.74 | 13.26 | 19.68 | 22.64 |
| $LaNi_{0.6}Fe_{0.4}O_3$ | 56.82 | 54.16 | - | - | 14.40 | 14.27 | 9.14 | 8.85 | 19.64 | 18.60 |
| $LaNiO_3$ | 56.55 | 45.55 | - | - | 23.89 | 20.43 | - | - | 19.54 | 18.33 |

X-ray diffraction and IR spectra were recorded to assess the presence of the perovskite-like structures. All synthesized solids showed two bands in the medium IR region at 600 – 500 $cm^{-1}$ and 400 – 450 $cm^{-1}$ characteristics of the perovskite structure. Their positions are in good agreement with those reported in the literature [4, 6]. Table 2 gives values of the BET specific surface area measured after calcinations and the position of the IR bands. It is observed that $\upsilon_1$ decreases as Ni or Fe substitutes Co indicating that Co-O bond is stronger than Ni-O and Fe-O.
The observed surface areas between 20-57 $m^2/g$ are high compare to the usual values obtained for these solids [7], being higher for the binary $LaCoO_3$ and $LaNiO_3$ perovskites.

**Table 2. BET Specific Surface Area and IR bands Position**

| Solids | Surface Area ($m^2/g \pm 10\%$) | IR Bands $\upsilon_1$ ($cm^{-1}$) | $\upsilon_2$ ($cm^{-1}$) |
|---|---|---|---|
| $LaCoO_3$ | 57 | 600 | 425 |
| La $Co_{0.6}Fe_{0.4}O_3$ | 20 | 592 | 423 |
| La $Co_{0.6}Ni_{0.4}O_3$ | 24 | 524 | 405 |
| La $Ni_{0.2}Fe_{0.8}O_3$ | - | 567 | 402 |
| La $Ni_{0.4}Fe_{0.6}O_3$ | 21 | 573 | 430 |
| La $Ni_{0.6}Fe_{0.4}O_3$ | - | 513 | 436 |
| $LaNiO_3$ | 27 | 576 | 443 |

The X-ray diffraction pattern of the as-synthesized solids at room temperature revealed that the perovskite structure was the main phase detected for all solids. It was observed that for Co series $LaCo_{0.6}Fe_{0.4}O_3$ and $LaCo_{0.6}Ni_{0.4}O_3$ the more intense peaks are doublets, characteristic of the $LaCoO_3$ structure with formation of La-Co spinel type intermediates, confirming the existence of a more complex structure. As an example, the X-ray diffraction pattern of $LaCo_{0.6}Fe_{0.4}O_3$ as-synthesized, reduced and after reaction is shown in Figure 2. It is observed that after reduction, structural breakdown of the perovskite occur producing Co°, Fe° and $La_2O_3$. However, the mayor phases detected after reaction were $La_2O_2CO_3$ and $LaFeO_3$ perovskite. This $LaFeO_3$ perovskite phase was observed for all studied series after reaction, which is not surprising due to the oxidative character of the reactants in the reforming reaction ($CO_2$ and $O_2$) and to the facility of iron to oxidize ($\varepsilon_{Fe}{}^{o}$= -0.33V).
For Ni-Fe series, no doublets were detected; however, a progressive shifting of the bands to lower 2θ angles as Ni content increases occurred. This could indicate formation of Ni-Fe solid solutions in all proportions. Figure 3 shows broadening of the 2θ scale between 32° and 33.5° in which the shift is clearly observed for these structures.
The relative stability of the perovskites in reducing atmosphere was also fallowed by in situ X-ray diffraction and by TPR. The results for all solids are shown in table 3. As an example, the in situ X-ray diffraction pattern of $LaFe_{0.6}Ni_{0.4}O_3$ as-synthesized and reduced at different temperatures is shown in Fig. 4. It is observed that when the temperature is raised, cations in position B are reduced to metals while La in position A is transformed to $La_2O_3$.
For the Co series, the most drastic changes are seen between 300-500°C (Table 3), while for the Ni-Fe series, degradation of the perovskite start at 600-700°C. The perovskite is progressively degraded until approximately 800°C. For all solids, the observed phases after reduction at 800°C were: Co, Ni, Fe, and $La_2O_3$ and oxides of the B cations in small proportions, depending on the composition of the starting perovskite. Higher iron containing perovskites were more difficult to reduce.
A similar behavior was observed by TPR: in general, the perovskite reduction proceeds in two steps; the first between 500-570°C due to the reduction of $Ni^{3+}$ to $Ni^{2+}$ and $Co^{3+}$ to $Co^{2+}$ and the second between 582-650°C due to the reduction of $Ni^{2+}$ to Ni° and $Co^{2+}$ to Co° [7], figure 4. With the exception of $LaNi_{0.4}Fe_{0.6}O_3$ perovskite, a third peak also appears at ~750°C, attributed to the partial reduction of Fe in the perovskite structure.
Carreiro et al. [8] attributed the band observed between 720 – 770°C in Fe perovskites to partial reduction of Fe in the perovskite structure.

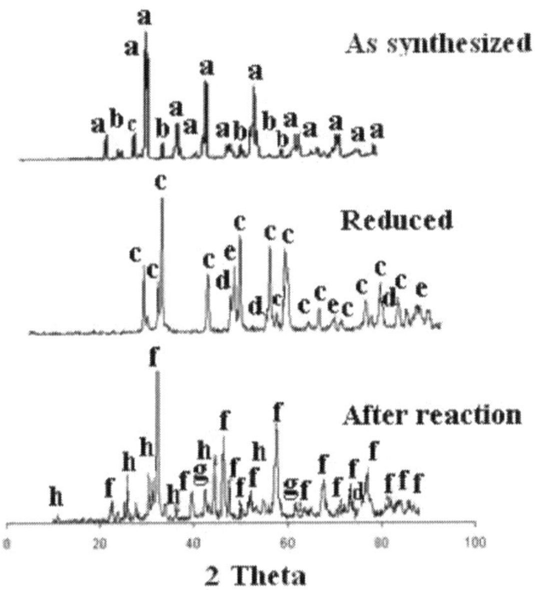


**Figure 2. XRD pattern of $LaCo_{0.6}Fe_{0.4}O_3$a: La $Co_{0.6}Fe_{0.4}O_3$; b: $Fe_2O_3$; c: $La_2O_3$; d: $Co^0$; e: $Fe^0$; f: $LaFeO_3$; g: CoO; h: $La_2O_2CO_3$**

The observed profiles for Ni-Fe with $x \geq 0.4$ and the progressive shift to higher temperatures as x increases is mainly due to the partial reduction of $Fe^{3+}$ to $Fe^o$, induced by the presence of $Ni^o$ particles.

The fact that there seems to exist an activated Fe reduction assisted by $Ni^o$ particles, favors Ni-Fe alloy formation after reduction. However, it is important to point out that alloy formation was detected mainly after reduction but not necessarily after reaction, where formation of an iron perovskite $LaFeO_3$ was observed, depending on the oxidation character of the reaction mixture.

The temperature of reduction and the presence of a third peak indicate that $LaNi_{0.6}Fe_{0.4}O_3$ and $LaNi_{0.2}Fe_{0.8}O_3$ are more stables in a reduction atmosphere than La $Ni_{0.4}Fe_{0.6}O_3$.

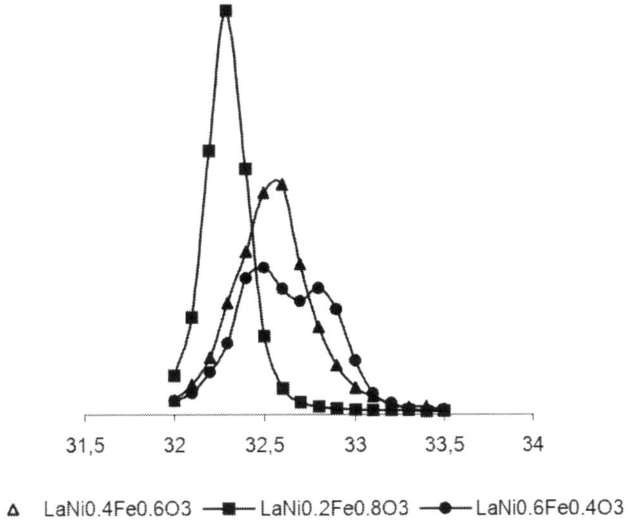


**Figure 3. Broadening of the XRD 2θ scale for $LaNi_xFe_{1-x}O_3$ series.**

It was also observed that compare to $LaNiO_3$ perovskite when Fe is introduced to the structure, both reduction peaks shift to higher temperature, which confirms the fact that Ni-Fe produces an alloy type interaction as observed in the X-ray diffraction pattern (Figure 5). $LaNi_{0.2}Fe_{0.8}O_3$ is the most stable perovskite of this series according to the in situ X-ray diffraction analysis (table 3).

**Table 3. In situ X-ray diffraction Analysis**

| Samples | Degree (2-theta) | T of Degradation (°C) | Phases after Reduction | Mean Size (nm) |
|---|---|---|---|---|
| $LaCoO_3$ | 32.920 | 300-400 | $La_2O_3$, Co°. | 13.6 |
| $LaCo_{0.6}Ni_{0.4}O_3$ | 32.611 | 400-500 | $La_2O_3$, Co°, Ni°, Co°-Ni°. | 27.7 |
| $LaCo_{0.6}Fe_{0.4}O_3$ | 32.868 | 400-500 | Fe°, Co°, Co°- Fe°. $La_2O_3$. | 20.8 |
| La $Ni_{0.2}Fe_{0.8}O_3$ | 32.630 | 700-800 | Ni°, Fe°, Fe°- Ni°. $La_2O_3$. | 34.8 |
| La $Ni_{0.4}Fe_{0.6}O_3$ | 32.696 | 700-800 | Ni°, Fe°, Fe°- Ni°, $La_2O_3$. | 35.9 |
| La $Ni_{0.6}Fe_{0.4}{}^{o}{}_3$ | 32.782 | 600-700 | Fe°, Ni°, $La_2O_3$, Fe°-Ni°. | 28.0 |
| $LaNiO_3$ | 32.439 | 600-700 | Ni°, $La_2O_3$, | 24.5 |

JCPDS-ICDD: Ni°(01-1272 ); Co° ( 01-1254 ); Fe° ( 01-1252 ) Ni°-Fe° (47-1417); Co°-Fe° (65-6829 ); $La_2O_3$ (05-0602)

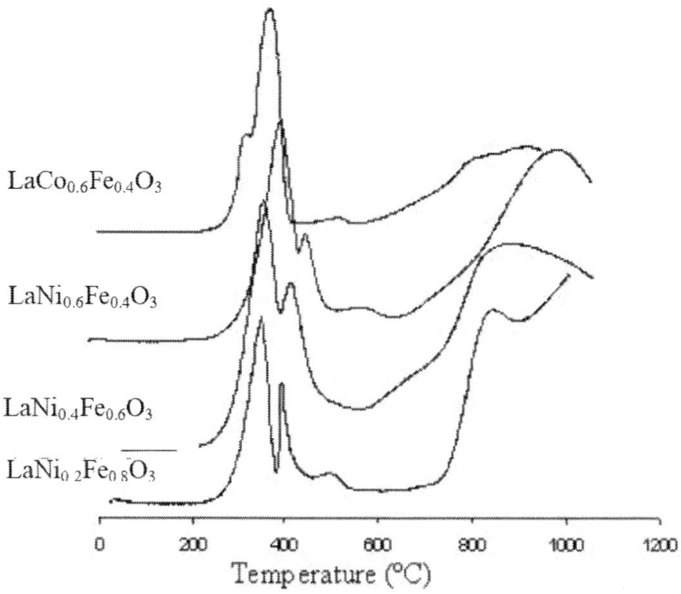


**Figure 4. TPR Profile of $LaCo_{0.6}Fe_{0.4}O_3$, $LaNi_{0.6}Fe_{0.4}O_3$, $LaNi_{0.4}Fe_{0.6}O_3$, $LaNi_{0.2}Fe_{0.8}O_3$ Perovskites**

The activity tests showed that Co containing perovskites ($LaCoO_3$ and $LaCo_{0.6}Fe_{0.4}O_3$) were not active in methane reforming at 750°C. However, a high conversion and activity to syngas was observed for these solids at 850°C (Table 4). Incomplete reduction of Co clusters at the lower temperature is claimed for this behavior [9].

For the combine reaction at 750°C, similar results were obtained for $LaNi_xFe_{1-x}O_3$ series, with $LaFe_{0.6}Ni_{0.4}O_3$ producing the higher $H_2$ rich syngas with a molar $H_2/CO$ ratio of 1.5, favorable for Fischer-Tropsch, MTBE and MTG process.

As seen in Table 4, on $LaNi_{0.6}Fe_{0.4}O_3$ and $LaNi_{0.4}Fe_{0.6}O_3$ perovskites $CH_4$ conversion was around 70%, while for the perovskite with a higher Fe content $LaFe_{0.8}Ni_{0.2}O_3$, selectivity's to $H_2$ and CO were quite close on Ni-Fe series.

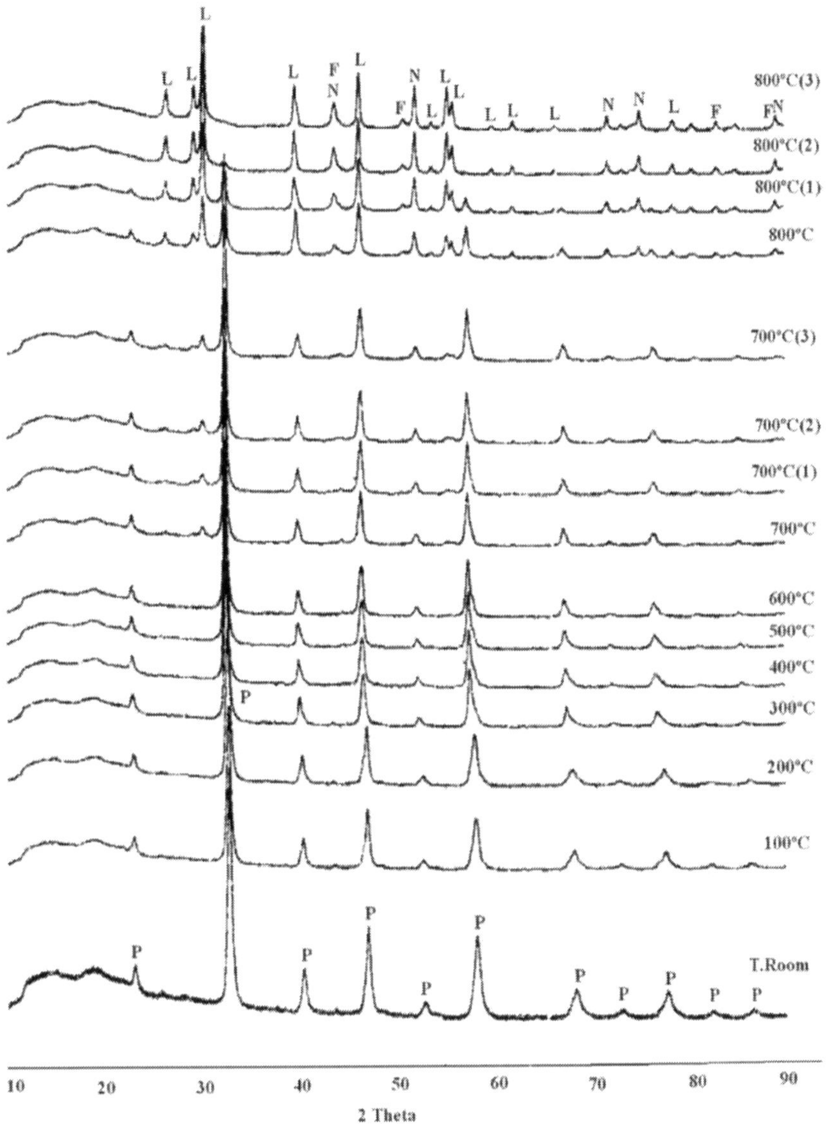


**Figure 5. In situ X-ray diffraction of $LaNi_{0.4}Fe_{0.6}O_3$. P: Perovskite $LaNi_{0.4}Fe_{0.6}O_3$, L: $La_2O_3$, F: $Fe^0$, N: $Ni^0$**

However, as shown in Figure 6, the higher Fe containing perovskite showed a progressive decrease with time in the selectivity to $H_2$. This behavior could be due to the production of CO and $H_2O$ via the reverse water gas shift reaction. This reaction will also explain the higher $CO_2$ conversion observed for this catalyst. Under the experimental condition used, $LaFe_{0.2}Ni_{0.8}O_3$ perovskite produce a $H_2/CO$ ratio near to 1, while a $H_2$ rich syngas with a $H_2/CO$ ratio of 1.50 is obtained on $LaFe_{0.6}Ni_{0.4}O_3$ perovskite.

**Table 4. Combine Reforming of Methane Using $CO_2$ and $O_2$**

| Sólido | $XCH_4$ % | $XCO_2$ % | $H_2/CH_{4\,(conv)}$ | $H_2/CO_{(molar)}$ |
|---|---|---|---|---|
| $LaCoO_3$ | 39 | -11 | - | - |
| La $Co_{0.6}Fe_{0.4}O_3$ | 17 | -139 | - | - |
| La $Co_{0.6}Fe_{0.4}O_3$* | 81 | 77 | 1.37 | 1.96 |
| $LaCo_{0.6}Ni_{0.4}O_3$ | 83 | 49 | 1.35 | 0.95 |
| La $Ni_{0.2}Fe_{0.8}O_3$ | 43 | 31 | 1.33 | 1.08 |
| La $Ni_{0.4}Fe_{0.6}O_3$ | 72 | 19 | 1.27 | 1.50 |
| La $Ni_{0.6}Fe_{0.4}O_3$ | 68 | 24 | 1.23 | 1.12 |
| $LaNiO_3$ | 87 | 46 | 1.32 | 1.16 |

tr =24 h, WHSV =24L/ h.g, $CH_4/CO_2$ =4, $CH_4/O_2$=2, W = 200 mg, Tr =750°C,
*Tr = 850°C, Tred = (700°C, 8h).

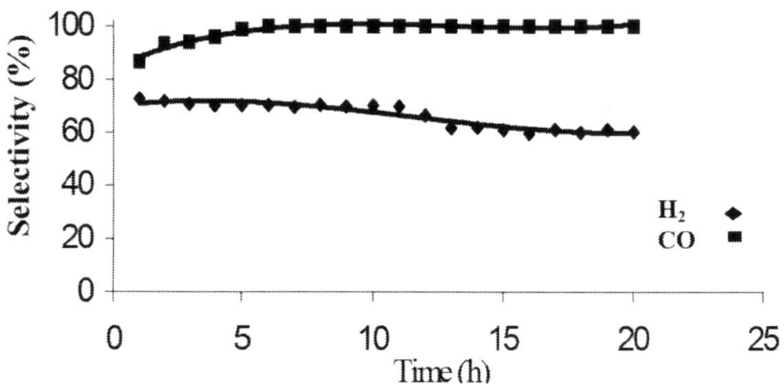


**Figure 6. - $LaNi_{0.2}Fe_{0.8}O_3$ Selectivity to $H_2$ and CO at 750°C, $CH4/CO_2/O_2$=2/0.5/1, VE= 24L.$h^{-1}$.$g^{-1}$**

H. Provendier et al. [10] on Fe-Ni perovskite show the stabilization of Ni by Fe. These authors propose that Fe controlled the reversible migration of Ni during reaction forming alloys which hinder the Ni sinterization. That effect is not clearly shown in our study, if it is occurring is in a lesser extension than the dispersive effect observed for $LaCo_{1-x}Ni_xO_3$ series. When the best catalyst of each series is compare higher conversions and selectivity to syngas are observed for Co-Ni than for Fe-Ni under the same experimental conditions. Also, as observe by X-ray diffraction during the reaction the catalyst is mainly formed by Ni dispersed in a mixture of $LaFeO_3$ and $La_2O_3$.

## 4. Conclusions

All synthesized solids showed perovskite type structure as evidenced by X-ray diffraction and IR. Under optimize experimental conditions $LaNi_{0.4}Fe_{0.6}O_3$ from Ni-Fe series show high activity and selectivity with a molar ratio of $H_2/CO>1$. TPR results show a higher stability to reduction and X-ray diffraction pattern evidenced the presence of a Ni-Fe solid solution. The $LaCo_{0.6}Fe_{0.4}O_3$ as Co precursor shows activity to the reforming reaction at 850°C. At these conditions excellent methane and $CO_2$ conversions with a molar ratio of $H_2/CO=2$.
For the $LaFe_{1-x}B_xO_3$ series, Fe under the reaction conditions is only partially reduced to Fe° remaining mainly as $LaFeO_3$ over which particles of metallic Ni is dispersed. In these systems Fe acts as Ni stabilizer controlling the Ni migration avoiding sinterization and coke formation.

## Acknowledgement

The authors thank Venezuelan FONACIT for its financial support through Agenda Petroleum Project N° 97-003739 and PICS PI 2003000024.

## References.

1. V. Szabo, M. Bassir, A. Van Neste and C. Kaliaguini, Appl. Catal. B: Environmental 43 (2003) 29.
2. Z. Zhong, K. Chen, Y. Ji and Q. Yan, Appl. Catal. A: General 156 (1997) 29.
3. M. R. Goldwasser', M. E. Rivas, E. Pietri, M. J. Pérez-Zurita, M. L. Cubeiro, L. Gingembre, L. Leclercq and G.Leclercq, Appl. Catal.A: General 255(2003) 45
4. E. Pietri, A. Barrios, O. Gonzalez, M. R. Goldwasser, M.J. Pérez-Zurita, M.L. Cubeiro, J. Goldwasser, L. Leclercq, G. Leclercq, Stud. Surf. Sci. Catal.,136 (2001) 381
5. H. P. Klug, L. E. Alexander, X-ray Difraction Procedures for Polycrystalline and Amorphous Materials, Wiley, London 1962,p.491
6. E. Pietri, A. Barrios, M. R. Goldwasser, M.J. Pérez-Zurita, M.L. Cubeiro, J. Goldwasser, L. Leclercq, G. Leclercq nd L. Gingembre, Stud. Surf. Sci. Catal.,30 (2000) 3657
7. J. M. D. Tascón, S. Mendioróz and L. González Tejuca, Z. Phys. Chem. 124 (1981)109
8. L. Carreiro, Y. T Qian., R. Kershaw, K. Dwingt and A. Wold, Mater Res. Bulletin, 20 (1985) 619
9. M. O'Connell, A. K. Norman, C. F. Hüttermann and M. A. Morris, Catal. Today, 47(1999) 123
10. H. Provendier, C. Petit, C. Estournes and A. Kiennemann. Appl. Catal. A: General 180 (1999) 163

# QATARGAS DEBOTTLENECKING PROJECT

*Muh Ilyasak*
*Engineering and Ventures Department*
*Qatar Liquefied Gas Company Limited*
*PO BOX 22666*
*Doha - Qatar*



*Prepared for Presentation at AIChE Spring National Meeting*
*4th Topical Conference on Natural Gas Utilization*
*New Orleans, LA, April 2004*

## Abstract

The Qatargas LNG plant consists of 3 LNG Trains, which were originally designed to produce and deliver 6 MTPA of LNG to Japan.

Trains 1 and 2 were commissioned in 1996, while the construction and start up of Train 3 was completed in 1998.

Qatargas intends to increase the production capacity of the plant by upgrading the process trains to produce over 3 MTPA of LNG per train to satisfy its future contractual obligations and capture new market opportunities.

The LNG train maximum capacity test was performed in November 1998, APCI simulation study confirmed that the Refrigeration and Liquefaction systems are able to produce LNG at rates thirty percent (30%) over design.

The study was launched to verify the capacity of the 32-inch sub-sea pipeline to transfer gas and condensate from offshore to onshore. It was concluded that the capacity of the line could be increased to the maximum of 1.6 BSCFD. This pipeline capacity limit is sufficient to run the Debottlenecked LNG Trains.

This paper describes how the Debottlenecking Project is implemented at Qatargas. It will discuss Feasibility study, Front End Engineering Design (FEED), Engineering Procurement and Construction (EPC).

## Introduction

The Qatargas LNG facility is located on the North-East coast of the state of Qatar, close to the international shipping routes with access to deep water (sea) required for transporting the product in large tanker ships.

The facility originally designed and constructed in the mid 90's is divided into two sections, namely:

a) The upstream facilities (offshore and onshore receiving facility).
b) The downstream onshore facilities (LNG plant).

The upstream system consists of the offshore production facilities, the submarine pipeline (sealine) and the onshore receiving facilities.

Each LNG train is originally designed to produce 2 MTPA of LNG delivered to Japan. Train 1 and 2 were commissioned in 1996, while the construction and start up of Train 3 was completed in 1998.

The Qatargas Project is fully integrated, i.e., Qatargas operates the entire production and LNG facilities from the wellheads to the loading berths.

Besides LNG, the Qatargas Plant also produces:

- Stabilized Field Condensate (North Field Condensate or NFC).
- Stabilized Plant Condensate (Qatar Return Condensate or QRC).
- Solid granulated Sulphur.

## Offshore Production Facilities

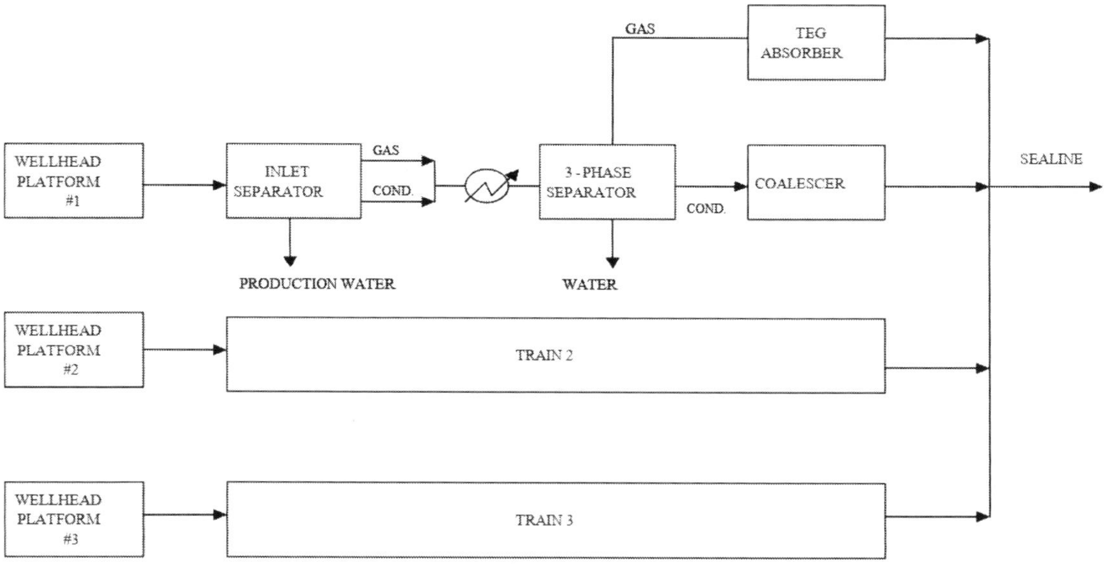


**Fig. 1 OFFSHORE PRODUCTION FACILITIES**

The gas and condensate are produced from the giant North Field Khuff reservoir with reserves in excess of 900TSCF.

The Offshore facilities include three wellhead platforms with a total of twenty producing wells and three gas and condensate processing trains. After separation of the production water, the gas and condensate are cooled by direct seawater cooling (Train 3) or a fresh water circulation loop (Trains 1&2), and separated again for removing the condensation water. The gas is directed to the TEG absorber for dehydration, and the condensate to a coalescer for removing the free water. The dry gas and condensate are mixed and transferred to the sealine.

The design capacity of Trains 1&2 offshore is 460MMSCFD, and that of Train 3 offshore is 485 MMSCFD, and they operate at 140 barg.

The 80 km long, 32" sealine has a design capacity of 1.4BSCFD and the delivery pressure to the Onshore Receiving Facilities is 80 barg.

**Onshore Receiving Facilities**

They include:

- One slugcatcher of 2,000 $m^3$ liquid capacity
- Three condensate stabilization trains
- Two off-gas compressors for recovery of the gas from the condensate strippers into the feed gas to the LNG trains

**LNG Plant**

It includes three identical LNG trains with a design feed gas capacity of 430MMSCFD each. The process includes:

- Traces Mercury removal on a Sulphur impregnated Carbon bed.
- Acid gas and mercaptans removal by the Sulfinol® process.
- Gas dehydration and traces Sulphur removal by adsorption on molecular sieve beds.
- Gas chilling by propane refrigerant, and NGL fractionation into ethane/propane/butanes (used as refrigerant make-up and recycled to LNG), and Plant Condensate.
- Gas liquefaction by the propane pre-cooled Mixed Refrigerant® process from APCI.
- Nitrogen removal from LNG.
- LNG storage in 4 x 85,000 $m^3$ tanks and Boil-off Gas recovery.
- Acid gas enrichment and Sulphur recovery.

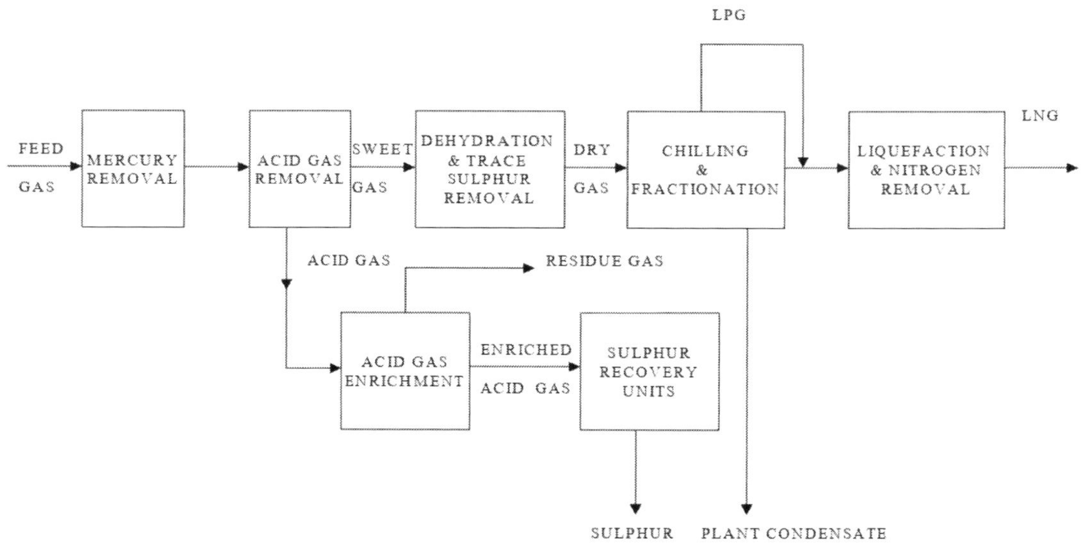


Fig. 2 LNG PROCESS

The propane refrigerant compressors are driven by one Frame 5C gas turbine (main compressor) and one 5MW electrical motor (booster compressor). The three MR compressors are driven by Frame 5C gas turbines.

Main Utilities are:

- Power generation (six Frame 6 turbo generators).
- Five MP steam boilers of 146TPH capacity each.
- Four water desalination units.
- Plant air and instrument air system.
- Two Nitrogen generation units.
- Fuel gas system.
- Flares system.
- Water treatment units.

## Debottlenecking Studies

After start-up and performance tests of the first two LNG Trains, Qatargas realized that the LNG rundown capacity could significantly exceed the design figure of 300TPH per train in winter and 275TPH in summer.

- A maximum capacity test of LNG Train 3 performed in November 1998 achieved an LNG production of about 380TPH but with some process restrictions.
- A simulation study completed by APCI in December 1998 indicated that the refrigeration and liquefaction facilities could achieve an LNG production of 390TPH

in winter conditions; however, it was identified that the refrigerant compressors and turbines would operate beyond their design limits.

In parallel to the LNG Debottlenecking studies, the maximum capacity and feasibility to Debottleneck the Offshore production facilities and sealine were also investigated.

- TotalFinaElf completed a sealine maximum capacity study in November 1998 and it was determined that the maximum transport capacity was 1.6BSCFD.
- Qatargas also awarded a feasibility study to Technip for debottlenecking the Offshore Trains 1&2 to 550 or 600MMSCFD, which was completed in March 1999 and concluded that the required modifications were relatively minor:
    - Installation of an Over Pressure Protection System for limiting the relief loads to the flares in case of operation upsets.
    - Upgrading of the three phase separators for improving the gas/liquid separation efficiency.
    - Replacement of some control valves to increase the sealine capacity.

Finally, the LNG Plant Debottlenecking and Reliability Study was awarded to Kellogg Brown & Root (KBR) in September 1999 with the following objectives:

- Determine the maximum sustainable capacity of the LNG Trains without any significant modification
- Identify the most appropriate process modifications for debottlenecking the LNG Plant to the sealine capacity limit of 1.6BSCFD
- Assess the impact of the Debottlenecking Project on the reliability performance of the LNG Plant.

The Study was completed in September 2000 after detailed evaluation of several process scenarios, and the conclusions were as follows:

- The maximum sustainable LNG rundown capacity of each Train (without process upgrading) was determined at 330TPH on yearly average, equivalent to 7.7MTPA LNG loaded.
- The debottlenecking case F was based on saturating the sealine capacity in both winter and summer conditions by upgrading the MR compressors and turbines, and replacing the main propane compressor and booster compressor (driven by a new gas turbine). This process scenario would allow increasing the LNG rundown capacity to 410TPH in winter conditions and 400TPH in summer, equivalent to 9.5MTPA LNG loaded.
- The reliability of the debottlenecked trains would be marginally affected (-0.2%) as compared to the performance of the existing trains.

## Offshore Facilities Debottlenecking Project

Based on the conclusions from the Debottlenecking Study, a FEED contract was awarded to Technip in October 1999 for debottlenecking the offshore Trains 1&2 from 460 to 600 MMSCFD capacity each.

The FEED package was issued in June 2000 and Qatargas decided to implement immediately the 3 critical modification projects:

- Over Pressure Protective System
- Upgrading of the 3-phase separators
- Replacement of gas pressure control valves and condensate level control valves to the sealine.

The three projects were successful completed in April/May 2002 during the Plant turnaround.

## Onshore Facilities FEED

Qatargas awarded the FEED contract for debottlenecking the Onshore Facilities (Receiving Facilities, LNG Plant and Utilities) to KBR in October 2000, which was completed in June 2001. The main modifications to be implemented were confirmed as follows:

- Upgrading of MR gas turbines from Frame 5C to 5D model, providing 13% additional power.
- Upgrading of MR compressors (new internals) to increase the flow capacity and match the additional turbine power.
- Replacement of the existing main Propane compressor by a larger capacity one (+38%).
- Replacement of the existing Propane booster compressor (driven by a 5MW motor) by a larger capacity compressor driven by a Frame 5C gas turbine.
- Replacement of the LNG product pumps.
- Addition of fourth boil-off gas compressor.
- Replacement of the MR/LP propane evaporator with a larger one.
- Addition of two new intermediate Propane flash drums.
- Upgrading of demisters in seven vessels.
- Upgrading of piping in Propane refrigeration unit.
- Replace De-Ethanizer trays.
- Modification of fuel gas system to improve fuel gas quality to LNG Trains.
- Replacement or upgrading of some control valves.
- Addition, replacement or upgrading of some pressure safety valves.

The new Propane system and the new MR system (case F) are shown in figure 3 and figure 4 respectively.

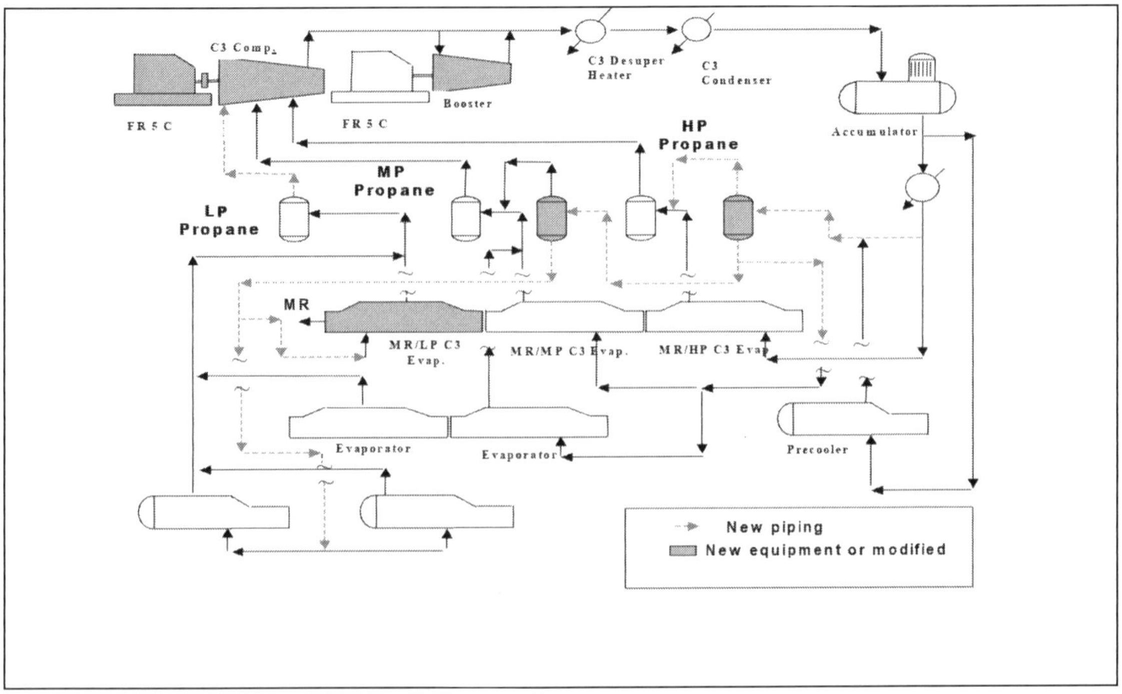


**Fig. 3 – New Propane System**

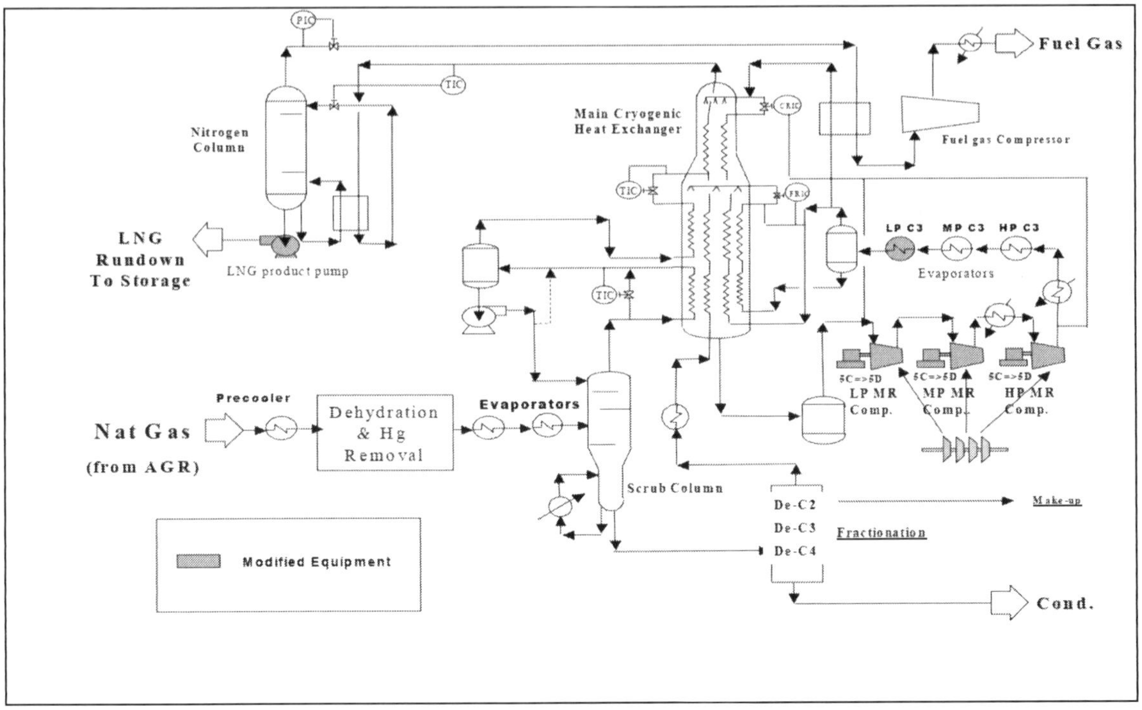


**Fig. 4 – New MR System**

As part of the FEED deliverables, a detailed constructability study was performed and it was determined that by maximizing the pre-construction activities the new equipment tie-ins and commissioning could be performed in 33 days as part of the train turnaround activities.

The LNG trains turnaround frequency is every three years, so one train is overhauled every year.

One key factor for achieving this objective was the decision to install the new main Propane compressor in the location of the existing booster compressor, which will be removed several months before the start of the turnaround. The new Propane booster compressor will be installed in the location of the existing main Propane compressor and driven by the existing Frame 5C gas turbine.

On this basis, it was determined that the project implementation schedule are as follows:

- Train 2 debottlenecking in September/October 2003.
- Train 3 debottlenecking in April/May 2004.
- Train 1 debottlenecking in April/May 2005.

In addition, it was decided to upgrade the Train 1 MR turbines in April/May 2002, as they were due for overhaul and this would bring some additional LNG production capacity to the Train.

**Turbines & Compressors Upgrade Contract**

In order to meet the fast-track project implementation schedule and to be able to upgrade MR Gas Turbines in Train 1 in 2002, Qatargas awarded supply and installation contracts to GE/Nuovo Pignone in March 2001, the Original Equipment Manufacturer for these machines. This contract was then reassigned to the selected EPC contractor.

The upgrading of the three MR gas turbines in Train 1 was successfully completed in May 2002, and the machines have run continuously since start-up, and allowed to increase the maximum LNG production from the train by about 9%.

**EPC Contract**

Following a pre-qualification procedure two EPC Contractors Joint Ventures were invited to bid and the EPC contract was awarded to Chiyoda / Technip Joint Ventures (CTJV) in October 2001.

**Debottlenecking Results**

Train 2 was the first train for which the entire debottlenecking modification completed in October 2003. The maximum sustainable LNG production achieved from Train 2 is 435 T/H, which exceeds the winter LNG design flow of 410 T/H by 6 %.

## Conclusion

Based on Train 2 performance, it has demonstrated that Debottlenecking LNG production targets could be achieved. The Debottlenecking Project will increase the LNG production to 9.5 MTPA loaded or 23% above the present maximum sustainable capacity of 7.7 MTPA.

The economic benefits to Qatargas are immense; the capital investment per tonnes of LNG to Debottleneck the existing LNG trains is significantly lower than the construction of a new train, and the additional output of LNG contributes to profits. This will place Qatargas in a good position to seize new market opportunities.

## CHLORIDE REMOVAL FROM AMINE SOLUTION IN TRAIN D

*Rahmat Minullah J and Nugroho Rachmadi*
*Process Engineering Section*
*PT Badak Natural Gas Liquefaction*
*Bontang, East Kalimantan Indonesia 75324*

*Prepared for Presentation at AIChE Spring National Meeting*
*4th Topical Conference on Natural Gas Utilization*
*New Orleans, LA, April 2004*

### ABSTRACT

Badak LNG/LPG plant has eight LNG trains, Trains A ~ H, with a total production capacity of 22.2 MTA of LNG and 1.0 MTA of Propane LPG and Butane LPG. Badak LNG/LPG plant uses Methyl Diethanol Amine (MDEA) solvent to remove the acid gas ($CO_2$) from the feed gas.

Due to a seawater ingress problem into the MDEA solution, the amine solution in Train D was contaminated by chloride ($Cl^-$) up to about 1200 ppm. Published literature and the solvent vendor advised PT Badak that the chloride impurities at this level might result in a potential stress corrosion cracking (SCC) of the stainless steel material in the system. Therefore, the chloride content need to be reduced to "a safe" figure of 100 ppm or lower as early as possible.

The chloride impurities removal has been carried out online by using a cation and anion ion exchange process. This process removal had reduced the chloride content from 1200 to 50 ppm, sodium content from 500 ppm to 50 ppm and also reduced the iron content from 15.0 to 2.5ppm. The ion exchange process was done online, the LNG train was not required to be shutdown, which prevented LNG production losses.

### INTRODUCTION

Badak LNG/LPG plant has eight existing Trains A ~ H with a total production capacity of 22.2 million ton/year of LNG and 1.1 million ton/year of LPG Propane and Butane. Trains A and B were initially started up in 1977; Trains C & D in 1983; Trains E, F, G & H in 1989, 1993, 1997 and 1999 respectively.

#### LNG Process

Liquefy the natural gas to Liquid Natural Gas (LNG) is done by using several main equipment, which it is called "train". The feed gas from the gas fields is introduced to Knock out drums to separate the liquid hydrocarbon and water and also to prevent liquid carry over to the $CO_2$ absorber. Carbon dioxide ($CO_2$) in the feed gas is absorbed by solvent of MDEA (methyl diethanol amine) with the strength of 40%-weight in the $CO_2$ Absorber. Water (moisture) carried over in the treated feed gas is adsorbed by molecular sieve in the dryers while the mercury is adsorbed in the Mercury Removal Vessel. The simplified process flow can be seen in figure-1.

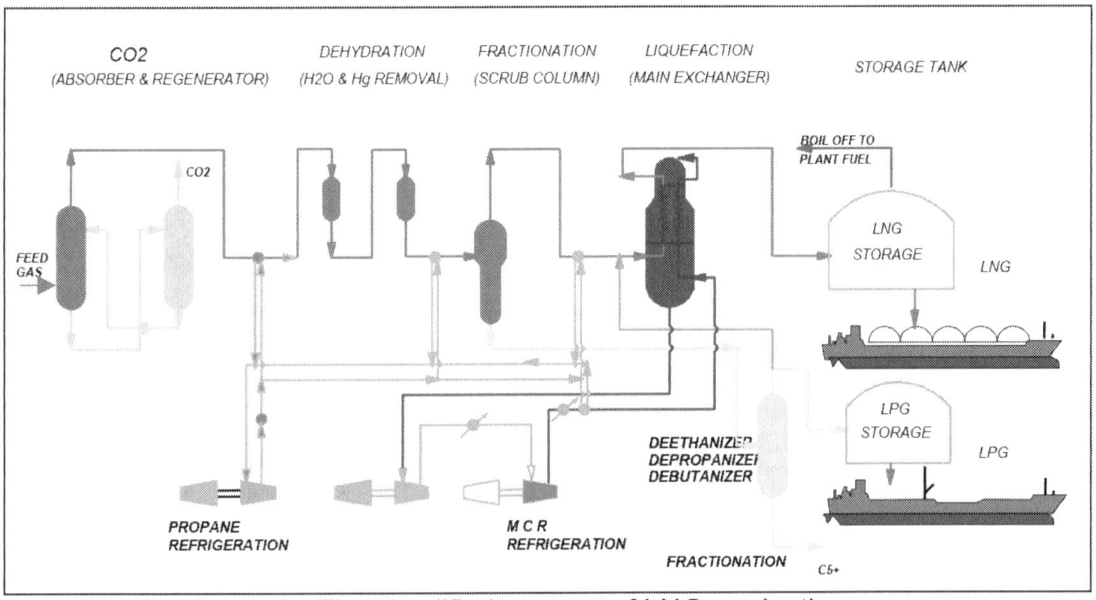


Figure 1. The simplified process of LNG production

The heavier component of the feed gas is stripped from methane as main component of LNG in the scrub column. After chilled by propane, the treated feed gas is then liquefied to LNG by multi component refrigerant in Main Heat Exchanger (MHE). The heavier component is separated in the fractionation columns to produce liquefied petroleum gas (LPG) commercial grade.

**The Process of $CO_2$ Removal**

The Carbon Dioxide ($CO_2$) Removal unit is designed to reduce $CO_2$ in the natural gas feed to a maximum of 50ppm(v) to prevent freeze-up in the liquefaction section. The process is based on the principle that amine will react chemically with $CO_2$ at high pressure near ambient temperatures. The acid gas is released from the solution at a high temperature & low pressure.

Feed gas enters the bottom of the absorber column; meanwhile lean amine solution enters the top of the column. As the amine solution moves downward through the column, it reacts and absorbs the $CO_2$ from the upward flowing feed gas. The treated gas from the top of the column is cooled in Overhead Cooler before flowing to the Overhead Separator. The treated gas is delivered to Drier for further processing. See figure-2.

Rich amine solution at about 61°C flows from the bottom of the absorber to Flash Drum. The rich amine stream pressure is reduced from 47.5 kg/cm$^2$a to about 7.0 kg/cm$^2$a in order to releases dissolved hydrocarbons in the solvent that is rejected to fuel gas system.

Rich amine solution from Flash Drum is preheated to about 105°C in Cross Exchangers with lean amine solution from the Regenerator and Regenerator Reboiler. The rich amine solution flows to Regenerator where it is operated at 1.25 kg/cm$^2$g to aid in releasing the carbon dioxide from the solution. The water vapor leaving the top of the Regenerator is condensed by Overhead Cooler and is recycled to Amine Regenerator. The $CO_2$ rich vapor is leaving the overhead and vented to the atmosphere.

**Cation and Anion Exchange**

The resin contains ionizable functional groups. These groups consist of both positively charged cation elements and negatively charged anion elements. However, only one of the ionic species is mobile. Typically, strong acid cation resin bead has ionic sites consisting of immobile anionic radicals and mobile hydrogen cation. The strong base anion resin bead has ionic sites consisting of immobile cationic radicals and mobile hydroxide anion. After the mobile anion/cation is exhausted, the regenerant solution is introduced to exchange exhausted mobile anion or cation. The common regenerant of cation is $H_2SO_4$ and anion is NaOH. The following are scheme of ions exchange in the resin:

$$\begin{bmatrix} Ca \\ Mg \\ 2Na \end{bmatrix} \begin{bmatrix} 2HCO_3 \\ SO_4 \\ 2Cl \end{bmatrix} + 2.R.H \Leftrightarrow 2.R. \begin{bmatrix} Ca \\ Mg \\ 2Na \end{bmatrix} + \begin{bmatrix} 2H_2CO_3 \\ H_2SO_4 \\ 2HCl \end{bmatrix} \qquad \text{Cation}$$

$$\begin{bmatrix} H_2SO_4 \\ 2HCl \\ 2H_2CO_3 \end{bmatrix} + 2R.OH \Leftrightarrow 2R. \begin{bmatrix} SO_4 \\ 2Cl \\ 2HCO_3 \end{bmatrix} + 2H_2O \qquad \text{Anion}$$

Seawater contains sodium chloride that dissolves perfectly in amine solution. These ions can be removed by ion exchanges process. Ion exchangers resin swaps the chloride ion with hydroxide ($OH^-$) and sodium ion by hydrogen ($H^+$). Sodium and chloride ions are "polar" cation/ anion. However, amine component or "methyl di-ethanol amine" (MDEA) and its activator are non-polar substance. Non-polar substances are difficult to be exchanged by cation/anion exchanger, while Polar ions are easily exchanged. Therefore, the selectivity of cation/anion exchanger toward "amine" versus "seawater" is the best process for decontamination in amine solution.

After Train E amine system was contaminated by seawater, several tests were carried out in Badak LNG Laboratory. Later, the test concluded that the ion exchange with the strong base anion resin was effective enough to remove the chloride ion less than 100ppm.

Commonly, the ion exchanger or demineralizer unit consists of two vessels (anion and cation bed) that equipped with valves, pump and piping for batch operation: a). In service step, contaminated solvent is introduced down flow to the beds. The polar ion adsorbed by the resin until the resin reaches the saturation point. b). 1st Rinsing step is to wash the remaining solvent in the bed and minimize the solvent losses. c). In regeneration step, regenerant is introduced co-currently into the bed. The regenerant exchanges the trapped contaminant ion with the regenerant anion/cation. d). 2nd Rinsing step is to wash the remaining regenerant before switch back to the step a). All steps are called one-step cycle.

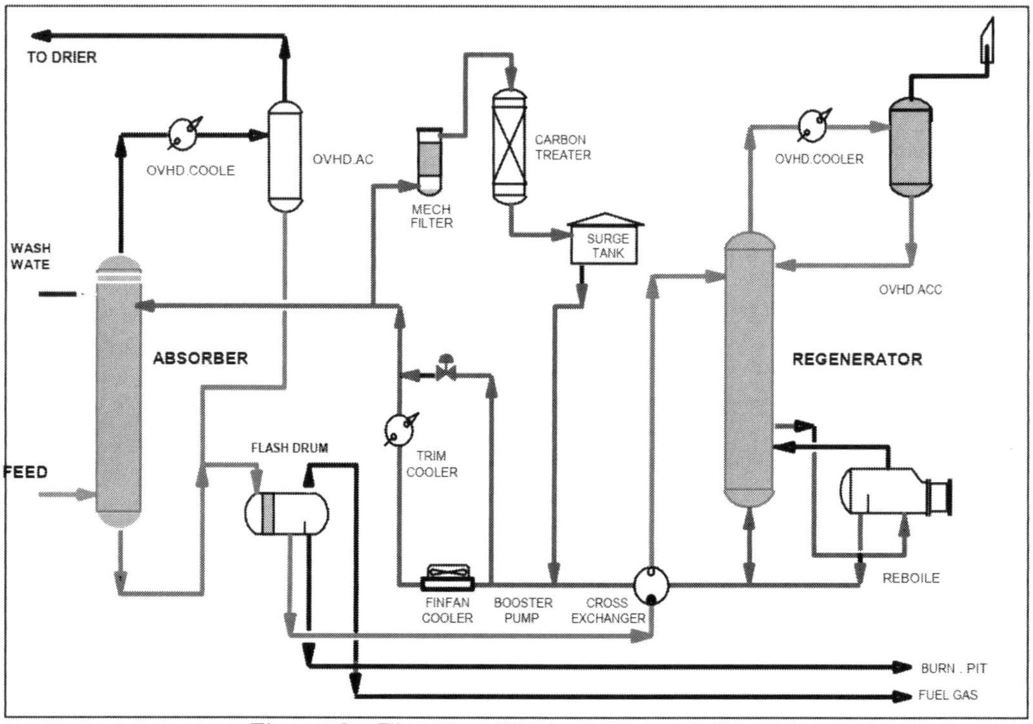


Figure 2. The simplified process of $CO_2$ Absorption

## DISCUSSION

### Background

Badak plant had experienced twice seawater ingress into the $CO_2$ Removal unit that contaminated the amine solvent. The first occurrence happened in 1991, the chloride concentration in amine was reaching 5,300ppm. The second one happened in 2002 and was reaching 1,200ppm. The seawater contains Chloride ion that it should be avoided in a $CO_2$ Removal unit due to a corrosion problem. Literatures advised that those levels are susceptible to a stainless steel material stress corrosion cracking (SCC). Most recommendation advice the maximum chloride ion should be lower than 100ppm. There is no single concern from process performance point of view. The process was running smoothly. No foaming incident, no $CO_2$ breakthrough.

### Chloride Removal Case in Train E

Seawater ingress occurrence in Train E in 1991 was caused by the tube leaks of overhead cooler amine regenerator that is a seawater heat exchanger. The weekly chloride laboratory sampling was alerting this ingression. Decision was made to dump all solution from the system and replace with the new solvent. This condition had driven by the fact that $CO_2$ removal unit was in heavy corrosion problem before this incident happens. Total $400m^3$ contaminated solvent was stored in the temporary tank for decontamination process later.

Based on this pilot result, Badak LNG carried out the decontamination process. It was considered to use the excess capacity of demineralizer unit of steam condensate make up preparation for chloride removal. As the chloride removal was the main purpose, only anion

exchange bed was used. Since the demineralizer unit had been installed, the purchasing of resin was not required. Two tanks were used, the first was contaminated amine tank as the supply and the second was intermediate tank as the final product. The cycle process consists of 4 steps: service, amine rinse, regeneration and regeneration rinse that called as one cycle.

**Initial Regeneration;** Two times regeneration was carried out at the first time to ensure that anion bed was completely exchanged so that the unit was ready to use.

**Chloride Removal Process;** The process was divided into four phases batch. In the first and third batch, the contaminated amine was introduced from temporary tanks (T1 and T2) to anion bed. After chloride removal process, the purified amine routed to a temporary tank (T3) by valves adjustment. In the second and fourth batch, amine from contaminated tank was entered back to anion unit and delivered back to the temporary tank. The process scheme can be seen in the figure-3. The processing time is about 138 kg of chloride/day with circulation of 50m3/hr.

**Amine Rinsing;** After the resin had reached the breakthrough point indicated by the amine effluent conductivity; the service was stopped. The amine rinsing was carried out to wash away the remaining amine in the bed.

**Anion Bed Regeneration;** The regenerant use 4%-wt caustic solution that prepared from 40 % caustic diluted by condensate water. This solution is pumped to the anion bed pneumatically by nitrogen. Prior to regeneration, the resin was preheated to 55°C in order to achieve highest resin regeneration efficiency. The produced waste regenerant was further neutralized by acid solution prior to disposal.

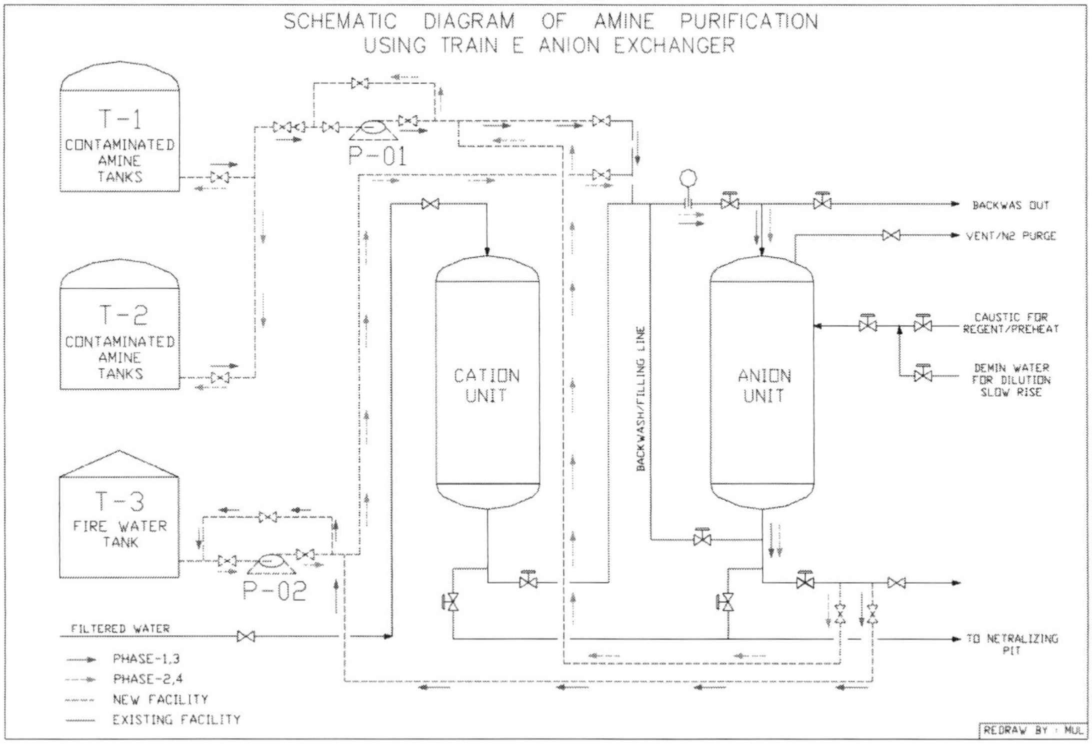


Figure-3. The simplified process of Chloride Removal in Train E

**Rinsing after Regeneration;** After completion of the regeneration step, the unit was rinsed by using the fresh condensate in order to remove the remaining regenerant in the bed.

### Summary of Train E Chloride Removal & Avoid Future Recurrence

The batch process was in four phases, 21 cycles and 31 regeneration within 16 days. In the first phase, chloride was reduced from 5,321 to 755ppm, second phase from 755 to 177ppm, the third phase from 177 to 42ppm and fourth phase from 42 to 32ppm. This decontamination process resulted the amine concentration decreased from 41% to 31.5%. To avoid seawater ingression due to overhead cooler tube leaks, the seawater pressure is reduced below the process pressure by adjusting the seawater inlet valve. Online conductivity monitoring was also installed as early warning system.

### Chloride Removal Case in Train D

Seawater ingress occurrence in Train D in 2002 was happened due to a non-continuous source. This case was different from the Train E case that the chloride level was built up steadily over the time. The root in Train D was quite complex. The most possible scenario was caused when the clean water sewer (CWS) underground that contained the seawater from the amine pump cooler return was lifted up due to flooding during heavy rain. Seawater entered the amine underground line through infiltration from the amine funnel located in the floor and accumulated in underground amine sump tank. It was entering the system during amine make-up preparation in the sump tank. As the result, 1200ppm chloride level was presence suddenly in $740m^3$ system and no more chloride content increase. This problem was known by monthly lab sample monitoring.

**Solution**; Badak LNG considered two alternatives in removing the chloride ion, either use the previous system or hire the third party service company. Considering the location of Train D from the nearest demineralizer unit, unavailability of big spare tanks and the requirement of online decontamination processing as well as cost and effectiveness consideration; Badak LNG decided to use the third party equipment & service. These movable equipments were ready to be dispatched and could be placed closely to the plant.

**The Process;** Principally, this chloride removal process uses "Ion Exchange" method that is similar approach of Train E case. The resin bed exchanges the chloride ion with $OH^-$ supplied by caustic (NaOH) regenerant. Beside chloride, the cation impurities such as Fe and Na were also removed by cation bed that was exchanged with $H^+$ from the sulphuric acid ($H_2SO_4$) during regeneration.

**Chloride Removal Processing;** The process was carried out online. The contaminated amine was tapped to the movable cation-anion exchanger beds and discharge end was tapped to the upstream carbon bed. The processing time is about 36.5 kg of chloride/day with circulation of $18m^3/hr$. When the resin was exhausted, the resin was regenerated. The scheme of chloride removal unit can be seen at figure-4. The anion bed exchanges all type of anion (included $CO_2$ and $H_2S$), therefore lower lean amine loading and lower $CO_2$ content in amine solution should be maintained. During this processes the lean amine loading was maintained at 0.01~0.015 mol $CO_2$/mol amine.

**Amine Rinsing Step;** The resin breakthrough was indicated by the amine effluent conductivity monitoring system. At first, the rinsing water was taken from the fresh condensate line. Half of the rinsing water was returned to amine system (inlet Carbon Treater) and the second half was directed to the temporary rinse tank for the next rinsing use. The next rinsing process (2nd cycles and afterward), the first-half rinsing water need was taken from the temporary rinse tank and it returned to the amine system. The second-half of the rinsing water need was taken from the fresh condensate line, returned to temporary rinse tank and so on. By doing this, the water consumption for rinsing and wastewater disposal to environment was less.

**Anion Bed Regeneration;** The anion regenerant is use 4% caustic solution that produced from 40% liquid caustic soda (NaOH) diluted with steam condensate. Waste regenerant was neutralized by acid solution prior to disposal to environment.

**Cation Bed Regeneration;** The cation regenerant is use 6% sulfuric acid solution that produced from 98% sulfuric acid diluted with steam condensate. Waste regenerant was neutralized by the caustic solution prior to disposal to the environment.

**Rinsing After Regeneration;** After completion of the regeneration step, the unit was rinsed by using the fresh condensate to remove the remaining regenerant. The next cycle would be continuing after completion of regeneration rinsing step. The arrangement of service, amine rinse, regeneration, and regeneration rinse were called with cycle. The finish time each step was determined based on conductivity monitoring.

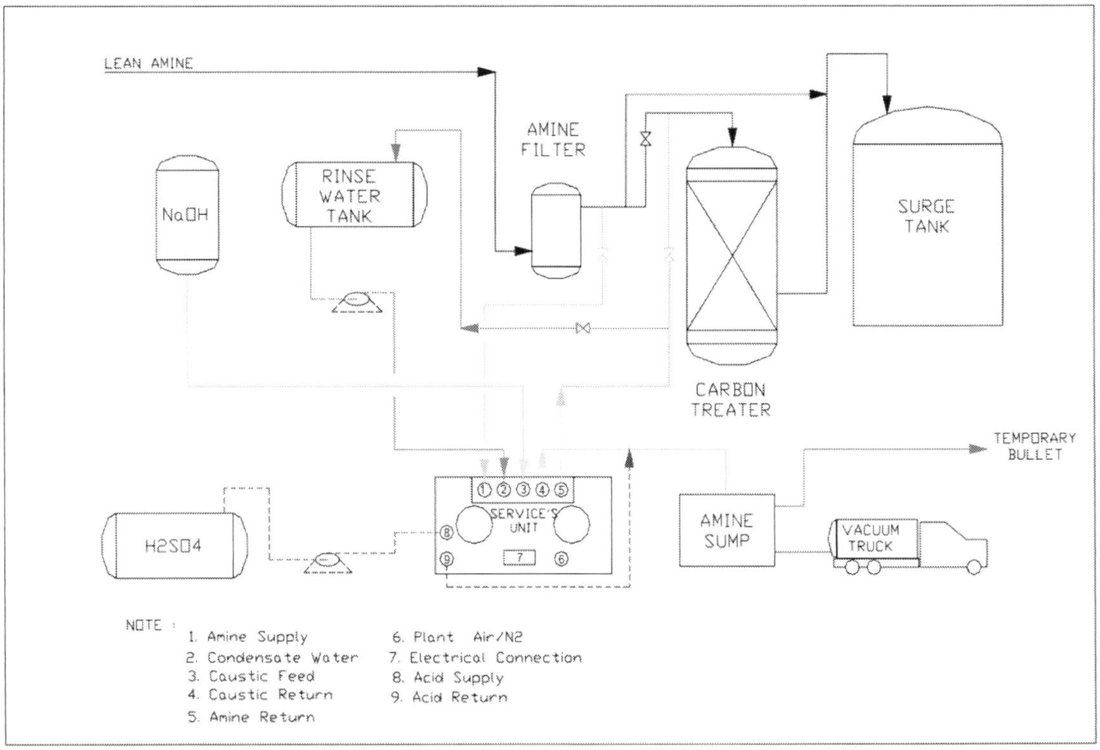


Figure 4. The simplified process of Chloride Removal in Train D

## Summary of Train D Chloride Removal & Avoid Future Recurrence

Chloride decontamination of 1,200 to 50ppm was finished within 24 days. Total processing cycle was 84. In addition, the sodium content has reduced from 500 to about 50ppm and iron content from 15.0 to 2.5ppm. To avoid the seawater ingress reoccurrence, the amine funnel construction was improved and the operator was requested to ensure the chloride presence in sump tank prior to the amine make up.

## CONCLUSION

1. Chloride decontamination of amine solvent can be done using ordinary demineralization ion exchange. In Train E, Badak LNG experience reduced the 5,321ppm level to 37ppm in 21 cycles within 16 days. The solvent loss is 28% of the inventory. The operating cost was USD 112/kg ion removed.

2. Chloride decontamination of amine solvent in Train D successfully reduced from 1,200ppm to 50ppm in 84 cycles within 24 days. The operating cost was USD 245/kg ion removed. The solvent losses was 0.2% of the inventory.

## REFERENCES


1. Betz, Handbook of Industrial Water Conditioning, Ninth-edition, 1991, Pennsylvania.

2. PT Badak, Contaminated Amine Recover Project (internal report), June 30, 1993, Technical Department.

3. PT Badak, Report of Train D Plant-1 Chloride Removal Project (internal report), August 6,2003, Process Engineering Section, Technical Department.

# INTEGRATED LIQUIDS RECOVERY TECHNOLOGY IMPROVES LNG PRODUCTION EFFICIENCY

*Hank M. Hudson, P.E., John D. Wilkinson, P.E., Kyle T. Cuellar, and Michael C. Pierce*

*Ortloff Engineers, Ltd.*
*415 W. Wall, Suite 2000*
*Midland, Texas 79701*
*http://www.ortloff.com*

*Prepared for presentation at the 2004 Spring National Meeting, New Orleans, Louisiana*
*April 26, 2004*



*Based on a paper previously presented at the 82nd Annual Convention of the Gas Processors Association, March 11, 2003, San Antonio, Texas, copyright held by author.*

**Abstract**

Most natural gas is transported from the wellhead to processing plants and thence to consumers in high pressure gas transmission pipelines. At remote locations, however, liquefying the natural gas for transport is increasingly common. The much lower volume of liquefied natural gas (LNG) relative to gaseous natural gas can reduce transportation costs by allowing delivery using cargo ships or transport trucks instead of pipelines.

Regardless of the method used to liquefy the natural gas stream, removal of a significant fraction of the heavier hydrocarbons is generally required before the stream is liquefied. The reasons for this hydrocarbon removal step are numerous, including control of the LNG heating value and the value of these heavier hydrocarbon components as products in their own right. Unfortunately, little attention has been focused heretofore on the efficiency of this step.

For more than 30 years, Ortloff has been developing technology for recovering liquids from natural gas that offers higher recovery, better efficiency, greater simplicity, and better reliability than other available processes. This natural gas liquids recovery technology can be extended to allow efficient liquids recovery in conjunction with liquefaction of the natural gas. Careful integration of the hydrocarbon removal step into the LNG liquefaction process can produce both LNG and a separate heavier hydrocarbon liquid product using significantly less energy than reported for current processes.

This paper presents the recovery and efficiency performance of the new liquids removal technology when applied to typical natural gas liquefaction cycles. Several examples illustrate the performance advantages of the new technology.

## Introduction

In many locations around the world, large deposits of natural gas have been discovered where the potential production capacity far exceeds the capacity of the consumers in those regions. For those regions where there is no existing pipeline infrastructure for transporting the natural gas, liquefaction of the natural gas for subsequent transport in cargo ships or trucks can be an economical means for getting the natural gas to market. A number of large scale plants for producing liquefied natural gas (LNG) are currently under construction or in the planning stages for this purpose.

Most LNG contracts specify a range of acceptable heating values for the LNG sold into a particular market. In most cases, this requires that a certain fraction of the heavier hydrocarbon components found in the natural gas be removed prior to liquefaction so that the LNG does not exceed the upper limit for the heating value. Some natural gases also require removal of the heavy ends to prevent operating problems in the liquefaction cycle, such as freezing of aromatic hydrocarbons at low temperatures.

Ortloff has been developing technology for recovering liquids from natural gas for more than 30 years. Throughout this time, the governing principles for Ortloff technology have been higher recovery, better efficiency, greater simplicity, and better reliability than other processes available to gas processors. As a result, the Ortloff processes for liquids recovery are generally acknowledged as the state-of-the-art around the world.

The removal of the heavier hydrocarbons from the natural gas is considered to be a feed conditioning step in most existing LNG plants, and is not integrated into the overall liquefaction scheme beyond the use of some of the high-level refrigeration from the LNG cycle. During the past few years, much of the development effort at Ortloff has focused on extending our natural gas liquids recovery technology to allow efficient liquids recovery when integrated with the liquefaction of the natural gas.

Although we expected there was room for improving the efficiency of the liquids removal step in the LNG cycle, we found that improving the liquids recovery efficiency was only part of the benefit. In fact, the overall efficiency of the **total** LNG production cycle could be significantly improved through careful integration of the hydrocarbon removal step into the LNG liquefaction process. This allows not only production of both LNG and a separate heavier hydrocarbon liquid product using significantly less energy than reported in the literature for current processes, but has the added benefit of producing a high-quality co-product to generate a second revenue stream for the facility. This paper presents the results of several case studies conducted by Ortloff to determine the recovery and efficiency performance of the new liquids removal technology when applied to typical natural gas streams.

## Current Liquids Removal Technology

The processing scheme used in most LNG plants is shown in block flow fashion in Figure 1. After removal of $CO_2$ and any sulfur compounds, the natural gas is dehydrated before entering the liquids removal section. Heavier hydrocarbons are removed from the natural gas using high-level refrigerant to provide the cooling needed to condense the liquids, and the residual gas is then liquefied using high-level and low-level refrigerant.

Figure 2 shows a typical example of the liquids removal section used in many LNG plants. The natural gas stream is cooled using high-level refrigerant to condense a portion of the stream. The cooled stream is then directed into a "wash tower" where the condensed liquid serves as reflux to absorb heavier hydrocarbons from the natural gas. The wash tower often includes a reflux condenser to minimize loss of the heavier components and a reboiler to minimize the quantity of lighter hydrocarbons co-absorbed in the tower bottoms. A second tower is typically used to strip light ends from the wash tower liquids so that only LPG (liquefied petroleum gas, $C_3$+ hydrocarbons) or condensate ($C_4$+ hydrocarbons) is extracted from the natural gas. The light ends stripped from the liquids are cooled and then recombined with the natural gas leaving the wash tower to become the feed for the liquefaction section.

There are several aspects of this wash tower operation that compromise the efficiency of the liquids removal step. Among these are:

- The wash tower operates at high pressure, causing poor separation efficiency.
- Poor separation efficiency requires further fractionation of the liquids removed.
- Rich inlet gas is a poor reflux stream for absorbing heavier hydrocarbons.
- All of the refrigeration comes from external sources, increasing power requirements.
- The external reboil heat further increases the refrigeration load.

At high pressure, the relative volatilities between the lighter components (methane and lighter, ethane and lighter, or propane and lighter, depending on the hydrocarbons to be removed) and the heavier components are not as favorable as they are at lower pressures. Consequently, the heavier hydrocarbons leaving as the bottom product of the wash tower generally contain a significant amount of lighter components that have been co-absorbed. These light ends must be stripped from the liquid product so that the liquid can be stored at a reasonable pressure, necessitating further fractionation of the wash tower bottoms to remove the light ends from the liquid. As a further consequence of the poor relative volatilities, the fractionation column must use refrigeration to generate reflux for the top of the column to avoid excessive losses of the heavier hydrocarbons.

All of the refrigeration for the wash tower is provided from an external source, typically a portion of the high-level refrigerant used for the initial cooling of the natural gas and for the pre-cooling of the low-level refrigerant in the LNG cycle. This adds to the overall power consumption of the LNG plant. In addition, the reboil heat added to the system in the wash tower and the fractionation column elevates the temperature of the light ends leaving the fractionation column, requiring still more refrigeration to cool the light ends before they are recombined with the natural gas flowing to the liquefaction section.

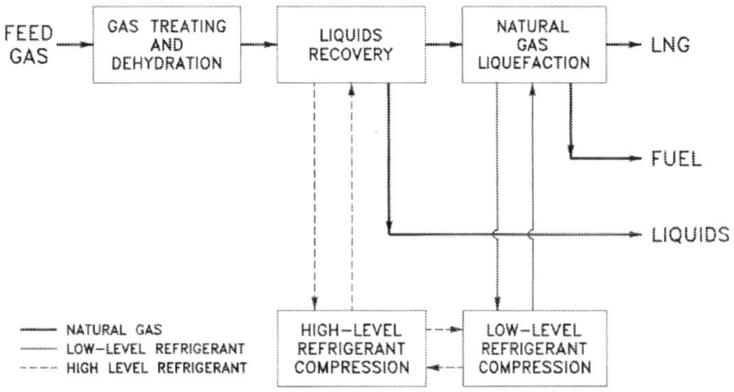


Figure 1 — Typical LNG Plant Block Flow Diagram

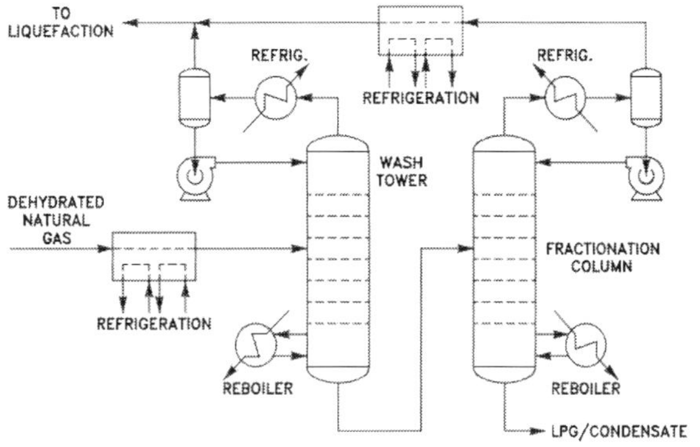


Figure 2 — Typical Liquids Recovery Section

## Natural Gas Liquids Recovery Technology

Ortloff has developed a number of processes for efficient recovery of liquids from natural gas and other gas streams. [Wilkinson et al. 2002] These processes offer many advantages over other processes available to gas processors, most notably: higher recovery, better efficiency, greater simplicity, and better reliability.

These processes achieve higher product recoveries through the use of reflux stream(s) for the recovery column that are appropriate for the liquids to be recovered. This allows overcoming the vapor-liquid equilibrium effects that limit the recovery potential of many competing processes. Some of these processes feature dual reflux streams for use when very high product recoveries are desired.

These liquids recovery processes provide better efficiency than other processes through the use of split-vapor process technology. This design concept separates the gas streams used for power recovery from those used to provide tower reflux, so that each can be optimized separately. The typical performance improvement is on the order of 30% greater efficiency compared to first-generation technology. As a result, retrofitting existing plants to use this technology has been quite popular. [Lynch and Pitman 1996]

Separating the power recovery and reflux components of the process also helps simplify plant operations since it minimizes the interaction between these two aspects of the process design. This makes process optimization much easier than for many other processes, particularly for plants that are subjected to variable feed conditions and/or product economics. These processes deal with variation in the richness of the feed gas quite well, allowing much more consistent product recoveries so that the heating value of the plant residue gas remains stable despite changes in the plant feed gas.

The capabilities of some of these liquids recovery processes are summarized in Table 1. As can be seen, some processes are best suited for NGL recovery (natural gas liquids, generally $C_2$+ hydrocarbons), some for LPG recovery, and some can be used efficiently for either. These processes have been used in gas plants ranging in size from as small as $0.3 \times 10^6$ $Sm^3/D$ [10 MMSCFD] to as large as $60 \times 10^6$ $Sm^3/D$ [2,100 MMSCFD]. These sizes bracket the gas processing capacity of all existing LNG trains and all of those currently in the construction or planning stages. For reference, a feed rate of $7 \times 10^6$ metric tonnes per year (t/Y) to an LNG plant equates to about $23 \times 10^6$ $Sm^3/D$ [800 MMSCFD] of natural gas entering the plant.

Table 1 — Liquids Recovery Processes

| Process | | $C_2$ Recovery | | $C_3$ Recovery | |
|---|---|---|---|---|---|
| | | Ultra-High | High | Ultra-High | High |
| CRR | **C**old **R**esidue **R**eflux | ✓ | | ✓ | |
| RSV | **R**esidue **S**plit-**V**apor | ✓ | | ✓ | |
| RSVE | **R**esidue **S**plit-**V**apor w/ **E**nrichment | | ✓ | | ✓ |
| GSP | **G**as **S**ubcooled **P**rocess | | ✓ | | ✓ |
| SFR | **S**plit-**F**low **R**eflux | | ✓ | ✓ | |
| SRP | **S**upplemental **R**ectification **P**rocess | | ✓ | ✓ | |
| IOR | **I**mproved **O**verhead **R**ecycle | | | ✓ | |
| SCORE | **S**ingle **C**olumn **O**verhead **RE**cycle | | | ✓ | |
| CORE | **C**ascade **O**verhead **RE**cycle | | | ✓ | |
| OHR | **O**ver**H**ead **R**ecycle | | | | ✓ |

## Integrated Liquids Recovery Technology

A block diagram for the Ortloff integrated LNG and liquids recovery process is shown in Figure 3. Depending on the particular process configuration used in the liquids recovery section and the desired recovery level for the co-product liquid stream, the treated and dehydrated natural gas feedstock is cooled using a combination of high-level refrigerant, a slipstream of low-level refrigerant, and/or process streams, followed by processing for liquids recovery. The resulting methane-rich residue gas then flows to the liquefaction section at relatively low temperature and relatively high pressure to be liquefied to form the LNG stream.

For most of the process configurations used in the liquids recovery section, a slipstream of low-level refrigerant is used to supply a portion of the process cooling. For some of the configurations, one or more of the process streams generated within the liquids recovery section can be used to cool low-level refrigerant rather than supply process cooling directly. At lower recovery levels there is generally a surplus of cooling available from the liquids recovery process, which is used to supplement the pre-cooling of the low-level refrigerant.

This integrated processing scheme offers several notable advantages over the liquids recovery technology used in current LNG plants:

- Liquids recovery occurs at intermediate pressure, providing better separation efficiency.
- Better separation efficiency eliminates the need for supplemental fractionation.
- Better reflux streams are utilized, minimizing the loss of the desired heavier hydrocarbons.
- Power recovery is used to generate refrigeration within the liquids recovery process.
- A second high-quality liquid product is recovered for sale, enhancing project economics.
- Liquid recovery can be tailored to match the circumstances of each particular project.

By performing the liquids recovery at an intermediate pressure, the relative volatilities between the lighter hydrocarbons and the heavier hydrocarbons are much more favorable. This makes it possible to capture the desired heavier hydrocarbons with minimal co-absorption of the lighter hydrocarbons, eliminating the need to provide a supplemental fractionation column like that required by the wash tower technology of Figure 1. The separation efficiency is further improved by properly matching the

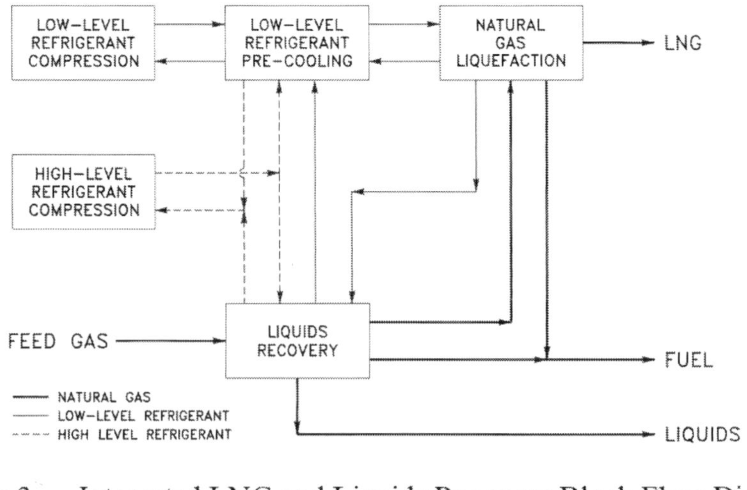


Figure 3 — Integrated LNG and Liquids Recovery Block Flow Diagram

composition of the reflux stream(s) to the hydrocarbons to be captured so that vapor-liquid equilibrium effects do not limit the recovery of the heavy hydrocarbons.

Since the separation is accomplished at an intermediate pressure, work expansion machines (i.e., turboexpanders) can be used to generate part of the refrigeration for the liquids recovery process. This has two very desirable effects on the overall process efficiency. First, the refrigeration provided by work expansion directly eliminates external refrigeration that would otherwise be required to condense the co-product liquid stream. Second, the recovered power can be used to partially re-pressurize the methane-rich overhead gas, so that condensation of the gas stream in the liquefaction section requires less refrigeration. The combination of these factors significantly reduces the amount of external refrigeration required in the LNG plant.

The composition of the co-product liquid stream from the liquids recovery section can be matched to the circumstances of a particular LNG project by selecting the appropriate processing scheme. In locations that have a market for ethane, an NGL product can be produced from the liquids recovery section to feed ethylene plants, etc. If there is no market for ethane, an LPG product can be produced instead to supply the local chemical, heating, and/or fuels markets. Or, if the only need is to control the heating value of the LNG, a condensate product for the local liquid fuels market can be produced. And, for locations where future development may create a market for lighter liquids, or where demand for products fluctuates, processes suitable for variable liquid co-product production can be selected. In all cases, the liquid co-product is controlled to meet the appropriate specification for hydrocarbon liquid streams (i.e., $C_1$:$C_2$ ratio for NGL streams, $C_2$:$C_3$ ratio for LPG streams, etc.).

## Case Studies

To illustrate the advantages of integrating the liquids recovery section with the liquefaction section of an LNG plant, several case studies are presented in the remainder of this paper. For the purposes of these case studies, it was assumed that one gas turbine drives the high-level refrigerant compressors and the second turbine drives the low-level refrigerant compressors. In order to balance the load between the two gas turbines and fully utilize their available power, it was further assumed that both gas turbines are also coupled to starter/helper motor-generators to allow sharing the load equally between the two drivers by generating electricity with the more lightly loaded turbine to supplement the power provided by the more heavily loaded turbine.

With the two gas turbines electrically coupled in this fashion, it is possible to drive the other major rotating equipment in the facility (pumps, overhead compressors, etc.) using some of the electric power generated by the one gas turbine. Similarly, the two hydraulic turbines (on the high pressure mixed refrigerant stream and on the LNG stream) can drive generators to supplement the power produced by the gas turbines. Thus, the production rate for each case studied was adjusted to fully use the power available from the two gas turbines to drive all of the rotating equipment in the liquids recovery and LNG liquefaction sections. (If the other rotating equipment is instead driven by a separate source of power, the increase in power available for refrigeration would allow increasing the production rate accordingly.)

Fuel gas for the gas turbines driving the refrigerant compressors was assumed to be withdrawn from one or more methane-rich streams in the liquids recovery section, prior to feeding the liquefaction section of the plant. The fuel gas withdrawal rate was set by the power requirements of the compressors, assuming a turbine fuel rate of 11.32 MJ/kW-H [8,000 BTU/HP-H]. Any fuel gas consumed by other plant systems (such as regeneration gas heaters, auxiliary boilers, and/or amine reboilers) was not included when determining the fuel gas withdrawal rate.

Case Study 1 — C3MR LNG Cycle

This first case study is based on liquefaction of a natural gas stream similar to that described previously in the literature. [Nibbelke et al. 2002] The LNG liquefaction section uses a mixed refrigerant to supply the low-level refrigeration, with propane refrigeration used for the initial chilling of the natural gas and for pre-cooling of the mixed refrigerant. This is commonly referred to as a C3MR cycle. The pertinent parameters for the study are summarized in Table 2.

The liquid co-product produced for each of the cases in the study varied depending on the desired level of liquids removal. For cases producing $C_2$+ NGL, the $C_1$:$C_2$ ratio in the liquid was controlled to a typical NGL specification. For cases producing $C_3$+ LPG, a $C_2$:$C_3$ ratio consistent with producing commercial-quality propane in downstream fractionation was used. Similarly, for cases producing condensate, a $C_3$:$C_4$ ratio consistent with producing commercial-quality butane in downstream fractionation was used.

The LNG production rate for each case examined was set by assuming that two General Electric MS7001EA gas turbine drivers provide the refrigeration for the liquefaction cycle, with the power available based on 28.5°C [83.3°F] ambient air. No power recovery from the waste heat in the gas turbine exhaust was assumed. (Approximately 30-35 MW of power per gas turbine could be generated using an absorption refrigeration system, a closed-loop hydrocarbon refrigeration cycle, or a simple steam cycle, **but no credit has been taken for this in the specific power consumption values given later**.)

Commercial-quality propane refrigerant in a closed vapor-compression refrigeration cycle was used to supply high-level cooling for the liquids recovery section and pre-cooling of the low-level mixed refrigerant. All of the simulation models used a relatively simple propane system with three pressure levels for the propane evaporators (process chillers). The pressure levels for a particular simulation were set by the specific cooling requirements of the process selected.

The mixed refrigerant streams used in the simulation models are multi-component mixtures of nitrogen, methane, ethane, and heavier hydrocarbons. The particular composition (as well as the number of components) was adjusted for each particular simulation to achieve a good match between the condensation of the process streams and the evaporation of the mixed refrigerant stream. For most of

Table 2 — Process Parameters for Case Study 1

| | | | |
|---|---|---|---|
| Natural Gas Feedstock | | Heat Sink for Cooling | |
| Component | Mole Percent | Cooling Medium | ambient air |
| Nitrogen | 0.12 | Temperature | 28.5°C [83.3°F] |
| Methane | 86.05 | | |
| Ethane | 6.40 | Refrigeration Drivers | |
| Propane | 4.25 | Gas Turbine | GE Frame 7 |
| i-Butane | 1.18 | Number Used | 2 |
| n-Butane | 1.17 | Power per Turbine | 76,266 kW |
| i-Pentane | 0.31 | Waste Heat Recovery | none assumed |
| n-Pentane | 0.30 | | |
| Hexane Plus | 0.22 | Liquefaction Cycle | |
| Total | 100.00 | C3MR (3-stage propane cycle) | |
| | | | |
| Temperature | 22.0°C [71.6°F] | | |
| Pressure | 57.7 bar(a) [837 PSIA] | | |

the process configurations used in the liquids recovery section, a slipstream of mixed refrigerant was used to supply a portion of the process cooling. At the lower recovery levels there is generally a surplus of cooling available from the liquids recovery process, which was used to supplement the pre-cooling of the mixed refrigerant.

The degree of subcooling of the LNG stream was set so that the temperature of the LNG after the pressure reduction step (to storage pressure) was slightly below the bubble point temperature of the stream and no flash vapor was predicted. In practice, a gas recovery system will be needed to capture the boil-off vapor from the LNG storage tank(s), but this detail was not included in the process simulations for this study.

Six different cases were modeled with varying levels of co-product liquid recovery. Two NGL recovery cases were run, one for ultra-high (95%) ethane recovery and one for high (85%) ethane recovery. Similarly, two LPG recovery cases were run, one for high (95%) propane recovery and one for medium (88%) propane recovery. Two condensate recovery cases were run, one for high (92%) butane recovery and one for low (72%) butane recovery. The results of these cases are summarized in Table 3. (The annual production rates given in this table and all of the tables that follow are based on operating 8,760 hours per year. Although it is typical to de-rate the production capacity, the assumed on-stream factor for LNG plants appears to vary considerably from company to company. For ease of reference, no such de-rating was included in the study results.)

In the article cited, the data presented give values ranging from 290 kW-H/t to 336 kW-H/t for the specific power consumption of an LNG plant using a C3MR cycle. As shown in Table 3, the specific power consumptions of the integrated designs are 20% to 35% lower than this range.

Figure 4 displays these results in graphical form, giving the liquid production rate as a function of the LNG higher heating value (HHV). The nearly constant total liquid production, regardless of the degree of heavier hydrocarbon removal, is not as surprising when you consider that all of the gas must be liquefied as either LNG or liquid product. In past installations, however, the processes used for liquids recovery have not been particularly efficient and have not allowed for good heat integration with the liquefaction cycle. The new liquids removal technology addresses both of these limitations, allowing efficient liquids production regardless of the makeup of the co-product liquid stream.

Table 3 — Results for Case Study 1

| | | NGL Product | | LPG Product | | Condensate Product | |
|---|---|---|---|---|---|---|---|
| | | Ultra-High | High | High | Medium | High | Low |
| $C_2$ Recovery | | 95.0% | 85.0% | --- | --- | --- | --- |
| $C_3$ Recovery | | 100.0% | 96.4% | 95.2% | 87.9% | --- | --- |
| $C_4$ Recovery | | 100.0% | 98.9% | 100.0% | 99.9% | 92.5% | 71.7% |
| $C_5$+ Recovery | | 100.0% | 99.8% | 100.0% | 100.0% | 99.6% | 98.8% |
| Feed Gas | $10^6$ t/Y | 6.23 | 6.38 | 6.19 | 6.24 | 6.06 | 6.24 |
| LNG | $10^6$ t/Y | 4.11 | 4.30 | 4.68 | 4.77 | 5.16 | 5.42 |
| Liquids | $10^6$ t/Y | 1.82 | 1.77 | 1.20 | 1.17 | 0.60 | 0.52 |
| Total Liquid | $10^6$ t/Y | 5.92 | 6.07 | 5.89 | 5.94 | 5.76 | 5.94 |
| LNG HHV | $MJ/Sm^3$ | 37.8 | 38.1 | 39.7 | 39.9 | 42.3 | 42.7 |
| Specific Power | kW-H/t | **225.5** | **219.9** | **227.0** | **225.0** | **232.0** | **225.0** |

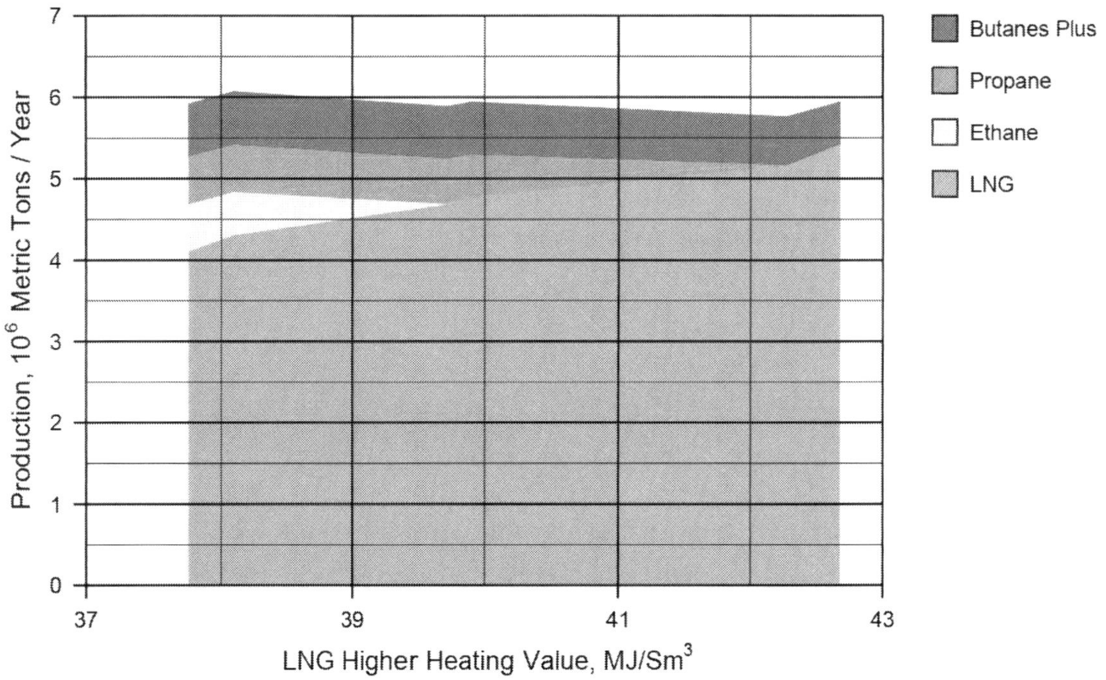


Figure 4 — Liquid Production versus LNG Heating Value for Case Study 1

<u>Case Study 2 — DMR LNG Cycle</u>

The second case study is based on liquefaction of the same natural gas stream as that in Case Study 1. For this case study, however, the LNG liquefaction section uses a second mixed refrigerant rather than propane to supply the initial chilling of the natural gas and the pre-cooling of the low-level mixed refrigerant. This is commonly referred to as a DMR (dual mixed refrigerant) cycle. The pertinent parameters for the study are summarized in Table 4.

Table 4 — Process Parameters for Case Study 2

| | | | | |
|---|---|---|---|---|
| Natural Gas Feedstock | | | Heat Sink for Cooling | |
| Component | Mole Percent | | Cooling Medium | ambient air |
| Nitrogen | 0.12 | | Temperature | 28.5°C [83.3°F] |
| Methane | 86.05 | | | |
| Ethane | 6.40 | | Refrigeration Drivers | |
| Propane | 4.25 | | Gas Turbine | GE Frame 7 |
| i-Butane | 1.18 | | Number Used | 2 |
| n-Butane | 1.17 | | Power per Turbine | 76,266 kW |
| i-Pentane | 0.31 | | Waste Heat Recovery | none assumed |
| n-Pentane | 0.30 | | | |
| Hexane Plus | 0.22 | | Liquefaction Cycle | |
| Total | 100.00 | | DMR (dual mixed refrigerant cycle) | |
| | | | | |
| Temperature | 22.0°C | [71.6°F] | | |
| Pressure | 57.7 bar(a) | [837 PSIA] | | |

| Table 5 — Results for Case Study 2 | | | | | | | |
|---|---|---|---|---|---|---|---|
| | | NGL Product | | LPG Product | | Condensate Product | |
| | | Ultra-High | High | High | Medium | High | Low |
| $C_2$ Recovery | | 95.0% | 85.0% | --- | --- | --- | --- |
| $C_3$ Recovery | | 100.0% | 96.4% | 95.2% | 87.9% | --- | --- |
| $C_4$ Recovery | | 100.0% | 98.9% | 100.0% | 99.9% | 92.5% | 71.7% |
| $C_5$+ Recovery | | 100.0% | 99.8% | 100.0% | 100.0% | 99.6% | 98.8% |
| Feed Gas | $10^6$ t/Y | 6.62 | 6.82 | 6.54 | 6.64 | 6.42 | 6.62 |
| LNG | $10^6$ t/Y | 4.39 | 4.62 | 4.96 | 5.09 | 5.49 | 5.76 |
| Liquids | $10^6$ t/Y | 1.93 | 1.89 | 1.27 | 1.24 | 0.63 | 0.55 |
| Total Liquid | $10^6$ t/Y | 6.32 | 6.51 | 6.23 | 6.34 | 6.12 | 6.31 |
| LNG HHV | MJ/Sm$^3$ | 37.8 | 38.1 | 39.7 | 39.9 | 42.3 | 42.7 |
| Specific Power | kW-H/t | **211.4** | **205.1** | **214.4** | **210.9** | **218.4** | **211.8** |

Six cases with the same levels of co-product liquid recovery as Case Study 1 were modeled. The only difference was the use of a simple mixed refrigerant cycle to supply the high-level cooling in these simulations. The results of these cases are summarized in Table 5 and shown graphically in Figure 5. Again, the liquid production rate is nearly constant regardless of the degree of heavy hydrocarbon removal. For comparison, the values for the specific power consumption computed from the data given in the previously cited article ranged from 259 kW-H/t to 290 kW-H/t for an LNG plant using a DMR cycle. As shown in Table 5, the integrated designs have specific power consumptions that are 16% to 29% lower than this range.

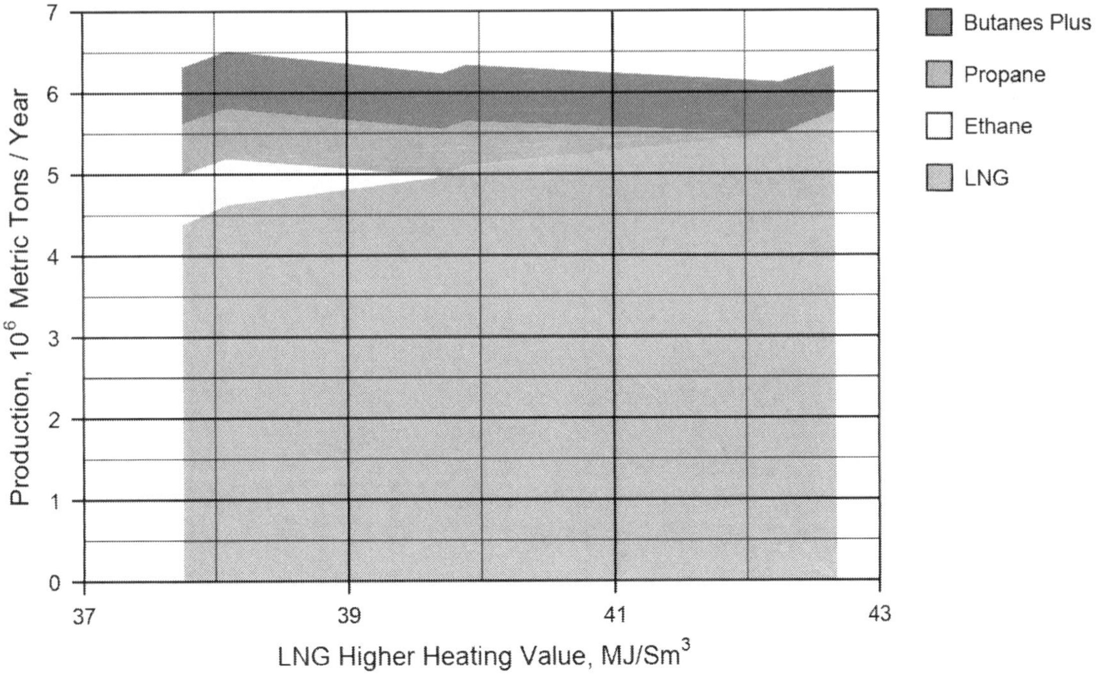


Figure 5 — Liquid Production versus LNG Heating Value for Case Study 2

Case Study 3 — DMR LNG Cycle with Flash Gas Recovery

Case Study 3 is based on a leaner natural gas stream than the first two Case Studies which also contains a significant fraction of inert gases (nitrogen and lighter). A DMR cycle was used for this Case Study, with the addition of a flash gas recovery system to capture the residual vapor from the final LNG flash-down and compress it to sufficient pressure so that it can used as fuel gas in the LNG plant. The pertinent parameters for the study are summarized in Table 6.

The LNG production rate for each case examined was set by assuming that two General Electric MS7001EA gas turbine drivers provide the refrigeration for the liquefaction cycle, with the power available based on 38°C [104.4°F] ambient air. As before, no power recovery from the waste heat in the gas turbine exhaust was assumed, **and no credit has been taken for this in the specific power consumption values given later**.

The degree of subcooling of the high pressure LNG stream was adjusted for each case so that the flash gas formed during the pressure reduction step (to storage pressure) was slightly less than the fuel requirements for the gas turbines driving the refrigerant compressors. This flash gas was compressed to a suitable pressure for use as turbine fuel, then mixed with a small supplemental stream of methane-rich gas from the liquids recovery section as needed to balance the turbine fuel requirements. In practice, a gas recovery system will also be needed to capture the boil-off vapor from the LNG storage tank(s), but this detail was not included in the process simulations for this study.

Six different cases were modeled with varying levels of co-product liquid recovery. Two NGL recovery cases were run, one for ultra-high (95%) ethane recovery and one for medium (80%) ethane recovery. Similarly, two LPG recovery cases were run, one for ultra-high (99%) propane recovery and one for medium (85%) propane recovery. Two condensate recovery cases were run, one for high (90%) butane recovery and one for low (70%) butane recovery. The results of the cases are summarized in Table 7.

Table 6 — Process Parameters for Case Study 3

| | | | | |
|---|---|---|---|---|
| Natural Gas Feedstock | | | Heat Sinks for Cooling | |
| Component | Mole Percent | | Cooling Medium | ambient air |
| Nitrogen / Helium | 3.91 | | Temperature | 38°C [100.4°F] |
| Methane | 87.20 | | | |
| Ethane | 5.50 | | Cooling Medium | cooling water |
| Propane | 2.02 | | Temperature | 35°C [95.0°F] |
| i-Butane | 0.34 | | | |
| n-Butane | 0.54 | | Refrigeration Drivers | |
| i-Pentane | 0.17 | | Gas Turbine | GE Frame 7 |
| n-Pentane | 0.14 | | Number Used | 2 |
| Hexane Plus | 0.18 | | Power per Turbine | 71,260 kW |
| Total | 100.00 | | Waste Heat Recovery | none assumed |
| | | | | |
| Temperature | 25.5°C | [78°F] | Liquefaction Cycle | |
| Pressure | 65.5 bar(a) | [950 PSIA] | DMR with flash gas compression | |

| Table 7 — Results for Case Study 3 | | | | | | | |
|---|---|---|---|---|---|---|---|
| | | NGL Product | | LPG Product | | Condensate Product | |
| | | Ultra-High | Medium | Ultra-High | Medium | High | Low |
| $C_2$ Recovery | | 95.0% | 80.2% | --- | --- | --- | --- |
| $C_3$ Recovery | | 100.0% | 96.4% | 99.2% | 84.9% | --- | --- |
| $C_4$ Recovery | | 100.0% | 98.9% | 100.0% | 99.9% | 90.5% | 70.4% |
| $C_5$+ Recovery | | 100.0% | 99.8% | 100.0% | 100.0% | 99.2% | 95.4% |
| Feed Gas | $10^6$ t/Y | 5.91 | 6.00 | 5.95 | 6.02 | 5.90 | 6.08 |
| LNG | $10^6$ t/Y | 4.32 | 4.48 | 4.86 | 4.97 | 5.11 | 5.32 |
| Liquids | $10^6$ t/Y | 1.07 | 1.00 | 0.58 | 0.54 | 0.28 | 0.24 |
| Total Liquid | $10^6$ t/Y | 5.39 | 5.48 | 5.44 | 5.51 | 5.39 | 5.56 |
| LNG HHV | MJ/Sm$^3$ | 37.2 | 37.5 | 38.9 | 39.1 | 40.2 | 40.4 |
| Specific Power | kW-H/t | **231.3** | **227.7** | **229.5** | **226.7** | **231.8** | **224.7** |

On average, the specific power consumptions in Case Study 3 are about 8% higher than those for the corresponding cases in Case Study 2. This is partly due to the leaner natural gas stream (which generally requires more refrigeration per unit of liquid production) and partly due to the power required for flash gas compression. This, together with the lower power available from the Frame 7 turbines due to the higher ambient air temperature, results in lower liquid production as shown in Figure 6. Nevertheless, the specific power values are still significantly lower than those reported in the literature, on the order of 10% to 23%.

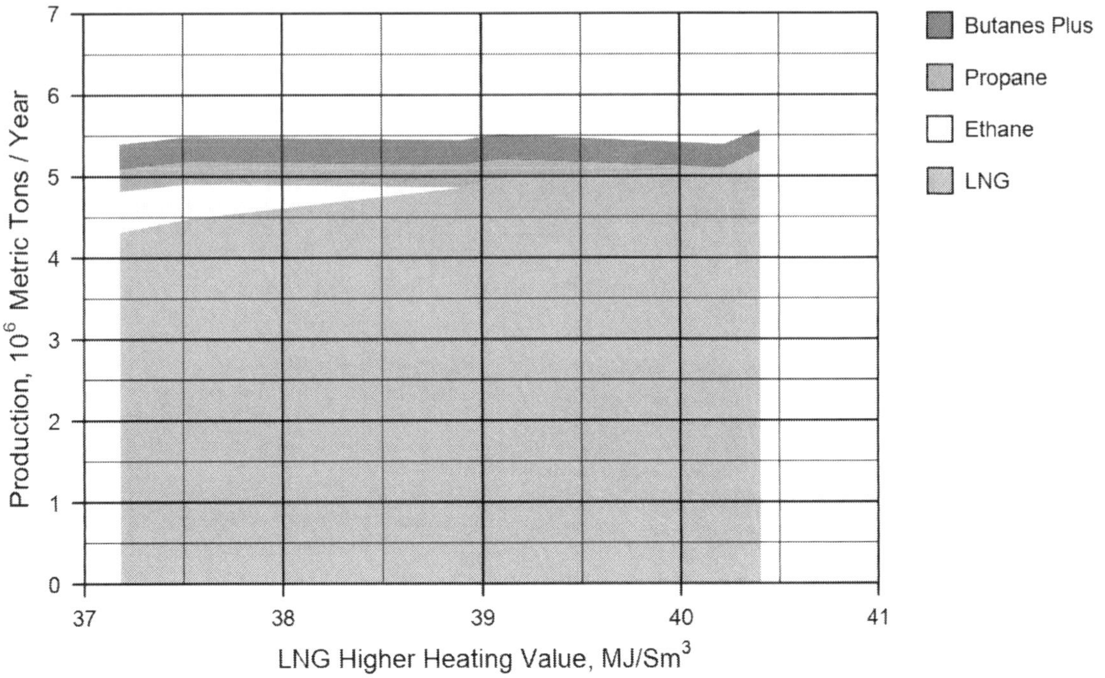


Figure 6 — Liquid Production versus LNG Heating Value for Case Study 3

## Conclusions

The value for the specific power consumption of LNG plants is reported in the literature to range from 259 kW-H/t to 336 kW-H/t. We believe that with proper process selection and heat integration between the liquids recovery and the liquefaction steps, a considerable reduction in the power consumption can be accomplished.

As shown in the Case Studies presented in this paper, the specific power consumption using fully integrated liquids recovery technology ranges from **205** kW-H/t to **233** kW-H/t depending on the type of liquid co-product produced in tandem with the LNG and the type of liquefaction cycle used. These values are based on relatively simple C3MR and DMR liquefaction cycles, and could be further reduced using more sophisticated cycles, cascade refrigeration, or some of the other more recent developments in liquefaction technology.

The benefits of integrating the liquids recovery technology with the LNG liquefaction cycle are three-fold:

(1) Lower specific power consumption means that larger train sizes are possible for a given driver selection. For instance, a 6.8 million metric ton liquefaction train is possible using a pair of Frame 7 drivers. With inlet air chilling on the gas turbines, the train size could be as large as 7.6 million metric tons.

(2) The co-product liquid stream can be tailored to optimize the cash flow for a particular project by producing an NGL stream, an LPG stream, or a condensate stream to serve as feedstock for chemical manufacturing, domestic heating fuel, and/or liquid fuel refineries.

(3) Changes in feed gas composition have much less impact on the liquefaction section because the liquids recovery section can maintain its removal efficiency over a wide range of compositions.

Process configurations are available to provide maximum efficiency for a given product, or to allow efficient adjustment of the product recovery levels if the product consumption and/or economics may be variable. The process efficiency and heat integration provided by this liquids recovery technology allows choosing the appropriate hydrocarbon liquid recovery level with minimal impact on the overall power consumption.

This liquids removal technology can be easily integrated into any type of natural gas liquefaction cycle. This means that all of the liquefaction cycles in use today can benefit from this technology, and that the choice of liquefaction cycle can be made independently of the recovery level desired in the liquids recovery section. This degree of flexibility should help to simplify the design and optimization of new LNG plants. It is also likely that existing LNG plants can be retrofitted to use this technology to raise the liquids recovery efficiency, increase plant throughput, or both.

## Literature Cited

Lynch, J. T. and Pitman, R. N., "Texas Plant Retrofit Improves Throughput, $C_2$ Recovery", Oil & Gas Journal, June 3, 1996, pp. 41-48.

Nibbelke, R., Kauffman, S., and Pek, B., "Liquefaction Process Comparison of C3MR and DMR for Tropical Conditions", Proceedings of the 81st GPA Annual Convention, Dallas, Texas, March 11-13, 2002.

Wilkinson, J. D., Hudson, H. M., Cuellar, K. T., and Pitman, R. N., "Next Generation Processes for NGL/LPG Recovery", Hydrocarbon Engineering, May 2002, pp. 77-84.

## HOW TO SECURE THE PROJECT TIME SCHEDULE TAKEN INTO ACCOUNT THE EXTREME COLD WINTER CONDITIONS?

***Piet Van Den Bossche, Director Project Development***
***Tractebel Gas Engineering Belgium SA***
***Brussels, Belgium***

***Prepared for Presentation at AIChE Spring National Meeting***
***4th Topical Conference on Natural Gas Utilization***
***New Orleans, LA, April 2004***

## ABSTRACT

The Snøhvit LNG Project will be realised on the island of Melkoya, in the proximity of Hammerfest in Northern Norway. This area has specific ambient winter conditions, where the freezing temperature together with the high Northern wind, gives a wind chill factor up to - 45°C.

As such, outside works are become unfeasible from end of October until end of March.

The project for the storage ( 4 full containment storage tanks ) and send – out (3 different sets of loading arms ), of the liquefied natural gas products ( LNG , LPG and condensate ), was awarded to Tractebel Gas Engineering – Belgium SA on the 16$^{th}$ July , 2002, while the commissioning is scheduled for August to October 2005.

With respect for the extreme winter conditions, the project management decided:

- To optimise the outside works during the summer period.
- To create the possibility to work inside the tanks during the winter periods.
- To minimise the site erection works by prefabricating in Belgium the piping racks in modules.
- To integrate the equipment in skid modules
- To slip form erect the concrete walls at the storage tanks.

Weekly workshop meetings and monthly combined project review meetings are guiding the project to a common objective: completion of the project within time and budget.

**Snøhvit Project Introduction**

The Snøhvit project embraces the first export facility for liquefied natural gas (LNG) in Norway and Europe. Huge volumes of gas deeps beneath the Barents Sea will be piped ashore, cooled down and shipped by special carrier to Spain and the USA. Shipment starts in 2006.

Snøhvit is the largest ever industrial project in northern Norway's Finnmark county, costing about NOK 45 billion overall. Costs associated with LNG ship construction come in addition.

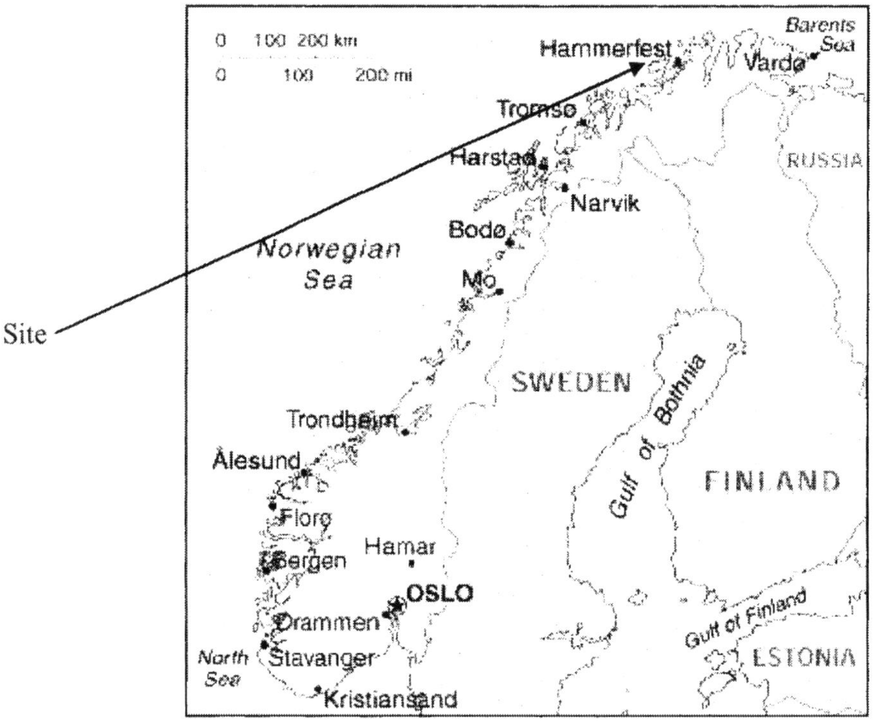


**Slide 1. Location north Norway**

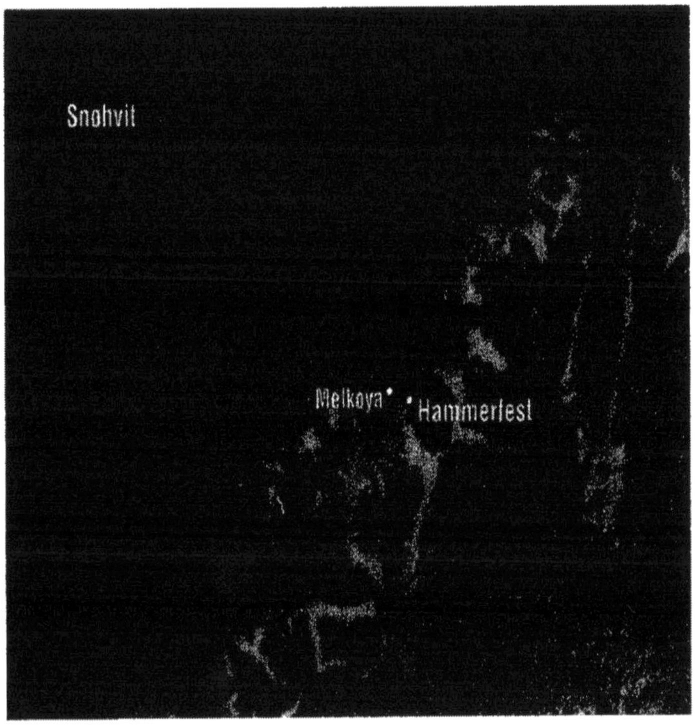


**Slide 2. Gas production system to export terminal**

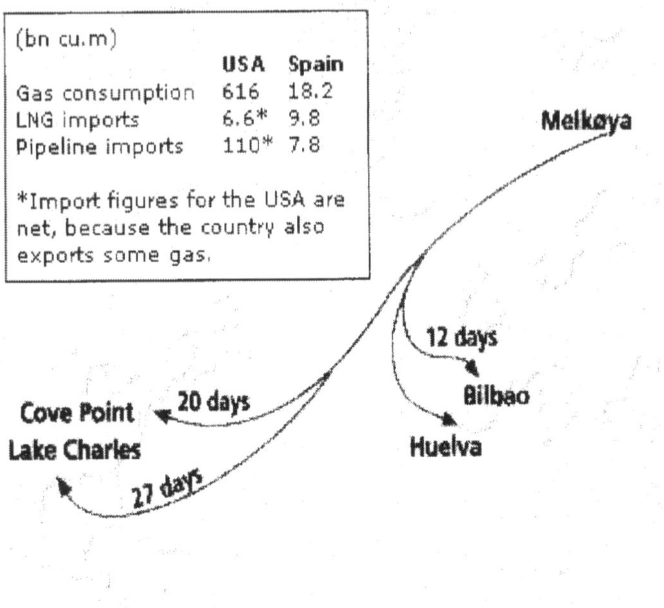


**Slide 3. LNG export**

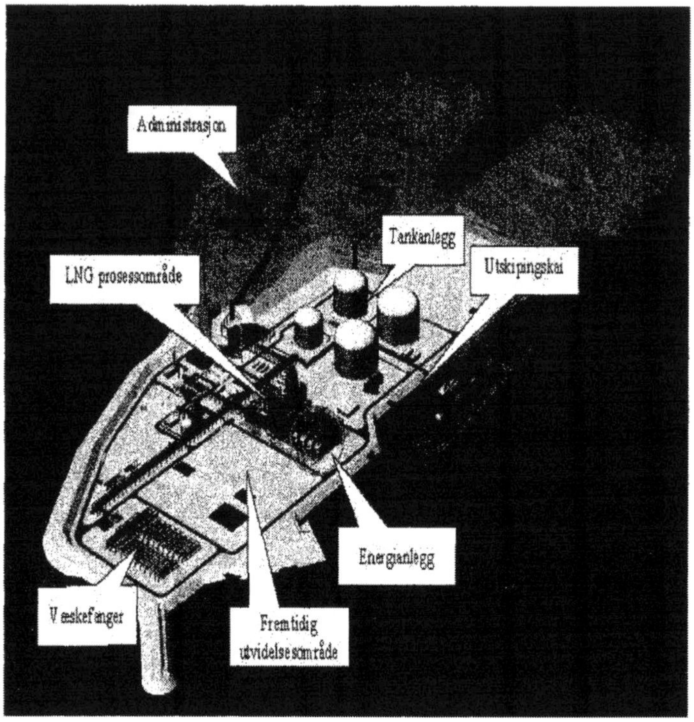


**Slide 4. Overall arrangement**

**Storage and Loading Project Presentation**

The project for the storage (4 full containment storage tanks) and send-out (3 different sets of loading arms), of the liquefied natural gas products (LNG, LPG and condensate), was awarded to Tractebel Gas Engineering – Belgium SA (TGE) on the 16th July, 2002 for the Consortium TGE, Fabricom and Entrepose.

The major items of the project are:

- 2 x 125.000 $m^3$ full containment LNG Tank.
- 75 000 $m^3$ condensate Tank.
- 45 000 $m^3$ LPG Tank.
- Jetty with 3 Loading Systems (1/product).
- Metering Systems.
- LNG HP – Send Out to Liquefaction Plant.
- LNG BOG Blower.
- External Flare and Incinerator.
- All interconnecting piping, electrical and instrumentation works.

LNG Export Terminal for Snøhvit License Group :

- **Operator** :
STATOIL for Snøhvit License Group
- **Contractor** :

Tractebel Gas Engineering Belgium SA for Consortium

Tractebel, Fabricom, Entrepose (TFE)

**Slide 5. Project presentation**

**Slide 6: Project 3D model**

- 2 x 125. 000 m$^3$ LNG Tanks
- 75.000 m$^3$ Condensate Tank
- 45.000 m$^3$ LPG Tank
- Jetty with 3 Loading Systems
- Metering Systems
- LNG HP-Send Out to Plant
- LNG BOG Blower
- External Flare and Incinerator

***Slide 7. Main systems***

**Overall Snøhvit Project Schedule**

The Engineering work for the Snøhvit project by Linde started early 2002. The Site preparation and blasting works on the Melkoya Island started mid 2002.

The Mechanical completion of the plant is scheduled for August 2005, followed by the commissioning activities, while the first gas production is planned early 2006.

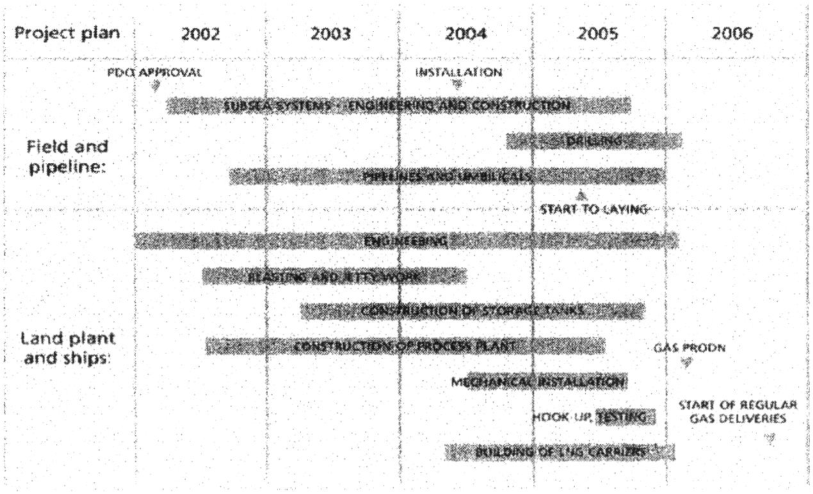


**Slide 8. Overall Schedule**

Slide 9. Melkoya island location

Slide 10. Export terminal

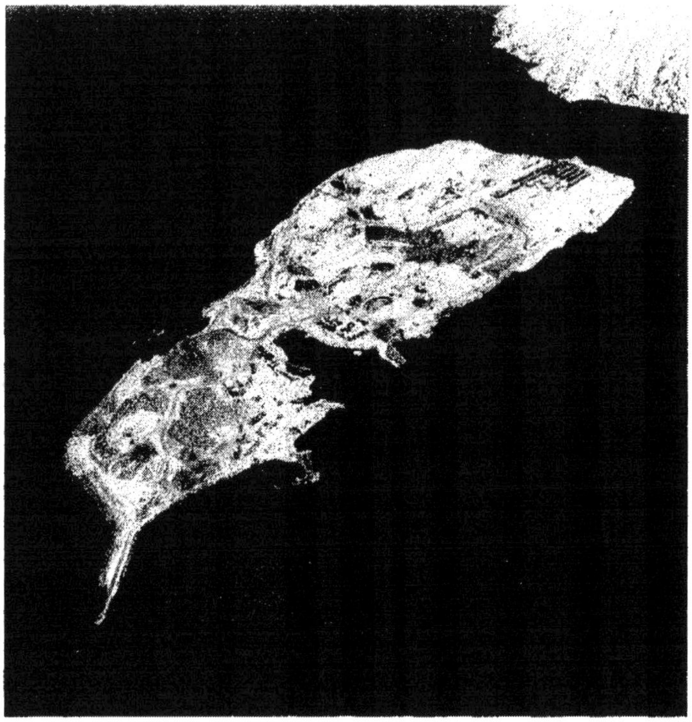

**Slide 11. Melkoya – April 2003**

**Storage and Loading Project Schedule**

The Contract was awarded to TGE on July 16$^{th}$, 2002. Specific key contractual dates of the contract are:

- Contract award: 16 July 2002.
- Blasting plans issued: 01 October 2002.
- Site mobilization: 01 April 2003.
- LNG tank 1 ready for commissioning: 03 August 2005.
- LNG tank 1 ready for operation: 17 August 2005.
- Plant fully ready for operation: 31 October 2005.

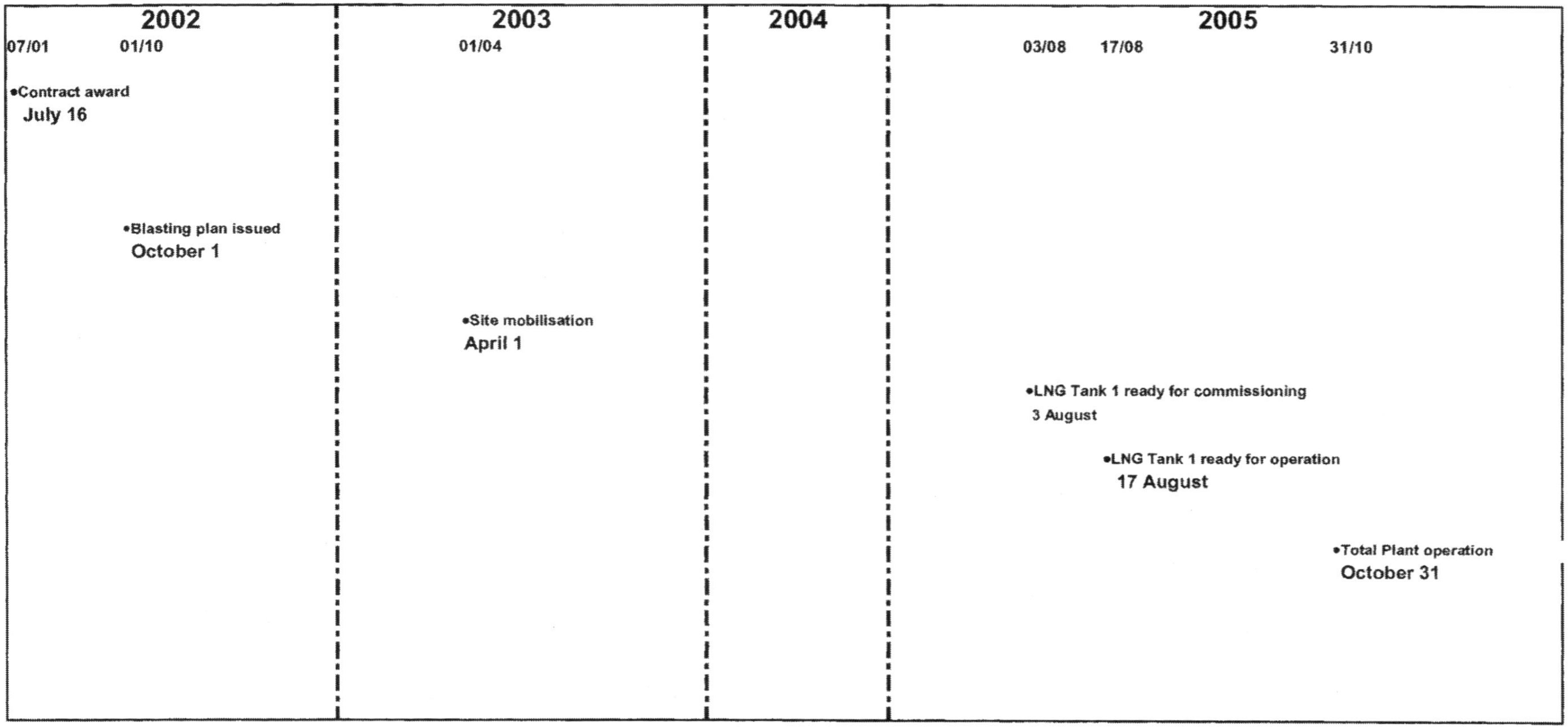


**Slide 12: Project schedule key dates**

**HOW TO SECURE THE PORJECT SCHEDULE TAKEN INTO ACCOUNT THE EXTREME COLD WINTER CONDITIONS.**

The Snøhvit LNG Project will be realised on the island of Melkoya, in the proximity of Hammerfest in Northern Norway.

This area has specific ambient winter conditions, where the freezing temperatures of-5°C to -15°C and the high winds blowing from the North, with velocity up to 25m/s, gives a wind chill factor up to -45°C.

Snow drift around the installed construction areas creates an additional obstacle. As such, outside works are become unfeasible from end of October until end of March.

**Constraints of Normal Construction Schedule.**

Within a normal construction schedule, it takes about 28 months between starting on site with the LNG Tank raft and the moment LNG Tank is ready for commissioning.
In other words, starting on April 1$^{st}$ 2003, leads to completion on August 1$^{st}$, <u>but:</u>

- With working all over the year, even in winter time.
- Preparing the roofs during winter and lifting the roof in May 2004, within the high-winds season. (see slide 13).

Within a normal construction schedule, site piping erection works starts ± 18 months after contract award, with detail engineering isometrics completed 24 months after contract award.

This involves starting erection works in January 2004, in full winter period, and complete the piping works during the winter period of 2005.

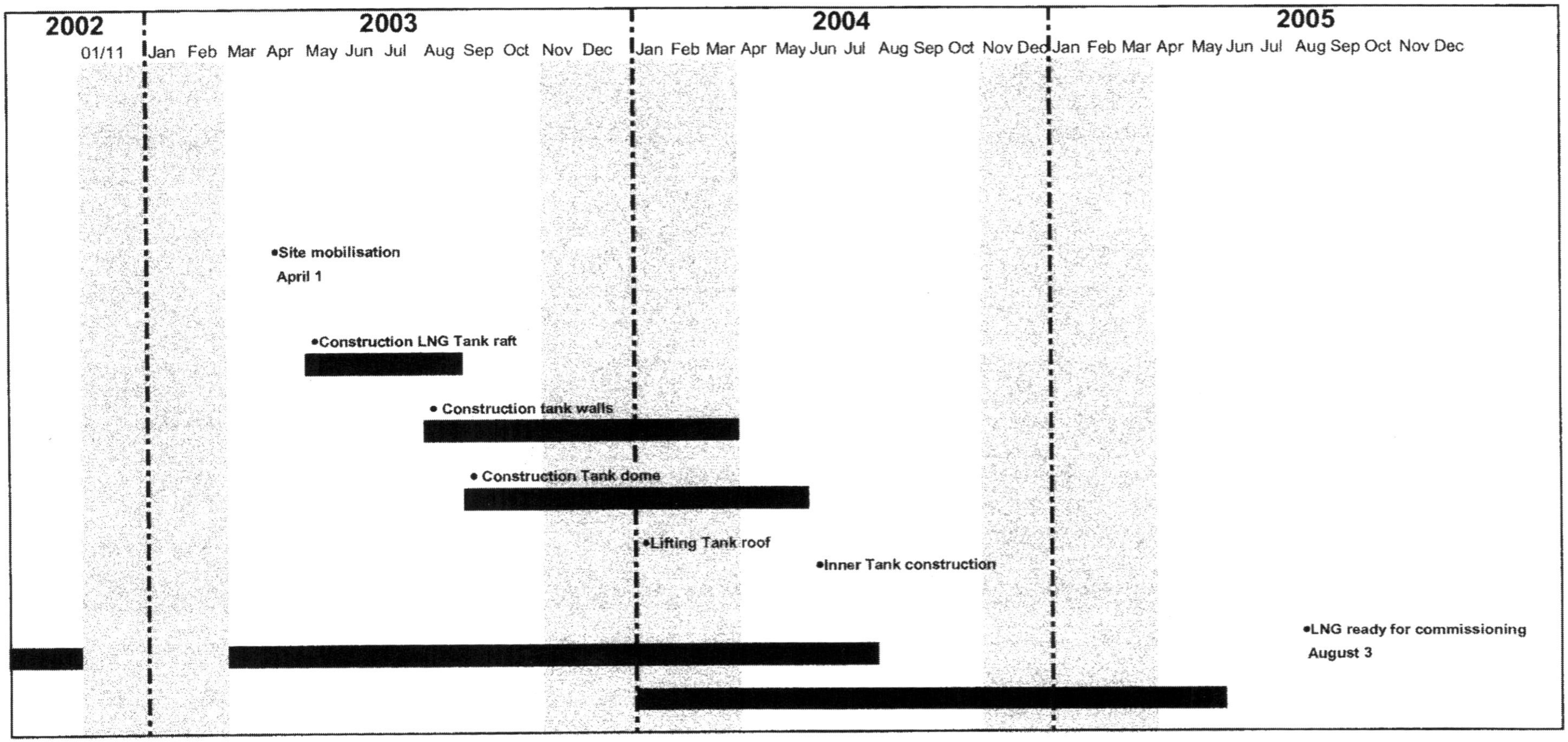


**Slide 13: Normal LNG - project schedule key dates**

The same applies for the erection of electrical and instrumentation works, where the activities could start on February 2004 but for quality reasons would be interrupted during the winter period in 2004 and 2005.

**Specific Snøhvit Project Construction Schedule.**

With respect for the extreme winter conditions, the project management decided:

- To minimise the construction works at site.
- To optimise the outside works during the summer period (April - October).
- To slip form and erect the concrete walls of the storage tanks,
- To prefabricate, transport and install the tank roofs in summer period, in order to create workable conditions for inside works in the tanks during winter period.
- To minimise the site erection works by prefabricating in Belgium the piping racks in Belgium.
- To integrate the equipment in skid modules.

In order to meet the project time schedule including the intermediate contractual milestones following specific measure were taken:

- Blasting of the island and site clearance (under Client control) from July, 2002, until end of March 2003).
- Site mobilisation on April 1$^{st}$, 2003.
- Fast track detailed design performed in 12 months in order to start site activities for the tanks and prefabrication works. (modelisation of pipe racks and equipment ships).
- Accelerate the procurement activities for the bulk material (piping, insulation, electrical heat tracing, instrumentation) in order to meet the production schedule of the modules.
- Realization of concrete works. (foundation and wall) for 3 tanks plus steel roofs before the winter period of 2003 -2004.
- Begin inside tank works for the metallic part during the winter period 2003-2004.
- Start prefabrication of piping skids in 2003, in order to have a first shipping campaign to site in spring 2004.
- Install <u>all</u> piping and equipment skids before the winter of 2004.
- Precommissioning and mechanical complete the plant in spring 2005.
- The LNG tanks hydro tested in June 2005.
- Plant fully commissioned in October 2005.

Those specific Snøhvit aspects where leading to an overall project as presented in slide 14.

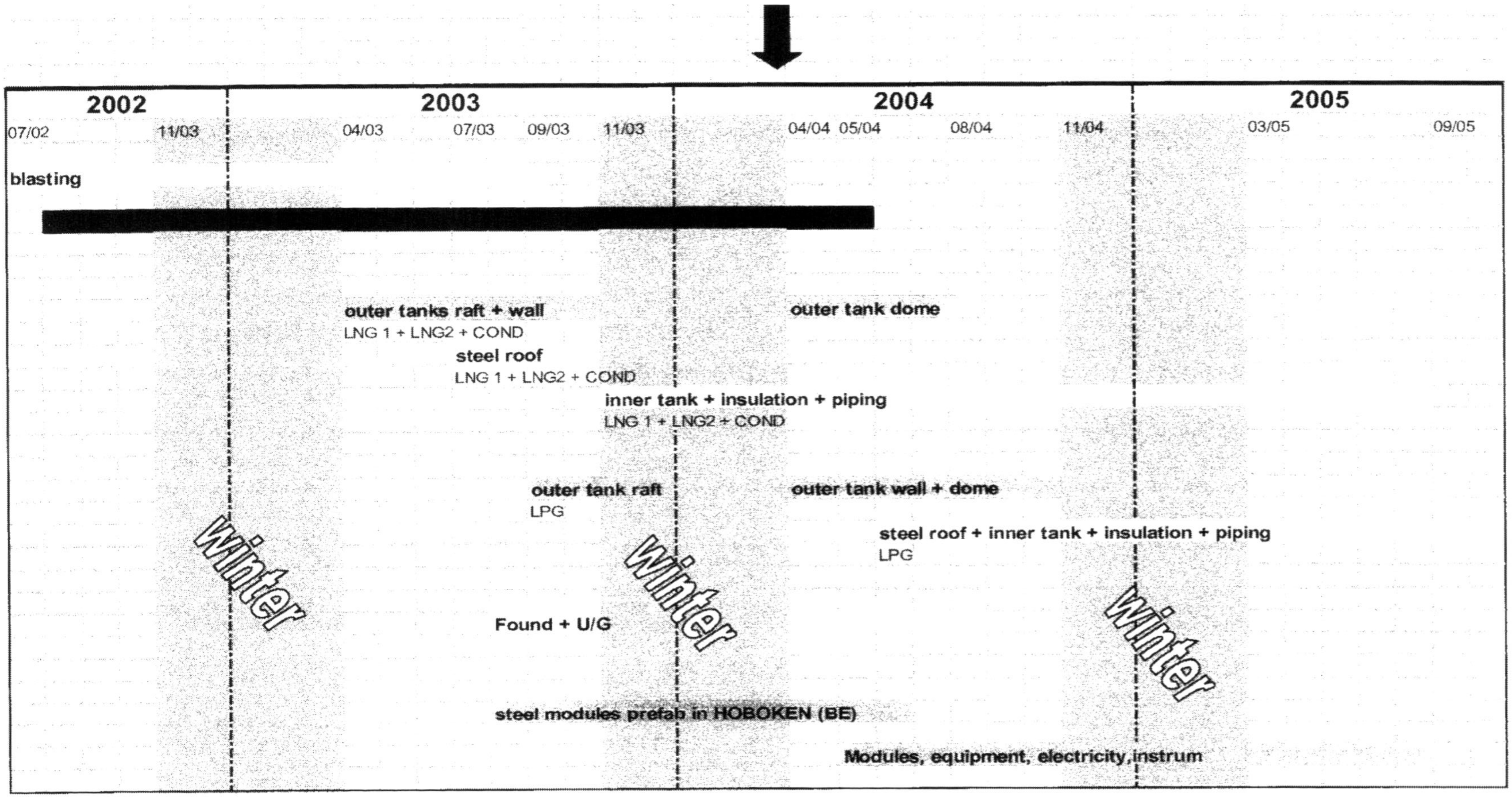


**Slide 14. Snøhvit specific aspects**

**Actual Project Status**

Before the winter of 2003 TGE realised the following activities on site:

- Closure of 3 tanks, (2 LNG/ 1 condensate) by end of October 2003.
- Raft erection of 4th tank (LPG)
- Concrete foundations for 25 % of piping modules.

Slide 15 is presenting the actual project schedule for the tanks, realized and scheduled for 2004.

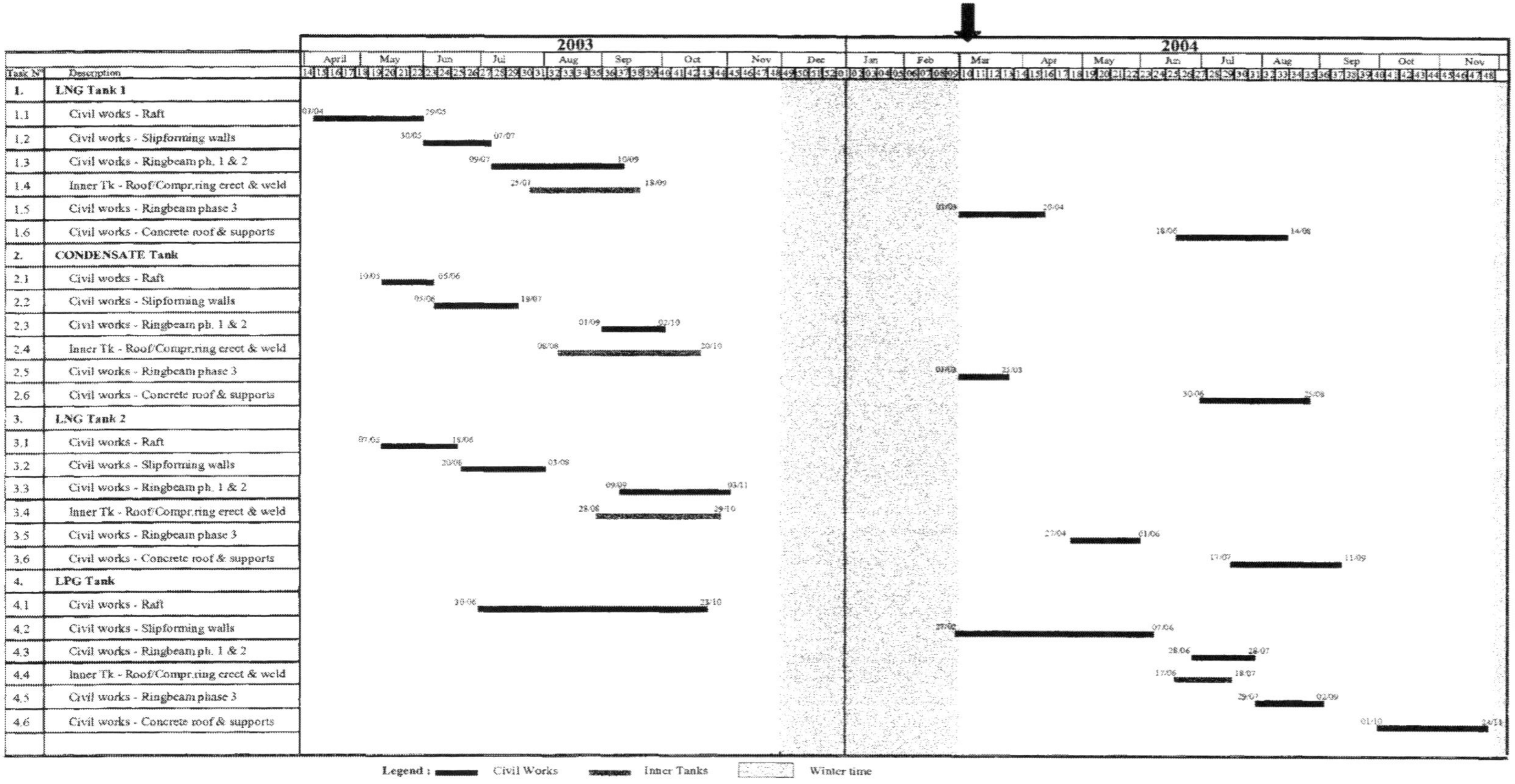


**Slide 15. Snøhvit project construction schedule for the tanks**

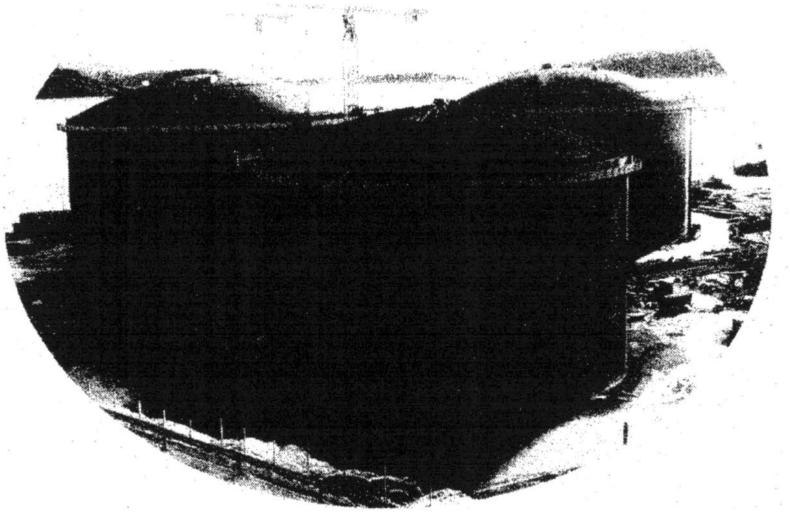

**Slide 16. Tanks completed**

Today, TGE has achieved already a 30 % progress on the inside works of the LNG tanks and 45 % on the condensate tanks.

Foundations works for the remaining piping modules will start on April 1st. Outer tanks concreting works for the LPG will start April 15th in order to have the dome installed by July 2004.

First campaign with 20 modules is ready for shipment and site erection works will start by end of June.

TGE refers to attached pictures of the piping modules which are finalised and ready for shipment.

**Slide 17. Jetty Module**

**Slide 18. Pipe Rack Module**

**Slide 19. Stair Tower**

## CONCLUSION

Taking into account this specific ambient weather conditions and having in mind that the Snøhvit project is a fast track project the only way to achieve the project schedule was to address a specific execution plan regarding tank construction and site erection works.

The decision taken by Contractor during the bidding stage to proceed in this way was facilitating the Contractor to mobilize the project team and structure immediately after contract award.


## Acknowledgments

The Client, fully supporting the effort to meet the target deadline, integrating his project team into Contractor's office task force during the engineering phase, and is today integrated within Contractors fabrication workshops offices of the modules in order to anticipate on all potential quality, interfaces and timing aspects.

Weekly workshops meeting and monthly combined project review meets where and when guiding the project to a common objective: completion of the project within time and budget.

# AN EFFORT TO IMPROVE PLANT THERMAL EFFICIENCY IN BADAK LNG PLANT

*Archiaston Musamma*
*PT Badak NGL Co.*
*Bontang, East Kalimantan - Indonesia*

*Prepared for Presentation at AIChE Spring National Meeting*
*4th Topical Conference on Natural Gas Utilization*
*New Orleans, LA, April 2004*

## ABSTRACT

PT. Badak Natural Gas Liquefaction, one of the first LNG plants in the South East Asia region has been in operation for 25 years, developing valuable experience in the operation of an LNG process. The first two trains were started-up in 1977, followed by six others trains in 1983, 1989, 1993, 1997 and 1999. All trains were designed with steam turbine drivers, seawater cooling, propane pre-cooling and multi component refrigerant as the final stage of cooling process. Thermal efficiency of Steam Turbine driven system is unique compared to an LNG plant that uses Gas Turbine drivers.

Obviously, the thermal efficiency target starts from the process design, however, the effort continues day-to-day for continual improvement. There are many contributing process variables that play a roll in developing the overall efficiency result. In the Badak LNG Plant, thermal efficiency improvement is a main focus now and for the future. In the past, concern of thermal efficiency was secondary behind "the plant debottlenecking capacity" that was driven by the growth demand of LNG. Capacity enhancements put the bottlenecked equipment at the edge of curve and away from their best efficiency. At turn down capacities during low LNG demand, the mode of operation should be changed effectively to meet the objective of plant thermal efficiency optimization. This needs a distinctive strategy, as every LNG train or piece of equipment has its own characteristics, in spite of the same design.

## 1. INTRODUCTION

**Company Profile;** Badak LNG plant is located in Bontang, East Kalimantan, Indonesia. Owned by Pertamina (The Indonesian State Owned Oil & Gas Firm). The plant is operated by PT. Badak NGL Co., an Indonesian Operating Company, 55% by Pertamina, 20% BP Indonesia, 15% by JILCO (Japan Indonesian LNG Company), 10% by Total Indonesie. The total capacity with eight trains is 22.4 million ton per year (MMTPA) of LNG; 1.0 million Ton of LPG and 11.0 million barrel of Condensate per year.

**Plant Capacity & Retrofit History;** Today operating capacity of Trains A~D are 2.64 MMTPA/train, 146% of the nameplate design after refrigerant compressors and drivers retrofit. The Train E~G are 2.86 MMTPA/train, 121% of the nameplate design after retrofitting the major refrigerant compressors. Train H is 3.1 MMTPA, 106% of the nameplate design, excess from the original design capacity.

**Plant Thermal Efficiency (PTE);** The focus of improvement moves to the efficiency, after safety and reliability maintained. In the past, concern of efficiency or Plant Thermal Efficiency (PTE) is the secondary behind "the plant debottlenecking capacity" that was driven by the growth demand of LNG. The capacity enhancements put the bottlenecked equipment operate at the edge of curve and away from their best efficiency.

**Benchmarking;** In 1999 & 2000, Badak LNG includes in the key performance comparison of World LNG plants conducted by third party. One of the compared performances is PTE. The comparisons reveal that Badak LNG PTE is 89.7% versus the world average of 90.8% and the best is 92.8%. Other performance parameters are the operating cost, reliability index, etc. that Badak LNG stays on top position. Plant Thermal Efficiency finding advice Badak LNG to puts thermal efficiency as the new target for the improvement.

## 2. OBJECTIVE

The objective of this presentation is to share the lessons learned in identifying, planning and executing the PTE improvement.

## 3. PLANT THERMAL EFFICIENCY OF BADAK LNG

**Design Concept;** Badak LNG Plant was built with the design concept of the first Train in 1977. The next trains in general are carbon copies with some operation lessons learnt. The basic philosophy of using the previous design was the proven record of reliability, use the common spare part to reduce the stock items, faster project construction and simplicity. This is with the price of slightly lower plant thermal efficiency than the common new LNG design with the gas turbines driver.

**Operating Condition Contribute PTE:** There are several factors in daily operation that contribute to the losses and decreasing the PTE:

- **Plant Trip;** Plant trip release the MCR inventory to the flare. The start up process requires derime process to sweep away the moisture and cool down process. The short plant trip and fast start up release 800,000Nm$^3$/hr is 0.04% PTE. Schedule shutdown release about 2,000,000 Nm$^3$ gas that equivalent of 0.1% of PTE.
- **MCHE Tube Leaks;** The MCHE tubes leak is a problem that most of LNG trains have been experiencing this problem. The leak rate below 100,000Nm$^3$/day could be recovered by sending it to the fuel gas system. If unscheduled shutdown for repair is not possible during high LNG demand; the losses could be as high as 0.04% of PTE.

**PTE in the Last 3 Years;** PTE of Badak LNG in the past 3 years is shown on the Figure-1. The pattern was contributed by several factors, such as Plant Reliability (every shutdown there is a losses), Production Load Factor (there is an optimum efficiency at certain % of running capacity), feed gas pressure (the lower the feed gas the lower the available JT effect, the higher refrigeration load require), etc.

**Figure - 01**

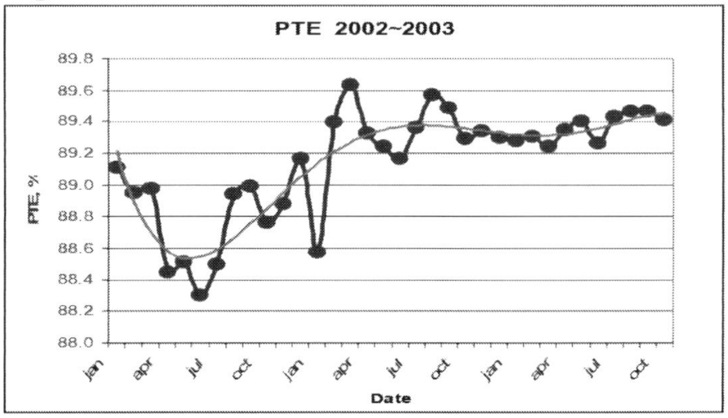


**The Structure of Energy Balance;** Typical Badak LNG energy and volume balance is reflected by design, feed gas composition and daily operations.

**Table - 1**

TYPICAL VOLUME / ENERGY BALANCE

| SUBJECT | % vol |
|---|---|
| LNG Production | 80.39 |
| LPG production | 1.31 |
| Condensate Production | 0.85 |
| Lean Gas to Domestic Usage | 0 |
| Boiler Fuel Consumption | 10.87 |
| CO2 Vented to Atmosphere | 5.15 |
| Total MCR and C3 Make-Up | 0.04 |
| Marine Flared Gas | 0.03 |
| **Processed Flared Gas and losses** | **1.36** |
| TOTAL | 100 |

| % energy | SUBJECT |
|---|---|
| **89.0** | Overall Plant Thermal Eff. |
| 10.5 | FUEL(from fuel gas product) |
| 0.26 | Flared |
| 0.09 | MCR Make Up |
| 100.0 | Total Output |

## 4. BENCHMARKING & PROSPECT FOR IMPROVEMENT

**Benchmarking:** Benchmarking is an important tool for positioning and determining the realistic objective. However, the information was limited and general. Therefore, evaluation was needed.

### 4.1. Standardization of the Process Different

The benchmarking data was expanded into a specific comparison such as how much energy utilization of Gas Turbine vs. Steam turbine, how much PTE would be affected if there are different of feed gas pressure, the $CO_2$ content in the feed gas, ratio of "LNG to LPG" product, and other possible un-standardize / un-normalized condition that should be translated into the same condition. The final different of post normalization could be interpreted as the opportunity derived from the benchmarking that referred from the plant that mostly the latest one. This becomes the objective of PTE improvement for Badak LNG. To be noted that the process condition and equipment design of each location is unique, it carry inherent factor that affect efficiency. Some of the factor could be beyond of the control.

The process different between Badak LNG and the benchmark LNG plant were checked in more specific item as the following.

**Feed Gas Pressure**

The feed gas pressure contained the cold energy when it flashed to the lower pressure (JT Effect). The higher the feed gas pressure, the lower the refrigeration needs to produce the same LNG production. The average World LNG Plants data is 62 barg versus PTB 47barg. For the same LNG production, higher feed gas pressure reduces the total compressor refrigeration requirement by 25,000HP. This subsequently reduces the steam usage. The PTE correction if PTB have similar feed gas pressure is +0.5%.

**CO2 Content in Feed Gas**

The higher the carbon dioxide contained in the feed gas, the higher the heat energy required to release CO2 in Acid Gas Removal Unit. This increase the steam need in reboiler and the pumping energy for amine circulation. The $CO_2$ in feed gas of World LNG Plants is 3.3% versus PTB 5.2%. For the same LNG production, less $CO_2$ in feed gas reduce the steam reboiler consumption. The PTE correction if PTB have similar CO2 content is +0.02%.

**Steam Turbine vs. Gas Turbine**

The World LNG Plants main compressors are driven by the Gas Turbines. PTB are driven by the Steam Turbines. The gas turbines follow the Brayton Cycle (Air Compressor- Combustor- Turbine), while the Steam Turbine follows the Rankine Cycle (Boiler-Steam Turbine-Condensate Pump). The efficiency of gas turbine is 29~32%. The efficiency of steam cycles about 22~26%. The PTE correction if Badak LNG have been used the gas turbine is +2.1%.

**The Ratio of LNG to 2nd Products**

As the benchmarking cover the PTE of the overall plant operation, including the side products such LPG and condensate. The side product ratio to LNG becomes important matter to be check. LPG product require a fractionation effort compare to the LNG that more the refrigeration effort. The energy use to liquefy LNG is higher than LPG fractionation. Therefore, the ratio of these two products should be compared in a standardized value. The LNG ratio to LPG of the World LNG Plants is 67.1% versus PTB 72.3%. For the same LNG production, the World LNG Plants requires 7% less refrigerant horsepower than PTB LNG. The PTE correction if PTB have similar LNG ratio is +0.9%.

**Overall PTE Result After Theoretical Normalization:** After data normalization the PTE can be compared. Total PTE value correction is +3.5%. Therefore, Badak LNG PTE would be about 93.3% if normalized to a similar condition. This shows that Badak LNG PTE issue was a design concern rather than a daily operating. However, there is still opportunity to improve from the operating condition that will be discussed briefly.

**The Prospect forImprovement**

The above global mapping "viewpoint" was further evaluated in more detail information.

**Marine Flare Study**

**Marine Flare;** Marine flare is the LNG vapor during ship loading that can't be recovered due to recovery system limitation or because of excess fuel gas supply over the fuel gas demand.

**Ship Vapor Return;** Fuel gas supply from LNG ship vapor return is uneven case that depends on loading or no loading, the number of simultaneous ship loading, ship temperature, number of boil off gas compressor in operation, and the LNG loading rate. In addition, the fuel balance is an un-steady state condition that is depends on the LNG rate of production, LNG tanks level and LNG temperature delivered from the trains.

**The Facts;** The fact of the existing operation was handicaps by following circumstances:

a) Low pressure can be delivered from the Ship LNG i.e. 1.2 kg/cm$^2$.a. before the ship requesting to open the marine flare. By design, the boil off recovery piping pressure drop was design at 1.9kg/cm$^2$.a.
b) Withdrawal the residue liquid pump in quenching drum was sized too large. It was for the maximum scenario of three simultaneous loading, while most of the case, the loading is following pattern on the table-1.

**Opportunity & Challenge;** The above pattern indicates there is enormous opportunity to recover the Ship BOG. At the same time, it is a challenge to find the optimum configuration considering the pattern is highly irregular. By the statistic, 75% of total marine flare could be recovered to replace 75% of fuel gas make up.

**Table-1**

| | DOCK OCCUPANCY | |
|---|---|---|
| **DOCK** | **% of Time** | **REMARKS** |
| **Dock-1** | 14.9 | One only |
| **Dock-2** | 19.7 | One only |
| **Dock-3** | 11.7 | One only |
| **1-2** | 5.4 | Two simultan |
| **1-3** | 2.5 | Two simultan |
| **2-3** | 4.1 | Two simultan |
| **1-2-3** | 0.1 | Three simultan |
| **No Loading** | 41.6 | |
| **Total=** | 100.0 | |

**Figure-1 (Dynamic of Marine Flare & Make Up)**

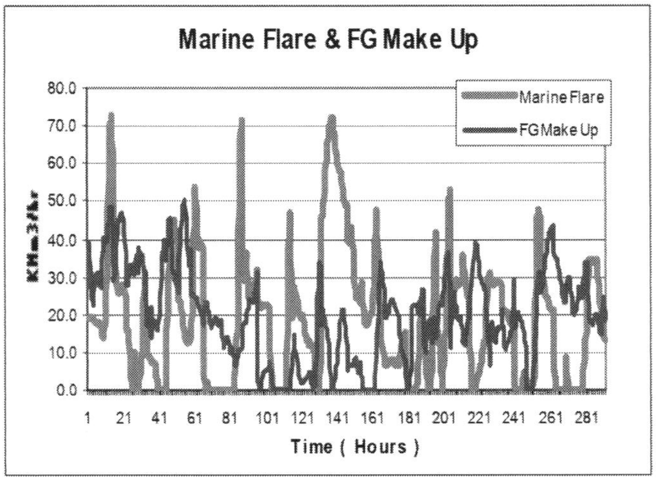


**Flaring;** There are some points that the recovered ship BOG would be flared during low demand of fuel gas makes up. This cannot be avoided, as the nature of centrifugal compressor should be run flat as much as possible, due to a short operating range. Therefore, during low fuel gas make up, there is still fuel gas excess. However, most of the time the recovered ship BOG would still be utilized to reduce the fuel gas make up.

**Solutions;** To install additional shore side blowers to boost the pressure from docks to the quenching drum. The liquid withdrawal of the quenching drums equipped by ejectors to improve the range ability of liquid withdrawal from the quenching drum that cannot be covered by the pump operating range. The ejector motive fluid uses the fuel gas boil off compressor discharge. The realistic PTE improvement is estimated as high as 0.29%. This is equivalent of USD 7.0 million/year.

## The Optimum Excess Air in Boiler

Most of the fuel gas use in Badak LNG is Boilers; little fuel gas use for two 12MW power generator gas turbines. Historical data shows that the boilers operate at high air fuel gas ratio (20-80%). The best performance test record indicates that boiler can run at 15% excess air at 80% load. The prospect of boiler efficiency is up to 2% higher if the excess air meets the ideal performance test condition.

The recent performance test indicates that the excess air could be maintained at 17%. If all boilers are maintained fix at this point, the individual boiler efficiency can be improved from 84.7 to 85.4%. The overall contribution of PTE improvement is 0.07%. This is equivalent of USD 1.7 million/year. This needs an extra day-to-day operational care.

## Load Optimization in LNG Trains to gain higher PTE

Load optimizations of LNG trains were come from of the facts that the energy uses to produce LNG is depends on the rate of production. The process historian provides a real statistic data at wide range of operation.

Highest production is trade off with the efficiency. C3-MR compressors are the main energy user in the system; therefore it dictates the plant efficiency. The highest compressors efficiency is not located at the highest production rate. In general, the most Trains were running at their best efficiency when the production rate at about 85% of the maximum capacity. For boilers, the turn down capacity was quite flat. It stays at highest efficiency at 40~100% of maximum load. Boiler nature differs from the common rotating machines.

**Best Efficiency of Single Trains;** Figure –3 derived from the steam required in C3-MR steam turbine driver to produce the LNG. The best efficiency is the highest ratio of LNG to steam. For example, Train H reaches the highest efficiency at 750m$^3$/hr that is 87% of it maximum sustains capacity. Running at higher or lower than 750m$^3$/hr would result at slightly lower efficiency. The efficiency curve pattern of other trains are quite similar, the only difference is the point of peak efficiency due to train size difference and a temporary individual operating performance at a given period.

**Figure-3 (Operating Range & Efficiency)**

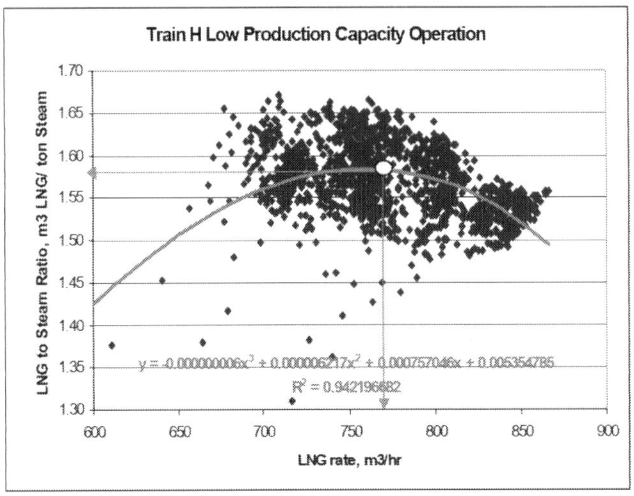


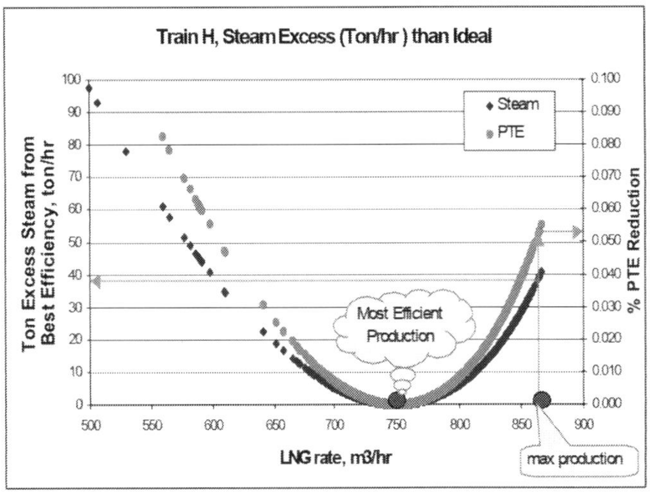

**Optimum Efficiency of All Trains;** By managing the best efficiency of each train as much as possible at fix total production rate at turn down target, the hypothesis best PTE picture can be developed and selected.

**Example of Turndown Operation at 87% Maximum Sustain Production;** There is opportunity to gain slightly efficiency improvement during turn down operation. Once example at 87% of the maximum production, the Train production adjustment at their individual best efficiency would produce 0.04% better overall PTE rather than running each train at equal flat production. This is equal to USD 100,000/month.

**Table-2**

| Scenario | % Load Production at 2800m3/hr total | | | | Total | Steam Use | LNG to | % PTE |
|---|---|---|---|---|---|---|---|---|
| | Train E | Train F | Train G | Train H | Prod | Turbine Comprs | Steam ratio | to max prod. |
| | | | | | m3/hr | ton/hr | | |
| Proportional | 87 | 88 | 88 | 88 | 2800 | 1822 | 1.54 | 0.07 |
| Optimal | 99 | 85 | 85 | 83 | 2800 | 1808 | 1.55 | 0.11 |
| | | | | | | | Gain = | 0.04 |

## 5. SUMMARY

- **Marine Flare;** The marine flare recovery improvement has a prospect to improve Plant Thermal Efficiency by 0.34% by installing booster compressors to fuel gas. The approximate energy saving is USD 9 million/ year and the estimate capital investment required is USD 3 million.
- **Maintains Boiler Excess Air Capacity at 17%;** Maintaining the boiler best efficiency all the time will increase the boiler average efficiency from 84.7 to 85.4%. The overall contribution of PTE improvement is 0.07%. or USD 1.7 million/year cost of energy. This require elaborate care of day-to-day operation.
- **Turndown Operation Management of LNG Train;** Manage and operate at the best individual train efficiency during turn down operation. This improves Plant Thermal Efficiency as highas 0.04% or equivalent with USD 80,000 of cost energy saving every month. This require elaborate care of day-to-day operation.

## 6. CONCLUSION

- There is significant room of improvement to save the energy and increase the efficiency from LNG ship boil off gas return. Those need a capital investment that the return is quite attractive.
- Some efficiency improvement needs elaborate care of day-to-day operation and it should be consistently follow to gain and maintain a piece by piece energy saving.

## 7. REFERENCES CITED


- Plant Thermal Efficiency Update, Archiaston Musamma, 2000, TRM 2000, Paris, France.
- PT. Badak Plant Thermal Efficiency, Archiaston M, Process Eng. memo to GM, 2001 no. 256/BM30/2001-344.
- Plant Thermal Efficiency Study, PTB Proc. Eng. & LAPI ITB, 2001
- Loss Terminology in Plant Thermal Efficiency, Archiaston M, Process Eng. Memo to MFG Mgr, no. 277/BM30/2001-344

## Lurgi's Gas To Chemicals (GTC®): Advanced technologies for natural gas monetisation

*Harald Koempel, Waldemar Liebner, Matthias Wagner*
*mg engineering - Lurgi AG*
*Lurgiallee 5, D-60295 Frankfurt am Main, Germany*

Prepared for Presentation at "AIChE Spring National Annual Meeting (April 25-29, 2004)
Session: Advanced Gas Conversion and Hydrogen Production"



### Introduction

Why is an engineering company investing a big portion of its considerable R&D budget in technology development for the conversion of natural gas to other fuels and to chemicals? Why are more and more members of the oil and gas industry promoting this development? Main answers are:

- plentiful gas supply sources, available for about 130 years – independent of petroleum reserves
- environmental aspects and regulations, and
- "monetizing" the abundant natural gas reserves in remote areas as well as associated gas which normally is flared without benefit

Both, economic and environmental benefits from the use of natural gas are the driving forces and will support the continuous innovation regarding gas-based technologies.

Figure 1. Flared Natural Gas (2002)

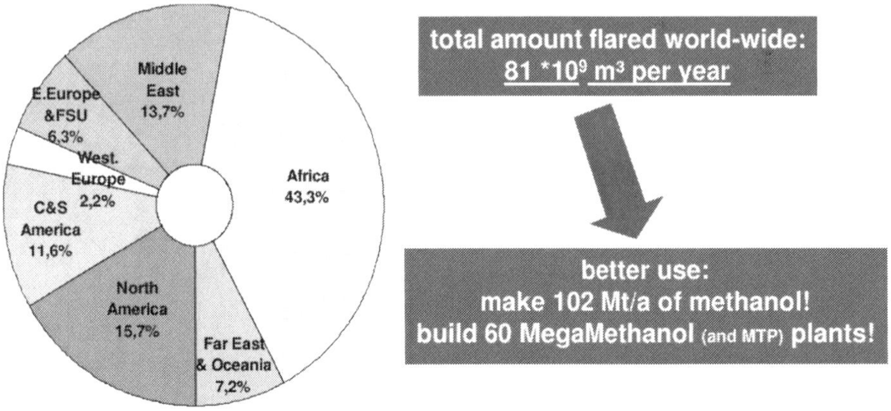


Source: Cedigaz, 2003

What first comes to mind with GTL is Fischer-Tropsch, the classic route from coal or natural gas to transportation fuels (synfuels). Lurgi on the other hand promotes methanol-based technologies for upgrading of natural gas to value-added products. These primarily would be DME (dimethyl ether), propylene, synfuels and "gas-based petrochemicals".

Since Lurgi introduced its new groundbreaking MegaMethanol® process for plants with a production of 5,000 tons of methanol per day and more, methanol will be available at a constant low price in the foreseeable future. This development has an enormous impact on downstream technologies for the conversion of methanol to more valuable products.

The first derivative of methanol in this context is DME which has a high potential as alternative to conventional diesel fuel as feedgas for gas turbines in power generation and as supplement to LPG. The next step is the use of methanol as feedstock for the production of olefins which is one of the most promising new applications. Lurgi's new Methanol-to-Propylene (MTP®) process presents a simple, cost-effective and highly selective technology. Both routes allow for the production of petrochemicals which then would be gas-based. Lurgi also proposes a methanol-based technology for production of synfuels which compares well with the FT-processes.

## Natural Gas – A Key Feedstock for the 21st Century

The total proven gas reserves amount to approx. 180 trillion cubic meters world-wide which translates into a gas reserve-to-production ratio, i.e. a gas reserve lifetime of 70 years. Furthermore, estimated additional gas reserves will cover a lifetime of 65 years more. [Cedigaz 2003] Compared with the reserve lifetime of 41 years for petroleum and 230 years for coal, there is no doubt that natural gas will be a key fuel component in the 21st century.

However, currently a considerable portion of this reserve is wasted yearly: A brief look at Figure 1 "Flared Natural Gas" explains the main incentive for engineers and environmentalists as well to come up with novel ideas for the utilisation of this gas.

Existing technologies for natural gas conversion are shown in Figure 2: via the conversion to syngas, hydrogen and ammonia, Fischer-Tropsch products, methanol and DME are produced. Currently, the production of chemicals requires only around 5% of world gas consumption [Quigley and Fleisch 2000].

Figure 2. Uses of Natural Gas

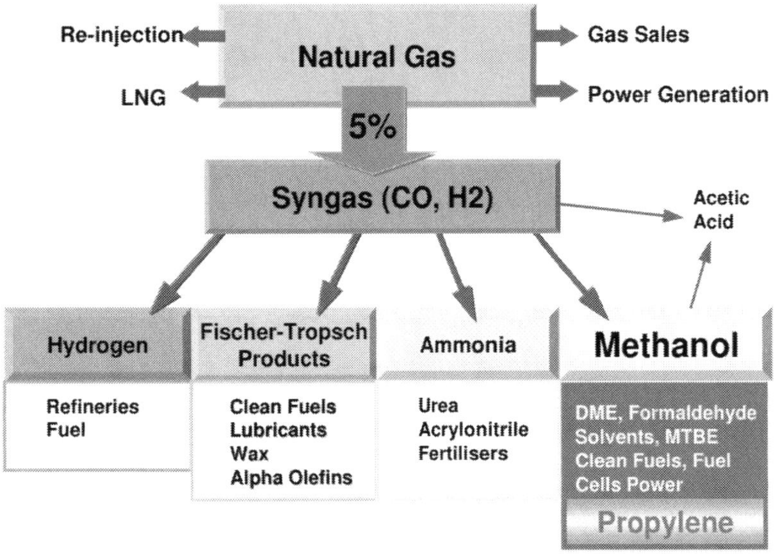


Lurgi AG focuses on new routes from $C_1$ to valuable products by combining a chain of proprietary Lurgi technologies that are based on low cost natural gas supply and large scale methanol plants. These are new DME and synfuels technologies and an exciting new process for the highly selective conversion of methanol to propylene.

## Lurgi MegaMethanol®: Basis for More Valuable Products

The term MegaMethanol® refers to plants with a capacity of more than one million metric tons per year, the actual "standard" size being $1.7*10^6$ t/a (equivalent to 5,000 t/d). To achieve such a large capacity in a single-train plant a special process design is required. For this reason Lurgi focused on the most efficient integration of syngas generation and methanol synthesis into the most economical and reliable technology for the new generation of future methanol plants [Streb and Göhna 2000].

The unique advantages of the Lurgi MegaMethanol® technology result in "ex-gate" methanol prices of about 65 $/t or less and make this process ideally suited as part of Lurgi's route from $C_1$ to propylene and others. This year two such plants of 5,000 t/d capacity will be started up: Atlas/Trinidad in summer and Zagros/Iran by year's end. Conceptual studies and engineering activities for MegaMethanol® plants with single-train capacities of up to 7,500 t/d and more have been successfully finalised making these plant sizes ready for commercialisation.

An environmental sidenote: 80 billion cubic meters of natural gas are flared or vented annually, see Fig.1 [Cedigaz 2003]. That amount would be sufficient to feed about 60 MegaMethanol® plants with a capacity of 102 million tons per year in total.

## DME - A Valuable Product From Methanol

Dimethyl Ether, DME, is industrially important as the starting material in the production of the methylating agent dimethyl sulfate and is used increasingly as an aerosol propellant. In the future DME can be an alternative to conventional diesel fuel or a feedgas for power generation in gas turbines. Both applications are based on large-scale production facilities in order to achieve an economic fuel price.

Traditionally, DME was obtained as byproduct of the high-pressure methanol synthesis. Since the low-pressure methanol synthesis was established, DME has been prepared from methanol by dehydration in the presence of suitable catalysts. The dehydration is carried out in a fixed-bed reactor. The product is cooled and distilled to yield pure DME.

A modification of the methanol synthesis would allow for co-generation of DME within the methanol synthesis loop. This technical path comprises two disadvantages. While dehydrating methanol, the water vapor content increases, thus enhancing the water gas shift reaction. By converting CO into $CO_2$, the quality of the synthesis gas deteriorates. The kinetics of the reaction of $CO_2$ and $H_2$ is slower than the one of CO and $H_2$. As a result, the synthesis catalyst volume and the recycle loop capacity have to be increased. In addition, due to its low boiling point a cryogenic separation is required in order to separate DME from the synthesis recycle loop.

As a result of these disadvantages of the co-generation of methanol and DME Lurgi favours the concept of generating DME from methanol by dehydration.

**Figure 3. DME Production by Methanol Dehydration**

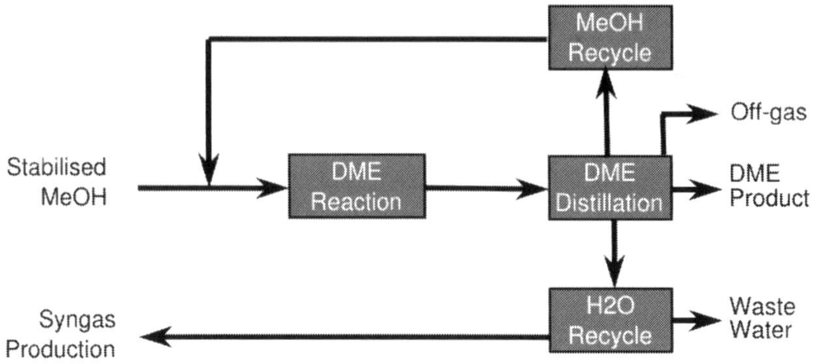


If a DME Unit is added to the MegaMethanol® plant, the distillation of methanol is reduced from a three-tower system to one tower at considerable savings. Figure 3 shows the simple and inexpensive flowsheet for the dehydration of methanol. In this process all types and qualities of DME can be produced. The different specifications for fuelgas, power generation or pure DME can be achieved just by varying size and design of the DME distillation towers.

The economics of the Lurgi DME Process are summarised in Table 1 assuming the following general set-up: Natural gas consumption and Product Value (EPC) are standardised on methanol equivalent (7050 t/d methanol capacity); DME product quality is at least 99.2 wt% DME; Natural gas consumption figures include energy demand for air separation and power generation; Total Fixed Cost include air separation, power generation and off-sites; Natural gas price assumed as US$ 0.5 / MMBtu; Depreciation is 10 % of Total Fixed Cost; Return on Investment (ROI) is set to 20 % of Total Fixed Cost; Operating cost for operator staff, plant overhead, maintenance labour and material are included.

All investment cost figures are budgetary estimates of +/- 20 % accuracy. Specific site conditions are not reflected with these numbers. The figures show the superb economics of MegaMethanol® in combination with a separate dehydration step.

Table 1: Economics of the Lurgi DME Process

| *Plant Type* | *Mega Methanol & Dehydration* |
|---|---|
| DME capacity | 5,000 t/d |
| Natural Gas Demand | 28.5 MMBtu / t MeOH<br>40.2 MMBtu / t DME |
| Total Fixed Cost (EPC) | 415 MM US$ |
| Cost of production | 93 US$ / t DME |

From all this it follows that DME, a traditional derivative of methanol, can be a promising alternative fuel for power generation, diesel, LPG or the manufacture of olefins when produced in large capacities. The production of DME by dehydration of methanol, i.e. in two steps, is more economic than a single-step synthesis as proposed in Japan and Denmark. This is discussed in detail in another paper in the olefin session of this conference [Liebner 2004].

## Propylene - An attractive product with high value

Demand growth of propylene till 2002 was above 8% and after that is projected at higher than 5%. Polypropylene is by far the largest and fastest growing of the propylene derivatives, and requires the major fraction of about 60 % of the total propylene. The increasing substitution of other basic materials such as paper, steel and wood by PP will induce a further growth in the demand for PP and hence propylene. Other important propylene derivatives are acrylonitrile, oxo-alcohols, propylene oxide and cumene. The average growth rate for propylene itself is estimated very conservatively to be 4.5 % per year for the next two decades.
How to satisfy this demand for propylene?

Currently, steam crackers and FCC units supply 66 % and 32 %, respectively of propylene fed to petrochemical processes. However, as FCC units primarily produce motor gasoline, and steam crackers mainly ethylene, propylene will always remain a by-product (e.g. 0.04-0.06 t/t of ethylene for steam crackers with ethane feedstock and 0.03-0.06 t/t, respectively of motor gasoline and distillates production for FCC units). Current forecasts indicate an increasing gap of propylene production that has to be filled by other sources. Lurgi's new MTP process directly aims to fill that gap.

## Lurgi's Methanol to Propylene (MTP®) Technology

Lurgi's new MTP® process is based on an efficient combination of the most suitable reactor system and a very selective and stable zeolite-based catalyst. Since the process has been described in detail elsewhere [Rothaemel and Holtmann, 2001], suffice it to say here that Lurgi has selected a fixed-bed reactor system because of its many advantages over a fluidised-bed. The main points are the ease of scale-up of the fixed-bed reactor and the significantly lower investment cost.

Furthermore, Süd-Chemie AG manufactures a very selective fixed-bed catalyst commercially which provides maximum propylene selectivity, has a low coking tendency, a very low propane yield and also limited by-product formation. This in turn leads to a simplified purification scheme that requires only a reduced cold box system as compared to on-spec ethylene/propylene separation.

With Figure 4 a brief process description reads: Methanol feed from the MegaMethanol® plant is sent to an adiabatic DME pre-reactor where methanol is converted to DME and water. The high-activity, high-selectivity catalyst used nearly achieves thermodynamic equilibrium. The methanol/water/DME stream is routed to the first MTP® reactor where also the steam is added. Methanol/DME are converted by more than 99%, with propylene as the predominant hydrocarbon product. Additional reaction proceeds in the second and third MTP reactors. Process conditions in the three MTP® reactors are chosen to guarantee similar reaction conditions and maximum overall propylene yield. The product mixture is then cooled and the product gas, organic liquid and water are separated.

**Figure 4. MTP®: Simplified Process Flow Diagram with production figures**

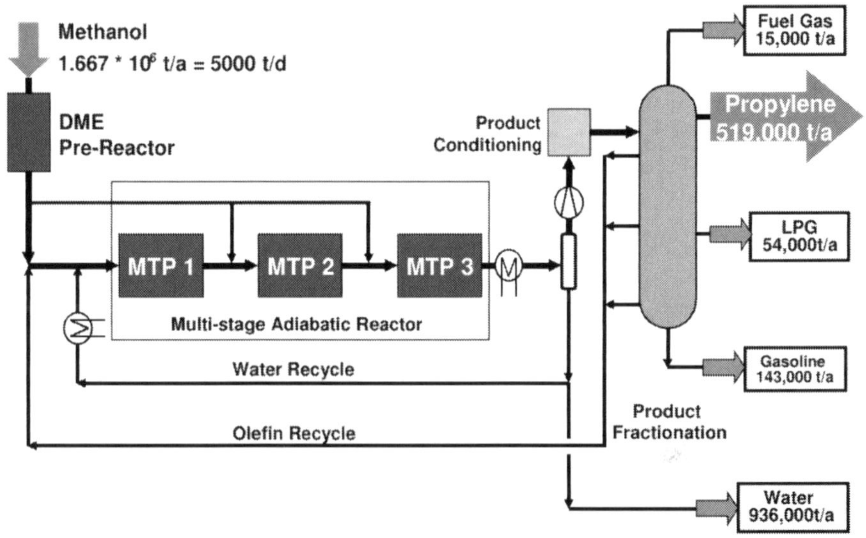


The product gas is compressed and traces of water, $CO_2$ and DME are removed by standard techniques. The cleaned gas is then further processed yielding chemical-grade propylene with a typical purity of more then 97%. Several olefin-containing streams are sent back to the main synthesis loop as an additional propylene source. To avoid accumulation of inert materials in the loop, a small purge is required for light-ends and the $C_4/C_5$ cut. Gasoline is produced as a by-product.

Water is recycled to steam generation for the process; the excess water resulting from the methanol conversion is purged. This process water can be used for irrigation after appropriate and inexpensive treatment. It even can be processed to potable water where needed.

An overall mass balance is included in Figure 4 based on a combined MegaMethanol® / MTP® plant. For a feed rate of 5,000 tons of methanol per day (1.667 million tons annually), approx. 519,000 tons of propylene are produced per year. By-products include fuel gas (used internally) and LPG as well as liquid gasoline and process water.

Further integration and optimisation of the total plant complex including syngas, methanol, propylene production and offsite facilities will again decrease the capital investment and production costs.

The technological status of MTP® in the areas of process and catalyst can be summarised as follows: The basic process design data were derived from more than 9,000 operating hours of a pilot plant at Lurgi's Research and Development Centre. Besides the optimisation of reaction conditions also several simulated recycles have been analysed. Parallel to that Lurgi decided to build a larger-scale demonstration unit to test the new process in the framework of a world-scale methanol plant with continuous 24/7 operation using real methanol feedstock. After a cooperation agreement with Statoil ASA was signed in January 2001 the Demo Unit was assembled in Germany and then transported to the Statoil methanol plant at Tjeldbergodden (Norway) in November 2001. Later in 2002 Borealis joined the cooperation.

The Demo Unit was started up in January 2002, and the plant has been operated almost continuously since then. As of September 8$^{th}$ 2003, the Demo Unit completed the scheduled 8000 hours life-cycle test. With that the main purpose of the test was achieved: to demonstrate that the catalyst lifetime meets the commercial target of 8000 hours on stream. Cycle lengths between regenerations have been longer than expected. Deactivation rates of the methanol conversion reaction decreased with operation time. Propylene selectivity and yields were in the expected range for this unit with only a partial recycle. Also, the high quality of the by-product gasoline and the polymerisation grade quality of the propylene were proven. For details see the parallel paper on the Demo Unit results [Liebner 2004].

The catalyst development is completed and the supplier commercially manufactures the catalyst.

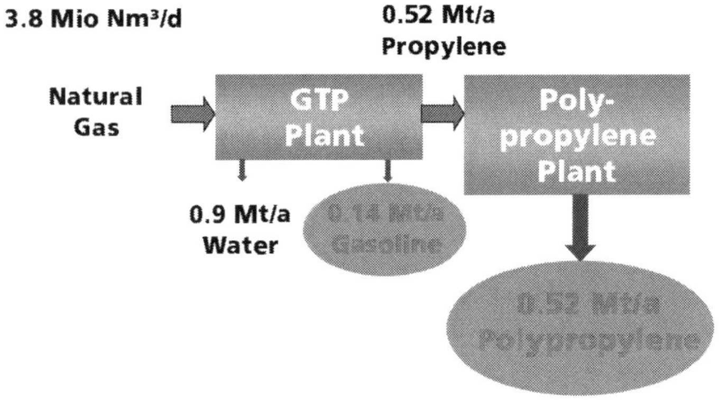

**Figure 5 Block Flow Diagram - PP Complex**

## GTP Economics

Since propylene by itself is more an intermediate than an end product, an economics estimate was performed for a complete natural gas to polypropylene complex. In this case of integrating a MegaMethanol® and a MTP® plant we designate the resulting unit as "Gas to Propylene", GTP®, as shown in Figure 5.

Thus, the economic assessment included the GTP route with a polypropylene unit for the production of a more saleable, higher-value end product. The cases presented here take into account a "difficult" region of remote/stranded gas as well as contingencies for the newly developed route. With that, the investment cost estimate is fairly high and still an attractive return can be expected as seen in Table 4.

Table 2 shows as Case A the production costs of intermediates and end product derived for the flow sheet of Figure 5. The corresponding rates of return are given in Table 4.
The remarkable facts here are the low production costs for propylene and for the end product polypropylene. These leave room for healthy profit margins. Given the facts that contingencies may be outweighed by additional integration savings or may not be needed in full, this route is seen as the most promising and most economic natural gas utilisation.

Case B as given in Table 3 shows the potential of the same complex to produce a significantly higher amount of gasoline, albeit at the corresponding lower propylene production rate. This was studied as part of a sensitivity and risk analysis. Even in this case far from the original design which is optimised for propylene yield, the rates of return as shown in Table 4 remain impressive.

Table 2 Production Cost, GTP/ PP Complex
CASE A: high propylene / low gasoline production

| **Table 2** | | GTP | PP |
|---|---|---|---|
| Capacity | t/a | 520,000 | 520,000 |
| Investment Cost EPC | Mio US$ | 565 | 165 |
| Owner's Cost incl. Capitalised Interest | Mio US$ | 113 | 33 |
| Feed Cost | US$ | Natural Gas<br>0.5 $/MMBtu | Propylene<br>$/t |
| Production Cost | US$/t | 210.1 | 261 |
| - Raw Materials | US$/t | 57.1 | 212.8 |
| - Utilities | US$/t | 10.8 | 6.8 |
| - Operation & Maintenance | US$/t | 29.6 | 8.6 |
| - Plant OVHD & Insurance | US$/t | 31.6 | 9.2 |
| - Depreciation | US$/t | 81.0 | 23.6 |
| Credit for by-product Gasoline | US$/t | -35.7 | - |
| Cost of Product at ROI = 0 | US$/t | 174 | 261 |

Table 3 Production Cost, GTP/ PP Complex
CASE B: low propylene / high gasoline production

| **Table 3** | | GTP | PP |
|---|---|---|---|
| Capacity | t/a | 440,000 | 440,000 |
| Investment Cost EPC | Mio US$ | 565 | 165 |
| Owner's Cost incl. Capitalised Interest | Mio US$ | 113 | 33 |
| Feed Cost | US$ | Natural Gas<br>0.5 $/MMBtu | Propylene<br>$/t |
| Production Cost | US$/t | 248.6 | 271 |
| - Raw Materials | US$/t | 67.6 | 215.2 |
| - Utilities | US$/t | 12.7 | 6.7 |
| - Operation & Maintenance | US$/t | 35.1 | 10.2 |
| - Plant OVHD & Insurance | US$/t | 37.4 | 10.9 |
| - Depreciation | US$/t | 95.8 | 28 |
| Credit for by-product Gasoline | US$/t | -71.9 | - |
| Cost of Product at ROI = 0 | US$/t | 177 | 271 |

It should be noted that a very low gasoline price of 130 US$/t has been considered in both cases. According to the high quality found by the Statoil refinery lab, higher prices would be justified. With these, the profitability of Case B would increase to nearly this of Case A. This robustness is based on the fact that with diminishing selectivity towards propylene automatically the gasoline yield rises and that never any detrimental by-products are formed. Additionally, all "unconverted" compounds are recycled as designed for anyway. In other words, the optimum propylene to gasoline ratio will depend on the relative value of the two products. The main product price -polypropylene's- was taken as 650 US$/t, a value slightly below the average since 1990 as quoted by CMAI; recent prices as quoted by ICIS-LOR being well in the 700dreds. With these moderate to low product prices and a reasonable investment cost contingency the economic evaluation presented here is purposefully conservative.

Table 4: ROI and IRR, GTP / PP Complex, CASE A, B

| **Table 4** | CASE | A: high propylene | B: high gasoline |
|---|---|---|---|
| Investment Cost EPC | Mio US$ | 730 | 730 |
| Owner's Cost incl. Capitalised Interest | Mio US$ | 146 | 146 |
| Feedstock Cost | US$ | Natural Gas 0.5 $/MMBtu | |
| Production Cost | Mio US$ | 154.3 | 150.7 |
| - Raw Materials | Mio US$ | 49.7 | 46.6 |
| - Utilities | Mio US$ | 9.1 | 8.6 |
| - Operation & Maintenance | Mio US$ | 19.9 | 19.9 |
| - Plant OVHD & Insurance | Mio US$ | 21.2 | 21.2 |
| - Depreciation | Mio US$ | 54.4 | 54.4 |
| Revenues | Mio US$ | 356.6 | 317.2 |
| - Gasoline (130 US$/t) | Mio US$ | 18.6 | 31.6 |
| - Polypropylene (650 US$/t) | Mio US$ | 338 | 285.6 |
| Return On Investment [1)] | ROI % | 23.1 | 19 |
| Internal Rate of Return [2)] before tax | IRR % | 25.1 | 20.6 |
| IRR on equity (30% of EPC + owner's cost) | IRRE% | 36.8 | 30.3 |

1) ROI estimate based on Chem Systems methodology, 2) IRR estimate based on COMFAR

## Lurgi's Fischer-Tropsch Experience

Historically, Lurgi was one of the developers of FT in the 1920-30ties. FT in the form of (fixed bed) ARGE-synthesis was commercialised in 1952 in Sasolburg, RSA. All five original reactors are still in operation. A sixth one was started in 1987 as capacity extension.

Modern FT reactor technology prefers slurry phase reactors, either tubular or fluidised bed. Lurgi has commercial experience in all these reactor designs. Also, Lurgi has designed all syngas production units of all currently operating industrial FT-plants: Sasol/Secunda, RSA, utilising coal gasification; Mossgas, RSA, - combined reforming of NG and SMDS Bintulu, Malaysia - partial oxidation of NG.

The syngas production route which among others is used for MegaMethanol® is offered by Lurgi as MegaSyn® and is available for FT syntheses also.

## Lurgi's Route to Transportation Fuels: MtSynfuels®

Given the economically highly attractive technologies of MegaMethanol® and MTP® as described above it nearly follows by itself to combine them with an industrially proven process for the conversion of olefins to diesel. A gas-based synfuels plant using this process, then named COD (derived from Conversion of low molecular weight Olefins to Diesel), was developed and built by Lurgi for Mossgas (today: PetroSA), RSA, in 1992 and is performing well since its start-up in 1993.

Remarkably, the industrial design was based on a scale-up factor of 3600 over the preceding demonstration plant. This basically was possible through the use of fixed-bed catalysis (on zeolite basis) which lends itself to easy scale-up. Other important process features are semi-continuous operation and a 98% conversion of $C_3$- and $C_4$- olefins.

The Lurgi route to synfuels, MtSynfuels® shown in Figure 6 is a combination of this type of process with MegaMethanol® and a simplified MTP®. Extensive engineering and estimating studies have been performed to prove the feasibility and economic viability of this new route. Table 5 shows the technical results, the product slate and Table 6 gives the economics as a comparison with an existing FT plant.

**Figure 6. Gas Refinery via Methanol - Lurgi's MtSynfuels®**

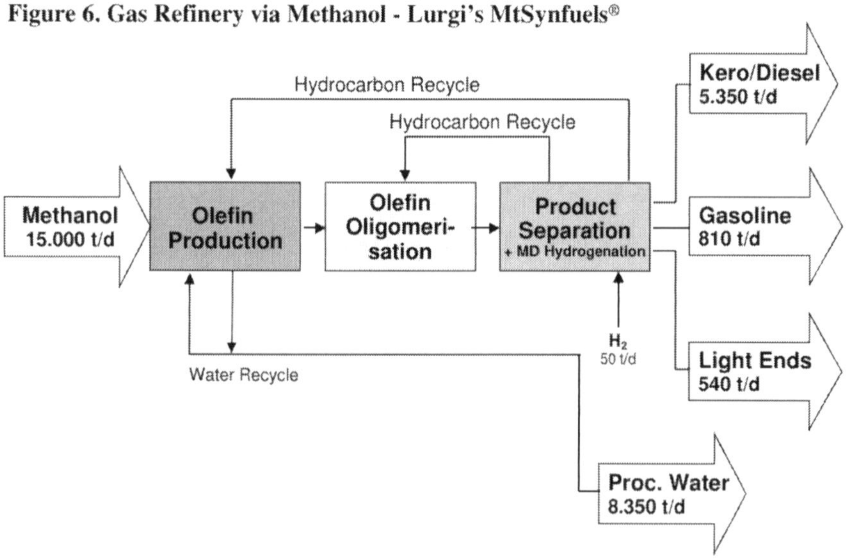


Table 5. Comparison MtSynfuels® vs. FT-Synthesis: Product Slate and Properties

| **Table 5** Product Slate | | | | MtSynfuels® | FT Synthesis |
|---|---|---|---|---|---|
| Naphtha : Kero+Diesel (max.) | | | | | 1 : 2.3 – 1 : 6 |
| Gasoline : Kero+Diesel | | | | 1 : 8 | |
| Product Properties | | | | | 3) |
| | Specification (Europe from 2005) | | | | |
| Gasoline | | | | | |
| -Aromatics | vol.% | max. | 35 | 11 | < 1 |
| -Benzene | vol.% | max. | 1 | << 1 | << 1 |
| -Sulphur | ppmw | max. | 50/10 1) | <<1 | < 1 |
| -Olefins | vol.% | max. | 18 | 6 | > 30 |
| -RON 2) | | | 91/95/98 | 92 | < 40 |
| -MON 2) | | | 82,5/85/88 | 80 | < 40 |
| Diesel | | | | | |
| -Polyaromatics | vol.% | max. | 11 | << 1 | < 1 |
| -Sulphur | ppmw | max. | 50/10 1) | << 1 | < 5 |
| -Cetane No. | | min. | 51 | >52 | > 70 |

1) Diesel with 10 ppmw sulphur has to be available on the market
2) RON / MON for Regular Gasoline / Euro-Super / Super – Plus
3) Properties of FT-naphtha

In an earlier study the authors [Koempel, Liebner, Feb. 2002] have summarised and compared proven and new FT-processes with Lurgi's alternative route MtSynfuels®. Tables 5 and 6 are taken from this paper. Their comparative economic evaluation is based on the following assumptions:

Plant location: Middle East; plant capacity: 50000 bpd products; NG Price: 0.50 US$/MMBtu
Depreciation: 10 % for ISBL, 5 % for OSBL; Return on Investment (ROI): 10 %;

Total Capital Investment includes total plant capital (ISBL+OSBL) plus 20 % for other project cost, year 2000; Cost of Production includes depreciation and 10 % ROI; Product Market Prices: Year 2000, adjusted to shipping costs from Middle East and applicable tariffs for the considered regions.

These tables show that MtSynfuels® compares well with existing FT plants. Admittedly it lacks full commercialisation, but so do most of the other FT processes discussed currently. In contrast to these, MtSynfuels® is proven in three of four steps with the demo unit for the third step (MTP®) having confirmed the lab results by a 8,000 hours test run.

Table 6. Comparative Economics - Cost of Production Estimate

| **Table 6** | | MtSynfuels® | existing FT[1] | |
|---|---|---|---|---|
| Total Capital Investment | [MM $] | 1,181 | 1,671 | |
| Total Plant Capital | [$/bpd] | 19,680 | 27,856 | |
| NG to Process (LHV) | [$/bbl] | 3.82 | 4.22 | |
| | [MMBtu/bbl] | 7.64 | 8.44 | |
| Cat. & Chemicals | [$/bbl] | 2.19 | 1.53 | |
| Utilities | [$/bbl] | 0.28 | 0.8 | |
| Cost of Production + ROI | [$/bbl] | 22.47 | 28.68 | |
| Market Prices [2] | | Western Europe | US Gulf Coast | Japan |
| - Gasoline | [$/bbl] | 34.75 | 31.42 | 29.70 |
| - Diesel | [$/bbl] | 28.88 | 26.51 | 33.21 |

[1] Chem Systems 2001
[2] Corresponding Crude Oil Price: abt. 21 $/bbl

**Figure 7. Gas-based Petrochemistry**

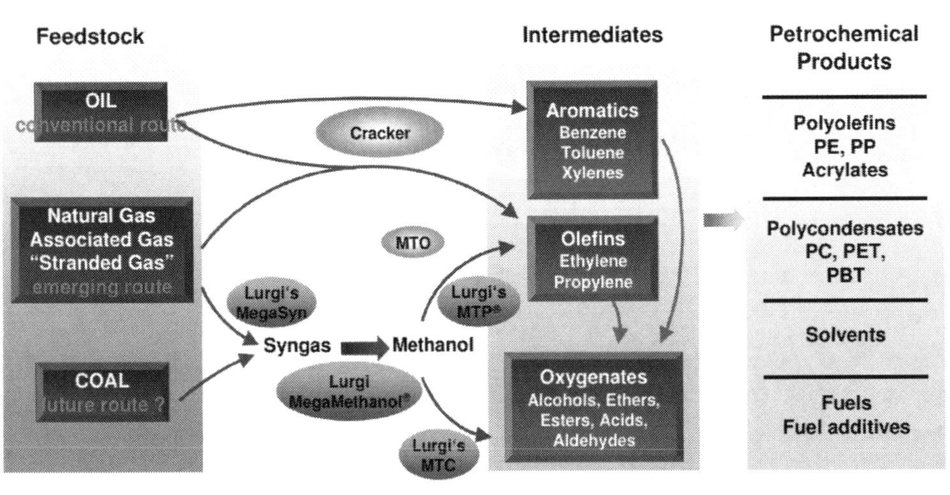


## From Gas to Petrochemicals

It has been shown above that propylene produced via MTP® competes well with cracker-derived product. In more general terms it develops that the chain of Lurgi's technologies described here provides an alternative route to petrochemicals. Almost all steps are technically proven and the economic competitiveness mainly depends on the natural gas price. This again follows from market pressures and the need or willingness to monetise gas reserves.

Figure 7 shows how the conventional cracker route from crude oil through olefinic and aromatic intermediates to highly valued petrochemical products is complemented -and replaced possibly- by "gas-to-methanol-and-others" processes. There is even the possibility to use coal as the primary feedstock for this methanol-to-petrochemicals route, an alternative seriously considered in China which lacks large oil or gas reserves but has an abundance of coal.

## Conclusions

There are abundant natural gas reserves providing low cost feedstock for methanol production and aiming at better use of natural resources especially in the case of associated gases being flared. DME and Propylene produced from methanol will increase the value of natural gas considerably and offer an exciting potential of growth and a high earnings level. Lurgi's MegaMethanol® technology brings down the net methanol production cost below US$ 50 per ton, wherever low cost natural gas is available. This opens up a completely new field for downstream products like DME, propylene and synfuels.

Based on simple fixed-bed reactor systems, conventional processing elements and operating conditions including commercially manufactured catalysts, Lurgi's MegaDME, MTP® and MtSynfuels® technologies provide attractive ways to "monetise" natural gas.

Driven by the excellent market prospects and additional environmental aspects, Lurgi has developed its own technology chains starting from natural gas via methanol to DME or propylene and polypropylene, based on the combination of highly efficient concepts at low investment costs. In the next step these concepts lead to gas-based refineries and gas-based petrochemicals.

Figure 8. Gas to Chemicals Processing Routes

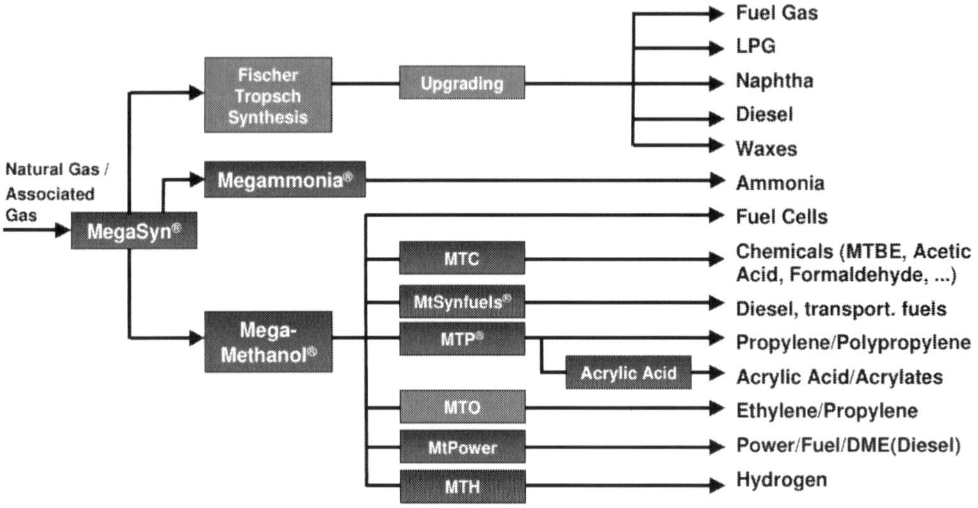


Figure 8 summarises the gas to chemicals routes. With the exception of FT and MTO which are offered as licensed technologies, all others depicted here are Lurgi proprietary technologies – a direct result of the high importance Lurgi always attached to gas and syngas conversion. MtPower depicts the utilisation of methanol and DME as energy carriers, made possible by the low production costs associated with the "Mega-plants".

Eventually, financial, strategic and political interests will determine the ultimate selection of any "gas-to-value" technology. The task of the engineering company is to provide as many attractive alternatives as possible to accommodate for all sorts of local conditions. With the technology portfolio described above Lurgi is up to this challenge.

## REFERENCES

**Cedigaz: "The 2002 Natural Gas Year in Review"**, April 2003, www.cedigaz.com

**Th. M. Quigley and Th. H. Fleisch:** "Technologies for the Gas Economy", EFI – Gas to Market Conference, San Francisco, October 11 – 13, 2000.

**S. Streb and H. Göhna: „MegaMethanol® - paving the way for new down-stream industries",** World Methanol Conference, Copenhagen (Denmark), November 8 – 10, 2000

**M. Rothaemel and H-D. Holtmann**: "**MTP, Methanol To Propylene - Lurgi's Way",** DGMK-Conference "Creating Value from Light Olefins – Production and Conversion", Hamburg, October 10 – 12, 2001

**H. Koempel, W. Liebner: "Gas to Liquids? Gas To Chemicals? Gas to Value!"**, ERTC Petrochemical Conference, Amsterdam, February 20-22, 2002

**M. Rothaemel, H. Koempel, W. Liebner: "Progress Report on MTP with focus on DME"**, AIChE Spring National Annual Meeting, New Orleans, April 25-29, 2004, Session: Olefins Production

# Syngas Production for GTL with Gas Heated Reforming - Efficient and Environmentally Friendly

**Jim Abbott**
**Bernard Crewdson**

Johnson Matthey Catalysts,
PO Box 88,
Billingham,
TS23 1XN,
United Kingdom



Prepared for presentation at the Spring AIChE Meeting, New Orleans, April 25-29, 2004.
Session 9007, Gas Conversion to Liquid Fuels and Chemicals. Unpublished.



**ABSTRACT**

Development in Gas to Liquids production technology continues to gather pace. The effective production of synthesis gas is central to achieving a process with a high thermal efficiency.

A disadvantage with many available syngas production technologies is that they consume excessive oxygen – feedstock is used up in producing heat and carbon dioxide rather than giving a high conversion to a balanced hydrogen / carbon monoxide mixture. The adoption of Gas Heated Reforming (GHR) changes this balance – the internal heat recycle dramatically cuts oxygen consumption and leads to significantly more efficient syngas production. In turn, this reduces the size of the plant for specific FT liquids' production. GHR syngas production can also be integrated favourably with the use of Gas Turbines to further enhance efficiency.

The best advantages can be achieved if the process is operated at a low steam to carbon ratio. The paper explores the beneficial features of using GHR in syngas manufacture for FT, and reviews the development progress towards operating the process at optimal conditions.

## Introduction

The main requirement of a process to make syngas feed for FT liquids production is to synthesise a poison-free gas with a composition where $H_2$:CO molar ratio is close to 2:1, with minimal un-reactive components, as efficiently and inexpensively as possible.

A number of different process routes are available today, using either Steam reforming, Catalytic or non-catalytic partial oxidation (CPO and POX), Autothermal reforming (ATR), Gas Heated Reforming (GHR) or a combination of these – each has its pros and cons.

Steam reforming alone gives many compositional disadvantages – including the excessive production of hydrogen and the levels of un-reacted methane and $CO_2$ in the syngas. ATR and GHR use an oxidant and hence produce more balanced gas compositions with the carbon content being mostly present as reactive CO. The use of high purity oxygen as oxidant is obviously preferable in terms of syngas quality and is assumed in this paper– air produces a syngas, which has 35-50% N2 content. The wider arguments around advantages and disadvantages of using air are outside the scope of this paper, which examines the relative merits of ATR and GHR using oxygen.

POX, by itself, gives a composition, which is too rich in CO, and therefore has to be balanced by producing a second, parallel stream. This, second stream uses steam reforming to obtain hydrogen or a hydrogen-rich syngas, which adds to the overall expense. CPO, without the use of steam, is perhaps the most elegant and theoretically ideal process, when practised using oxygen – however there is still a major technical challenge to overcome in achieving economic conversion and selectivity figures.

Capital cost is the single biggest input into the economic feasibility of a GTL project. Operating efficiency is an important factor in this when it impacts on the size of the process plant required to achieve a given plant capacity.

## Autothermal Reforming (ATR)

Currently, the most widely proposed syngas technology for GTL production is autothermal reforming (ATR). There are a number of variants of this basic process, some incorporating proprietary technology, but all comprise the partial oxidation of natural gas with steam and oxygen, or air, followed by equilibration over a catalyst bed, as shown below

In autothermal reforming, after the feed gas and steam mixture is preheated, it is combusted with an oxidant and passed over a catalyst to bring the steam-methane reforming reaction to equilibrium at a temperature of 975-1050°C (1790-1920°F). An adiabatic pre-reforming step may additionally be used to remove higher hydrocarbons from the feedstock – this has no net effect on the heat and mass balance. The ATR product gas is very corrosive in the temperature range of 400-750°C (750-1400°F) and must be cooled by raising steam in a waste heat boiler.

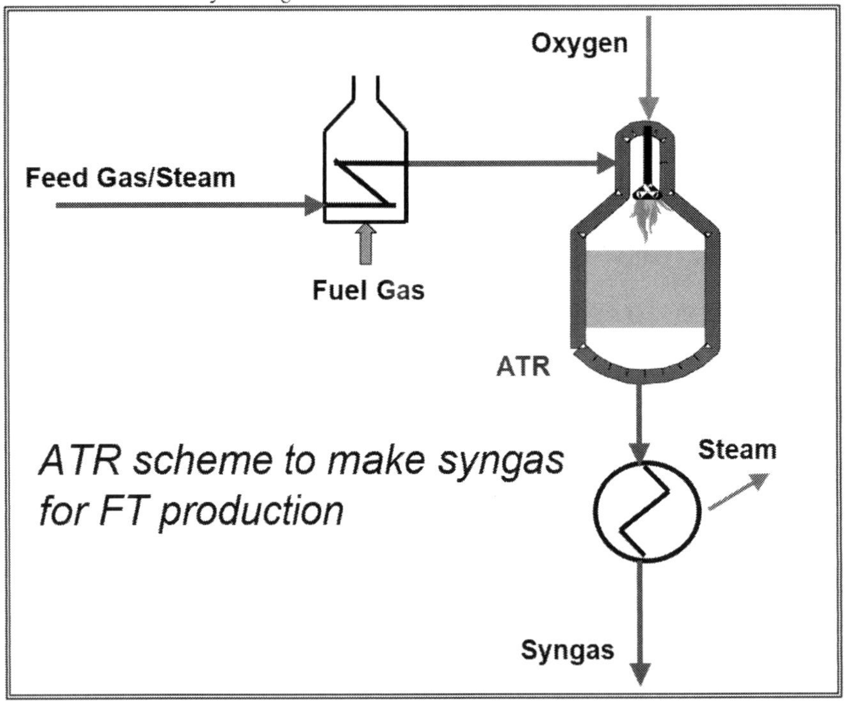

*ATR scheme to make syngas for FT production*

## Gas Heated Reforming (GHR)

The performance of a GTL plant using oxygen-based reforming can be substantially improved by coupling the oxygen-fired reformer with a Gas Heated Reformer (GHR). The GHR allows the recycle of high grade heat from the reformed gas directly back into the reforming process, reducing the oxygen requirement and eliminating the need to generate large quantities of high pressure steam from the process train. The configuration is shown below. A more detailed description of GHR technology can be found in reference 1. Gas Heated Reformers are now in service on three ammonia units, a methanol unit and two syngas research units.

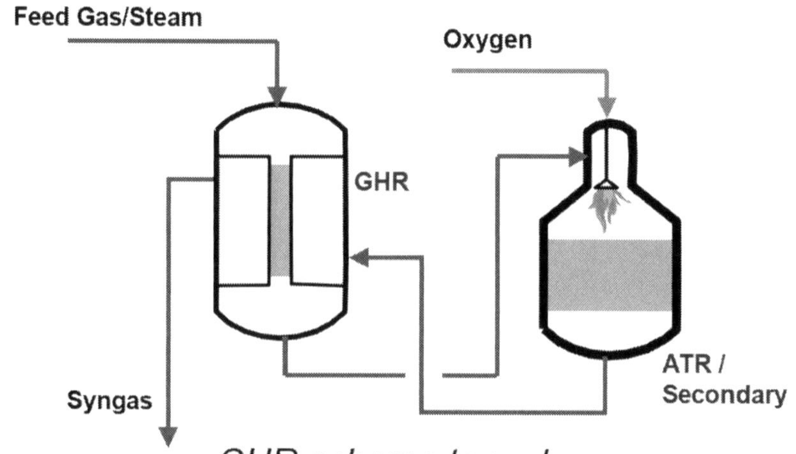


*GHR scheme to make syngas for FT production*

It is possible to configure a GHR/secondary combination in series or parallel with respect to the natural gas feed. In both cases, the full reformed gas flow is used to provide heat to the GHR tubes. JM Synetix have found the series configuration to be superior, both in terms of efficiency of conversion to CO and in terms of GHR size – this is covered in more detail in Reference 2.

The beneficial effect of heat recycle using GHR, in comparison to ATR is shown below, with molar gas quantities for GHR on top (in red) and those for ATR below (in black):

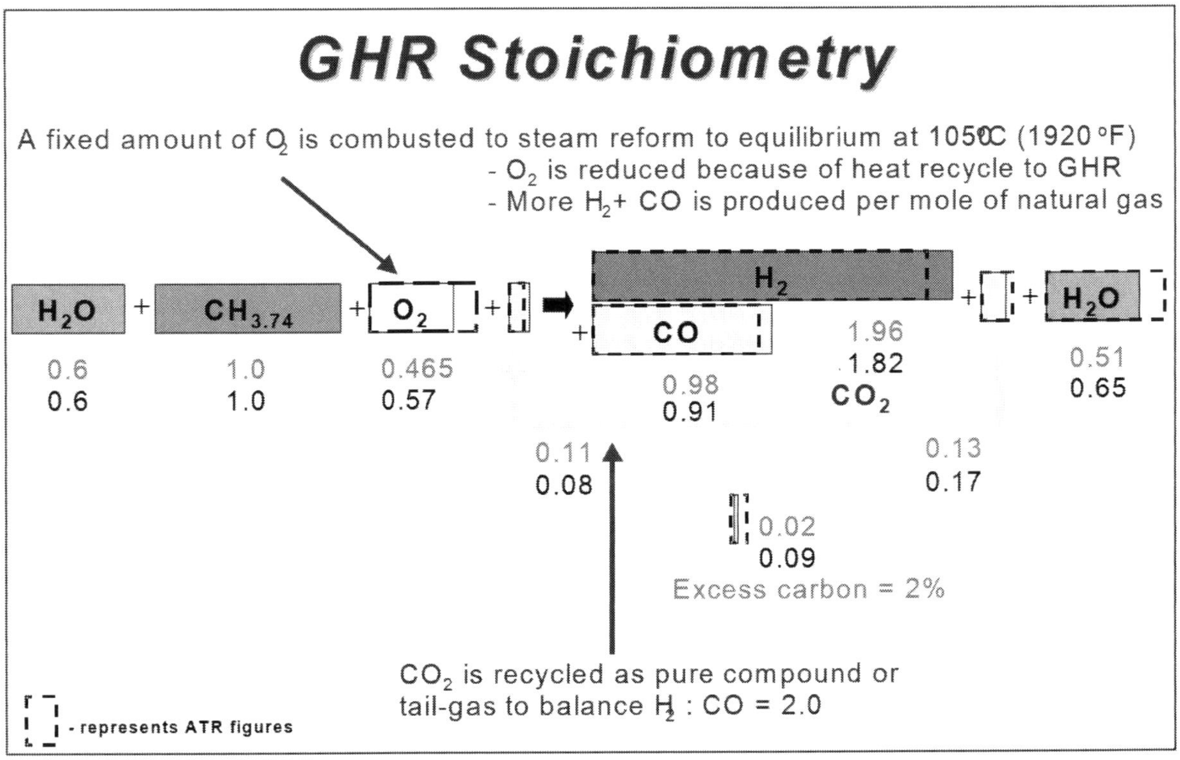


For a constant ATR inlet temperature and steam ratio (here 0.6) a fixed amount of oxygen is combusted to satisfy the requirement of heating the gases and steam reforming them to equilibrium at 1050°C, (1920°F). Of course, this gives a syngas, which is too lean in CO, ($H_2$: CO = 2.19) because $CO_2$ is not reactive for FT. It is possible to recycle $CO_2$ to the ATR and thereby achieve substantially the correct syngas $H_2$: CO ratio of 2.0.The balance shows that there is fundamental, in-built, process inefficiency. For every 100 units of carbon in the feed, 9 units of carbon must be discarded to achieve a satisfactory FT feed composition.

Looking at GHR on the same basis as the preceding ATR example (natural gas composition, steam ratio, ATR/secondary exit temperature) 18% less oxygen combustion is required to satisfy the heat balance and less valuable hydrogen is consumed, leaving more available for FT liquid synthesis. The scheme is now well-balanced - for every 100 units of carbon in the feed, only 2 units of carbon need to be purged, with almost 8% improvement in reaction efficiency to CO – this boosts the overall carbon efficiency in the GTL plant.

**GHR vs. ATR comparative performance**

The benefits of heat recycle ripple throughout the entire flowsheet, including both the process and facilities. This can be seen in the following table, which compares ATR and GHR flowsheet statistics at different steam ratios, configured for zero power export:

| Reformer | | ATR | GHR | GHR |
|---|---|---|---|---|
| Steam ratio | | 0.6 | 1 | 0.6 |
| Carbon eff. | % | 75.2 | 80.7 | 82.84 |
| Total energy | Gj/bbl | 9.61 | 8.95 | 8.72 |
| Oxygen | kg/bbl | 209.5 | 162 | 159.8 |
| CO2 emissions | kg/bbl | 120.2 | 86.9 | 75.4 |
| Water produced | kg/bbl | 180.5 | 153.1 | 152.1 |

The GHR-based process operating at a steam ratio of 0.6 achieves 9% lower total energy consumption than ATR. The maximum efficiency improvement (and the steam ratio at which it occurs) will depend on many factors, such as feed natural gas composition, FT conversion and selectivity, LPG recovery etc. There is still a 5%+ efficiency improvement for a GHR flowsheet, when operated at a steam ratio of 2.0. There is around a 25% reduction in oxygen capacity required for GHR, which is relatively insensitive to steam ratio.

The GHR heat-recycle results in shrinkage of utility requirements, and hence capital cost of utility equipment, as per the table below. In the ATR flowsheet, steam is produced involuntarily and a significant quantity is surplus to internal power requirements.

| Reformer | | ATR | GHR | GHR |
|---|---|---|---|---|
| Steam ratio | | 0.6 | 1 | 0.6 |
| HP steam | kgs/bbl | 867.9 | 273 | 225 |
| Power required | kWhrs/bbl | 161.9 | 130.4 | 128 |
| Air Cooling | kWhrs/bbl | 937 | 698 | 650 |

The cost savings in the utility areas combine with those for air separation to give a significant capital cost advantage for a GHR flowsheet over an ATR flowsheet at low steam ratio. At a steam ratio of 2.0, our analysis shows a GHR flowsheet to have a similar capital cost to an ATR flowsheet with a steam ratio of 0.6.

**GHR – Alternative Power Integration**

As noted before, the utilisation of GHR eliminates the involuntary steam requirement characteristic of an ATR flowsheet. This means that there is more flexibility for the flowsheet designer to choose how to balance heat and power. The diagram below shows this principle:

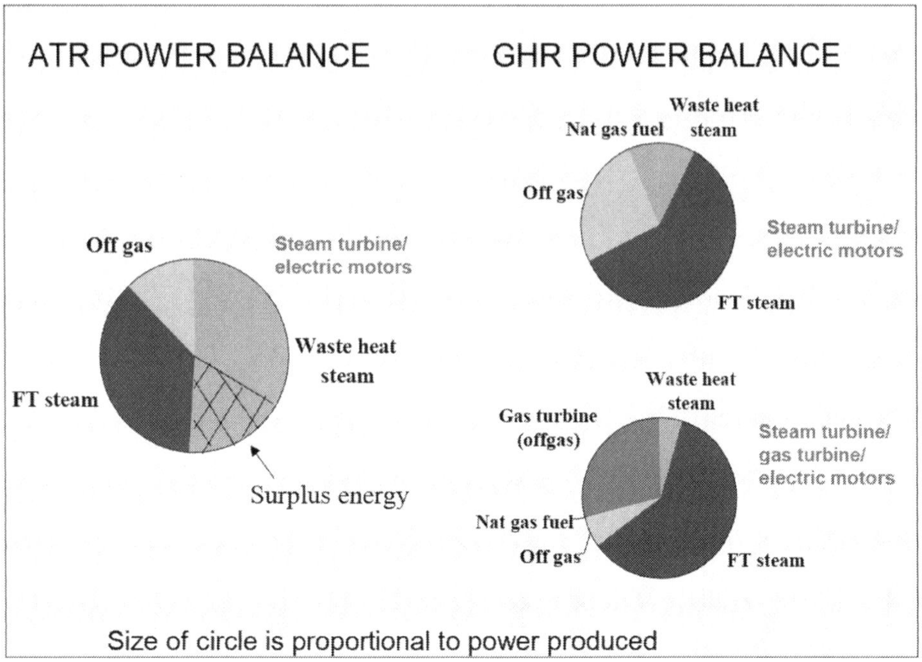


Power can come from steam generated within the process or by combusting waste gas from the process or additional natural gas fuel

For the ATR, it would be possible to generate power that is surplus to facility requirements – shown by the shaded area of the pi chart in the diagram. There is a slight energy shortage for the GHR flowsheet, when configured with steam turbines as the only power sources (either as direct compressor drives or when used to generate electricity for electric motor drives). This can be eliminated by the use of a gas turbine fired on offgas fuel, thereby increasing the overall natural gas efficiency, removing the need to fire additional natural gas as fuel. This would typically improve the overall efficiency by almost 2 percentage points. The actual integration of the gas turbine would depend on a number of factors, such as the size of plant, turbine frame size and relative magnitude of power requirements for different drive duties.

This flexibility is not readily available with an ATR flowsheet, where using a gas turbine would mean that there would be even more steam excess to power generation requirements.

| Reformer | | ATR | GHR | GHR |
|---|---|---|---|---|
| Steam ratio | | 0.6 | 0.6 | 0.6 |
| Gas Turbine ? | | No | No | Yes |
| Feed Carbon eff. | % | 75.2 | 84.56 | 84.56 |
| Total Carbon eff. | % | 75.2 | 82.84 | 84.56 |
| Total energy | Gj/bbl | 9.61 | 8.72 | 8.54 |
| CO2 emissions | kg/bbl | 120.2 | 75.4 | 66.4 |

**Use of Natural Gas Saturator**

The heat-recycle feature has yet another positive benefit, when the requirements for make-up water and waste-water treatment are considered.

There does not need to be a waste heat boiler in a GHR flowsheet. The gas leaving the shell of the GHR is typically at 500-570 $^{o}$C (930- 1060 $^{o}$F ), rather than at 950 $^{o}$C+ ( 1740 $^{o}$F+) for ATR, and can then be interchanged efficiently with the incoming feedstock to cool it by a further 100-150 $^{o}$C (180-270 $^{o}$F). A saturator can be used to humidify the natural gas by contacting it directly with a hot water stream in a circuit. This hot water stream derives its energy from the cooling syngas product stream. There is enough heat to raise substantially all the process steam for steam ratios of 1.0 or less and is thus the preferred means to do this. This diagram shows the arrangement.

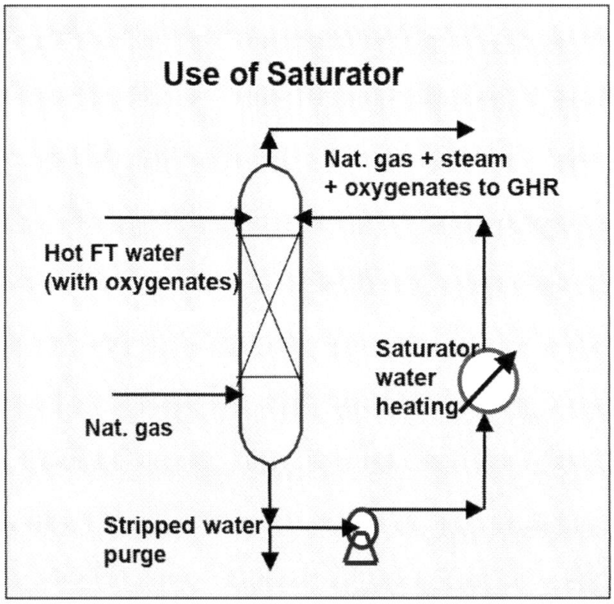


In contrast to a boiler, the saturator circuit can use process waste-water streams to provide its make-up water. A very large, waste-water stream in a GTL plant is the co-produced water from FT synthesis (especially when based on Cobalt catalysts). Another key feature is its composition – it typically contains around 1-2 wt% co-produced alcohols and other organic oxygenate compounds. This will add a large cost to the plant associated with cleaning up the water before it can be discharged or used beneficially.

By processing the FT co-produced water through the saturator in a GHR flowsheet, the waste -water organic load can be reduced by 98-99%, with a huge reduction in effluent treatment plant set-up and operating costs and a corresponding boost to plant efficiency. The alcohols are recycled to the GHR, where they are reformed back to CO and $H_2$.

However, for an ATR flowsheet there is no natural recycling path for the byproduct oxygenates. The choice is as follows:

a) Accept higher effluent treatment costs
b) Instal a topping, distillation column to treat FT condensate and recycle overheads.
c) Instal a condensate stripper using superheated process steam
d) Instal a natural gas saturator

Our analysis shows that options (b) and (c) do reduce the effluent burden, but not as effectively as using a saturator and require installation of extra equipment. Option (d) gives the same reduction and benefit, but also requires extra equipment and means that there is potentially even more steam that is excess to requirements.

**Development of GHR Technology**

JM Synetix's experience with Gas Heated Reformers goes back many years. It was developed initially by Synetix in the 1980s as part of its Leading Concept Ammonia (LCA) process and two plants were commissioned in England using the technology in 1988. In 1994, BHP Petroleum commissioned a methanol plant based on Synetix's GHR technology, using oxygen. The plant, which is now operated by Coogee Energy, is near Melbourne, Australia and was the first commercial demonstration of Gas Heated Reforming for methanol production. Further process and mechanical improvements have and continue to be made.

We have learnt much about the design and operation of these units including an in depth knowledge of the appropriate materials of construction. Gas Heated Reformers, and other systems operating in a synthesis gas environment, are susceptible to a form of corrosion known as metal dusting where components can rapidly deteriorate under the influence of components of the process gas. Metal dusting is complex and is influenced not only by the composition of the alloys and process gas but also by other operating conditions and design details of the system. Although small scale and laboratory test work can sometimes provide useful background information, testing, which fully replicates the proposed plant conditions, is essential in order to provide reliable design information.

JM Synetix's commercial experience with Gas Heated Reformers extends down to a steam ratio of 2.0 and today we can design GTL syngas units to achieve the 5% energy savings and other benefits described above. In order to get the maximum capital and efficiency benefits, associated with running at steam ratio of 1.0 or less, further demonstration of the metallurgy was needed and Synetix has designed a Materials Demonstration Unit (MDU), which is operated by its partner, Methanex in a development venture in New Zealand.

The New Zealand MDU comprises 2 Gas Heated Reformers with full size tubes, each coupled to a secondary reformer and associated equipment. The two separate trains operate under different conditions and are demonstrating materials suitable for both GTL and advanced methanol process conditions. The MDU is testing a number of different alloys simultaneously under conditions replicating those of full-scale commercial units. The objective of the test program is to provide long-term test data under commercial plant conditions thus giving a high level of confidence in the results of the work.

[Photograph deleted owing to file size]

**MDU Unit at Methanex site, NZ**

There were a number of problems in the early operation of the plant, mostly not related to the technology per se (many related to oxygen generation), but the plant has since run consistently and provided invaluable data. As anticipated, some materials have been found to be not suitable for the duty. The GTL train has now accumulated several months' operation at a steam ratio of around 1.0 and has also run at lower steam ratios. Testing will continue.

**Conclusions**

The fundamental heat-recycle feature of GHR technology brings wide-spread efficiency and environmental benefits to GTL production with respect to big reductions in gaseous and liquid effluents compared to alternate production routes.

Today, a GTL process, incorporating the Synetix GHR operating at a steam ratio of 2.0, offers significant efficiency and operating cost benefits at a capital cost comparable with an ATR -based process. At these process conditions the materials of construction are already established and are within JM Synetix's experience.

A demonstration program is underway to prove appropriate materials of construction for using GHR at low steam ratios, which will enable all the benefits of using GHR to be realised.

Reference 1- Efficient, Cost-Effective & Environmentally friendly Synthesis gas technology for Gas to Liquids production, BJ Crewdson + KJ Elkins, IBC GTL Conference, London, Feb 2002
Reference 2 - Fischer-Tropsch production Developments using Gas Heated Reforming Technology, PEJ Abbott + BJ Crewdson, GPA Europe conference, London, 21 Feb 2002

# Enhanced Heat Transfer Reformer (EHTR) Technology for GTL Applications

*John M. Repasky*
*Air Products and Chemicals, Inc.*
*7201 Hamilton Boulevard*
*Allentown, PA 18195*

*Robert T. Reader*
*Mid-America Commercialization Corporation*
*1500 Hayes Drive*
*Manhattan, KS 66502*



Prepared for Presentation at the AIChE Spring National Meeting
New Orleans, LA, 25-29 April 2004
Unpublished



## Abstract

Gas-to-Liquids (GTL) conversion of remote natural gas to hydrocarbon fuels is developing quickly. Syngas generation is the first step of the process and the largest portion of capital investment. A 100,000 bpd GTL site will require nominally 2500 MMscfd of 2:1 H2/CO syngas from multiple trains of some of the largest syngas equipment in the world. Enhanced Heat Transfer Reformer (EHTR) technology is designed to integrate with partial oxidation (POX) or auto-thermal reforming (ATR) to produce GTL syngas. The EHTR replaces a boiler to recover high-level heat from the primary syngas generator to produce additional syngas instead of producing steam. The new reactor configuration improves carbon utilization, reduces oxygen consumption, and reduces the syngas steam system. This paper provides an overview of the conventional syngas generation technologies POX and ATR before focusing on the integration of EHTR technology into the GTL flow sheet.

## Introduction and Background

Gas-to-Liquids (GTL) conversion of remote natural gas to hydrocarbon fuels is developing quickly. These mega-sites will require tremendous amounts of synthesis gas. A single site may require nominally 900 MMscfd of natural gas and 20,000 tpd of oxygen to produce 2500 MMscfd of syngas toward eventually producing 100,000 bpd of synthetic petroleum products for export. Site capital investment is targeted at nominally $20,000 per bpd of installed capacity resulting in $2 billion dollar investment. In its simplest form, the process may be divided into the three basic steps of syngas generation (including air separation units), Fischer-Tropsch synthesis, and product upgrading with the overall capital divided in the nominal ratio of 60 / 30 / 10, respectively, between the three. It is clear that lower cost syngas is the key to reducing costs of synthetic hydrocarbon fuels.

Partial oxidation is a benchmark technology for GTL syngas. Natural gas and oxygen react in the POX reactor to produce syngas with a hydrogen-to-carbon monoxide ratio of nominally 1.8 to 1. A specialized, and quite expensive, POX waste heat boiler produces significant steam as it cools the POX syngas. Finally, a separate steam-methane reformer is required to produce hydrogen rich syngas for blending to obtain the 2 to 1 ratio syngas required for Fischer-Tropsch Synthesis. Carbon dioxide is eventually vented.

Auto-thermal reforming is another benchmark syngas technology. Natural gas, oxygen, steam, and partially recycled carbon dioxide react in the ATR reactor. The amount of carbon dioxide recycle may be used to control the syngas $H2/CO$ ratio, and nominally 75% recycle yields the desired 2 to 1 syngas required for Fischer-Tropsch Synthesis. Again significant steam is produced as the syngas cools in the process gas boiler.

## Enhanced Heat Transfer Reformer Technology

Enhanced Heat Transfer Reformer (EHTR) is a tubular heat exchange reforming reactor designed to operate in parallel with a primary syngas generator. Natural gas and steam react via the steam-methane reforming reaction in the catalyst filled tubes of the EHTR to produce additional syngas. The endothermic heat of reaction is provided via convective heat exchange with syngas from a primary syngas generator.

Air Products has previously commercialized EHTR technology for the production of hydrogen in which an EHTR reactor is arranged in parallel with a fired steam-methane reformer (SMR). ATR and POX reactor configurations are preferred for GTL syngas production. In one example, natural gas and oxygen react in a POX reactor to produce nominally 1.8 to 1 syngas. Natural gas and steam react in a parallel EHTR to produce hydrogen rich syngas. High carbon conversion may be realized by complete recycle of all carbon dioxide from the Fischer-Tropsch synthesis section of the GTL site and nominally 75 / 25 natural gas feed split between the POX reactor and the EHTR yields the desired 2 to 1 syngas. Both the mass and heat balances of the two reactors are inherently optimized by their fundamentally coupling. The exothermic POX reactor produces a carbon monoxide-rich syngas which in turn provides the heat to drive the endothermic production of hydrogen-rich syngas in the EHTR unit. Furthermore, the high-level heat recovery in the EHTR yields a much lower syngas temperature to the process gas boiler. Boiler design is simplified and the size and cost of the steam system are significantly reduced.

Summarizing the benefits of EHTR heat exchange reforming: The EHTR recovers high-level heat for additional syngas production, on the order of 20 to 30 percent. The additional syngas production offsets duty in the primary reformer resulting in reduced size and cost of the primary reformer. When coupled with oxygen-based reforming technologies, the EHTR reduces oxygen consumption, and with that, the size and cost of air separation units (ASU). The high-level heat recovery to produce syngas greatly reduces the size and cost of the steam system, and steam production can be tuned to match the site needs. These reductions in duty on major equipment and the ability to integrate recycle streams make EHTR an environmentally friendly technology. The parallel arrangement of the EHTR makes it ideal and cost effective for plant expansions. Analogous to plant expansions, the EHTR may be used to de-bottleneck maximum train size of grass-roots GTL syngas applications. De-bottleneck reduces risk associated with scale-up of the primary syngas generator and leads to greater economy of scale and reduced capital for the overall facility.

**Process Evaluation and Economics**

Air Products has conducted process evaluation and economic analysis for the three GTL syngas processes POX+SMR, ATR, and POX+EHTR. The design basis produced 500 MMscfd of nominal 2:1 syngas for Fischer-Tropsch Synthesis, resulting in nominal 21,000 bpd of FT Liquid Products with approximately 60/40 split between diesel and naphtha. The 1996 study by Bechtel sponsored by the United States Department of Energy served as the basis for Fischer-Tropsch Synthesis design. Economics assumed natural gas at $0.50/MMbtu.

Comparing POX+EHTR against the POX+SMR benchmark within the syngas generation section of the site alone, 23% of the natural gas feed and 100% carbon dioxide containing recycle gas are processed through the EHTR unit to produce the target 2:1 H2/CO ratio syngas. The syngas production in the EHTR reduces the size of the POX reactor by 21% and results in 24% reduction in oxygen requirement from 4800 sT/D to 3600 sT/D. EHTR heat recovery provides a lower temperature syngas product. This allows simpler steam boiler design, reduces the overall steam system by 56%, and allows the export steam to be tuned to optimize the overall site. The chemical recuperation of carbon dioxide reduces the natural gas requirement by 6%, or stated another way can allow 6% increased F-T liquid production from a given natural gas supply. CAPEX is reduced by 28%, primarily driven by reduction in oxygen and the smaller steam system. OPEX is reduced by 4%, primarily driven by reduction in natural gas consumption. The net effect is 15% reduction in the overall cost of GTL syngas.

Comparing POX+EHTR against the ATR benchmark within the syngas generation section of the site shows similar results. Again, 23% of the natural gas feed and 100% carbon dioxide containing recycle gas are processed through the EHTR unit to produce the target 2:1 H2/CO ratio syngas. In this case, the oxygen requirement is reduced by 18% from 4400 sT/D to 3600 sT/D, the overall steam system is reduced by 58%, and the natural gas requirement is reduced by 6%. The economic results are also quite similar. CAPEX is reduced by 26%, again primarily driven by reduction in oxygen and the smaller steam system. OPEX is reduced by 10%. The net effect is a 16% reduction in the overall cost of GTL syngas.

Further benefit may be realized when the overall GTL site configuration, including fuel gas, steam, and power, is reconfigured around EHTR technology. In particular, reduction in excess steam production from syngas generation enables site power and steam requirements optimally supplied from a gas turbine. Gas turbine integration enables further optimization of the site fuel, steam, and power balances. For example, Fischer-Tropsch tail gas or hydrogen may be utilized as gas turbine fuel. The gas turbine HRSG can provide feed preheat for syngas generation, allowing ancillary fired heaters with their associated capital and operating expenses and environmental impacts to be eliminated. Nitrogen from the air separation unit can be integrated to provide ballast gas to the gas turbine. Relative to the 1996 Bechtel study, Air Products has shown 11% reduction in natural gas, 45% reduction on oxygen, improvement in LHV thermal efficiency from 58% to 69%, and improvement in carbon efficiency from 72% to 83%. Improvements in Fischer-Tropsch synthesis technology beyond the 1996 Bechtel base line, which most likely have already occurred, further improve the efficiencies.

## Commercialization

Air Products has previously commercialized EHTR technology for the production of hydrogen in combination with a fired steam-methane reformer (SMR).

EHTR syngas generation was selected as the economic process-of-choice to meet the zero steam export scenario at Sacramento, CA. The facility has operated at its design capacity for nearly 16 years with greater than 95% on-stream factor since its original start-up in 1986. Not a single shutdown has been caused by the EHTR. Mechanical inspection of the EHTR in 1998 found the unit to be in excellent condition, with no signs of metal dusting.

EHTR technology was installed to de-bottleneck the hydrogen plant at the Repsol Refinery in La Coruna, Spain in 2001. In this expansion case, custom internals were designed to accommodate syngas product exit at the bottom of the EHTR unit to match up directly with the existing process gas boiler. The unit provided 20% capacity expansion with minimal additional plant modifications.

Most recently, EHTR technology was incorporated into the grass-roots design of Air Products 40 MMscfd hydrogen facility New Orleans, LA. The Oct 2003 start-up was smooth and performance test has confirmed the unit to perform at design expectation.

## Metal Dusting

The potential for metal dusting corrosion is a critical issue in most all heat exchange reforming equipment and remains the primary technical challenge toward commercialization of EHTR in combination with POX or ATR for the production of syngas toward GTL application. In particular, the syngas and metal temperatures pass directly into the metal dusting regime as the syngas product cools on the shell-side of the EHTR. EHTR design for GTL is a more severe operating condition than existing hydrogen experience.

## Patent Ownership

Specific patent ownership for the application of EHTR technology to GTL and methanol has been transferred from Air Products and Chemicals, Inc. to Mid-America Commercialization Corporation.

## Conclusions

- EHTR technology offers significant economic benefit for GTL 2:1 syngas generation.
- Air Products has commercialized EHTR for hydrogen production under moderate severity metal dusting conditions in both grass-roots and plant expansion projects: Sacramento, 1986; Spain, 2001; New Orleans, 2003.
- Metal dusting remains the primary technical challenge toward commercialization of EHTR for GTL application.
- Specific patent ownership has been transferred from Air Products and Chemicals, Inc. to Mid-America Commercialization Corporation.

## References


Bechtel Corporation, Baseline Design/Economics for Advanced Fischer-Tropsch Technology, Topical Report VI: Natural Gas Fischer-Tropsch Case, Volume II, Plant Design and ASPEN Process Simulation Model, US DOE Contract No. DE-AC22-91PC90027, August 1986.

Allam, R.J. and R.J. Cotton. EP1197471(A1), 2000.

Wang, S., S. Nataraj, and J.M. Repasky. US2003/0162846(A1), 2002.

Repasky, J.M., D.M. Brown, S. Wang, and R.J. Allam. "GTL Syngas - The Potential for Heat Exchange Reforming", presentation at IBC GTL Conference, London, U.K., 27 Feb 2002.

Ratan, S. and C.F. Vales. "Improve Your Hydrogen Potential", Hydrocarbon Processing, March 2002.

Brown, D.M., Repasky, J.M., S. Wang, and R.J.Allam, "New Near-Term Technology for Reducing the Cost of Syngas in Remote Gasification", presentation at IBC GTL Conference, Singapore, 7-8 May 2002.

Repasky, J.M., D.M. Brown, S. Wang, and R.J. Allam. "New Synthesis Gas Technology for GTL Applications", presentation at CWC Global GTL Summit, London, 29-30 May 2002.

Repasky, J.M., D.M. Brown, and R.J. Allam. "Advances in the Production of Synthesis Gas and Hydrogen for Clean Fuel and Power Systems", presentation at IBC GTL Conference, Milan, Italy, 8-9 Jul 2002.

Brown, D.M., J.M. Repasky, and E.C. Heydorn. "New Syngas Technology for Methanol and DME Production", presentation at IBC GTL Conference, Houston, TX, 23 Oct 2002.

Brown, D.M., J.M. Repasky, and M.O. LeFevre. "Heat Exchange Reforming - An Efficient Route to Synthesis Gas", presentation at EFI General Meeting, San Francisco, CA, 30 Oct-1 Nov 2002.

Repasky, J.M., D.M. Brown, and M.O. LeFevre. "New Syngas Technologies for GTL, Methanol, and DME", presentation at IBC GTL Conference, London, U.K., 3-4 Feb 2003.

**Oxygenates Produced During Fischer-Tropsch Synthesis**

*Burtron H. Davis, Robert A. Keogh and Karuna Chaudhari*
*Center for Applied Energy Research*
*University of Kentucky*
*2540 Research Park Drive*
*Lexington, KY 40511*

Prepared for Presentation at the AIChE 2004 Spring National Meeting, New Orleans, LA
April 25-29, 2004, Gas Conversion to Liquid Fuels and Chemicals

Unpublished

## Extended Abstract


This presentation will report the effect of Group IA alkali promoters on a supported Fischer-Tropsch iron catalyst. The general formula for the catalyst is 100Fe/4.5Si/1.44X, where X is the Group IA alkali promoter. All runs were completed in a 1 liter CSTR. The start-up oil used in the run was a C-30 polyolefin oil. All runs were conducted at a temperature of 270°C, a pressure of 175 psig and a $H_2$:CO ratio of 0.67. Space velocity was changed from 3 to 40 Sl/hr/g catalyst to investigate the oxygenate yield and selectivity change with conversion. Identification of the individual oxygenate compounds was done with commercially available standards. Analysis of the aqueous phase was done using an Agilent 6890 capillary GC equipped with a TCD detector and a 30 meter HP-Plot Q column.

The Group IA promoted catalysts produced more total oxygenates in the aqueous phase when compared to the unpromoted catalyst. This comparison was made at a 50 mole% CO conversion. The dominant oxygenate class produced was linear alcohols. The carbon number of the alcohols ranged from 1 to 5 with ethanol having the highest concentration in the aqueous phase. The yield of total oxygenates and alcohols decreased in the following order: K > Cs > Rb > Na > Li > no promoter.

Aldehydes had the second highest yields in the aqueous phase produced using the Group IA promoters. Acetalaldehyde had the highest yields in the compound class. The carbon number distribution for the aldehydes varied with the promoter, but generally ranged from 2 to 5. The aldehyde yields in the aqueous phase decreased in the following order: K > Cs > Na > Rb ≈ no promoter > Li.

The total acid and ketone yields were approximately the same for the runs with the promoted catalysts. Only acetic and propionic acid were produced in the aqueous phase. The yields of acetic acid were significantly higher than those of propionic acid. The yields of total acids decreased in the following order: Cs > Rb > K > Na > Li > no promoter.

The addition of a promoter did not significantly increase the yields of ketones in the aqueous phase when compared to the ketone yields for the unpromoted catalyst. In addition, the ketone yields were essentially the same for all the Group IA promoters. Acetone, 2-butanone and 2-pentanone were the ketones produced by the Group IA promoters with acetone having the highest yields. Only acetone was produced in the aqueous phase for the unpromoted catalyst.

The oil plus wax ($C_4$ to ~ $C_{35}$) hydrocarbon fraction produced by the promoted catalysts also contained linear alcohols, acids, ketones and aldehydes. The alcohol yields, in general, were significantly greater when compared to the other oxygenates found in the hydrocarbon fraction. The alcohol carbon number varied with the promoter, but in general ranged from $C_2$ to $C_{14}$ with maximum yields in the $C_5$ to $C_7$ range. The highest yield and greatest carbon number distribution were produced by the cesium and rubidium promoted catalysts. In the hydrocarbon fraction of the unpromoted catalyst, only ethanol and n-propanol were detected in trace amounts.

# SUBMERGED MOTOR LNG PUMPS FOR SEND-OUT SYSTEM APPLICATIONS

*Steven G. Rush*
*Vice President, Sales and Service*

*Ebara International Corporation*
*Sparks, Nevada, USA*
*srush@ebaraintl.com*

*Prepared for Presentation at AIChE Spring National Meeting*
*4th Topical Conference on Natural Gas Utilization*
*New Orleans, LA, April 2004*

## Introduction

With the current growth in the LNG market worldwide, there is a high demand for additional LNG receiving terminals and regasification systems in many locations. These terminals, whether on-shore or off-shore, contain gas send out systems which use particular types of submerged, electric motor LNG pumps for LNG transfer and pressurization.

The type of pumps discussed here are a submerged, electric motor type, with an integral shaft, having the entire motor, bearings and all other components completely flooded with LNG.

Recently, various methods of transferring LNG from ship to tank storage for transfer later to a send out system, or transfer directly from ship into a regasification or send out system have been studied.

This paper will present the types of submerged motor LNG pumps used in send out system applications as well as the safety aspects and design of the equipment typically used. In addition, the various methods of transfer using submerged motor pumps will be presented, along with some aspects to consider for each method.

## Safety

The submerged electric motor pump (SEMP) has been used in LNG applications since the early 1960's. Since that time, the SEMP has been used in almost all LNG receiving terminals. The primary reason for this is that the design is inherently safe, compared to an external motor type pump with dynamic shaft seals.

The motor and hydraulic sections of the SEMP are directly coupled with a common shaft, submerged in the liquid, with no oxygen present. This means the motor is not located in the atmosphere in the hazardous area, and no rotating seals are required. The design almost completely eliminates the possibility of leakage of flammable gas into the atmosphere.

In addition to the safety aspects, since no coupling is required between the motor and pump sections, there are no alignment problems normally associated with pumps using a coupling. Also, since the complete assembly is submerged, the liquid acts as very effective sound insulation and the pumps operate very quietly.

## Basic Design

In the traditional, land-based LNG receiving terminal, the system consists of a storage tank, which contains retractable (removable) or in-tank type pumps, and a send out system, which contains vessel mounted type high-pressure LNG pumps and vaporizers. This type terminal also includes a jetty where the LNG carrier would dock and discharge its LNG into the on-shore storage tanks.

The pumps used in the on-shore storage tanks are submerged motor, retractable (removable) types, as shown in Figure 1. This type pump, sometimes also referred to as a "primary" pump, for a typical receiving terminal, will have a flow rate of approximately 200 to over 400 $m^3/h$. This pump would normally only require one or two impeller stages, since it only needs to transfer the LNG out of the tank and into the secondary system.

For vaporizer feed duty, a relatively high pressure is required due to the high pressure drop across the vaporizer. For this application, a multistage vessel mounted type pump is used which can produce pressures up to approximately 140 $kg/cm^2$. The multistage vessel mounted pump design is shown in figure 2.

For the primary transfer pumps, the motor voltage is normally 400 to 480 volts, three phase, but can also be made at higher voltages depending on site power supply. The secondary vaporizer feed pumps are normally higher power, which require from 4160 to 6600 volts. Both pumps can be manufactured for either 50 or 60 Hz power, depending on site requirements.

The primary pump is installed into the storage tank through a discharge "column" mounted inside the tank. At the base of the column is a suction valve, which is opened by the pump itself. As the pump is lowered into it, the valve opens, allowing the LNG to flow into the inlet of the pump. The pump has a seal located near it's base, which allows the discharge liquid to be pumped out the top of the pump and out the top of the discharge column.

FIGURE 1

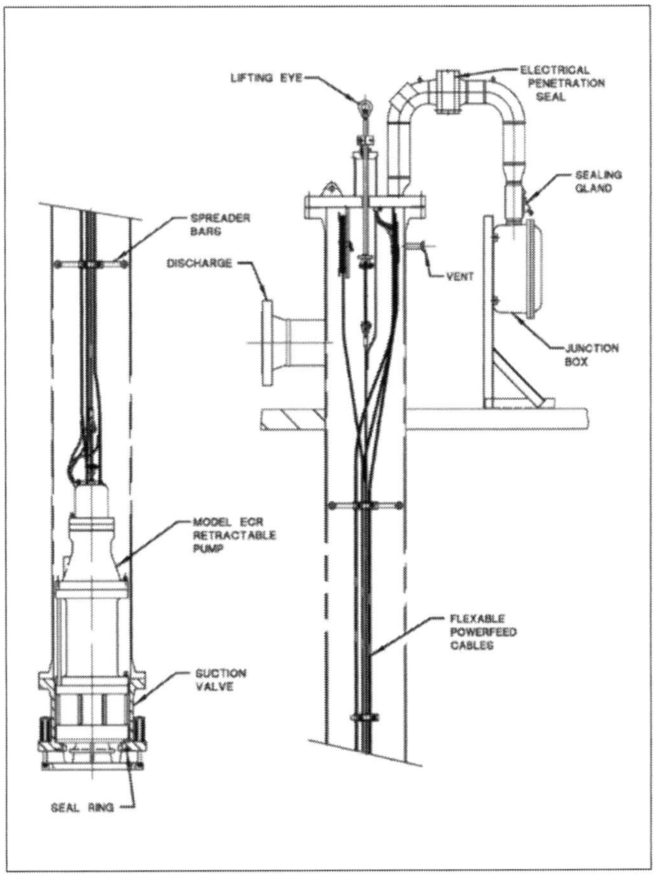


FIGURE 2

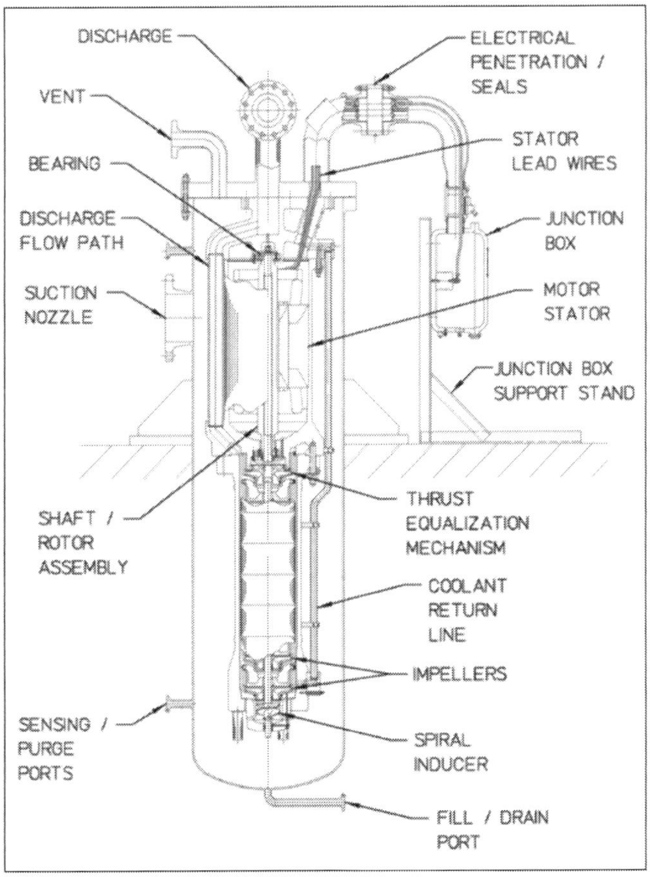


Since the primary pump is installed in the storage tank, which is already provided with a vent system, no other venting connection at the headplate or column is required, although venting of the column to equalize pressure to the main tank area is required prior to start-up. The heat from the pump primarily is transferred to the pumped fluid, with only a small amount of heat being transferred back to the liquid in the storage tank.

The high-pressure secondary pump is installed in its own self-contained suction vessel, with standard flanged inlet and discharge pipe connections. The vessel needs to be vented and is also provided with a vent nozzle at the top of the headplate. The vessel can be made to the ASME Section VIII code or to just about any pressure vessel code as required.

Apart from the piping connections, both pump types require power cables to be connected at their junction boxes located on the headplate.

## Primary Pump, Design Concerns

With an SEMP used for LNG, it is always important to keep in mind that LNG is normally handled close to its boiling point. Even just a small increase in temperature or a drop in pressure can cause the LNG to vaporize.

As previously noted, the primary pumps in the storage tank are used simply to transfer the LNG to the send-out system. Normally, two or three primary pumps are used depending on total flow requirements and the need for back-up or redundancy. From the point of view of the pump design, there is no particular spacing requirement between the pump columns. Normally, the column spacing will be dictated by the tank design and spacing of the piping, valves, etc, at the tank top.

Care should be taken in the tank design for the location of the inlet piping to the tank. If the inlet pipe is placed too close to the pump columns, the warmer liquid entering the tank can affect the NPSHR of the pumps when filling the tanks during lower level operation.

Another requirement is the need for column venting prior to start-up. Most columns will have discharge piping which can vented to the tank top to equalize pressure between the tank and column, and particular care should be taken to ensure there are no low spots which would trap liquid or prevent proper venting. This becomes very important at low liquid levels, where 2 meters of LNG is equal to only 0.09 $kg/cm^2$ pressure. If the column pressure is too high, the lower section of the column, including the pump can be completely void of liquid, which can create problems during starting.

## Secondary Pumps, Design Concerns

The secondary pumps are normally installed in a “bank” or row, with the suction nozzles all connected to a common suction header. The discharge lines are also normally connected to a common line, but each pump should have its own discharge control valve, as well as a by-pass or “kick-back” line to allow proper starting and operation. Discharge control is extremely important to allow proper start-up when operating several pumps in parallel.

With the secondary pump, it is also very important to recognize proper venting. Since the secondary pumps normally have much higher motor power, particular care must be taken to ensure any heated LNG or vapor is taken care of. Vent lines should always be rising as they leave the pump vessel and provide good venting back to a low pressure space. Many problems with secondary pumps over the years have been attributed to poor vent system design.

In the secondary pump system, a phase separator, or recondenser is normally installed in the suction area. This tank is typically used to allow the liquid to settle long enough to allow any vapor to be vented, and is also used to introduce LNG from the boil-off gas system to try and recover as much of the LNG as possible. The design of the system in this area should also be treated carefully to ensure that the LNG temperature is still well into the liquid phase as it enters the pump suction.

With the recent interest in off-shore or remote send-out systems, an important topic is the electrical supply to start the high-pressure send out pumps. Since the motors in these pumps are typically from 1000 to as much as 2300 kW, a large starting system is required. A cryogenic motor is a unique design, and the starting current required is approximately 6 ½ times the full load current. It is difficult to reduce this value because of the amount of torque required for starting a cryogenic motor.

To reduce the starting current, soft starters, autotransformers and variable frequency drives can be used with SEMPs, and have been used very successfully in many applications. However, proper setup of starting parameters in any current reduction type starting system is critical, and the pump manufacturer should be consulted to prevent problems.
In some offshore or remote locations, using the LNG carrier as the primary system, delivering the LNG directly to the secondary pumps has been discussed. It appears the biggest obstacle in this type system is the venting and boil-off from the send out system during operation. Using the boil-off gas to feed local gas turbine generators for power, or pressurizing the gas using compressors and feeding into the downstream gas system appear to be popular alternatives.

## Monitoring and Protection

Both primary and secondary SEMPs have vibration monitoring systems installed to be able to trend vibration. Both types use an accelerometer mounted to the outside of the pump casing (yes, submerged in the LNG). As can be imagined, this is a very difficult installation, and unfortunately, the reliability of these in-tank vibration systems is not very good.

For the in-tank primary pumps, there is little other alternative to monitor vibration, as the pump is too far from the top of the column to be able to measure anything meaningful. There are many in-tank SEMPs installed throughout the world with no vibration monitoring system that operated extremely well with no problems. Normally, the primary pump is a very reliable pump, with only one or two stages, and using a vibration system for monitoring is not really necessary. Monitoring the pumps by other means, such as trending motor amperage, can be done with good reliability.

For the secondary pumps, the accelerometer can be mounted on the headplate of the suction vessel, outside of the suction vessel. While this method does not provide as much sensitivity it can be used very well for trend monitoring, and has recently become the preferred method for mounting.

One of the best ways to monitor SEMPs is to simply use the trending capabilities of the DCS system. If motor amperage, flow and pressure are checked regularly and trended against one another, the performance of the pump can be verified. Over time, any apparent deterioration in performance is normally an indication that the unit is in need of overhaul.

For protection, the pumps should be protected with at least over and under current relays in the electrical system, connected to the motor power input. A loss in amperage is an indication of loss of suction pressure (cavitation), and an increase in amperage is an indication of a mechanical problem or a problem in the electrical wiring into the motor. In addition, a typical system will contain protection for low suction pressure or low liquid level, and low or out of specification flow and0or pressure. Newer DCS systems have excellent capabilities for creating acceptable operating parameters for SEMPs.

## Reliability

The reliability of a sendout system is extremely important. Most systems have several pumps operating at all times, with the need for pumps operating twenty four hours a day for long periods of time.

With LNG systems, mainly because of the fact that the LNG is typically handled close to its boiling point, there are several factors, including system design, which are very important to consider to ensure a safe, reliable system. When reviewing LNG pump failures over the years, the predominant root causes for failure are as follows:

- Cavitation due to poor system design or improper operation.
- Operation at flow rates away from the rated or best efficiency point for extended periods of time.
- Debris or contamination in the liquid.

Since LNG systems are normally clean, the first two points listed are normally the main cause for failure. Therefore, the system designer and operators need to be well trained and aware of the important issues surrounding design and operation to ensure a reliable send out system.

To ensure a successful send out system, it is important to consult with the SEMP pump manufacturer during the initial design or FEED stages, as well as when writing specifications for ordering equipment. As with any system, the result of proper system design should be a safe, simple, reliable and reasonably priced system.

## References


1. D. Cullen, J. Madison, *High Pressure Technology*, Hydrocarbon Asia, July/August 2001.
2. G. Louis Weisser, *Modern Submersible Pumps for Cryogenic Liquids*, World Pumps, January 1994.
3. D. Cullen, S. Rush, J. Madison, *Radial and Axial Diffusers for Submerged Electric Motor-Driven Pumps*, World Pumps, September 2000.

# A Case Study for Integrated LNG-to-Power: Technical Comparison of Available Cost Reduction Methods

*Joseph H. Cho, Ph.D., P.E.*
*Principal Technical Professional*
*Himanshu Patel, P.E.*
*Technology Consultant, LNG, GTL and Gas processing*
*David A. Coyle, P.E.*
*Technology Manager, LNG, GTL and Gas-Processing*
*Charles A. Durr*
*Technology VP, LNG, GTL and gas processing*

*Kellogg Brown & Root, Inc.*
*Houston, Texas 77002*

*Prepared for Presentation at the AIChE 2004 Spring National Meeting*
*4th Topical Conference on Natural Gas Utilization*



## ABSTRACT

Natural Gas that is imported as Liquefied Natural Gas (LNG) into an open power generation market competes against pipeline gas, liquid hydrocarbon fuels and coal. One of the key factors in its competitiveness is effectively reducing the delivered cost of electricity. This goal can be achieved by reducing the cost of gasifying LNG and thus optimize the economics of gas-to-power.

This paper presents a technical comparison of practical and viable processes that integrate LNG vaporization with power generation. Some of these processes are already in commercial development. The paper also presents new LNG integration processes for re-gasification and power generation, developed by KBR. These processes include extensive use of cooling water and waste heat recovery for LNG vaporization through its integration with gas turbine inlet air chilling. The paper provides a useful guideline for selecting the best integration scheme and for the evaluation of synergies.

# A Case Study for Integrated LNG-to-Power: Technical Comparison of Available Cost Reduction Methods

## INTRODUCTION

Liquefied Natural Gas (LNG) imports into the North American market are expected to increase over time in order to supplement declining gas production and to meet increasing demand. LNG imports will directly support both the natural gas pipeline and electric power markets. One of the key factors in the viability of an LNG import project lies in its ability to effectively reduce the delivered cost of its products, i.e., re-gasified LNG into the pipeline network and electricity into the power grid. In order to comply with the energy industry's requirements, many studies have been conducted to develop the integrated process between the LNG receiving terminal and the power plant.

A power plant must reject heat to condense steam and to cool equipment. In addition, power output can be increased by cooling the gas turbine inlet air, thus increasing the mass flow rate and improving the gas turbine compressor efficiency. On the other hand, an LNG terminal has a large storage of cryogenic energy, which has to be released in order to re-gasify LNG. Various methods have been applied to make use of this cold energy, which include cryogenic power generation, air liquefaction, solidification of carbon dioxide, deep freeze warehouse and storage, boil-off gas re-liquefaction, and seawater desalination.

If the seawater temperature is low or poor quality or unavailable due to environmental regulations, submerged combustion vaporizers (SCV) are commonly used for LNG re-gasification. The SCV uses 1.5 to 1.8% of natural gas for fuel, causing an increase in operating costs. The thermal integration of the two plants, supplying heat from the power plant to the LNG Receiving Terminal (RT) in a synergistic arrangement, results in significant benefits: reduction in fuel costs for LNG vaporization, reduction in emissions, and increase in power output. Currently available thermal integration can be categorized into the following:

1) LNG cold energy recovery
2) Heat recovery from the power plant
3) Combination of both heat and cold recovery

This paper classifies the commercially viable processes (patented processes including pending patents) into three categories based on their heat integration schemes. Technical advantages and economical aspects of each process are discussed. In addition, the technical advantages of the integrated processes developed by KBR are discussed.

## LNG COLD ENERGY RECOVERY

### *Gas Turbine Inlet Air Chilling*

Gas Turbine (GT) inlet air chilling is a commercially proven method to generate more power from a given machine. Figure 1 shows the gas turbine output (PG 7241 FA) [1] as a function of the inlet air temperature. The output can be increased by 10 % if the inlet air is chilled to 45°F from 75°F. This is the net increase in output after accounting for the increased pressure drop caused by the chiller and by the water removal vanes. Depending on site conditions inlet air chilling is usually economically attractive. GT inlet air chilling, which uses the LNG cold energy, has been proposed by different researchers [2-4]. The schematic diagram for GT inlet air chilling is shown in Fig. 2.

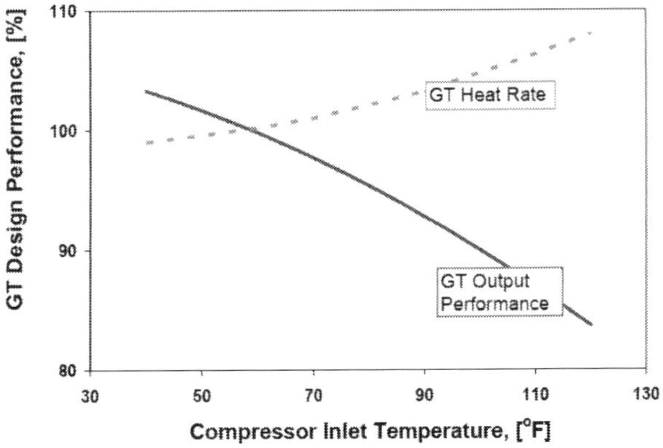


Fig. 1 - Inlet Temperature Impact on GT (GE PG7241FA) Performance [1]

Although the LNG can directly cool the GT inlet air, ice can be formed severely on the heat exchanger surface, which may rupture the tube, resulting in a potentially hazardous mixture at the GT inlet. To avoid this risk, a heat transfer fluid (HTF) loop (e.g. water-ethylene glycol solution) is recommended. Shimizu et al. [3] proposes intermediate fluid (propane) and HTF (brine or ethylene glycol water). The invention claims that this double fluid configuration can prevent the formation of ice on the heat exchanger with the cold LNG. It also claims that hazardous situation can be avoided when there is a leak of HTF on GT inlet exchanger.

With thermal integration, the design gas turbine inlet temperature is typically limited to 45°F. At ambient temperatures above 45°F, the power plant output decreases as shown in Fig. 1. At ambient temperatures below 45°F, the output of the power plant is similar to that of the non-integrated power plant, except for a slight penalty (<0.5%) for the inlet chiller coil pressure drop. The moisture content of the air must be considered because it directly affects the heat duty of the GT inlet air chillers. As the moisture content of the air increases, the amount of cold energy recovered from the LNG increases due to the increased condensing load on the chiller [5].

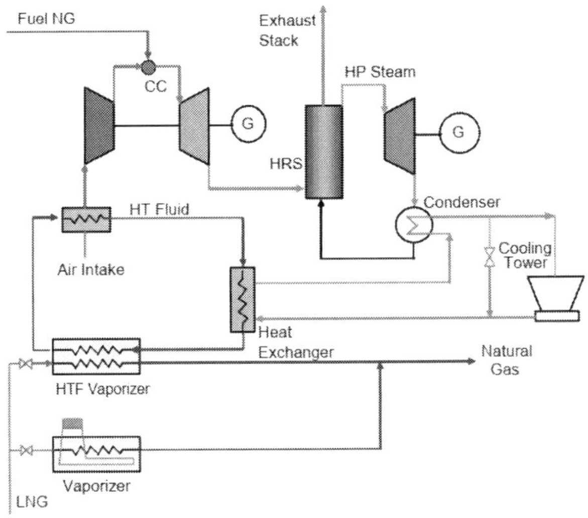


Fig. 2 - Typical Thermal Integration for GT Inlet Air Chilling

For a multi-purpose LNG RT, the amount of LNG vaporized is more than the amount of the fuel needed by the power plant. Therefore, the duty to cool down the GT inlet air is much less than the LNG re-gasification duty. The cold energy recovery by inlet air chilling is subject to ambient conditions and power plant load. As a result, different thermal integrations and alternate heat sources are required in order to ensure a stable send-out rate and temperature.

### *Cold Power Generation*

The most widely used and viable cold power generation methods are: Direct Expansion, Rankin Cycle, Combined Direct Expansion and Rankin Cycle, and Closed Loop Gas Cycle. The cold power generation is dependent on the gas send-out rate and on the pressure to which the vaporized LNG can be expanded. Typically, about 3 – 6 MW cold power generation units are under operation at several Japanese terminals.

Direct expansion of the re-gasified LNG or expansion of the operating fluids using the Rankin Cycle or a combination of both can directly generate electrical power at the LNG receiving terminal (Figs. 3 and 4). Several inventions related to the simple Rankin cycle are well described in Reference 6 through 8. The Rankin cycle can be divided into two types: Single component and mixed component fluid. At the initial development stage, Freon (R-22) or Ammonia was used.

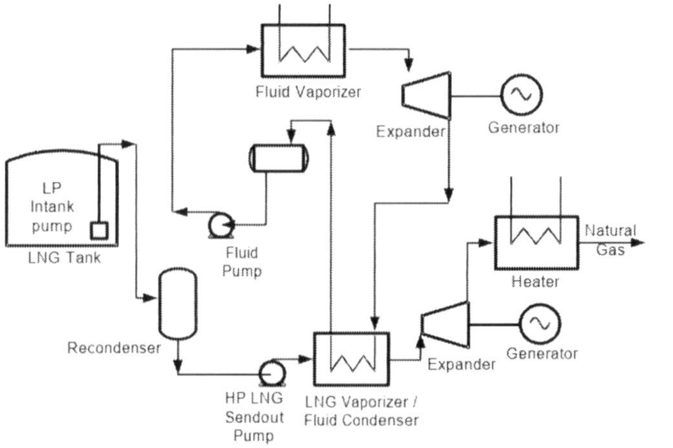


Fig. 3 - Schematic for Power Generation by Combined Direct Gas Expansion and Single Rankin Cycle

The direct generation of electrical power can also be achieved by a cold enhanced combustion recovery system with gas export using closed cycle gas turbine or an open cycle gas turbine or a combination of both. Only the closed cycle gas turbine option, where LNG is vaporized against a closed cycle gas turbine system, has been implemented thus far. Part of the re-gasified LNG is consumed in the fired heater where the high pressure closed cycle gas is heated prior to passing through a gas turbine expander. The low pressure gas from the expander is cooled initially in a recuperator against the flow to the fired heater and then it re-gasifies LNG in the vaporizer. The cold cycle gas from the vaporizer is recompressed and then reheated first in the recuperator and then in the fired heater. The expander turbine is used to drive both a compressor and an electric generator. Air, nitrogen, or helium is utilized as closed cycle gas resulting in the power generation up to 50 MW.

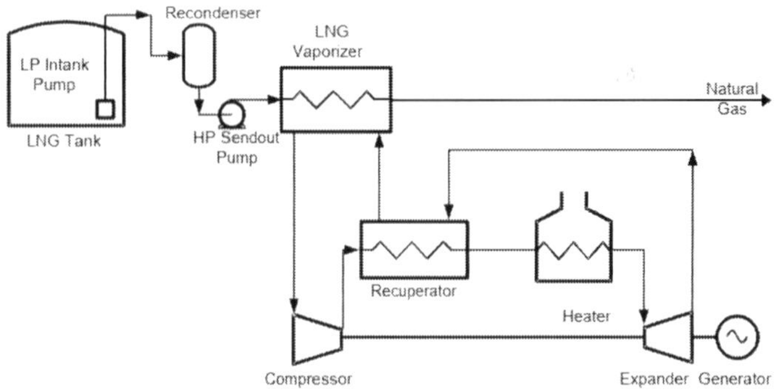


Fig. 4 - Schematic for Power Generation with Closed Cycle Gas Turbine

## HEAT RECOVERY FROM THE POWER PLANT

### *Once-Through Seawater*

The seawater used for steam condensing in the Combined Cycle Gas Turbine (CCGT) increases in temperature. Figure 5 illustrates two different thermal configurations for the heat recovery from the warm seawater. The use of seawater received from a power island for LNG re-gasification can reduce the amount of the required seawater ("Before Using" - Fig. 5A). The heat recovery

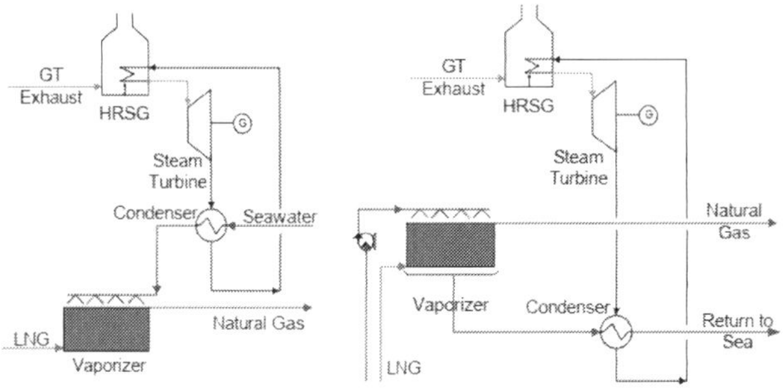


Before Using (A) After Using (B)

Fig. 5 – Integration Scheme with Once-Through Seawater

from the power plant can also make ORVs practical at a cold seawater location [9].

Conversely, the cold seawater from the LNG vaporizer(s) can be used for steam condensing. The advantages of the integration scheme as shown in Fig. 5b: "After Using (B)" are:

- A reduction in the amount of seawater intake (Normally, the amount of seawater is not determined by LNG vaporization duty, but the condenser cooling duty.);
- The enhancement of the power output by achieving a low condensing pressure with the cold seawater, which is colder than the seawater.

These types of thermal integration can share the seawater lift facility, which requires high capital costs. As a result, this integration leads to the reduced capital cost of the integrated project.

### *Low Pressure Steam*

Steam extracted from the steam turbine can be used as a thermal energy source to vaporize LNG in a dual heat SCV or a separate heater. In the event Low Pressure (LP) steam is not available from the power plant, SCVs can be operated by using the installed submerged burner.

Figure 6 illustrates the schematic diagram of a SCV modified to use LP steam as well as natural gas. This modified dual heat SCV reduces equipment costs, as well as space requirements, as compared to costs needed to install a separate SCV and heater. After the LP steam is condensed in the modified SCV, the condensate is returned to the power plant steam cycle. The performance penalty for this option is in the form of the reduced power output by the steam cycle; since LP steam is extracted from the power plant, less steam is available to the steam turbine to generate electricity. It is likely that the fuel cost savings for LNG re-gasification is more than the loss of power generation, resulting in overall economic improvement [5].

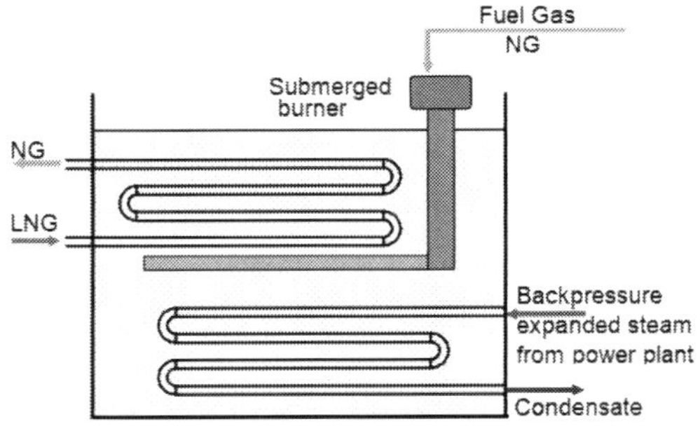


Fig. 6 - Dual Heat Source SCV

### *Utilization of Exhaust Gas from HRSG*

The exhaust gas produces steam at Heat Recovery Steam Generator (HRSG) and is vented through the stack. The temperature of exhaust gas in the stack is about 110°C. When the size of the gas turbine is small, generally HRSG is not installed. The exhaust gas temperature without HRSG is about 480°C. This means that the high valuable energy is exhausted. This exhaust gas can be used as a heat source for an absorbed type heating and chilling system for building cooling and heating (HVA/C).

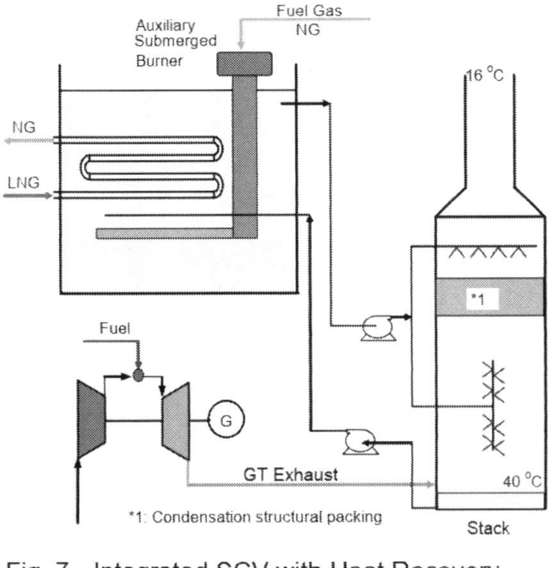


Fig. 7 - Integrated SCV with Heat Recovery from Exhaust Gas Stack

Distrigaz operating the Zeebrugee terminal (Belgium) commercially utilizes the hot turbine exhaust gas by extracting as much heat from it as possible in order to gasify LNG. Fig. 7 shows the schematic diagram of the innovative SCV integrated with the gas turbine in the Zeebrugee terminal. This concept dramatically increases its total energy utilization factor (106.8 %) [10].

## COMBINATION OF HEAT AND COLD ENERGY RECOVERY

An elaborate and theoretically sophisticated patent was awarded to Mandrin [2] in July 1977, and it expired in 1994. This patent claimed the combination of the following three process configurations (Refer to Appendix 1).

- GT inlet air super chilling (normal chilling plus methanol injection);
- Superheating regasified gas by the heat absorbed from a stack;
- Cold energy recovery from working fluid (Freon R-22).

This process is the basis of the current hypothesized thermal integration method. The process covers the cold power Rankin Cycle, heat recovery from the stack, and GT inlet air super chilling. However, this integration scheme has not yet been built. Since the process does not cover the heat integration with the steam condenser circulation water (CW) in a CCGT, this case study will focus on the thermal integration of LNG cold energy recovery and heat recovery from a CCGT with the steam condenser CW.

### *GT Inlet Chilling and Steam Condensing*

The heat recovered from the GT inlet air chilling is far less than the LNG vaporization heat duty. In order to meet the heat balance, other alternative heat sources are required for the LNG vaporization. Since a large amount of heat must be released from a steam condenser, the heat integration allows the use of this heat for LNG vaporization.

The concept behind the patented inventions (US Patent 5457951 and 6374591) involve heating up the HTF (after GT inlet air chilling) by heat exchange through steam condensing. The warmed HTF vaporizes cold liquid (LNG or LPG). The HTF is chilled by heat exchange through LNG cold liquid vaporization. The HTF absorbs heat from the air intake gas turbine while GT inlet air is chilled simultaneously. Heat recovery from the power plant and from the LNG cold energy supply is the same within the thermal integrated process. Again, HTF is warmed by the GT air chiller and is further heated-up by steam condensing.

Both processes propose that the LNG cold energy is used to cool the steam turbine during low ambient temperature conditions. Condenser cooling CW increases the steam turbine efficiency and output.

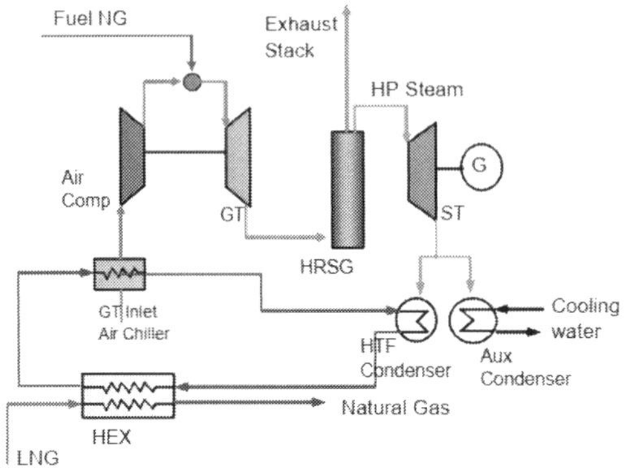

Fig. 8a - Conceptual Diagram of Direct Heat Recovery from Condenser [12]

Figures 8a and 8b show the direct condensing steam (as a result, warming of the HTF) and direct warm up of the HTF at a stack, respectively. The direct steam condensing concept has been applied to both patented inventions. The inventions claim that the auxiliary condenser will only be operated when the LNG terminal is not operating. The remainder of the heat released from the condenser was considered in the inventions. A system design needs to be refined to insure independent functionality of the LNG terminal and power island when the exchange system is not operable (e.g. maintenance, failure, etc.)

A recently patented invention [12] adds a stack heat recovery, which can be applied to the simple cycle. When the HTF heat recovered from the GT inlet air chilling is not enough for LNG vaporization, the HTF can be heated at the single cycle exhaust stack as shown in Fig. 8c. The heat recovery concept from the GT exhaust stack has been developed for the LNG regasification at Cove Point LNG RT and has been used since 1976. The GT inlet air chilling can be obtained by either single or double HTF closed loops. The modified HTF loop for LNG vaporization and GT inlet air chilling can be applied to both heat integrations, steam condensing and stack heating, as shown in Fig. 8c.

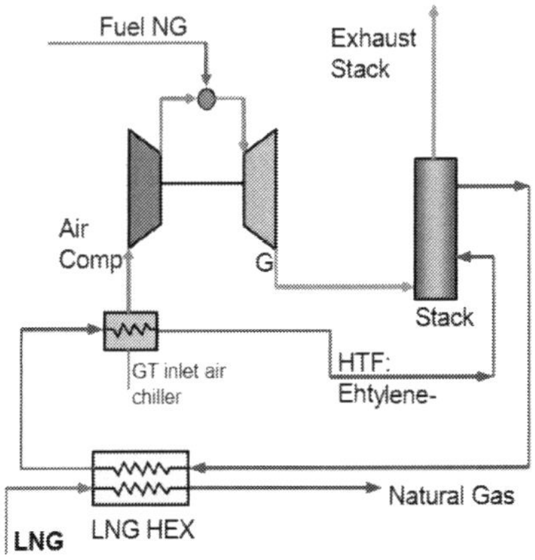

Fig. 8b - Conceptual Diagram of Direct Heat Recovery from GT Exhaust [12]

### *Steam Condenser Circulation Water*

A wet cooling tower or once-through seawater system is used to condense the steam in order to keep vacuum expansion in the steam turbine. The operating pressure and steam expansion efficiency in the steam condenser is dependent on the condensing system. Cold energy from the vaporizing LNG can also be used to cool steam condenser CW. In other words, the HTF can be heated-up by the steam condenser CW.

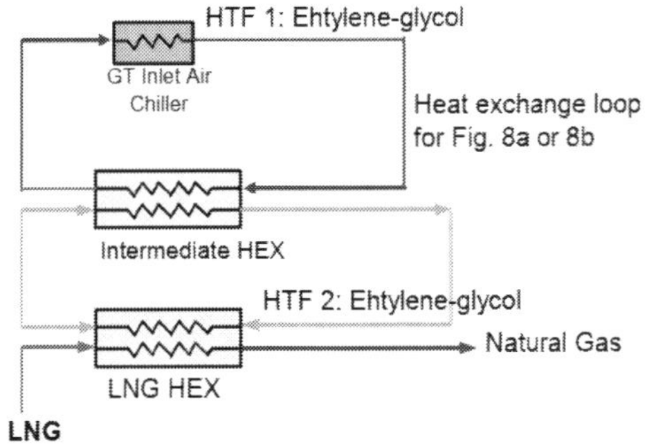

Fig. 8c - Conceptual Diagram of Double Heat Exchange Loop [12]

Another commercially available integration process is the GT inlet air chilling and heat gain from the steam condenser CW [13]. Since the heat obtained from the GT inlet air chilling is insufficient for the LNG vaporization, the heat balance is provided by the steam condenser CW. The basic concept in heating HTF is similar to the previous processes. However, this process obtains heat from the steam condenser CW, while the previous processes acquire heat from the heat exchange directly at steam condenser. The HTF is then heated to a desired temperature before it is returned to the LNG vaporizer.

The HTF is warmed-up by the GT inlet air chilling. The HTF is further warmed-up by the cooling water discharged from the once-through steam condenser or wet type cooler. Before the HTF is provided to the LNG vaporizer, two types of heat integration are available: HTF is heated-up by the steam condenser CW “Before” the condenser heat exchange or “After” the condenser heat exchange.

Figure 9a illustrates the HTF heated by the steam condenser CW, which is chilled before going into the condenser. The low CW temperature results in low condensing pressure as low as 50 mbara [14]. This improves the steam turbine power output. However, the integrated HTF loop for the GT inlet air chilling and heat absorbing from a condenser CW (Fig. 8) increases the flow into the GT inlet air chiller, resulting in higher frictional loss and increase in its size. The HTF flow control is also difficult.

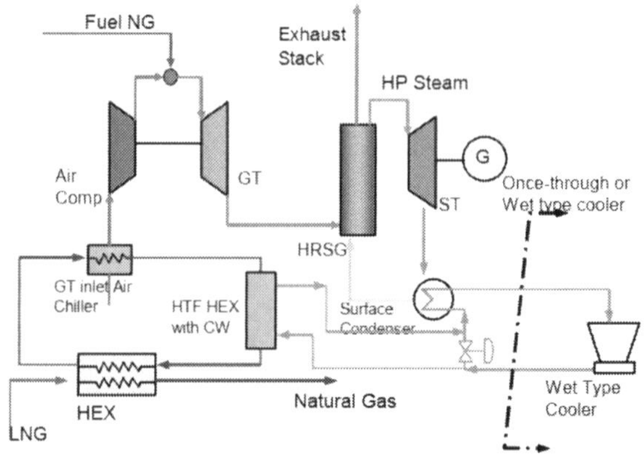

Fig. 9a - Heat Recovery from "Before" Condenser Circulation Water [13]

Figure 9b shows the HTF heated by the steam condenser CW after the condenser, which reduces the cooling load of a cooling tower. The advantage of this configuration compared to the Fig. 9a is the attainment of a high temperature of HTF. As a result, the HTF flow rate reduces.

Figures 9c presents the cooling of the condenser discharge, which adds heat to the HTF to assist in vaporization. With a once-through type steam condenser at the CCGT, LNG cold energy can be used to reduce the necessity of adding additional pumps to mix seawater, heated through the condenser, with unheated seawater in order to minimize thermal pollution. A similar process has recently been in operation at the AES Andres LNG RT and power, Dominican Republic.

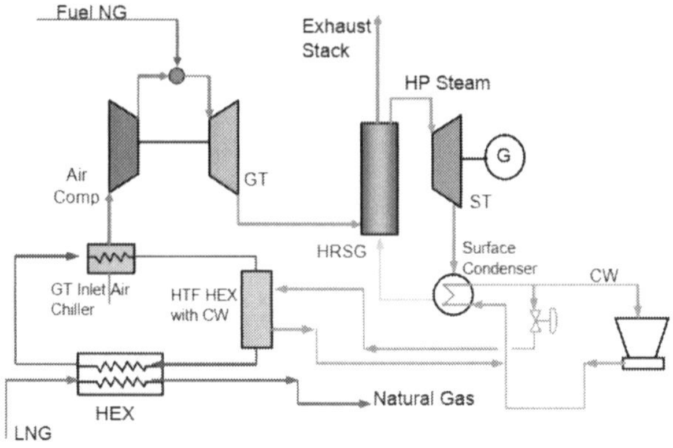

Fig. 9b - Heat Recovery from "After" Condenser Circulation Water [13]

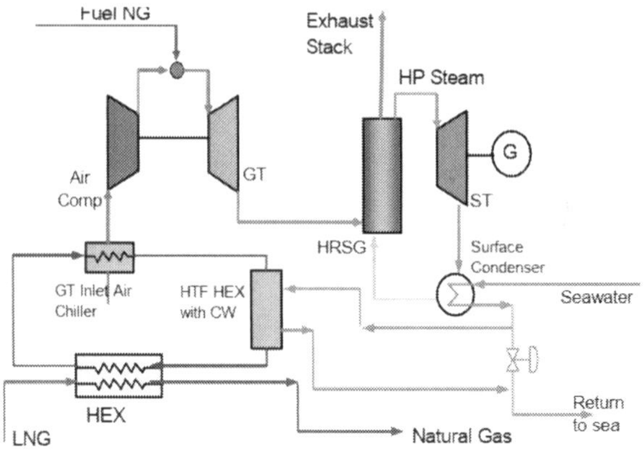

Fig. 9c - Heat Recovery from Once-Through Condenser Circulation Water [13]

### *GT Inlet Chilling and Super-Heating*

The condenser heat duty of 1,000 MW CCGT is about 2,000 MMBtu/h while the LNG vaporization heat duty is about 650 MMBtu/h for 1,000 MMsfcd sendout (equivalent to 7.5 mmtpa of LNG import). GT inlet air chilling is a function of ambient temperature and relative humidity. When ambient temperature and relative humidity decrease, the amount of heat available is reduced. The heat obtained from GT inlet air chilling is about 80 MMBtu/h for a frame 7FA GT when ambient temperature is 80°F and relative humidity is 70%, based on 45°F inlet air chilling. This heat is equivalent to the duty required to vaporize about 150 MMscfd.

A single HTF loop consisting of an HTF heat exchange with condenser CW is not suitable for multipurpose LNG RT, which provides regasified LNG to city gas as well as natural gas to CCGT fuel because the estimated HTF flow rate is high and a GT inlet air chiller becomes large. Utilizing additional vaporizers and superheaters (superheating heat exchangers) with the recovered heat from CCGT allows for easier controlling in the vaporization process than in an HTF heat-up with a condenser CW, as shown in Figs. 8 and 9.

The following configurations are new integration processes developed by KBR. The processes integrate GT inlet air chilling, heat recovery from the condenser CW, and superheating of the vaporized LNG with the condenser CW. The processes shown in Fig. 10 use an independent HTF for GT inlet air chilling. Since the GT inlet air chilling load varies depending on ambient conditions, the LNG inlet flow is controlled and the vaporized LNG is then warmed by a super heater (or a trim heater), whereby, heat is obtained from the condenser CW. For the most part, the LNG is vaporized by the heat recovered from the condenser CW.

The process configuration shown in Fig. 10a presents the "after condenser heat exchange" with the condenser CW, while Fig. 10b shows the "before condenser heat exchange". In Fig. 10a, the configuration reduces the cooling load with a wet type cooler. This configuration also reduces the cooling water requirement with the once-through type condenser. The cold NG is superheated by cooling water. Superheaters and LNG vaporizers can save heat loads of wet type coolers or they can reduce the amount of seawater for once-through type cooling systems. This is a practical integration of a high capacity multipurpose LNG RT and CCGT.

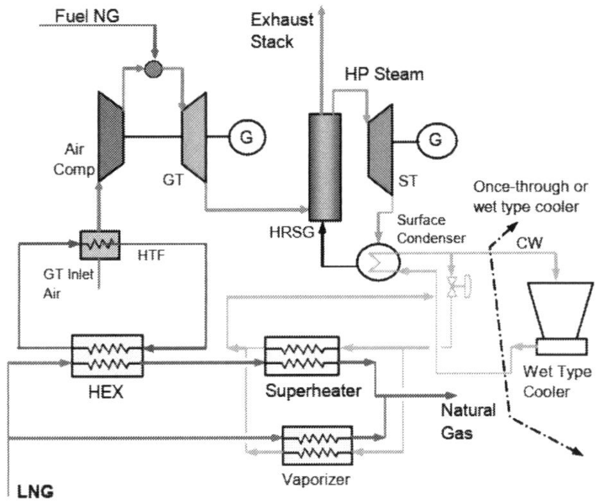

Fig. 10a - Integrated Super Heating and Vaporization with "After" Condenser Circulation Water

Fig. 10b configuration reduces the temperature of the steam condenser CW, resulting in an increase in steam power output due to the high steam expansion efficiency (a low vacuum expansion pressure can be achieved). The cold NG is superheated by cooling water. The required CW can also be reduced.

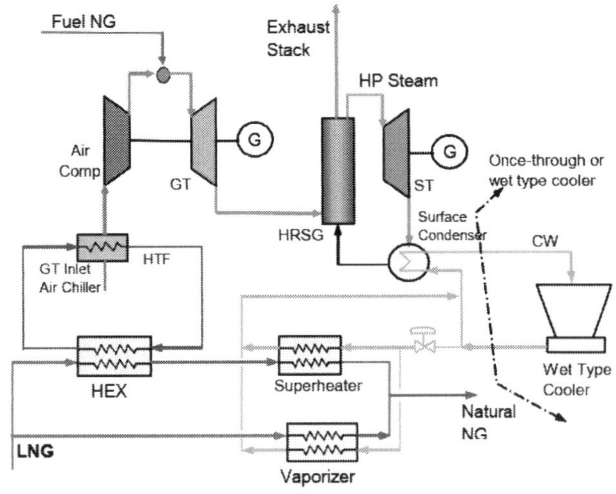

Fig. 10b - Integrated Super Heating and Vaporization with "Before" Condenser Circulation Water

The process configuration, shown in Fig. 10c, consists of an HTF loop for the GT inlet air chilling, double tube bundle vaporizers, and superheater(s). The HTF is heated by the CW.

As a result, as the temperature of CW decreases, steam turbine output increases. The double bundle heat exchanger design prevents freezing of the HTF. The HTF (normally an ethylene glycol-water solution or water) is heated by the GT inlet air chilling. The HTF vaporizes the intermediate fluid inside the double tube bundle (Appendix 2). The vaporized intermediate fluid is condensed at the surface of the tube bundles while the LNG is vaporized inside the tube bundles (absorbing heat). The double tube bundle vaporizers have been used at Cove Point LNG terminal since 1976. This type of vaporizer is useful in the application of cold power generation because both processes of the intermediate fluid condensing and LNG vaporization can occur inside the LNG vaporizer. The cold vaporized LNG can then be warmed up to the design point by the condenser SW.

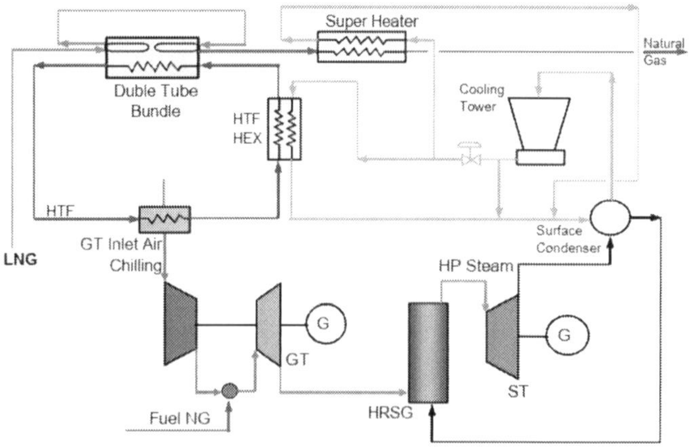

Fig. 10c - Integrated HTF Heating and Super Heating with Double Tube Bundle Vaporizer

## SYNERGY OF INTEGRATED LNG AND POWER

The benefits of integrated LNG regasification and power plant vary depending on project conditions as well as on the level of integration (or configuration). KBR is constantly advancing integration technology in LNG and Power to its highest level of efficiency, reliability and profitability. The synergies of this integration include:

- Reducing costs in construction and operation
- Optimizing the use of the LNG facility (storage & re-gasification)
- Increasing competitiveness in energy markets (gas & power)

Since previous studies have only revealed the qualitative synergies that cannot provide a project decision tool, KBR developed a specific value added simulation model for quantifying integration benefits with various parameters greatly affecting project economics. These involve LNG handling volume, power output, hourly send-out patterns, ambient temperatures and humidity variations, and various economic factors, depending on project conditions. This program allows KBR to both optimize the thermal integration between two plants and to quantify the synergy of integration. With an integrated project, the fuel consumption for LNG re-gasification is reduced by utilizing waste energy from the power facility. The integration also increases power generation by using the LNG receiving terminal as a low temperature heat sink. Reduced staffing with a combined project (estimated at 20-25%) results in reduced operating and maintenance costs.

KBR integration technology also provides the following benefits:

- Shared facilities
- Enhanced summer peaking capacity
- Intangibles

Each plant requires common facilities that can be shared between two plants when designed and developed simultaneously as an integrated project. Examples are seawater intake, seawater treatment, plant air, industry water, fire fighting, and a power generating system, as well as an emergency generating system. Sharing facilities between two plants is very important to maximize the benefits of an integrated project. The estimated cost savings due to sharing facilities is dependent on the capacity of both plants and the geological features of the location. Integration can boost electric power generation during the summer, when the electric power demand is high, by inlet air chilling (See Fig. 1).

In addition to the benefits quantified in the economic evaluation, the following benefits result from integration:

- Single point of contact for a contractor
- Integrated Engineering & Construction
- Reduced emissions (per unit of energy exported)
- Economy of scale

## CONCLUSIONS

The following are the conclusions of this case study:

The GT inlet air chilling is a commercially proven method that generates more power from a given machine. It is a viable integration method that uses the LNG cold energy. In order to maximize the synergy in thermal integration between an LNG RT and a power plant, combination of thermal integrations is required.

The LNG provides cold energy to the HTF as a result of GT inlet air chilling. In other words, the HTF obtains heat from GT inlet air chilling. The heat balance can be provided by heat exchange with direct steam condensing, HTF heating by a steam condenser CW, or superheating / additional vaporizers.

In relation to the direct heat exchange at the steam condenser, some limitations do exist on the power output because of the heat balance between LNG vaporization and heat release from a steam condenser.

GT inlet air chilling can provide heat to the HTF, which is further heated to a desired temperature before it is returned to the LNG vaporizer. As the vaporization load increases, the HTF flow rates into the GT inlet also increases. As a result, equipment of the GT inlet air chilling increases, resulting in an increase in air friction into the GT compressor suction, as well as equipment costs.

The newly developed process by KBR provides the GT inlet air chilling and superheating the cold vaporized LNG by heat exchange with the steam condenser CW, resulting in a reduction of HTF flow into the GT inlet air chilling. This process does not increase GT compressor suction losses. The heat required for the additional LNG vaporization can be obtained from the CW. As a result, the developed process will be productive and economical efficient integration, which in turn contributes to its integration synergy.

## ACKNOWLEDGEMENT

The authors wish to thank KBR (Kellogg Brown and Root, Inc.) for the sponsorship of this study and for the permission to prepare and publish this paper.

## REFERENCES CITED

1. Smith, K.L. et al.: "Advanced Technology Combined Cycles", GE Power Systems GER-3936A, April 2000.
2. Mandrin, C.: "Process and Apparatus for Evaporating and Heating Liquefied Natural Gas," U.S. Patent No. 4,036,028, July 1977.
3. Shimizu, A., Utamura, M., Hoizumi, S., and Komatsu, H.: "Gas Turbine Intake Air Cooling Appratus," U.S. Patent No. 5,626,019, May 1997.
4. "Bibliograph of Turbine Inlet Cooling (TIC) Publications," Turbine Inlet Cooling Association, www.turbineinletcooling.org/resources/ticbiblio.pdf.
5. Cho, J.H., Ebbern, D., and VanSickle, R.: "Economic Benefits of Integrated LNG-to-Power", presented at the AIChE 2003 National Meeting, New Orleans, 30 March 2003.
6. Ooka, I., and Hisazumi, Y.: "System for Generation of Electricity by Utilization of Heat Exchange Between Liquefied Natural Gas and Intermediate Heat Medium," U.S. Patent No. 4,320,303, March 1982.
7. Nozawa, R.: "Liquefied Natural Gas-Freon Electricity Generation System," U.S. Patent No. 4,422,298, December 1983.
8. Nozawa, R.: "Liquefied Natural Gas-Refrigerant Electricity Generating System," U.S. Patent No. 4,429,536, February 1984.
9. Durr, C., Coyle, D., Patel, H., and Cho, J.H.: "Advanced Design of LNG Import Terminals", presented at the AIChE 2003 National Meeting, New Orleans, 30 March 2003.
10. Gerard, A.: "40 MW Combined Heat and Power (CHP) Plant at Zeebrugge LNG Terminal: Innovations and Performances" Proceeding, 17th Congress of the World Energy Council, Houston, Texas, September 13-18, 1998.
11. Johnson, P.C. and Toombs, A.E.: "Improved Liquefied Natural Gas Fueled Combined Cycle Power Plant," U.S. Patent No. 5,457,951, October 1995.
12. Johnson, P.C. and Toombs, A.E.: "Liquefied Natural Gas Fueled Combined Cycle Power Plant and a LNG Fueled Gas Turbine Plant," U.S. Patent No. 6,374,591 B1, April 2002.
13. Wen, H., Narula, R.G.: "Method Apparatus for Vaporizing Liquid Natural Gas in a Combined Cycle Power Plant," U.S. Patent No. 6,367,258 B1, April 2002.
14. Chiu, C.H.: "LNG Receiving Terminal with a Combined Cycle Gas turbine (CCGT) Power Plant", Proceeding, Eleventh International Conference on Liquefied natural gas, Birmingham, UK, July 3-6, 1995.

## Appendix 1: Schematic Diagram of US Patent 4,036,028

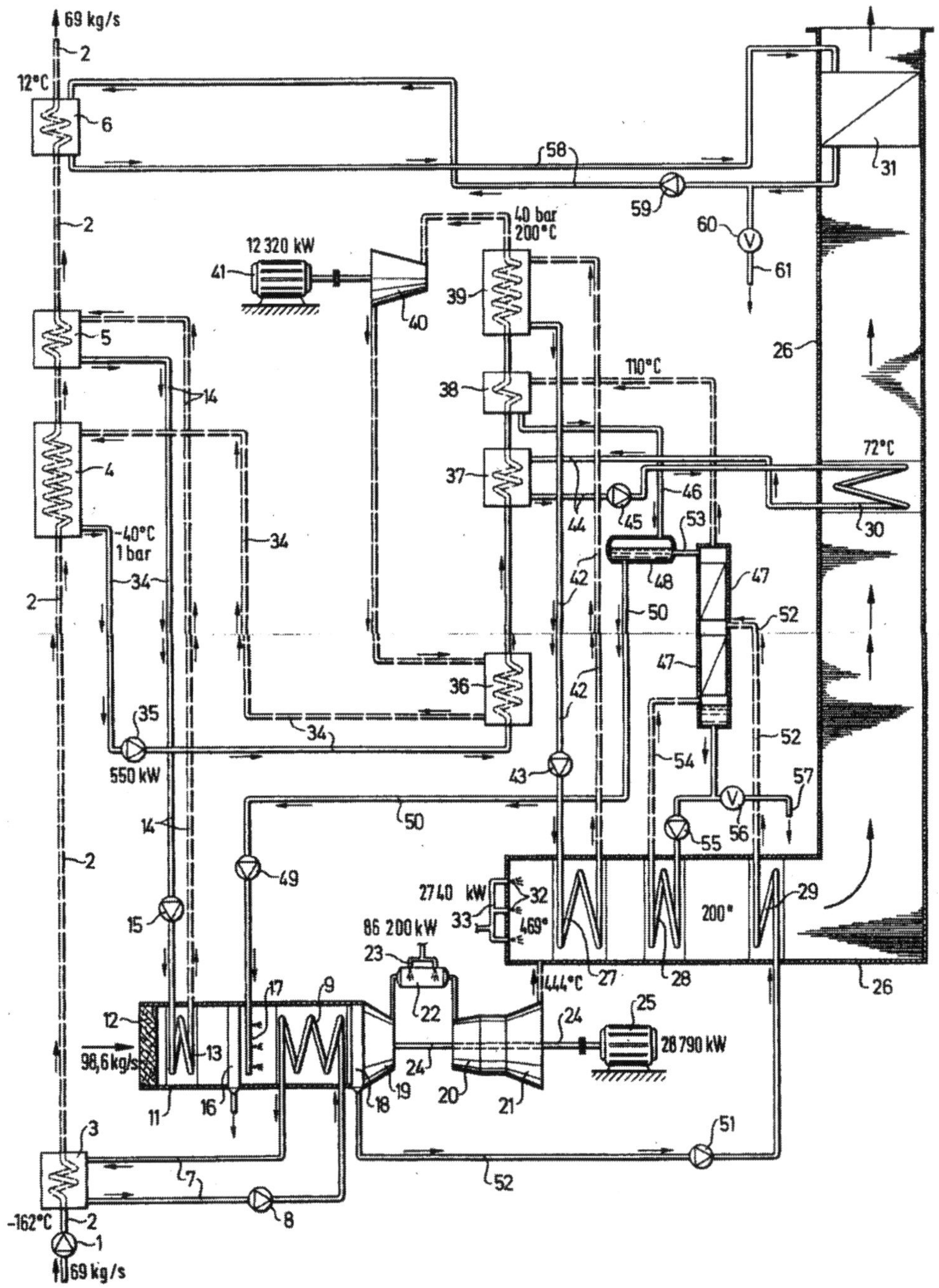


## Appendix 2: Schematic Drawing of Double Tube Bundle Vaporizer

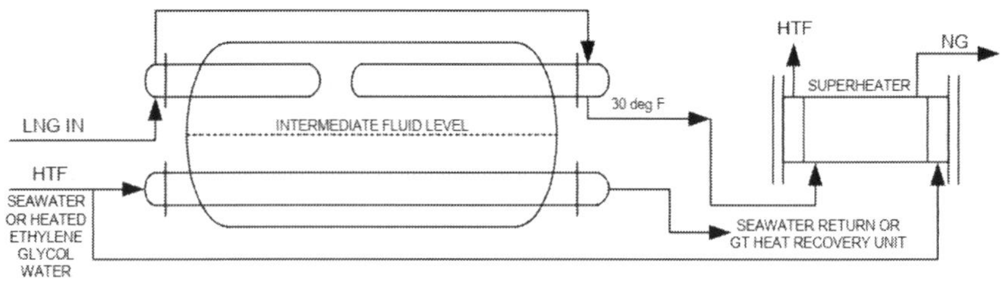

# Two-Phase LNG Expanders Replace Two-Phase Joule-Thomson Valves

*Chen-Hwa Chiu*
*Sr. Technology Advisor, LNG*
*ChevronTexaco, USA*
*Chen-Hwa.Chiu@ChevronTexaco.com*

*Hans E. Kimmel*
*Vice President R&D*
*Ebara International Corporation, USA*
*hkimmel@ebaraintl.com*

*Prepared for Presentation at AIChE Spring National Meeting*
*4th Topical Conference on Natural Gas Utilization*
*New Orleans, LA, April 2004*

## INTRODUCTION

As part of the liquefaction process of natural gas, the condensation phase requires the gas to be at a very high pressure, after which the pressure is reduced by expansion across a Joule-Thomson valve. During this pressure reduction process, the condensed fluid is flashed isenthalpically across the valve and into a phase separator vessel [1]. These pressure drops across a Joule-Thomson valve create very high turbulent friction losses, wasting energy as frictional heating of the fluid.

In contrast, a two-phase expander achieves the same pressure reduction while shaft power is generated from a near-isentropic expansion process. This keeps the frictional heating in the process to a minimum [2]. Thus, the substitution of two-phase flow expanders in place of Joule-Thomson valves in refrigeration applications increases overall system efficiency and can generate electric power in lieu of previously wasted two-phase energy [3].

In addition to the condensate phase of LNG production, two-phase expanders may also be used in cryogenic distillation processes. Dissolved nitrogen in pressurized LNG degrades the quality of the liquid. Expansion of low grade pressurized LNG across a two-phase expander extracts nitrogen from the LNG by vaporization while reducing the temperature of the remaining liquid. This Nitrogen rejection process is essentially cryogenic distillation, which upgrades the quality of the LNG while increasing overall production through temperature reduction [4].

## DESIGN METHODOLOGY

Two-phase expander design concepts essentially follow existing single-phase turbine and expander technology. The hydraulic energy of the pressurized fluid is converted to electric energy by first transforming it into kinetic energy, then into mechanical shaft power and finally to electrical energy through the use of an electrical power generator [2].

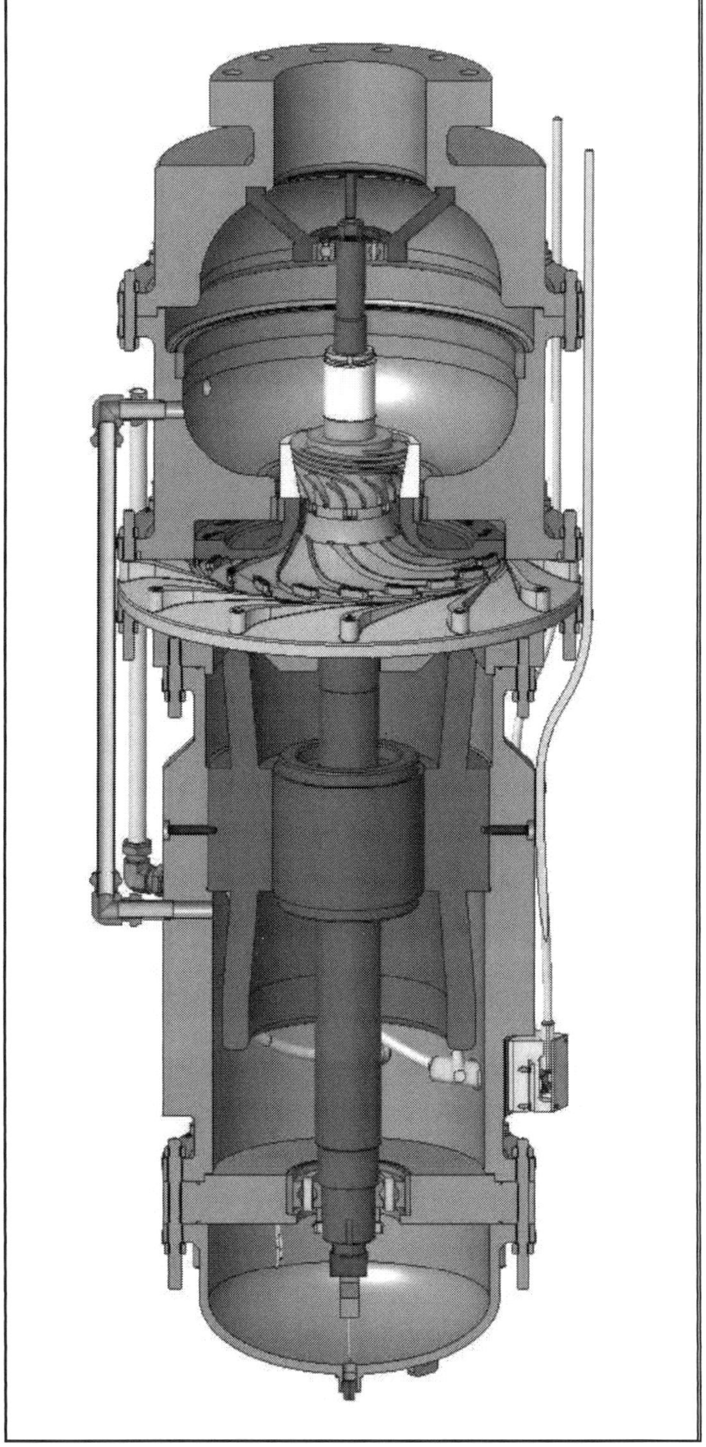

**Figure 1. Ebara Two-Phase Expander Cross Section**

Electric power generation in cryogenic technology consists of two basic concepts. First, the generator may be an air cooled machine, mounted outside the cryogenic liquid and coupled to the expander as described by T. Bond [3], or the generator may be submerged in the cryogenic liquid and mounted integrally with the expander on a common shaft [4]. In this second application, the cryogenic induction generators must use insulation systems specifically developed for use in submerged windings. These have been found to have significantly superior dielectric and life properties than the systems commonly use on air-cooled machines. [5].

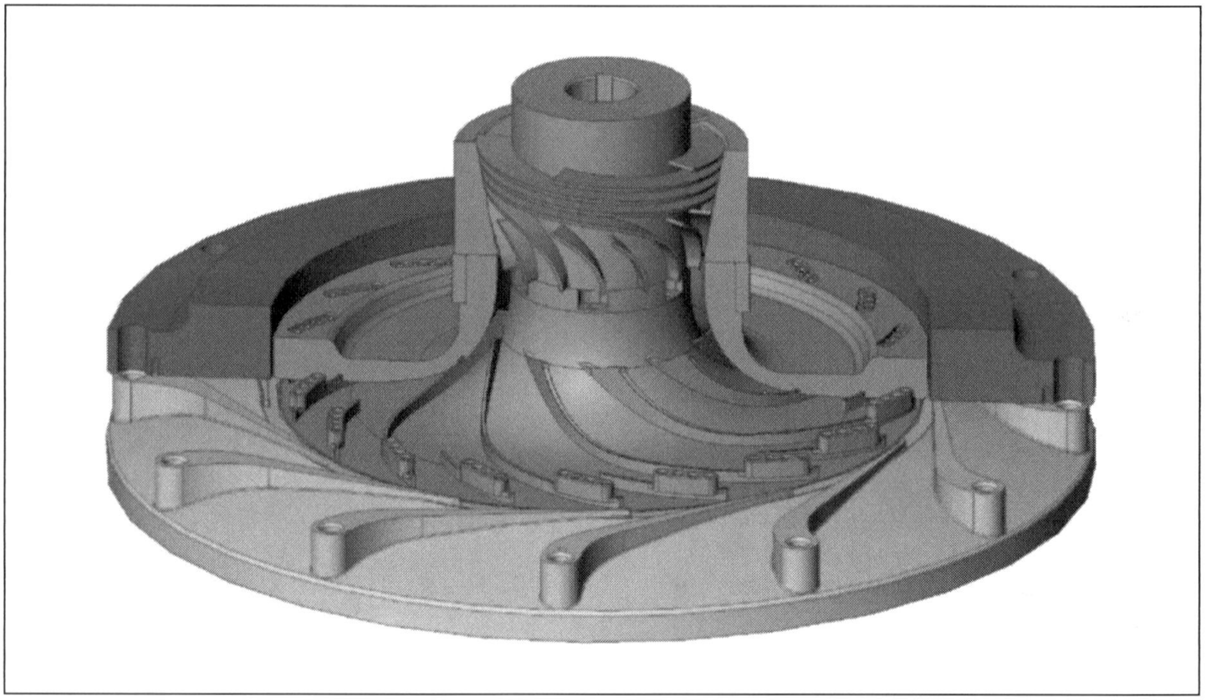

**Figure 2. Two-Phase Hydraulic Runner Assembly**

Figure 1 depicts the cross section of a typical Ebara International Corporation cryogenic two-phase submerged expander. As may be seen, the induction generator and the hydraulic expander are mounted on a common shaft. The expander consists of a nozzle ring generating the rotational fluid flow, a radial inflow reaction turbine runner and a two-phase jet exducer. Figure 2 depicts an enlarged cross section of the two-phase hydraulic runner assembly. Figure 3 shows the mounted jet exducer from the top.

To generate symmetric flow conditions, Ebara's two-phase expander is designed with a vertical rotational axis. Expanders with horizontal rotational axis [1,2,3] generate asymmetric flow conditions can result in higher vibration levels. In addition, the flow direction of the Ebara two-phase expander is upward to take advantage of the buoyant forces of the vapor bubbles, to stabilize the flow and to minimize flow induced vibrations

[6]. The hydraulic assembly is designed for continuously decreasing pressure to avoid any cavitation along the two-phase flow passage [7].

**Figure 3. Two-Phase Jet Exducer**

## FIELD EXPERIENCE USING TWO-PHASE FLOW EXPANDERS

To upgrade LNG by extracting undesired Nitrogen, two Ebara two-phase expanders were installed at the Krio Polish Oil & Gas Nitrogen Rejection Plant (Figure 4) in Odolanow, Poland. Figure 5 shows the two-phase expander assembling and Figure 6 the expander installation.

Figure 4 Nitrogen Rejection Plant in Odalanow, Poland

Figure 6 Expander Installation

Figure 5 Assembling the Expander

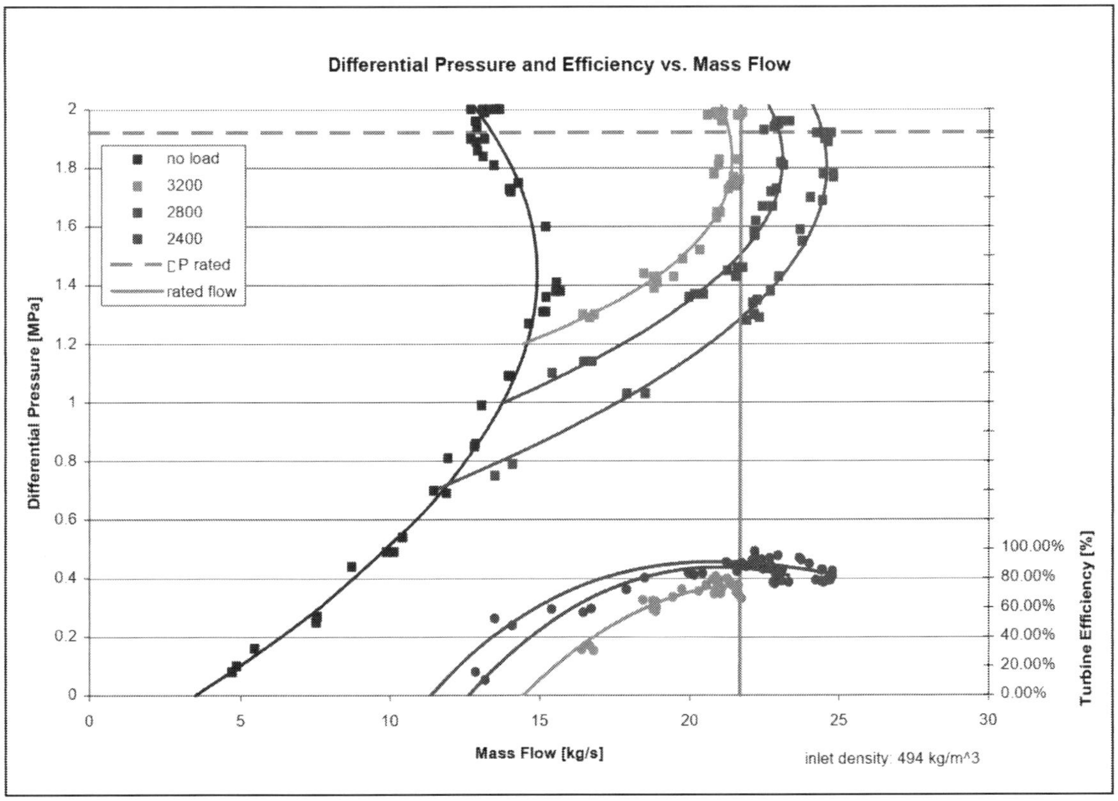


**Figure 7. Two-Phase Hydraulic Performance**

Two-phase expanders operate at variable speeds in order to adjust to the changing mass flows and pressure conditions of the plant. Figure 7 presents the hydraulic performance of the two-phase expander with differential pressure and efficiency versus mass flow for different rotational speeds.

Efficiency is defined as the ratio of electrical power generated divided by the hydraulic power input. Hydraulic power input is the product of mass flow and differential pressure. The solid vertical red line depicts rated mass flow and the horizontal dotted line indicates rated differential pressure.

Figure 8 presents the tested and measured points from Figure 7, but revised to depict differential pressure and efficiency versus volumetric flow. The specific volume of fluid changes during the expansion process across the expander and is shown in Figure 9 as a quadratic interpolation between the inlet and outlet specific volume. The horizontal dotted line in Figure 8 indicates the rated differential pressure. The solid red parabolic curve in Figure 8 indicates the rated volumetric flow. As may be noted, the rated volumetric flow increases with increasing differential pressure due to the expansion of the two-phase fluid.

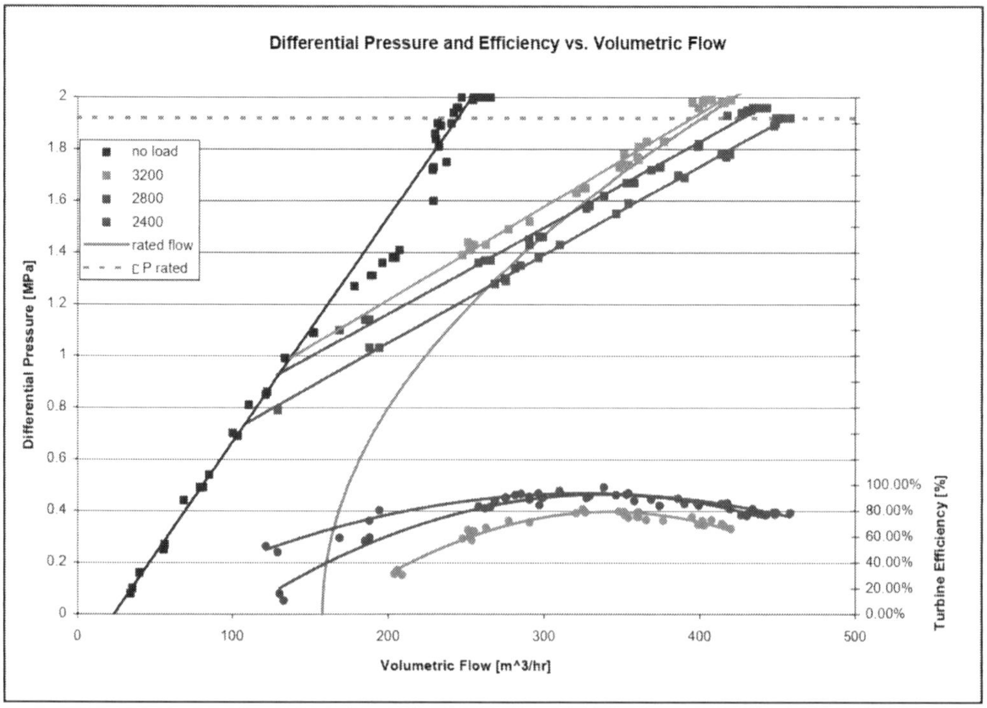


**Figure 8. Two-Phase Hydraulic Performance**

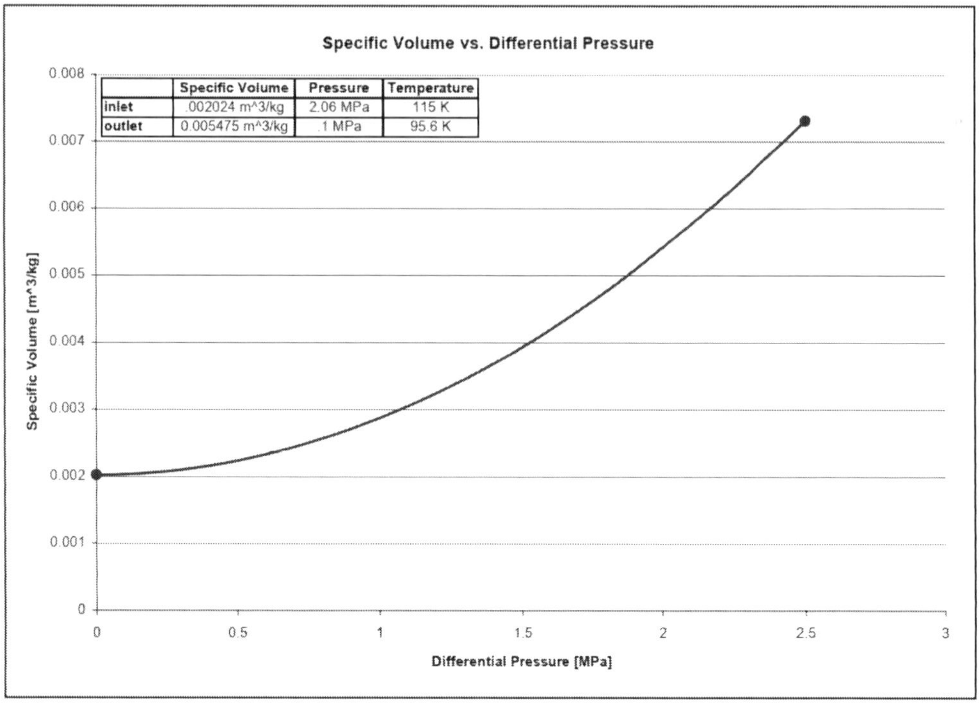


| | Specific Volume | Pressure | Temperature |
|---|---|---|---|
| inlet | .002024 m^3/kg | 2.06 MPa | 115 K |
| outlet | 0.005475 m^3/kg | .1 MPa | 95.6 K |

**Figure 9. Two-Phase Flow Specific Volume**

## BENEFITS ANALYSIS

The ultimate objective of all improvement efforts in liquefaction process technology is to increase economic benefits while simultaneously reducing the environmental impact of LNG production. Replacing inefficient valves with power recovery turbines for single-phase or two-phase expansion reduces the related plant power input by factors ranging from 3 to 5 times the expander's power output.

It is significantly more efficient to cool the LNG stream using two-phase expanders instead of single-phase expanders or other devices. Figure 10 presents the LNG temperature drop versus the power output for the previously described two-phase expander. As may be seen, the cooling effect on the LNG stream is directly related to the power output.

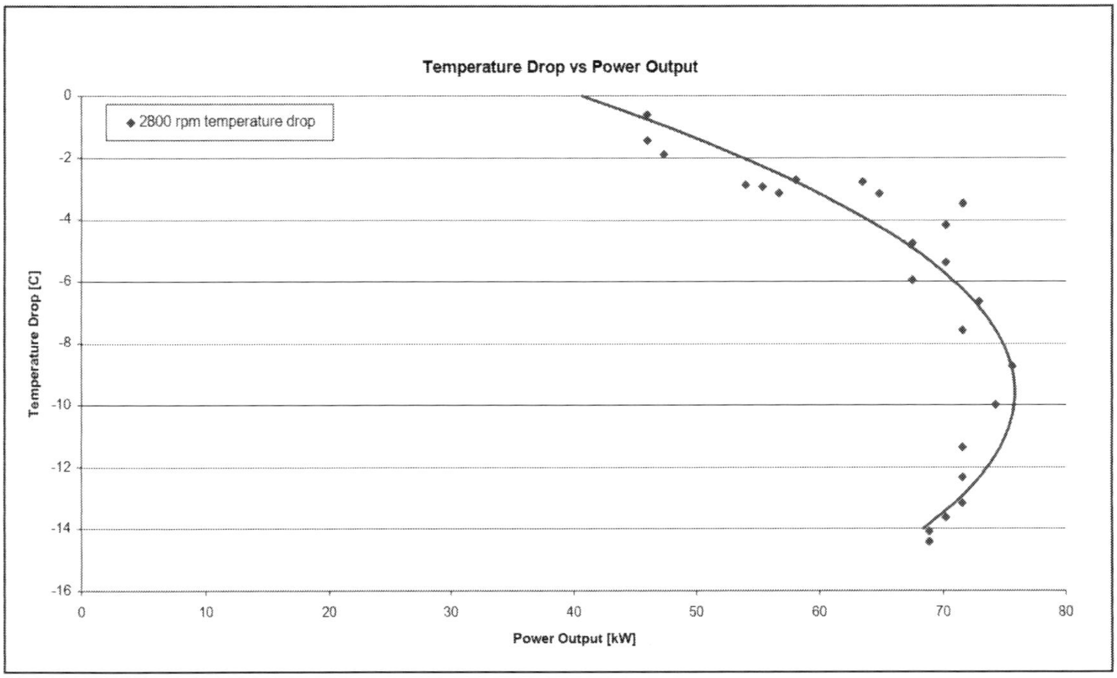


**Figure 10. Cooling Effect of Two-Phase Expansion**

## SUMMARY

Ebara two-phase expanders have been operating at the nitrogen rejection plant in Odolanow, Poland in excess of one year. Throughout that period regular inspections have shown no incipient failures in bearings or materials, vibration levels have been less than 20% of API 610 allowable limits and the ability to produce high grade LNG using less power and in an environmentally friendly manner has been conclusively demonstrated.

## REFERENCES CITED

1. Ross, Greg; Davies, Simon; Vislie, Geirmund; Hays, Lance; "Reductions of Greenhouse Gas Emissions in Oil and Gas Production and Processing by Application of Biphase Turbines", 1996, www.mpptech.com/techpp/tech_home.htm
2. Hays, Lance, "History and Overview of Two-Phase Turbines", International Conference on Compressors and Their Systems", Institution of Mechanical Engineers, London, 1999.
3. Bond, Ted, "Replacement of Joule-Thomson Valves by Two-Phase Flow Turbines in Industrial Refrigeration Application", 2000, www.mpptech.com/techpp/tech_home.htm
4. Chiu, Chen-Hwa; Kikkawa, Yoshitsugi; Kimmel, Hans E.; Liu, Yu-Nan; "New Cryogenic Two-Phase Expanders in LNG Production", Third Topical Conference on Natural Gas Utilization, AIChE 2003 Spring National Meeting, New Orleans, Louisiana, USA
5. Shively, R.A. and Miller, H., "Development of a Submerged Winding Induction Generator for Cryogenic Applications", in Proceedings of the IEEE Electrical Insulation Conference, Anaheim, California, 2000.
6. Gebhart, Benjamin et al.; "Buoyancy-Induced Flows and Transport" Hemisphere Publishing Corporation, New York, 1988, ISBN 0-89116-728-5
7. Hsu, Peter; Evrensel, Cahit A.; Kimmel, Hans E.; "Cavitation-Free Cryogenic Two-Phase Expanders", CAV 2003, Fifth International Symposium on Cavitation, Osaka, Japan, November 2003

## BIBLIOGRAPHY

1. Boom, R.W. et al.; "Experimental Investigation of the Helium Two Phase Flow Pressure Drop Characteristics in Vertical Tubes", Proc. ICEC 7, pg 468-473, 1978
2. Elliott, D.G.; Weinberg, E; "Acceleration of Liquids in Two-Phase Nozzles", Technical Report no.32-987, Jet Propulsion Laboratory, California Institute of Technology, Pasadena, California, USA, 1968
3. Filina, N.N.; Weisend II, J.G.; "Cryogenic Two-Phase Flow: Applications to large-scale systems", Cambridge University Press, 1996, ISBN 0-521-48192-9
4. Vislie, Geirmund; Davies, Simon; Hays, Lance; "Further Developments of Biphase Rotary Separator Turbine", Paper presented at IBC Separation Systems Conference, May 1997, Oslo, Norway.
5. Perlmutter, M.J.; Kimmel, H.E.; Chiu, C.H.; Paradowski, H.: "Economic and Environmental Benefits of Two-Phase Expanders", Proceedings LNG 14, 21-24 March 2004, Doha, Qatar.
6. Finley, Christopher D.; "Continuously Transient Operation of Two-Phase LNG Expanders" Fourth Topical Conference on Natural Gas Utilization, AIChE 2004 Spring National Meeting, New Orleans, Louisiana, USA.

# Development of High Speed Pumps

*Motoyasu Ogawa, Research Project Engineer*
*Nikkiso Co Ltd,*
*16-2, Noguchi-Cho, 2-Chome, Higashimurayama, Tokyo 189-8520, Japan*

*Bill Haesloop*
*Marketing Director*
*Nikkiso Cryo Inc,*
*4661 Eaker Street, Las Vegas, Nevada, 89031, USA*

*Prepared for Presentation at AIChE Spring National Meeting*
*4th Topical Conference on Natural Gas Utilization*
*New Orleans, LA, April 2004*

## Introduction

Nikkiso Co Ltd and Nikkiso Cryo Inc have developed a number of variable frequency driven high speed submerged electric motor pumps for cryogenic service more specifically for LNG service. The advantage of high speed operation is reduced size and cost.

Three models of high speed cryogenic pumps were developed for specific applications. Interestingly, they have found to be applicable pumping services that were not anticipated at the time of initial development.

Examples of development of these three pumps and their performance are presented together with results of full scale cryogenic tests using LNG.

## Development history of High Speed Cryogenic Pumps

Nikkiso decided to develop a series of high speed submerged electric motor pumps for cryogenic hydrocarbon liquefied gas service in the year 2000. As part of that decision, Nikkiso developed the Vehicle Fueling Pumps in 2001, (See AIChE Spring Meeting 2002 for presentation); the Micro Turbine Fuel Feed Pumps in 2002 and High Speed Vaporizer Feed Pumps in 2003.

The pump specifications and flow range of these units are presented on Table 1 and graphically on Figure 1 where they are compared to the performance that would be expected from the same hydraulic configurations operating a 2 pole motor speed at 60 Hz input frequency.

The range enclosed with the black line in Figure 1 is a typical performance range in Nikkiso cryogenic pumps. For the high speed LNG pumps developed, it is characteristic that the pumps have smaller capacity and higher discharge head than the typical pump.

Table 1 Pump Specification

| | Vehicle Fueling Pump | Gas Turbine Feed Pump | High Speed Vaporizer Feed Pump |
|---|---|---|---|
| Model | 62020-P15R | 62007L3-P15R | 60528L8-P200F |
| Capacity | 1-24 (m3/h)<br>4-105 (gpm) | 1-10 (m3/h)<br>4-44 (gpm) | 8-24 (m3/h)<br>35-105 (gpm) |
| Design Head | 240 (m) / 790 (ft) | 450 (m) / 1480 (ft) | 2370 (m) / 7780 (ft) |
| Motor Output | 15 (kW) / 20 (BHP) | 15 (kW) / 20 (BHP) | 200 (kW) / 270 (BHP) |

## Vehicle Fueling Pumps (VFP)

The VFP (Vehicle Fueling Pump) was originally developed for installation in LNG Vehicle Fueling Stations for fueling buses and large trucks powered by diesel style engines using LNG as fuel. The VFP is installed in its own suction vessel which is usually insulated by vacuum jacketing to control heat leak due to the large differential temperature between LNG and ambient air of about 380 °F Δ (difference between -320 °F and +60 °F)

The VFP must enable a wide range of flows to accomplish fueling of 1, 2 or 3 vehicles simultaneously each being fueled at 30 gpm and 100 psi. The way in which this wide range of flows is met is by the use of a variable frequency drive for the input power to the submerged electric motor of the pump. This allows operation at a variety of speeds as required to best suit the flow requirement.

In addition to being able to satisfy a wide variety of duty requirements using variable frequency input power, the operation at higher speeds will permit the higher pressure duties to be met without using additional numbers of impellers in series thus making the pump significantly shorter with resultant reduced cost. The final pump design is quite compact (Figure 2).

Even though the hydraulic design of the VFP pump was derived from the well proven high speed Nikkiso hermetically sealed "non-seal" pumps in ambient temperature service, the packaging of the entire unit was modified to suit installation in a cryogenic pump. This involved new and novel features as part of the overall development program. The mechanical design was evaluated using Rotor Dynamic Analysis and Finite Element Analysis (FEM). The hydraulic performance was verified by building a prototype unit and conducting performance tests with water followed by LNG tests at cryogenic conditions. Based on the prototype test results, the production unit was built and performance verified with LNG.

The VFP has already found a number of applications that were not originally anticipated. The unit has been installed as an LNG feed pump for the low flow high pressure pump as part of the El Paso Energy Bridge Regasification ship test program and is being considered as part of a number of LNG re-liquefaction processes where small low flow, high pressure pumps are required. Further, it is also in use to feed LN2 to a shell and tube heat exchanger used to control LNG temperature at a cryogenic test facility.

## Gas Turbine Feed Pumps

Recently, the Japanese Electric Power Utility Code was amended and a number of the electric power markets were liberalized to allow individual users to generate electric power for their own usage. Many of these independent power generators are driven by mini gas turbines (2 to 20 megawatt). LNG is used a fuel for these mini-turbines and the feed flow rate is smaller and head higher than the design parameters for the VFP.

The steps in development of this pump were essentially the same as described above for the VFP, that is, using proven components from the Nikkiso non seal pump product line where possible, except that a new impeller hydraulic configuration was designed because one was not available from the non seal pump series that had a specific speed as low as required for this application. The impeller and diffuser design were accomplished using Computational Fluid Dynamics (CFD) analysis (Figure 3) and proven with prototype water and LNG testing. The pump consists of the 3 (three) stages low specific speed pump and is driven by the same submerged electric motor that is used for the VFP.

The effect of multistaging also had to be confirmed in the water test with the prototype pump as only a single stage was analyzed by CFD. The outcome of the water tests confirmed the CFD analysis prediction within the accuracy ranges and the production pump was completed based on the prototype test results. The completed pump is very small in comparison with the usual pumps (Figure 4).

Finally, performance under cryogenic conditions (using LN2) was proven with the production pump at the Nikkiso test facility in Japan. The developed pumps were first shipped in August 2002 and have been in satisfactory service since delivery. After that, a number of pumps were delivered to assorted clients with micro-turbine generators throughout Japan.

## High Speed Vaporizer Feed Pump

The High Speed Vaporizer Feed Pump was developed as a low flow high pressure pump for initial pipeline pressure buildup (line packing) in a shipboard regasification system. This delivery system is for off shore delivery of high pressure gaseous methane to an onshore pipeline system thus eliminating the need for onshore LNG storage tanks

The regasification system has been installed in a number of membrane type LNG carriers and this enables the LNG Vessel to discharge regasified LNG to a sub-sea pipeline from to an offshore mooring buoy to an onshore gas pipeline.

The pump had to be designed to endure under the severe acceleration and motion condition as shown Table 2.

Table 2 Acceleration Condition (g = Gravitational Constant)

| Regas Operations Condition | | Transit Seagoing Condition | |
|---|---|---|---|
| Vertical acceleration | 0.4 g | Vertical acceleration | 0.8 g |
| Transversal acceleration | 0.2 g | Transversal acceleration | 0.5 g |
| Longitudinal acceleration | 0.1 g | Longitudinal acceleration | 0.6 g |

The acceleration that the pump will experience due to motion of the ship could affect the balancing of the thrust loads on the bearings that is automatically performed by the Nikkiso balance drum system. Nikkiso has developed a computer program to calculate the bearing loads of the cryogenic submerged electric motor pumps which considers all internal flows, local pressures, fluid velocities and thrust loads acting on the surfaces of the rotating assembly. These fields were analyzed by entering the effect of the combined accelerations on these loads and numerically solving the differential equations together with a finite element analysis. Each dimension of the balance drum system was evaluated and determined by using this computer program to assure that the bearing loads are maintained at acceptable levels under all flow conditions with these levels of accelerations imposed.

The vaporizer feed pump is required to raise the LNG pressure to high levels (100 barg or 1450 psig for the regasification ship application) the normal approach using a centrifugal pump is to operate at 3600 rpm (2 pole electric motor synchronous speed using a 60 Hz power supply). A pump operating at this speed would require 20 impeller stages for this pressure rise.

However if the pump is operated at higher speeds the pressure produced per stage would be increased by the speed ratio raised to the $2^{nd}$ power. Using a power input frequency of 120 Hertz as controlled by a variable frequency drive, the pump synchronous speed is raised to 7200 rpm and the pump only requires 8 impeller stages to attain the 100 barg discharge pressure. This make the pump physical size only half the size of a comparable unit at normal 2 pole motor speeds (3600 rpm). Refer to Figure 5 for size comparison.

A rotor dynamic analysis including electrical and unbalanced magnetic force was performed to confirm the robust design and technical integrity because the high speed vaporizer feed pump is larger, has more stages, and significantly more power than the previously developed VFP and micro turbine feed high speed pumps mentioned above. The analyses were based on a complete model including all impellers, the motor rotor, and the rotating balance drum assembly and included a Stability Analysis and an Unbalance Response Analysis of the integrated system.

The rotor shaft in an induction electrical motor must be stiff enough to counter the unbalanced magnetic forces due to any offset between the magnetic center of the stator and the center of rotation of the rotor without excessive deflection.

Although it intuitively seems that the motor rotor shrink fitted to the motor shaft would significantly stiffen the assembly, the deflection of versus load were physically measured for the actual rotor/shaft assembly and it was shown that there was no meaningful increase of the stiffness of the rotor/shaft assembly over the motor shaft alone. The motor shaft diameter could not be increased to increase its stiffness without adversely affecting the magnetic characteristics of the motor so the motor shaft was shortened to bring the anti friction bearings closer together and reduce the span between the supports which raised the rotor/shaft assembly stiffness to an acceptable level.

Full performance string testing of the high speed pump and its variable frequency drive using LNG at full speed and power was carried out at Nikkiso's cryogenic pump test facility in Las Vegas, Nevada.

Following these factory performance tests and inspections, the Pump and its variable frequency drive were delivered to the client's specially constructed full scale regasification ship simulator in Trussville Alabama. The test platform was hydraulically powered to induce all the ship motions and accelerations expected at maximum sea conditions. The entire vaporization system (pumps, surge drum, sea water vaporizers and all piping, controls, valves, etc) was mounted on this platform and operated at all expected conditions while under the motions of the platform.

The high speed 200 BkW vaporizer feed pump (as well as two 1100 BkW Nikkiso supplied 3600 rpm 11 stage main vaporizer feed pumps) were fitted with proximity probe to determine shaft (rotor assembly) movement while operating under all these conditions. Pump performance and shaft movement was measured for the full range of platform motions as well as no motion. In the results, it was confirmed that the platform movement and accelerations did not influence the performance of the pump and the axial load balance system properly controlled the bearing loads under all the motion conditions.

## Conclusion

The technology of high-speed submerged electric motor pumps in cryogenic applications was very clearly demonstrated with three separate high speed pump developments ranging from 1.0 m3/h (4.4 gpm) though 25 m3/h (110 gpm) and powers from 15 BkW to 200 BkW. The major advantage of high speed pump operation is significantly reduced size and cost. Nikkiso plans to continue development of this high speed pump technology to include higher power and larger flow pumps in the future.

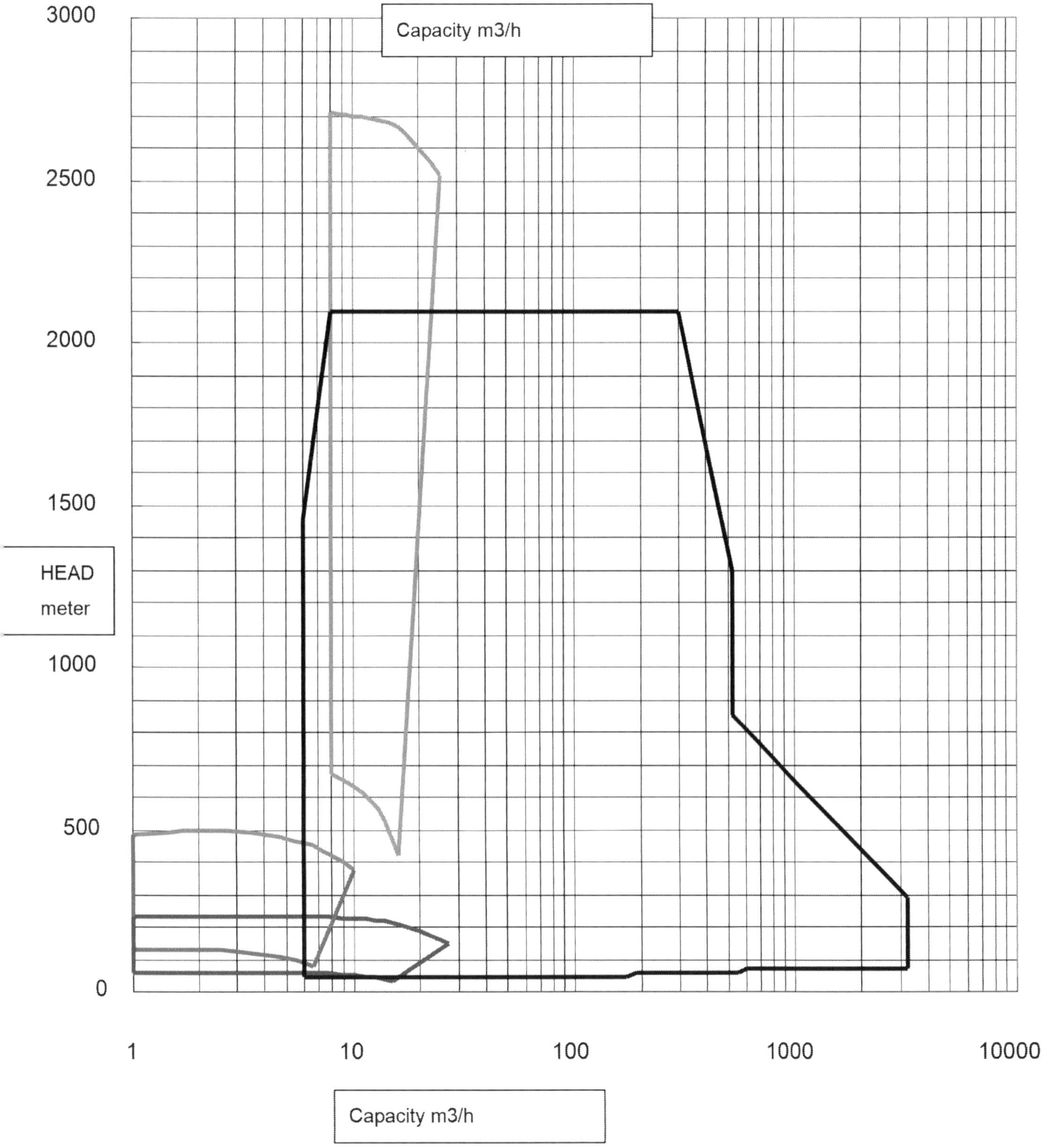

Capacity m3/h
3000
2500
2000
1500
HEAD
meter
1000
500
0
1
10
100
1000
10000
Capacity m3/h


Figure 1

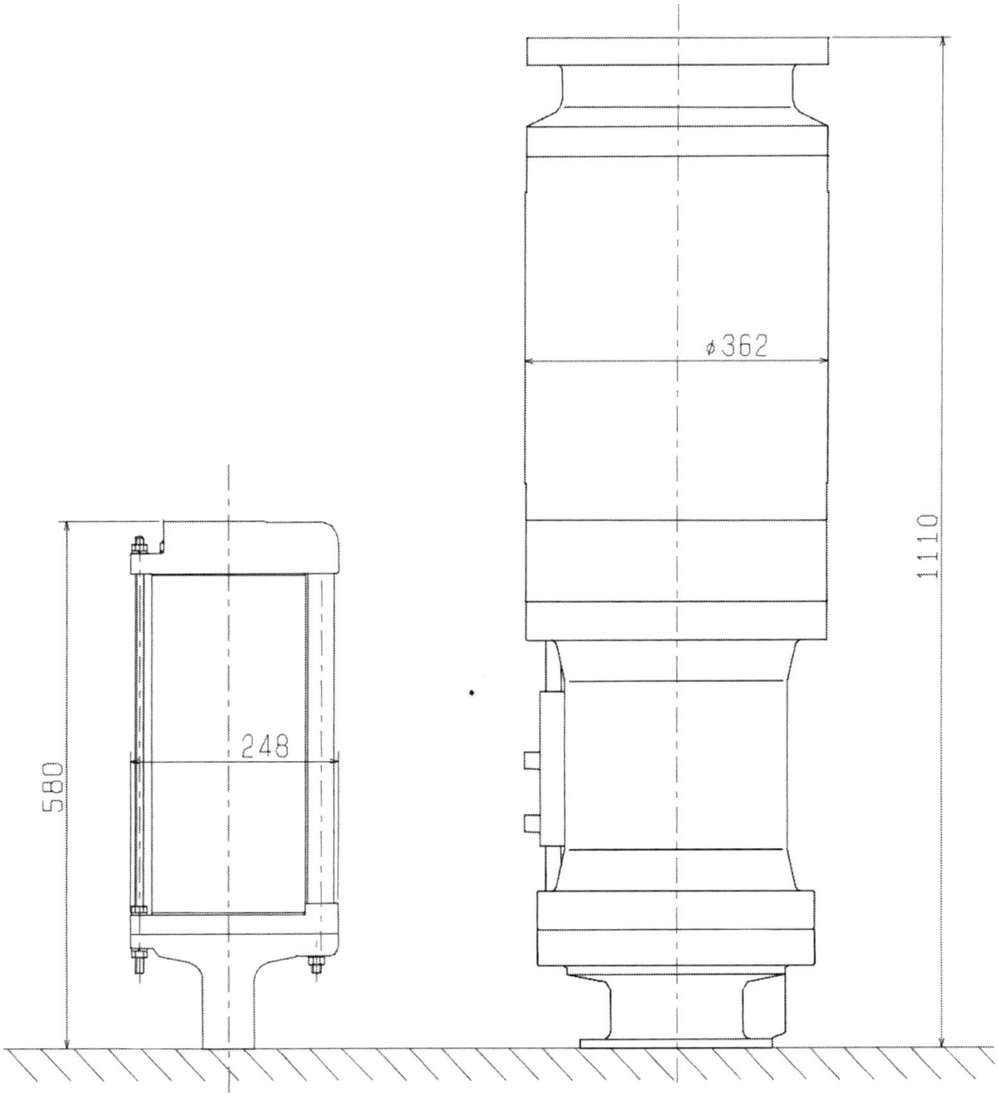


**Figure 2**

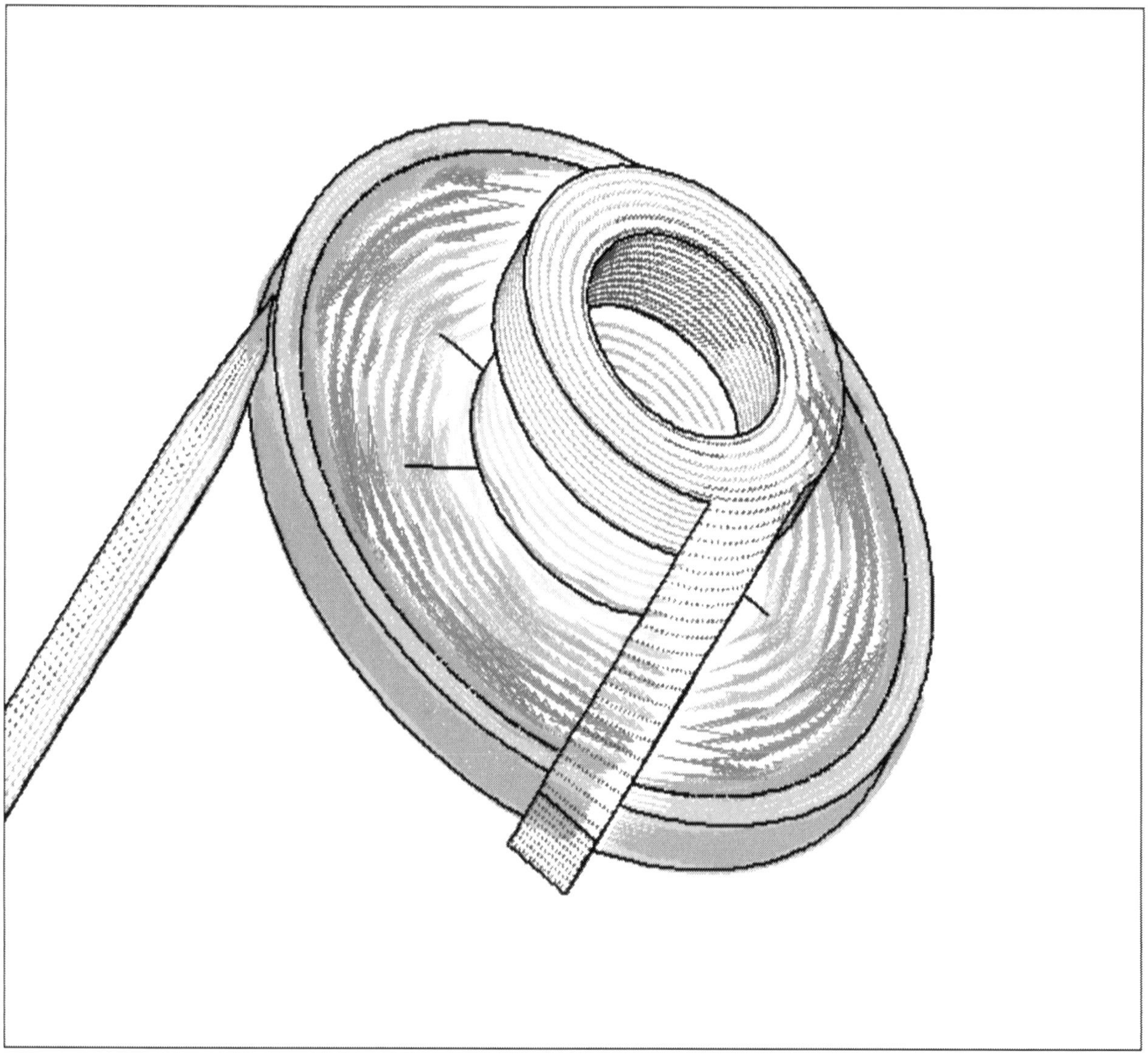

**Figure 3 CFD MODEL**

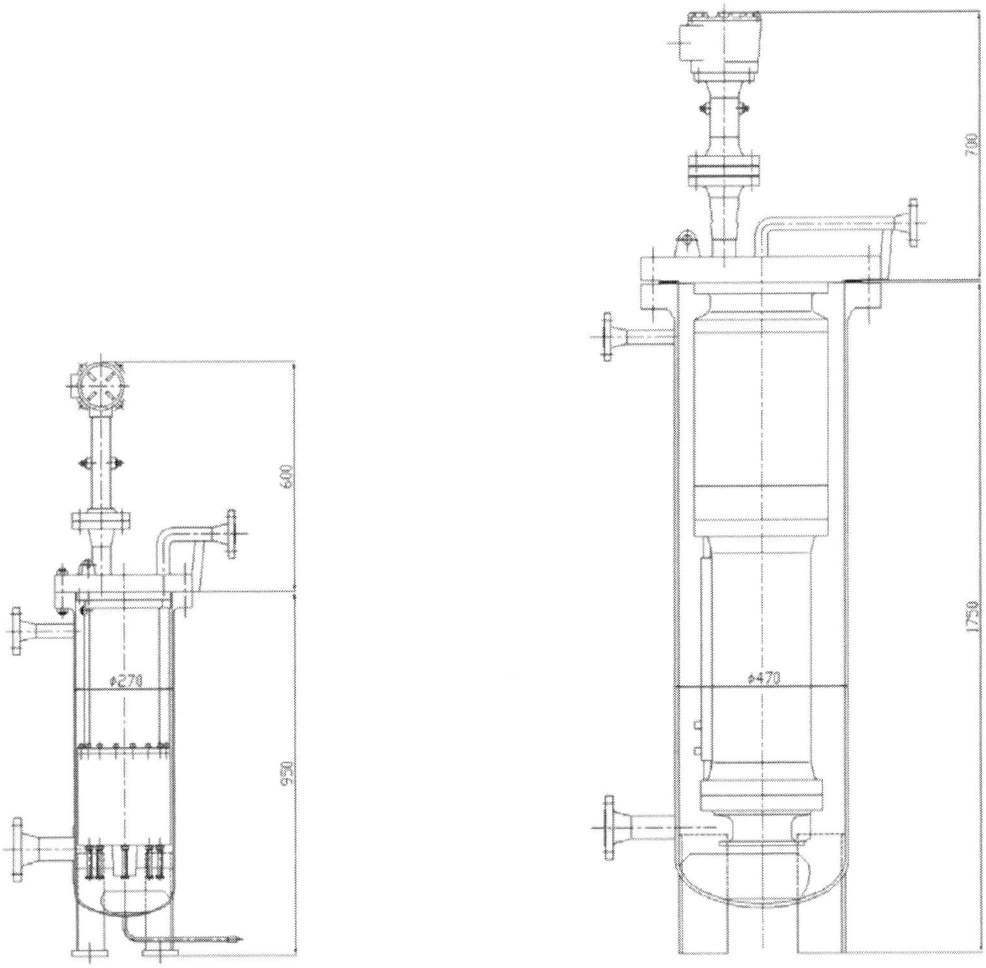


Gas Turbine Pump

Usual Pump

Figure 4

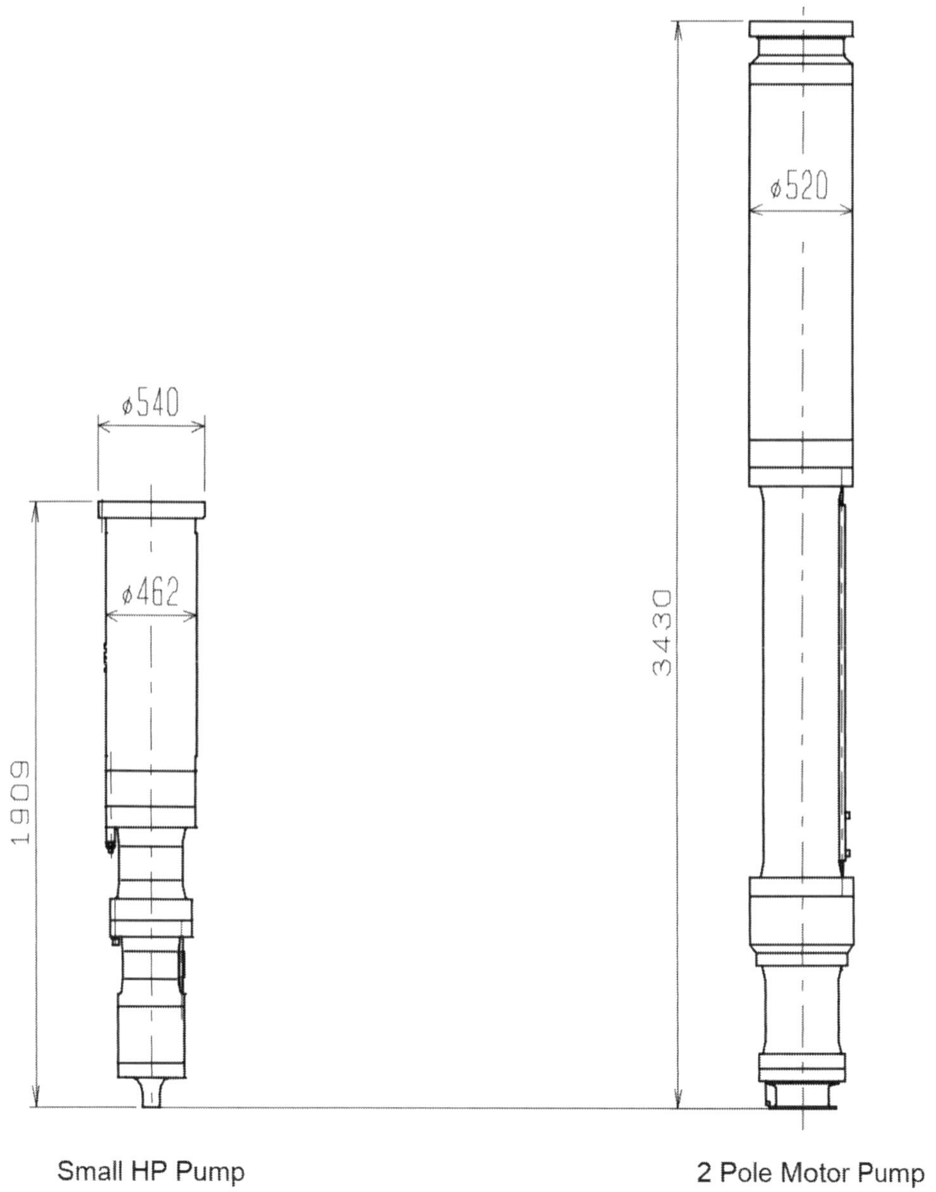


**Figure 5**

FIGURE 6

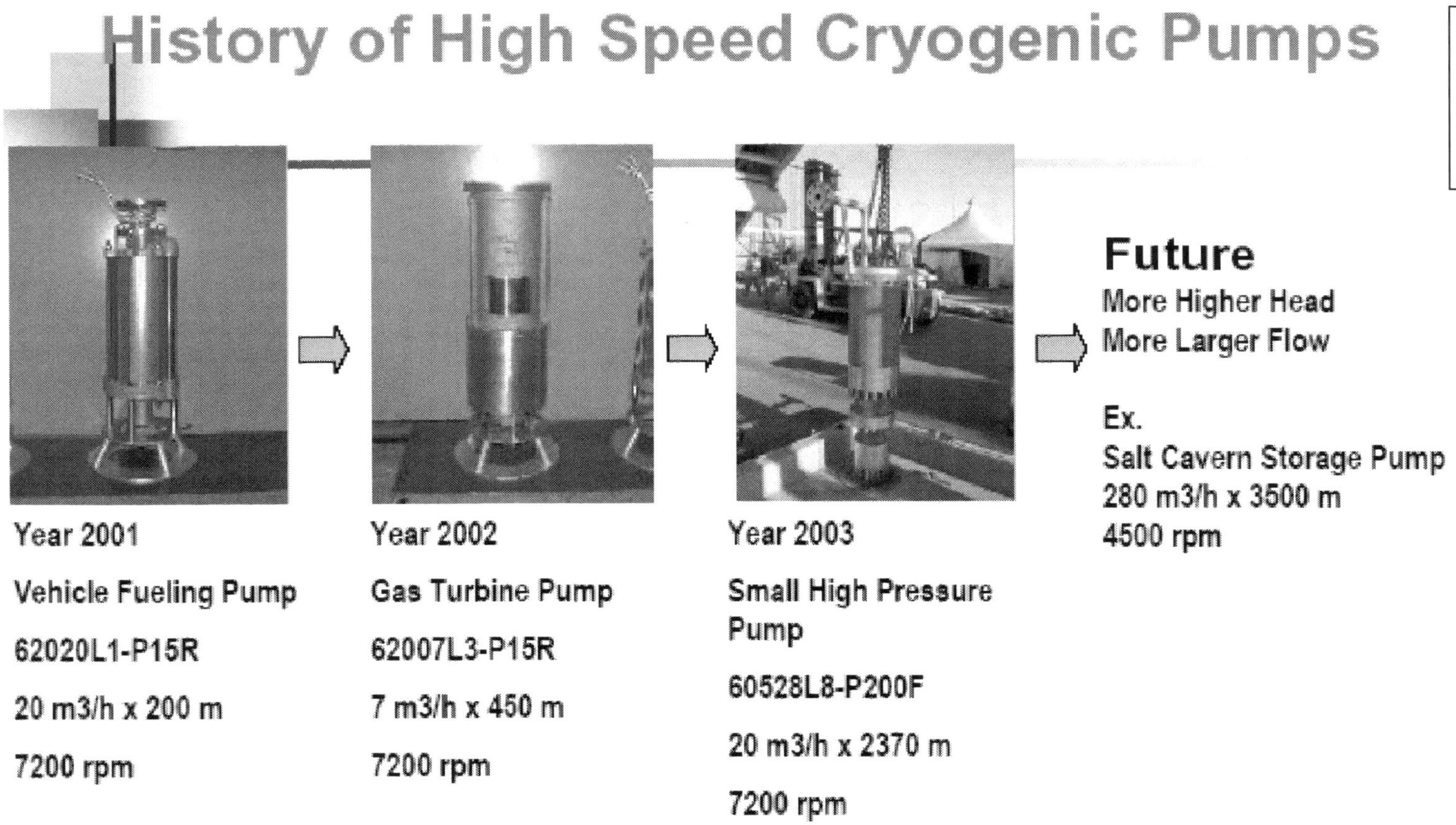
History of High Speed Cryogenic Pumps
Year 2001
Vehicle Fueling Pump
62020L1-P15R
20 m3/h x 200 m
7200 rpm
Year 2002
Gas Turbine Pump
62007L3-P15R
7 m3/h x 450 m
7200 rpm
Year 2003
Small High Pressure Pump
60528L8-P200F
20 m3/h x 2370 m
7200 rpm
Future
More Higher Head
More Larger Flow
Ex.
Salt Cavern Storage Pump
280 m3/h x 3500 m
4500 rpm

**FIGURE 7**

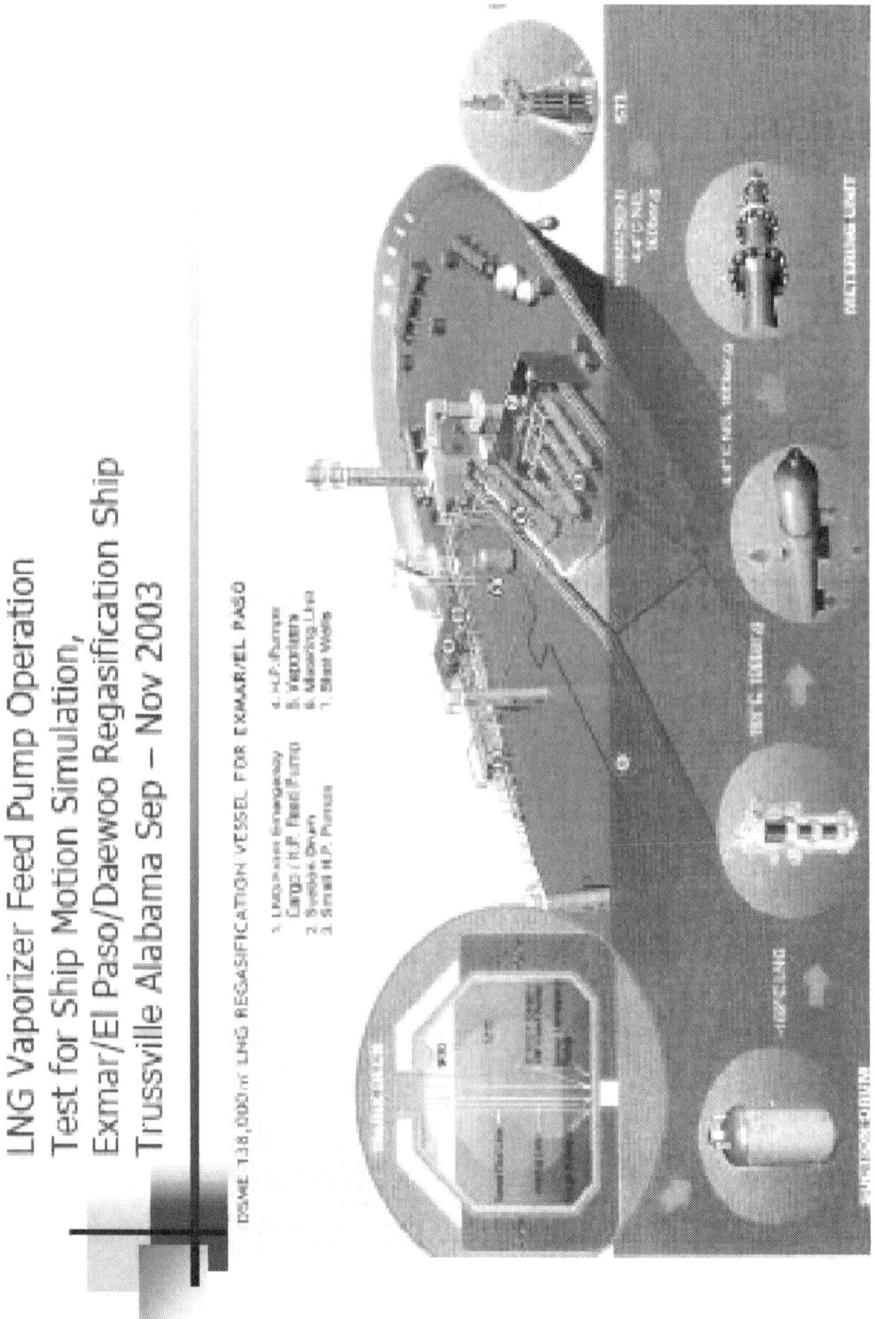
LNG Vaporizer Feed Pump Operation
Test for Ship Motion Simulation,
Exmar/El Paso/Daewoo Regasification Ship
Trussville Alabama Sep – Nov 2003

FIGURE 8

FIGURE 9

**FIGURE 10**

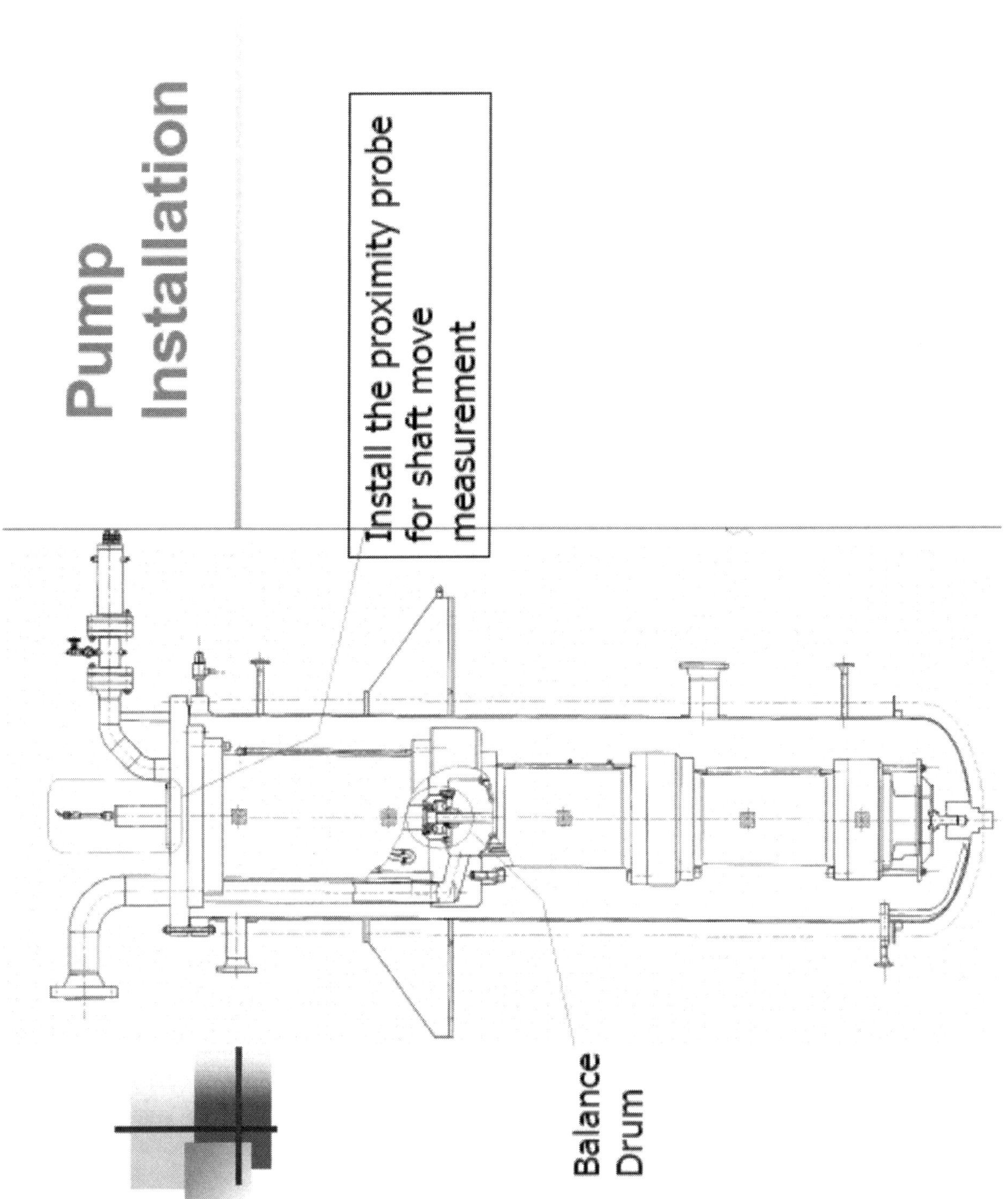
Pump Installation
Install the proximity probe for shaft move measurement
Balance Drum

FIGURE 11

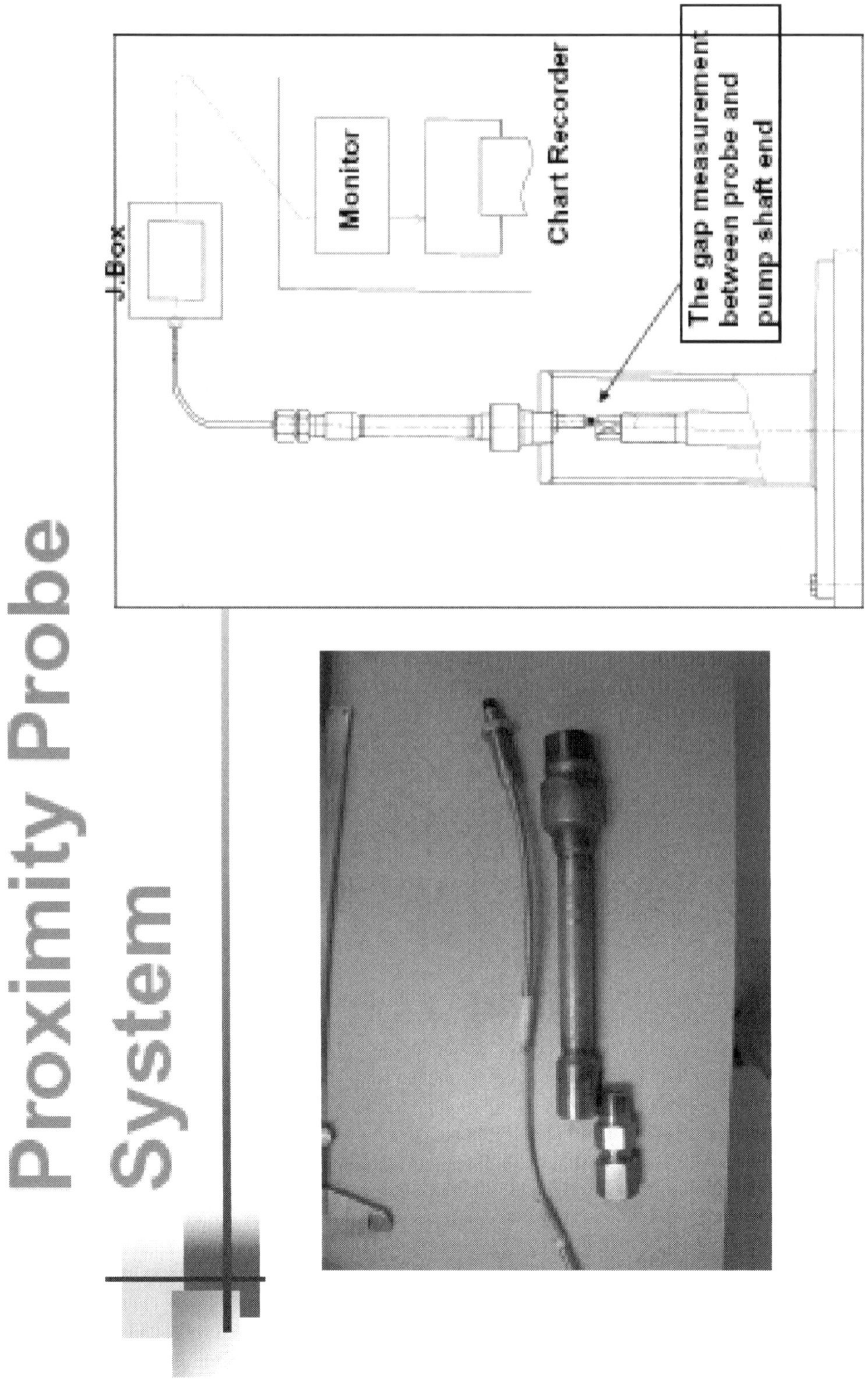

Proximity Probe System
J.Box
Monitor
Chart Recorder
The gap measurement between probe and pump shaft end

FIGURE 12

# Gap Measurement Results

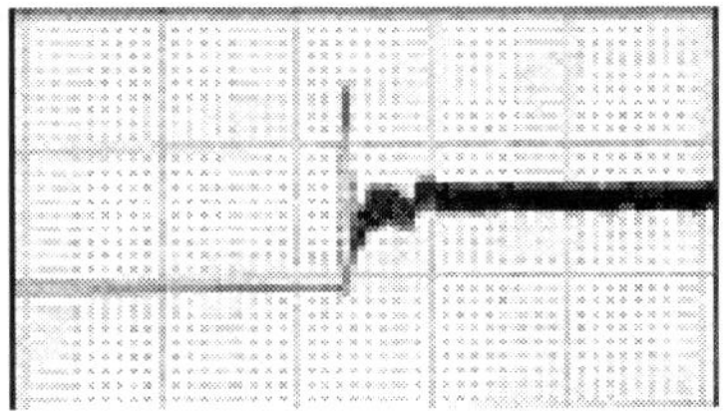

**Platform No Motion**

**Pump Start Up**

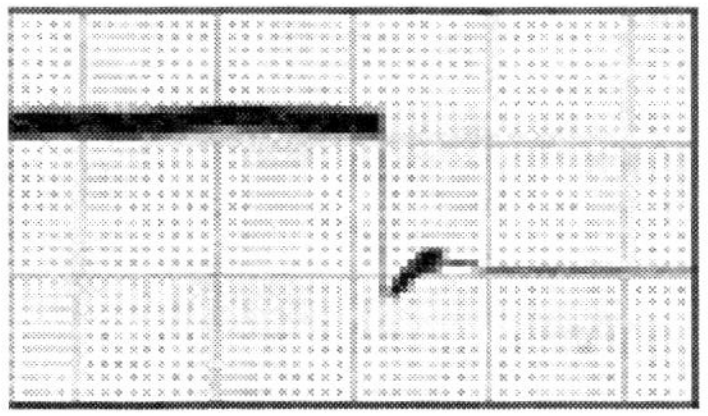

**Platform No Motion**

**Pump Stop**

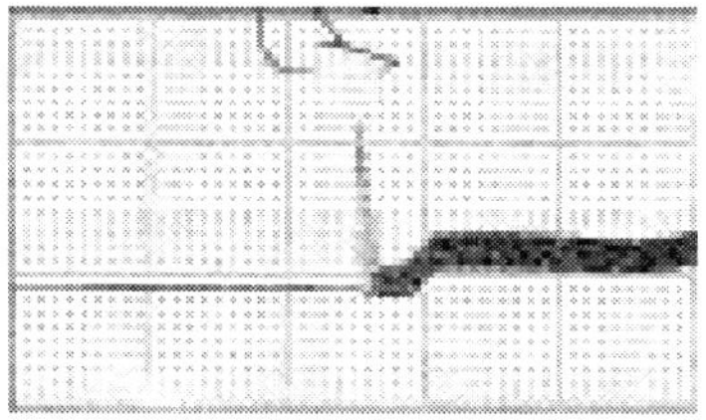

**Platform Inclined Position 9deg**

**Pump Start Up**

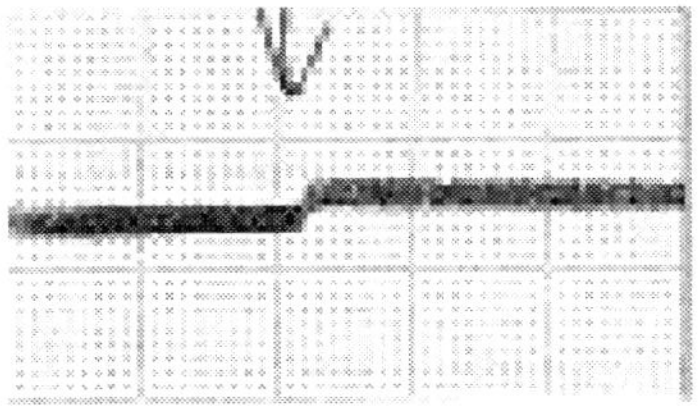

**Platform**
**+/-4.0deg Roll 9sec**
**+/- 6.8deg Pitch 11sec**

**Flow Change**

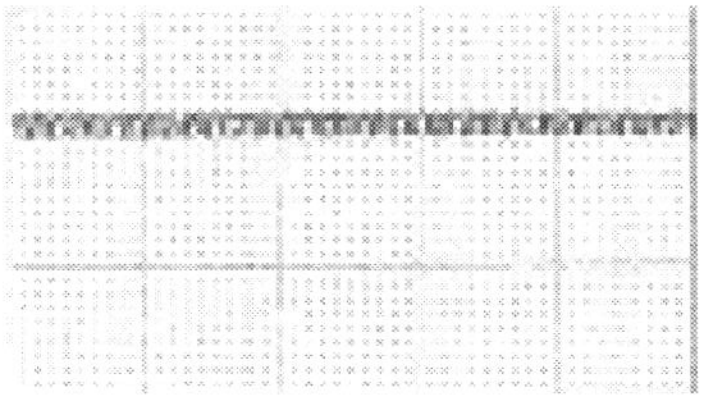

**Platform**
**+/-6.0deg Roll 6.3sec**
**+/- 6.8deg Pitch 6.3sec**

**90% Flow**

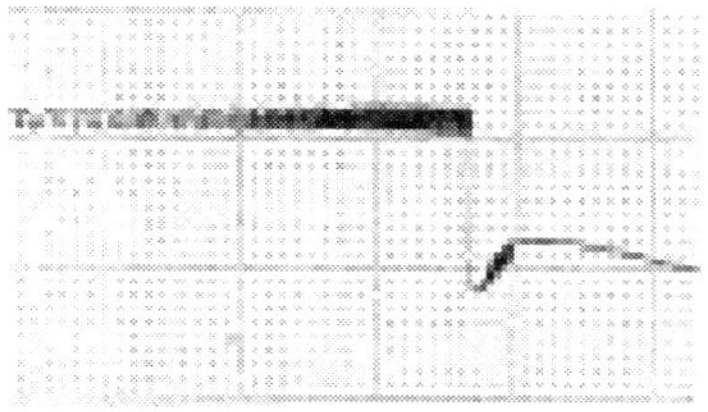

**Platform Inclined Position 9deg**

**Pump Stop**

**FIGURE 13**

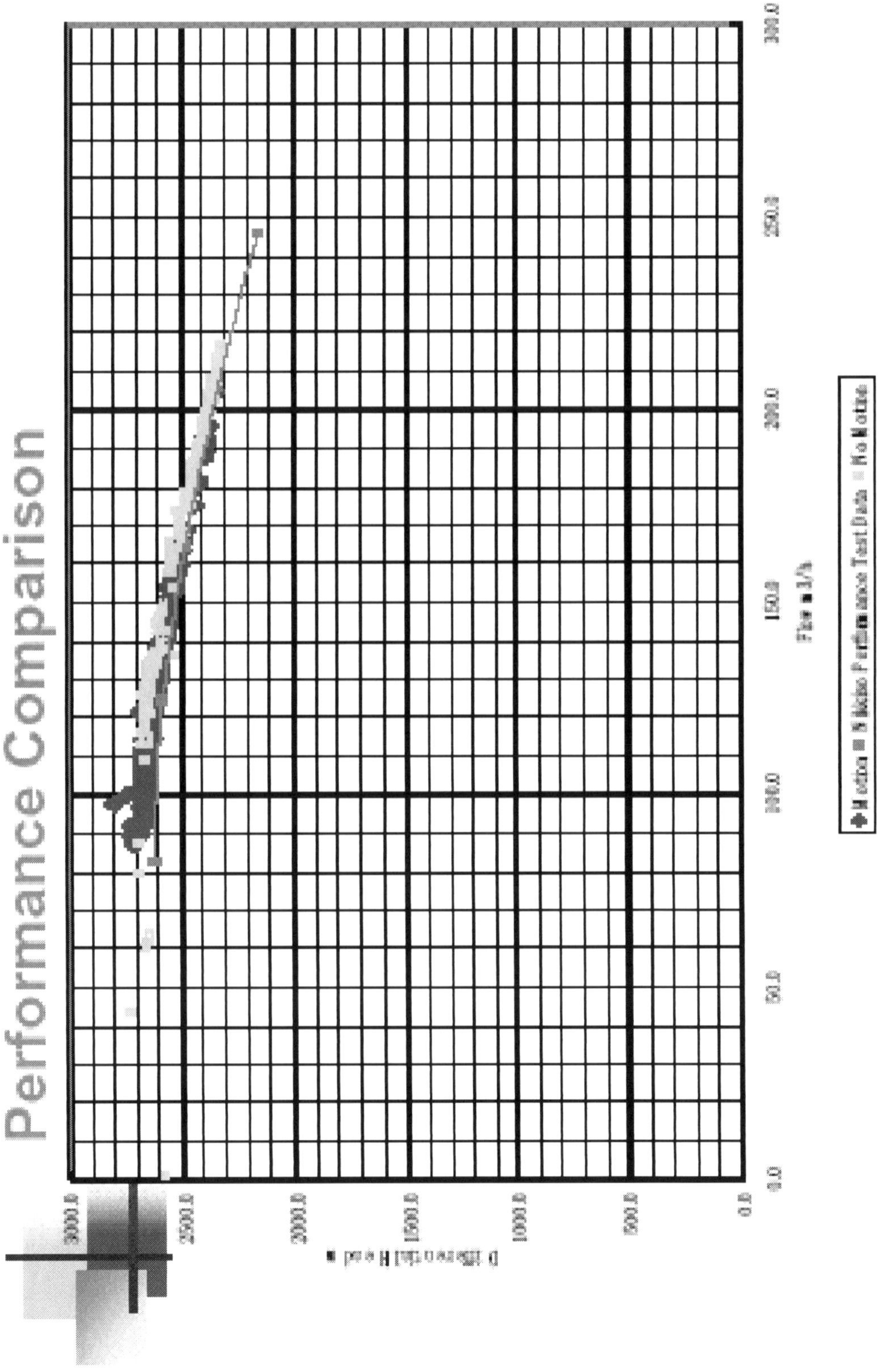
Performance Comparison
0.0
500.0
1000.0
1500.0
2000.0
2500.0
3000.0
0.0
50.0
100.0
150.0
200.0
250.0
300.0

# Current Trends in Hydrogen Plant Design

**Eric Wagner, Vinay Khurana, Peter De Geest**

**Technip USA Corporation**
555 West Arrow Highway
Claremont, California, 91711
ewagner@technip.com

Prepared for Presentation at Session 72, Conversion of Natural Gas to Hydrogen, at the AIChE National Spring Meeting, April 27, 2004, New Orleans, Louisiana

Abstract

The basic chemistry employed in the production of hydrogen has remained unchanged for many years. Nonetheless, the process plant designed to produce hydrogen continues to evolve and improve. This paper presents the current trends of hydrogen plant design and will focus on the steam reformer, by-product steam production, modularization and $NO_x$ reduction. Novel technologies and potential future directions will also be addressed.

## 1.0 Introduction to Hydrogen Production

There are a variety of ways to produce hydrogen ranging from electrolysis of water to gasification of heavy hydrocarbons. However, steam reforming of natural gas and naphtha continues to be the economical and therefore the most predominant path for large production facilities of hydrogen.

## 2.0 The Hydrogen Plant Flowsheet

The hydrogen plant can be divided into the following processing steps as represented in Figure 1: Feed Purification, Steam-Hydrocarbon Reforming, Water-Gas Shift Reaction, Hydrogen Purification, Heat Recovery and Steam Production.

### 2.1 Feed Purification

Hydrocarbon feeds typically contain impurities such as sulfur and chlorides which can poison the catalysts employed in the reforming and shift sections of the hydrogen plant and therefore need to be removed. The feed purification section is shown in Figure 2. A reactor converts organic sulfur and chlorides to $H_2S$ and HCl in the presence of a small amount of recycle hydrogen; these compounds are then removed over a chlorine guard catalyst vessel and a zinc oxide (ZnO) bed. Unsaturated hydrocarbons in Refinery Off Gases (ROG) are saturated in the presence of hydrogen in the hydrodesulfurization reactor. If the level of olefins is high in the ROG, a cooling mechanism is required to mitigate the exotherm encountered in the reactor.

### 2.2 Steam Reforming

The Steam Reformer (or primary reformer) is the heart of the hydrogen plant. It consists of multiple long tubes (commonly 4 inches in diameter and 40 feet in length) filled with nickel-based catalyst and situated in a rectangular or cylindrical firebox. Various configurations of the reformer exist such as side-fired, bottom-fired and the most common top-fired configuration shown in Figure 3. In the catalytic reforming step, steam reacts with the hydrocarbon to form synthesis gas, a mixture of carbon monoxide, carbon dioxide and hydrogen via the following reactions:

$$CH_4 + H_2O = CO + 3\,H_2 \qquad \text{(Methane-Steam Reaction)}$$
$$CO + H_2O = CO_2 + H_2 \qquad \text{(Water-Gas Shift)}$$

For hydrocarbons heavier than methane, the main reaction goes to complete conversion, represented as:

$$C_xH_y + xH_2O = xCO + (x+y/2)H_2$$

The overall process is highly endothermic requiring large amounts of heat input into the reformer. The actual steam used in the operation of the hydrogen plant is in excess of the stoichiometric values given above. Typical steam to carbon ratio range from 2.8 to 4.0 mole/mole, depending upon hydrocarbon feed and catalyst employed.

## 2.3 Shift Reaction

To increase the hydrogen yield, the syngas exiting the reformer is further reacted in an adiabatic shift reactor illustrated in Figure 4. The high temperature effluent from the reformer is cooled and then fed to one or more shift reactors where CO and steam selectively react to form hydrogen and CO2; the shift catalyst is inactive to the methane-steam reaction. The shift reaction is highly exothermic resulting in a significant temperature rise over the catalyst bed. Depending upon the temperature of operation, the shift reactor is referred to as High Temperature Shift (~ 640 F), Low Temperature Shift (~ 400 F) or Medium Temperature Shift (~ 580 F).

## 2.4 Hydrogen Purification

After cooling the shift reactor effluent, the hydrogen product is recovered in a pressure swing adsorption (PSA) unit which consists of a molecular sieve material that preferentially adsorbs hydrogen from the other components. The beauty of the PSA unit is that hydrogen is recovered at high pressure and at high purity (typically 99.9% or greater). Recovery of the hydrogen varies depending upon the number of PSA beds but is typically in the range of 85-90%; the optimum recovery occurring in the range of 300 – 400 psig. The PSA purge gas stream contains the remaining un-recovered hydrogen and other components ($CH_4$, CO and $CO_2$) and exits the PSA at low pressure used as the primary fuel in the reformer. A schematic of the PSA is given in Figure 5.

## 2.5 Heat Recovery and Steam Production

The thermodynamics of the steam reforming reactions are such that high temperatures are required in order to obtain sufficient conversion of methane. Reformer exit temperatures are typically in the range of 1550 – 1650 F. In order to heat the process gas to these temperatures, fuel is burned in the reformer radiant box and the resulting flue gas is collected at high temperatures and used as a convective heat source for pre-heating the reformer feed as well as by-product steam production in the reformer convection section, as illustrated in Figure 6. The amount of by-product steam can be reduced by incorporating combustion air preheat or the Pre and Post Reformers discussed below.

## 3.0 Current Trends in Hydrogen Plant Design

In addition to the basic hydrogen plant described above, there are several configurations that are increasingly employed today in hydrogen plant designs.

## 3.1 Pre-Reformer

A Pre-Reformer is an adiabatic reactor placed before the primary steam reformer that partially reforms some of the hydrocarbons before they enter the primary reformer. Use of a Pre-Reformer will reduce the size of the primary reformer in a grass roots facility or allow expanded hydrogen production if added to an existing plant. The Pre-Reformer contains a high-nickel content catalyst with high activity for low-temperature (pre)-reforming. The Pre-Reformer effluent is then re-heated before entering the primary reformer as shown in Figure 7.

Disadvantages of the pre-reformer are catalyst sensitivity to steaming and deactivation compared to the primary reforming catalyst. Advantages are improved feed flexibility, increased reformer inlet temperature because of fewer higher hydrocarbons, as well as lower by-product steam production and lower $NO_x$ formation which result from a smaller primary reformer.

## 3.2 Post-Reformer

The Enhanced Heat Transfer Reformer (EHTR) shown in Figure 8 is a combined reactor and heat exchanger that uses the high level heat contained in the primary reformer effluent as the heat source to reform a slip stream of feed; heat transfer is via convection rather than radiation. Like the Pre-Reformer, the EHTR will reduce the size of the primary reformer for a grass-roots design or allow more hydrogen to be produced in a retrofit application.

Disadvantages of the EHTR are cost and a higher methane slip (unconverted methane). Advantages are lower by-product steam production and lower $NO_x$ formation which result from a smaller primary reformer.

## 3.3 High Pressure Stripping

Some hydrogen plants now incorporate High Pressure Steam Stripping of process condensate. Ammonia formed in the reformer tubes from nitrogen in the hydrocarbon feed collects in process condensate after cooling of shift effluent. In typical deaerators very little ammonia is removed from process condensate and ultimately ammonia is released in the steam drum and finds its way into the overall facility's steam and condensate system via export steam. Deposition of bicarbonate and corrosion of copper containing metals are concerns.

The recent trend in the industry is to limit ammonia in steam to levels down to 1 ppm. There are several ways to achieve this goal. An attractive scheme is stripping of process condensate in a high pressure stripper using saturated steam from the steam drum and returning steam from the stripper to the reformer as process steam. The main benefit of the system is that the potential ammonia problem is confined to the hydrogen plant.

Another option can be to have segregated steam systems for process and export steam, the latter being generated using high quality boiler feed water without process condensate.

## 3.4 Modular Design

In order to reduce field labor and improve project schedule, hydrogen plants commonly incorporate some degree of Modular Design which means that certain sections of the plant are fabricated and pre-commissioned at fabrication shops and then connected together at the site of the plant. Even the reformer and convection section can have certain sections pre-assembled in shops and hydrotested before shipping to the field. Significant piping quantities and instruments can be installed in a controlled environment such as a fabrication shop as opposed to field installation and erection which can be adversely affected by weather and local labor conditions.

## 3.5 Power Integration

An interesting means to produce power along with hydrogen is to integrate a gas turbine into the primary reformer as illustrated in Figure 9. In this configuration, natural gas is burned in the turbine to produce power. The hot turbine exhaust (~1000 F) contains considerable excess oxygen even after combustion; this stream can be conveniently used as combustion air for the steam reformer. Provisions must be made so both the reformer and the turbine can operate safely and efficiently even if the other is not in operation. Use of the gas turbine integration improves the thermal efficiency of the process by typically five percent, which results in fewer $NO_X$ and $CO_2$ emissions.

## 3.6 Computation Fluid Dynamics

Computational Fluid Dynamics (CFD) is the science of predicting fluid flow, heat transfer, mass transfer, chemical reactions, and related phenomena by solving mathematical equations that represent physical laws, using a numerical process. Most reformers designed today utilize these sophisticated CFD tools to model the burners, reformer firebox, flue gas tunnels, convection section and combustion air flow in the penthouse for induced-draft reformers. The results of these studies will influence the mechanical layout of the reformer section resulting in a reduction in installed cost as well as fewer operational problems.

## 4.0 Environmental Considerations

The biggest environmental concern from the hydrogen plant is that of $NO_X$ formation. Ultra-Low $NO_X$ burners can achieve levels of 25-40 ppm depending upon combustion air pre-heat. In order to achieve $NO_X$ levels down to 10 ppm and below, Selective Catalytic Reduction (SCR) de-$NO_X$ reactors can be installed in the convection section. Ammonia, either anhydrous or aqueous, is injected into the flue gas and reacts with the $NO_X$ to form nitrogen.

## 5.0 Summary

As long as light hydrocarbons remain the preferred feedstock for hydrogen production, the basic layout and unit operations of the hydrogen plant will remain the same. Nonetheless, within this framework, the process plant will continue to evolve and improve. In the future, more improvements are expected. Progress towards a hydrogen based economy and increased use of fuel cells will certainly drive chemical engineers to proactively innovate the ways in which hydrogen is produced, both economically and safely.

## Figure 1 SMR based Hydrogen flowsheet process steps

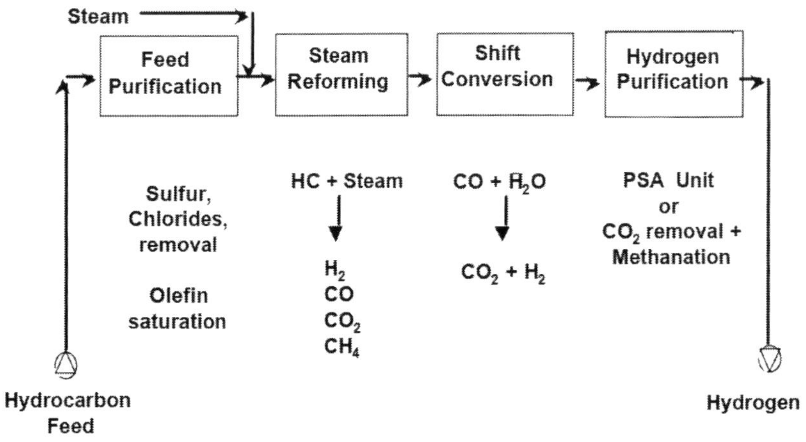


AIChE Spring Meeting , 2004 Technip 0

## Figure 2 Typical Feed Purification Configuration

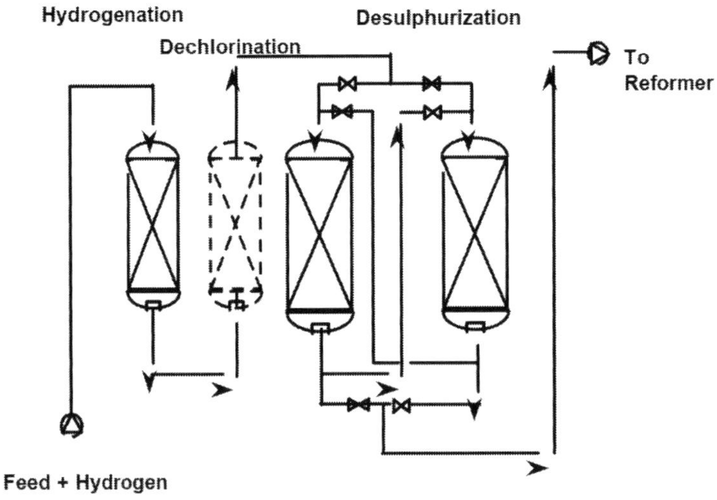


AIChE Spring Meeting , 2004 Technip 1

## Figure 3 Top-fired large reformer

## Figure 4 Shift Conversion configurations

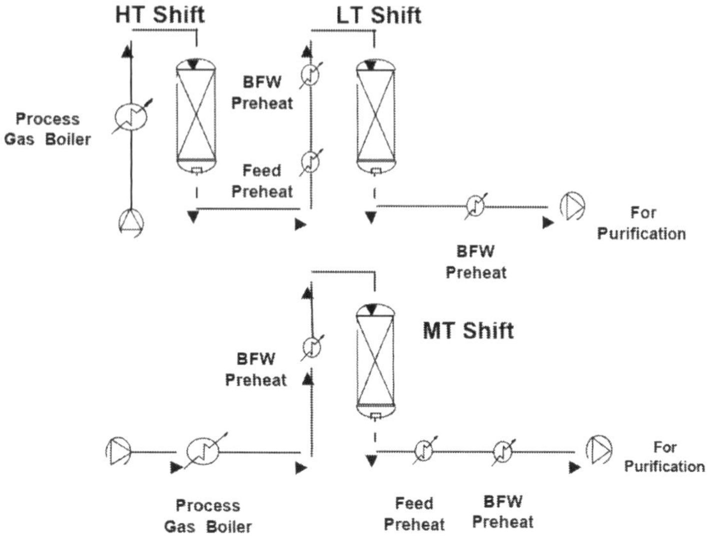

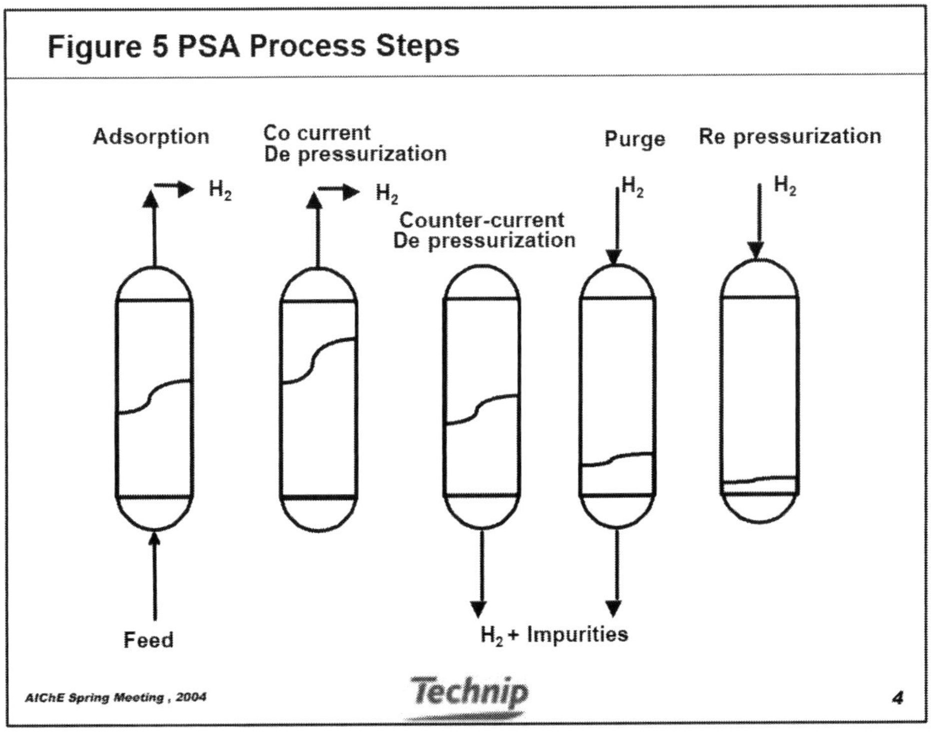
Figure 5 PSA Process Steps
Adsorption
Co current
De pressurization
Purge
Re pressurization
H2
Counter-current
De pressurization
Feed
H2 + Impurities
AIChE Spring Meeting , 2004
Technip
4

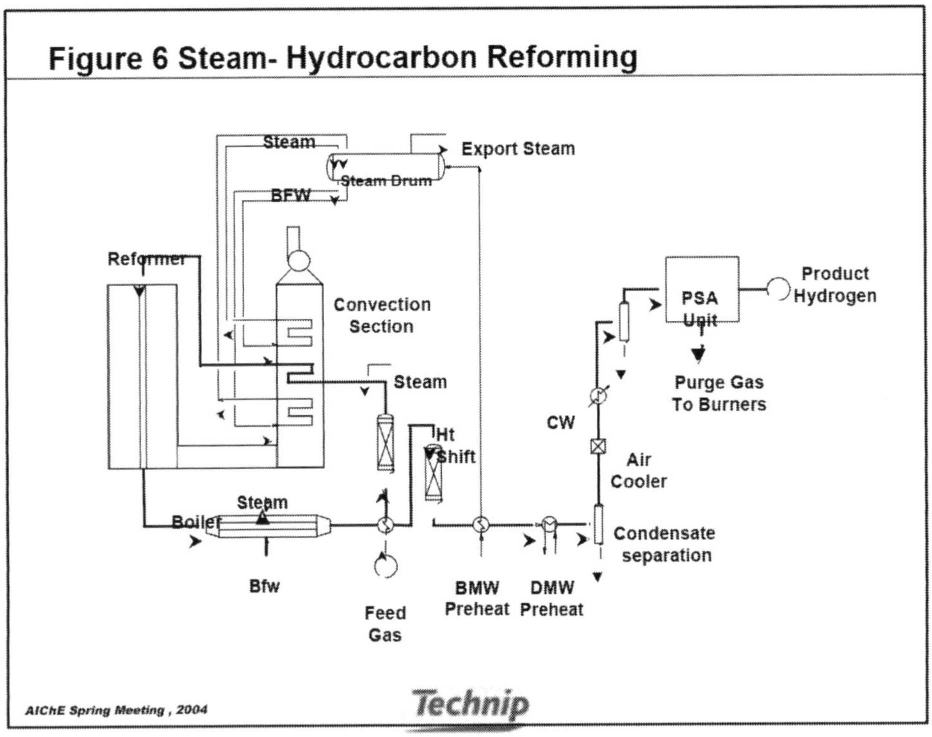
Figure 6 Steam- Hydrocarbon Reforming
Steam
Export Steam
Steam Drum
BFW
Reformer
Convection
Section
Steam
Ht
Shift
Steam
Boiler
Bfw
Feed
Gas
BMW
Preheat
DMW
Preheat
CW
Air
Cooler
Condensate
separation
PSA
Unit
Product
Hydrogen
Purge Gas
To Burners
AIChE Spring Meeting , 2004
Technip

## Figure 7 Pre-reformer Integration

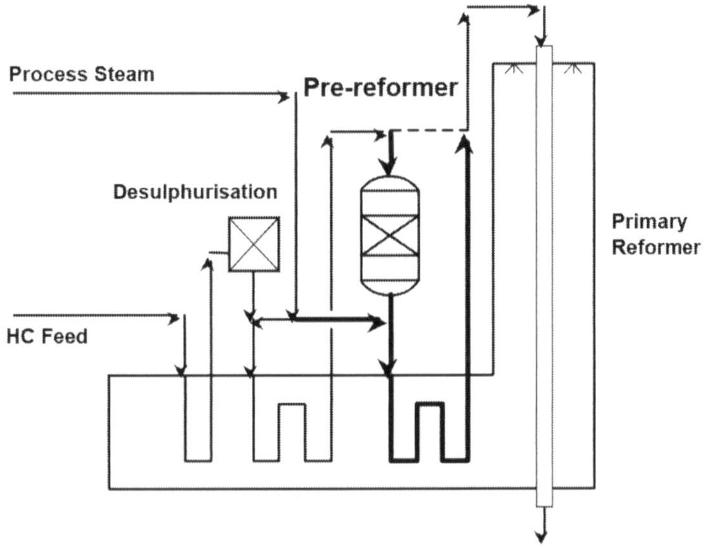




## Figure 8 As built EHTR

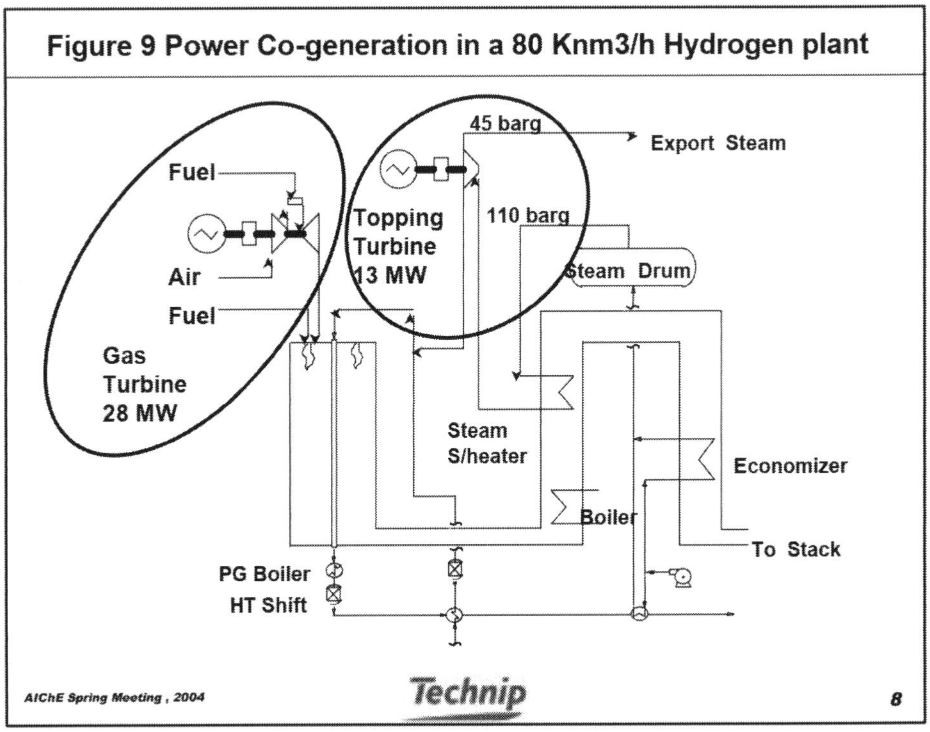
Figure 9 Power Co-generation in a 80 Knm3/h Hydrogen plant
45 barg
Export Steam
Fuel
Topping
Turbine
13 MW
110 barg
Steam Drum
Air
Fuel
Gas
Turbine
28 MW
Steam
S/heater
Economizer
Boiler
To Stack
PG Boiler
HT Shift
AIChE Spring Meeting , 2004
Technip
8

**Efficient Production of Purified Hydrogen Via Low Temperature Water-Gas-Shift Reaction**

**Devinder Mahajan**[1, 2]

[1]Energy Sciences and Technology Department, Brookhaven National Laboratory, Upton, New York 11973-5000 USA
[2]Department of Materials Science and Engineering, Stony Brook University, Stony Brook, New York 11794-2275 USA

**Key words**: Hydrogen production, sing-site catalysts, organometallic complexes, homogeneous catalysis, fuel cells.

Prepared for Presentation at the Symposium on Advanced Gas Conversion and Hydrogen Production, 2004 AICHE Spring National Meeting, New Orleans, LA April 25-29, 2004
Session: Natural Gas Utilization

## Abstract

Methane derived hydrogen contains copious amount of carbon monoxide the presence of which is detrimental to catalysts used in PEM fuel cells. We have designed a single-site organometallic complex that dissolves in liquid phase to efficiently catalyze water-gas-shift (WGS) reaction under very mild conditions. The catalyst has been evaluated in batch mode by dissolving in a methanol/water mixture to process synthesis gas with high turnover numbers at temperatures as low as 80°C under pressures < 1 MPa to yield a gas mixture that is essentially devoid of CO. The thermodynamic basis and the application of this technology will be discussed.

## Introduction

Fuel cells are considered the next breakthrough technology for transportation sector and power generation to arrest global warming [1, 2]. Fuel cells utilize hydrogen ($H_2$) as a fuel that produces no carbon dioxide ($CO_2$). Moreover, economical $H_2$ production and storage methods are sought to make desired "Hydrogen Economy" a reality in the coming years [3]. $H_2$ can be produced by steam reforming of methanol (equation 1) over Cu-Zn type catalysts [4] or partial oxidation of methanol [5] (equation 2) though a copious amount (> 20 ppm) of carbon monoxide (CO) is also produced. However, a modified Cu-Zn catalyst has been reported [6] to yield CO-free $H_2$ (equation 3).

$$CH_3OH + H_2O \leftrightarrow 3\,H_2 + CO_2 \qquad \Delta H = +49.4\ \text{kJ.mol}^{-1} \qquad (1)$$
$$CH_3OH + \tfrac{1}{2}O_2 \leftrightarrow 2\,H_2 + CO_2 \qquad \Delta H = -192.2\ \text{kJ.mol}^{-1} \qquad (2)$$
$$2\,CH_3OH + \tfrac{1}{2}O_2 + H_2O \leftrightarrow 5\,H_2 + 2\,CO_2 \qquad \Delta H = -142.8\ \text{kJ.mol}^{-1} \qquad (3)$$

However, low space-time-yields (STY) remain a challenge. Fuel cells face several technological challenges that must be addressed before commercialization can be realized [1, 2]. For example, proton exchange membrane (PEM) fuel cells utilize a platinum based catalyst that is readily poisoned by carbon monoxide. To extend the platinum catalyst life, two approaches are utilized. These are: 1) use of various platinum alloys to make the catalyst more CO tolerant [7] and 2) a water-gas-shift (WGS) step (equation 4) catalyzed by heterogeneous catalysts requiring expensive high temperature treatment [8] to reduce the carbon monoxide concentration to acceptable levels.

$$CO + H_2O \leftrightarrow CO_2 + H_2 \qquad \Delta H = -39.4\ \text{kJ.mol}^{-1} \qquad (4)$$

The forward Reaction 4 is attractive in that every mole of CO removed also results in concomitant generation of $H_2$ from $H_2O$. Transition metal complexes are also known to catalyze WGS reaction in a homogeneous mode [9] but a number of problems including incomplete CO removal, attrition of the expensive catalyst used or low STY pose a challenge. We have initiated a study to develop a WGS catalyst system. Our strategy is to first use formate decomposition as a fast screening method. We describe here our results of such a study in which the catalyst operates in a homogeneous mode and is an efficient formate decomposition catalyst at low temperatures.

## Experimental

A commercial batch unit, obtained from AE Zipperclave, was used for batch screening of the catalyst formulations. It was a commercial stainless steel 0.5 L reactor fitted with a magnedrive and modified for use in this study. The reactor was equipped with a six-blade impeller to ensure efficient gas/liquid mixing. Ports were provided for loading reactant gases, sampling vapor and liquid phases. The batch CSTR unit ratings were as follows: stirrer = 0 - 2000 rpm, T = 298 – 723 K, P = 100 - 14000 kPa.

## Results and Discussion

A number of transition metals are known catalyze WGS reaction in a homogeneous mode in which water serves as a primary solvent. But the catalyst integrity is at issue and the reaction rates are too slow to be of commercial interest [9]. The mechanism that produces $H_2$ in these systems has been thoroughly investigated and established. A typical mechanistic cycle involves activation of CO at the metal center and the nucleophilic attack of hydroxide ion on the coordinated CO yields a hydroxycarbonyl complex. The decarboxylation produces a metal dihydride that on elimination of $H_2$ completes the catalytic cycle. Metal formates also serve as intermediates in these reactions and their decomposition can be used for rapid screening of potential WGS catalysts.

We designed several base-activated transition metal catalyst systems and evaluated these for formate decomposition activity. The run conditions were as follows:

Baseline batch run conditions
Reactor = 0.5 L stirred reactor (CSTR)
Solvent = 50% Methanol/50% Water (volume/volume)
Catalyst formulation = Per Table 1
Catalyst loading = 2-3 wt%
T = 413 K
P = 1000 - 3000 kPa.
Inert gas ($N_2$) = 1000 kPa

---

The solution phase consisted of water/methanol mixtures in which the reaction took place. The catalyst under these conditions dissolved readily and was very likely in homogeneous mode, i.e., it was a single-site catalyst. The results of our study are listed in Table 1. The following observations are noteworthy:

1. Ni, Ru, Rh, Co, Pt, all showed formate decomposition activity but the rates were too slow as indicated by the reaction time.
2. The extent of formate decomposition was indicative of the reaction rate.
3. The reaction was faster with catalyst containing ligands (L) where L is a N-donor or P-donor.

**Table 1. Decomposition of Potassium Formate Catalyzed by Metal Complexes**

| Run No. | Catalyst formulation | Final Gas Analysis $H_2$ (mmol) | Final Gas Analysis CO (mmol) | Time, min | % $KHCO_2$ Decomposition |
|---|---|---|---|---|---|
| 1 | ----- | 1 | - | 140 | 2 |
| 2 | Ru-1 | 35 | - | 80 | 70 |
| 3 | Ni-1/L | 5 | - | 120 | 10 |
| 4 | Ru-2/L | 47 | - | <5 | 94 |
| 5 | Rh-1/L | 50 | - | <30 | 80 |
| 6 | Co-1 /L | 2 | - | 55 | 4 |
| 7 | Pt-1/L | 43 | - | <10 | 86 |
| 8 | Fe-1 /L | 3 | - | 30 | 6 |
| 9 | Ru-3/L | 50 | - | <2 | 100* |

*T = 413 K all other runs were at 373 K.

These preliminary results are promising for developing a viable system for commercial application. We are now evaluating the Ru-based catalyst system in a continuous mode for WGS reaction.

## Conclusions

The presented results show that of all the metals evaluated, Ru complexes activated by a base are highly active single-site transition catalysts that operates at low temperature and low pressure of $T < 150^oC$ and $P \sim 2000$ kPa, respectively and achieves almost complete CO removal via WGS reaction in a reaction time of less than a few minutes. *This is made possible by utilizing low operating temperatures that thermodynamically allow the reversible exothermic Reaction 4 to proceed almost completely to the right.* Further tests are underway to evaluate the commercial potential of this promising technology.

## References

[1]. Fuel Cell Technology Handbook. Gregor Hoogers, editor, CRC Press, Inc., I$^{st}$ edition (2002).
[2]. Fuel Cell Systems Explained. James Larminie and Andrew Dicks. John Wiley & Sons, 2$^{nd}$ edition (2003).
[3]. National Hydrogen Energy Roadmap, U.S. Department of Energy, November 2002.
[4]. Breen, J.P. and Ross, J.R.H. Catal. Today 51 521 (1999).
[5]. Cubeiro, M.L. and Fierro, J.L.G. J. Catal. 179 150 (1998).

[6]. Velu, S., Suzuki, K., and Osaki, T. Chem. Commun. # 23 2341 (1999).
[7] Mukerjee, S., Lee, S.J., Ticianelli, E.A., McBreen, J., Grgur, B.N., Markovic, N.M., Ross, P.N., Giallombardo, J.R., and De Castro, E.S. Electrochem. and Solid State Lett. 2(1) 12 (1999).
[8] Thomas, C.L. in "Catalytic Processes and Proven Catalysts", Academic Press, New York, 1970.
[9] Ford, P.C. Acc. Chem. Res. 14 31 (1980).

# COMBINED STEAM AND DRY REFORMING IN LOW TEMPERATURE PLASMA

T. Hoang, H. Le, L. L. Lobban, and R. G. Mallinson

School of Chemical Engineering and Materials Science
Institute for Gas Utilization Technologies
University of Oklahoma
100 East Boyd, Room T335 Norman, Oklahoma 73019

Prepared for Presentation at Gas to Chemicals Session for 2004 AIChE Spring
April 12, 2004



March 2004

Unpublished



## INTRODUCTION

Natural gas, with huge proven reserves in stranded and remote areas, has been increasingly received worldwide attention as an important alternating chemical feedstock to produce other chemicals and fuels. Methane is the principle constituent of natural gas and carbon dioxide is the major greenhouse gas and frequently present in natural gas. Any success in utilizing them will signify attainment of double objectives of reducing the greenhouse gases and the cost for separating system. A possible utilization of methane with carbon dioxide is production of hydrogen and acetylene. Hydrogen fuel has been considered as a potential fuel of future. Much research on producing hydrogen from natural gas has been conducted. Since 1998[11], The US Department of Energy (DOE) considers its hydrogen program a strategic program and has embarked on an aggressive research effort to develop "breakthrough" technologies to extract hydrogen from natural gas. Non-thermal plasmas, one of those innovative technologies also called a cold plasma and its non-equilibrium properties, high electron temperature and low bulk gas temperature, have been extensively investigated because of its low temperature operation and its potential low energy consumption.

This paper discusses the combined dry (R.1) and steam (R.2) reforming of methane to produce hydrogen and acetylene using an AC non-thermal plasma reactor, that allows low energy consumption (~8eV/molecule C) and high hydrogen selectivity (>80%) over a long period. Thermodynamics study has shown that the presence of water in dry reforming helps increase the conversion of methane[12] and due to the unique way of feeding water into reactor, the discharge could be stably maintained over a wide range of experiment parameters.

$$CH_4 + CO_2 \Rightarrow 2CO + 2H_2 \quad \Delta H=247\ kJ\ mol^{-1} \qquad R.1$$

$$CH_4 + H_2O \Rightarrow CO + 3H_2 \quad \Delta H=206\ kJ\ mol^{-1} \qquad R.2$$

## EXPERIMENTAL

A corona discharge generated by an AC power system with a point-to-point electrode configuration is shown in Figure 1. The top electrode was a 1/8-inch stainless steel tube in which de-ionized liquid water was introduced into the reactor by a Kd Scientific syringe pump model 100. The bottom electrode was a stainless steel rod. The gas gap (GG) between the two electrodes was always set at 15 mm unless otherwise specified. The reactor's quartz tube inside diameter (ID) was 10 mm. The feed gases, methane and carbon dioxide, were well mixed and flowed axially downward inside the reactor. Product gases exited through the bottom tube electrode in the case of the hollow bottom electrode study (Fig. 1). All the experiments were operated at atmospheric pressure. The power system has been described previously elsewhere[6,10]. Temperatures measured by Infrared were less than 600 K.[10]

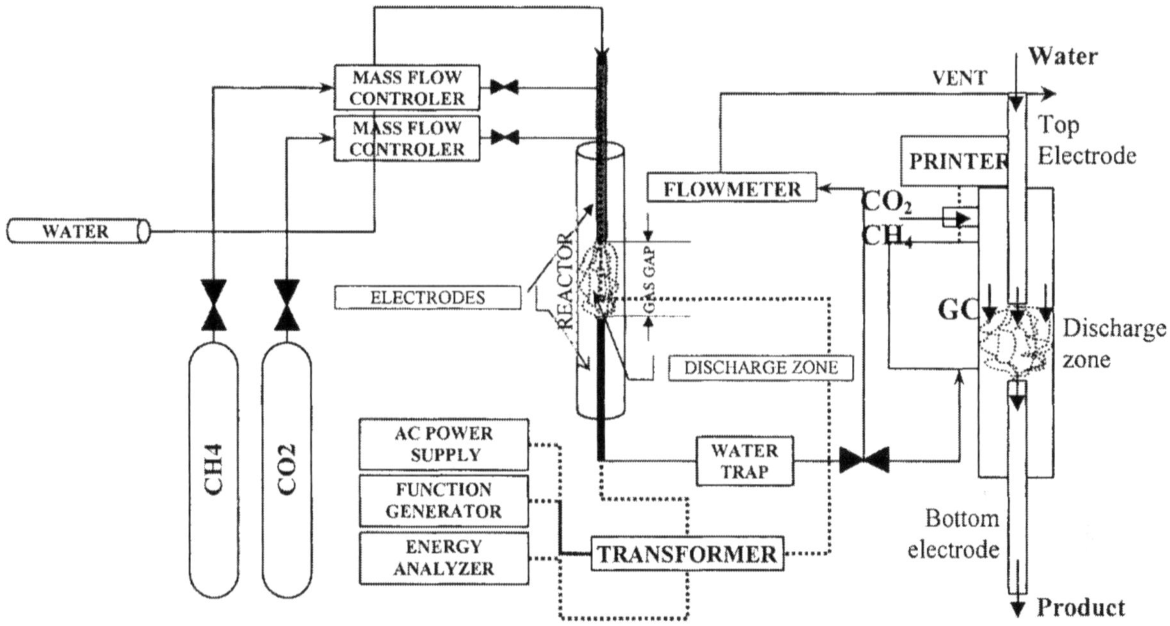


Figure 1 – Plasma reactor schematic and hollow electrode configuration

The product selectivities were calculated based on both methane and carbon dioxide. The electrical energy consumption, eV/molecule of C converted, was based on the total input power and the total amount of $CH_4$ and $CO_2$ reacted. Likewise, the term "eV/mol $H_2$ produced" is the energy consumption to produce a molecule of hydrogen.

## RESULTS

Corona discharges have electron energies at about 6eV which is high enough to dissociate $CH_4$, $CO_2$ and $H_2O$ in the feed stream because those species have bond energies ~5eV. Although other reports[6,13] suggest that pure methane cannot dissociate in a corona discharge due to high appearance potentials of methane ions (>12 eV), it might not be true because only a small degree of ionization of gases occurs inside corona discharge.[14]

| | | | | | | | | |
|---|---|---|---|---|---|---|---|---|
| $H_2O$ | + | e | ⇔ | H | + | OH | + e | R.3 |
| OH | + | e | ⇔ | H | + | O | + e | R.4 |
| $CO_2$ | + | e | ⇔ | CO | + | O | + e | R.5 |
| $CH_4$ | + | e | ⇔ | $CH_3$ | + | H | + e | R.6 |

These radicals, O, OH, and H, then abstract hydrogen from methane to form methyl radicals (CH3). Consequently, methane conversion is significantly improved in corona discharge.

| | | | | | | | |
|---|---|---|---|---|---|---|---|
| $CH_4$ | + | O | ⇨ | $CH_3$ | + | OH | R.7 |
| $CH_4$ | + | H | ⇨ | $CH_3$ | + | $H_2$ | R.8 |
| $CH_4$ | + | OH | ⇨ | $CH_3$ | + | $H_2O$ | R.9 |

Several experimental series have been conducted to study the effect of each parameters:

*Gas gap*

By alternating gas gap, one can study the effects of the electric field and residence time on power consumption and methane conversion as well as the product selectivities. The experimental conditions for the series of gas gap experiments were 200 sccm feed flow rate (varying residence time), 37.5 mole % $CH_4$, 37.5 mole % $CO_2$, and 25 mole % $H_2O$, 200 Hz, 10 mm ID reactor. The gas gap between the two electrodes was varied from 5.5mm to 15mm. The input power was also varied to maintain the discharge. As the gas gap increases, the feed conversions increased which could be due the increase in the residence time (from 0.13 seconds to 0.35 seconds) and the input power (from 8.5W to 24W). In methane conversion using a non-equilibrium plasma, the energy consumption usually increases with increasing feed conversion due to lower probability of the highly energetic electrons activating a previously un-reacted feed molecule at higher conversions. In this case, even though the conversions increase, the energy consumption remains relatively stable around 8 eV/molecule of C converted, which is 50% lower than previously reported.[6] It has been experimentally observed that as the gas gap increased, the visible discharge volume, resembling an oval shape, increased as well. The reason for the energy consumption not increasing is likely because increasing the gas gap decreases the amount of gas bypassing the discharge space.

The hydrogen selectivity increased from 58% to 70% with increasing gas gap. The acetylene selectivity also shows the same trend. However, the ethylene selectivities decreased slightly from 10% to 5%. The ethane selectivity was very low, less than 5%. Because the H balance, which is the amount of hydrogen in the reacted methane accounted for in the gas phase products, is less than 100%, at this $CO_2$ mole fraction, net water production instead of consumption is inferred. The feed and the vapor product (in this case separated from the liquid with a cold trap) of the experiment with 15mm gas gap were also analyzed by Mass Spectroscopy to detect methanol, ethanol, and any other heavy species in the product. No methanol (MW =32) or ethanol (MW=46) were observed. This result shows that under these experimental conditions, organic oxygenates are not produced in the AC corona discharge.

*Feed Composition Study*

The feed of combined steam and dry reforming under AC discharge has been studied to find the best composition in term of power consumption and methane conversion. Residence time was kept constant at 0.35s. Three series of experiments with different mole fractions of $CO_2$: 7.5%, 15% and 25% were conducted at 300Hz. In each series, the mole fractions of water ranged from 12.5% to 37.5%.

As shown in Figure 2, the conversion of methane and carbon dioxide increases at higher mole fraction of water in the feed. The lowest power consumption is 9eV per molecule C converted and 7eV per molecule of hydrogen produced. The highest conversion of 45% was obtained at the maximum mole fraction of water 37.5% and $CO_2$ 25% with the slightly higher power consumption 10eV/mol C. The results suggest that adding more water and $CO_2$ into the feed improve the conversions. This is consistent with the results in steam reforming. In steam reforming, the consumed power could reach 10eV but only 19% conversion of methane was obtained[9]. The maximum conversion in steam reforming was 80%; however, the power consumption is also as high as 40eV[9]. Therefore, the presence of $CO_2$ in the feed has reduced the power consumed by the system. It can be explained that besides the dry reforming, $CO_2$ has provided more O and OH radicals in the system. Thus, the more $CH_x$ (x=1,2,3) radicals were produced because of more hydrogen abstraction by the O species. As a result, higher conversion and lower power consumption were observed.

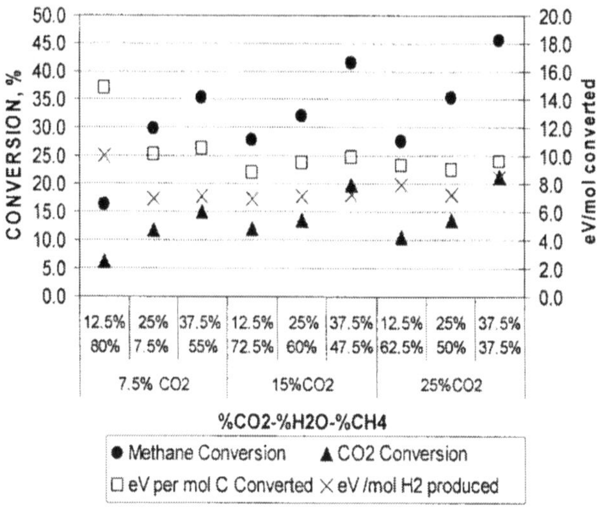


Figure 2 – Conversion and power consumption

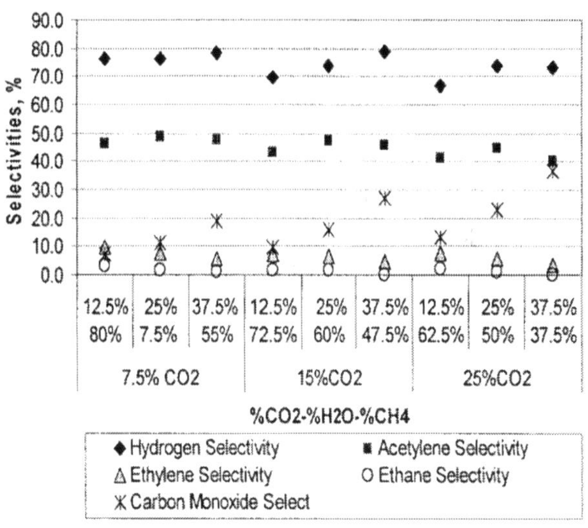


Figure 3 – Product selectivities vs. feed compositions

The distribution of products at different feed composition was also analyzed and was shown in Figure 3. Adding more water increased $H_2$ and CO selectivity, while decreasing ethylene and ethane selectivities. It is noted that the selectivity of ethane is very low in most cases and there are no trace of it at high mole fraction of water. These trends were due to the increasing conversion of methane and $CO_2$ when adding more water. Furthermore, from the pathway of plasma reactions, ethane is a primary product. At higher concentrations of water and $CO_2$, more hydrogen abstracting species were produced and thus more ethane and ethylene were converted. Hydrogen and acetylene selectivities were around 80% and 45% respectively. The more $CO_2$

present in the feed, the more CO was produced. At the highest 45% conversion of methane, the highest 40% CO selectivity was obtained and all $C_2$ selectivities decreased, as expected.

*Electrode study*

To understand the bypassing that may occur during the reaction, hollow bottom electrodes have been studied. In this configuration, because gases exit through the bottom hollow electrode, they were mixed better than in the case of coaxial exit gases through the bottom of the reactor. In addition, the bulk gases must pass through the discharge zone. Therefore, in this configuration, there may be less bypassing than in the case of gases exiting through the bottom of the reactor. A series of experiments were conducted with three types of electrodes: a steel rod, a ¼-inch steel tube and a 1/8-inch steel tube. The total flow rate was kept the same for each experiment at 200 sccm with 75 sccm $CH_4$, 50sccm $CO_2$ and 75 sccm $H_2O$. Other parameters also kept constant: 10mm reactor ID, 15 mm gas gap, 25 W power input and 300Hz frequency.

By changing the bottom electrode from a solid rod to a hollow tube, conversions of methane and carbon dioxide increase about 5%. The visible discharge shape and stability look better than in the case of the solid rod electrode. Product distribution changed with an interesting trend. Acetylene selectivity decreases by changing the bottom solid rod electrode to the hollow tube electrode. However, carbon monoxide selectivity acted in a different manner. The smaller the hollow tube diameter, the more CO was produced. That the gas products, which are generated after entering the discharge zone, mix below the discharge zone and re-enter the discharge zone to exit through the bottom hollow electrode. In other words, the residence time in this situation was a little longer due to mixing. These results were consistent with the pathway study,[10] because most CO was produced from the oxidation of carbon that was formed from decomposition of $C_2H_2$. In addition, CO is also produced from $CO_2$ and this can explain higher conversion of carbon dioxide in hollow electrode cases.

In conclusion, the hollow bottom electrodes help discharge stability and increases the CO fraction in product. This CO has the potential for further $H_2$ production if the water gas shift reaction is taken into account. CO selectivity is about 60% in this case compared with 45% in the solid electrode's case. The energy consumption decreased from 10eV to 9eV per molecule carbon converted in this experimental condition. The lowest promising energy consumption could be 5eV per molecule hydrogen produced if the water gas shift reaction is included.

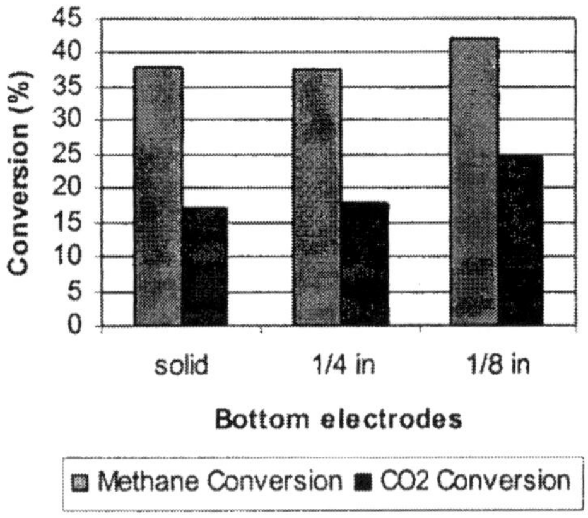


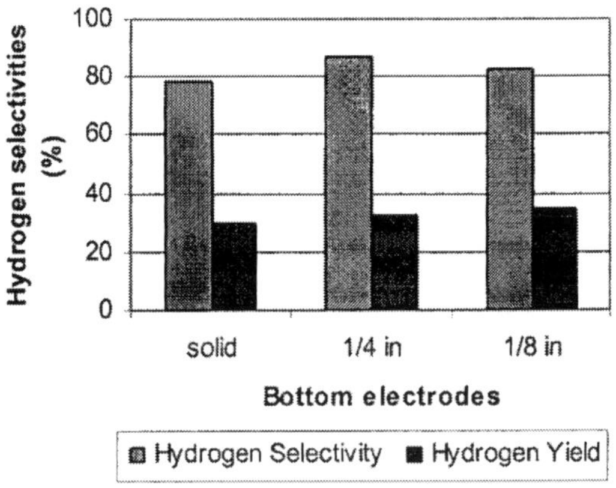


Figure 6 – Methane conversion and Hydrogen selectivities vs. bottom electrodes

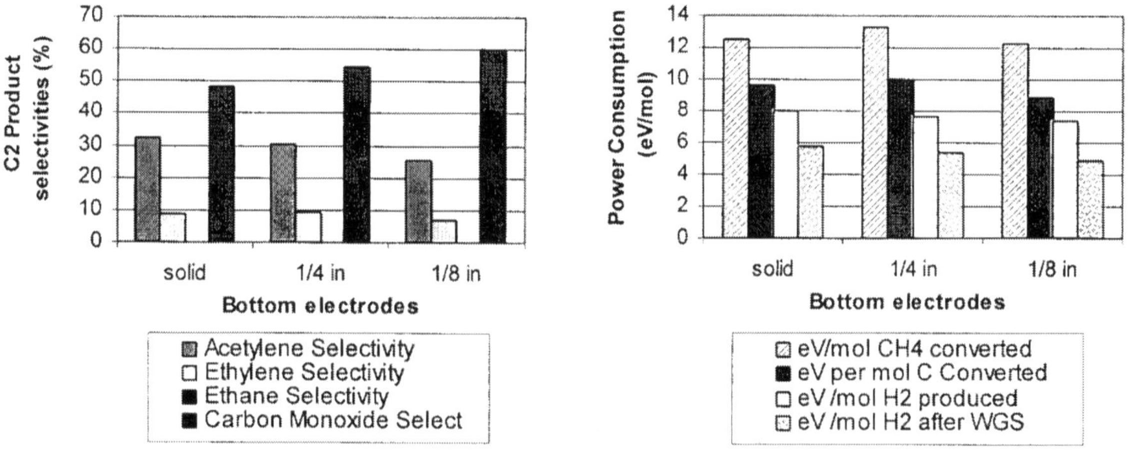


Figure 7 –$C_2$ selectivities and Power consumption vs. bottom electrodes

## SUMMARY

In the combination of steam and dry reforming, H, OH, and O radicals play an import role in activating methane molecules by abstracting hydrogen from $CH_4$, prolonging the chain reactions. O and OH radicals also help convert coke to carbon oxides, preventing the hydrogenation of coke to produce methane and removing solid coke particles out of the discharge space. Besides $CO_2$ reacting in dry reforming, it was also an additional source to produce reactive O species. As a result, the methane conversions in the steam reforming and dry reforming of methane are much higher than that in the pure methane discharge or steam reforming discharge under similar experimental conditions. The results indicate the improvement of power consumption in combined steam and dry reforming. At the same consumed power, higher conversion than steam reforming alone was obtained. The lowest power consumption is 8eV per molecule of C converted at experimental frequency 200Hz, which is 50% lower than the previous report in combined steam reforming and partial oxidation of methane[6]. The study of the bottom electrode configuration suggested an option to produce more $H_2$ and CO which can further reduce energy consumption to as low as 5eV per molecule $H_2$ produced (if water gas shift reaction is taken into account). In conclusion, combined dry and steam reforming under AC discharge is a promising method to produce hydrogen and may lead to new economical processes, especially when significant $CO_2$ is present in the gas.


### Acknowledgement

The authors would like to express their appreciation to ChevronTexaco, Inc. and PetroVietnam for their support to this research


### References


[1] Liu, C. J.;Xu, G. H.;Wang, T. M. Fuel Process Technol. 1999, 58, 119-134
[2] Malik, M. A.;Malik, S. S. J. Nat. Gas Chem. 1999, 8(2), 166-178
[3] Bradford, M. C. J.;Vannice, M. A. Catal. Rev.-Sci. Eng. 1999, 41 (1), 1-42.
[4] Li, Y.;Xu, G.;Liu, C.;Eliasson, B.;Xue, B. Energy & Fuels 2001, 15, 299-302
[5] Tsui, M.; Miyao, T.; Naito, S. Catal. Lett.2000, 69, 195.
[6] Supat, K.; Chavadej, S.;Lobban, L.; Mallinson, R. ACS Symposium Series, 2003, 852, 83.

[7] Yao, L.; Okumoto, M.; Nakayama, A.; Suzuki E. Energy & Fuels 2001, 15, 1295-1299.
[8] Li, Y.; Liu, C.; Eliasson, B.; Wang, Y. Energy & Fuels 2002, 16, 864-870.
[9] Luque J.;Kraus M. and Wokaun A.; Haffner K.; Kogeischatz U. and Eliasson B. J. Appl. Phys. 2003, 93(8), 4432
[10] Le, H., Master Thesis, University of Oklahoma, Norman, OK, 2003.
[11] Strategic Plan for DOE Hydrogen Program, US Department of Energy, Jan 1998 and its website http://www.doe.gov
[12] Cancino, J. F. Master Thesis, University of Oklahoma, Norman, OK, 2001
[13] Liu, C.; Marafee, A.; Hill, B.; Xu, G.; Mallinson, R.; Lobban, L. Oxidative coupling of Methane with ac and dc Corona Discharges. Ind. Eng. Chem. Res. 1996, 35, 3295. (22)
[14] Eliasson, B.; Kogelschatz, U. Nonequilibrium Volume Plasma Chemical Processing. IEEE Trans. Plasma Sci. 1991, 19, 1063. (36)

# OPTIMISATION OF VALUE -COST- PRICE IN INDIAN GAS MARKET

*SUDARSAN PAUL,*
*DEPUTY GENERAL MANAGER (LNG MARKETING)*
*BHARAT PETROLEUM CORPORATION LIMITED*
*13 WALCHAND HIRACHAND MARG, BALLARD ESTATE*
*MUMBAI - 400 001*
*INDIA*

*Prepared for Presentation at AIChE Spring National Meeting*
*$4^{th}$ Topical Conference on Natural Gas Utilization*
*New Orleans, LA, April 2004*

## ABSTRACT

Just as Coal powered the industrial revolution of the $19^{th}$ century and Oil fuelled the growth of the $20^{th}$ century, Gas is emerging as the preferred source of energy to power the economics of the $21^{st}$ century and India is no exception. High growth in population, increasing urbanization, rising incomes and standards of living of the people of India are expected to result in high and steady consumption growth in petroleum products and natural gas in the country.

From April 2004 commercial supply will start from Dahej LNG terminal of M/s Petronet LNG Limited. Sustained commercial operation will begin from last quarter of 2004 from Shell's Hazira LNG Terminal.

Indian market is looking for LNG from suppliers with softer terms and conditions some of which are mentioned below:

- Shorter duration contracts
- Pricing formula
    - Fixed price
    - Quasi fixed Price by setting a formula
    - Price with Floor and Ceiling
    - Negotiated ratio of fixed elements while lowering the ratio of the crude oil-linked portion.
- Flexible Terms and conditions

This rapidly growing energy market may evolve New Risk Distribution Model. The importers not only will demand that the suppliers bear some of the risks they have taken, but also that they may seek new profits in return for the additional risks they are going to assume by enrolling themselves into entire LNG chain.

More new LNG re-gasification terminals along with pipeline infrastructure are expected to be commissioned for providing terminalling, storage, re-gasification and transportation services nearer the demand centers. Thereby utility owners shall have direct access towards ownership of the gas under long, medium, short-term contracts and access of gas under spot purchase. The major utility owners are negotiating new generation contracts as "Terminalling, Re-gasification and Transportation Agreements (TRTA)".

Indian importers shall endeavor to create an optimal portfolio of LNG contracts by negotiating different pricing formula with different contractual terms. This will enable them to match demand patterns, customer needs and lower generating costs.

## Introduction

Just as Coal powered the industrial revolution of the 19th century and Oil fuelled the growth of the 20th century, Gas is emerging as the preferred source of energy to power the economics of the 21st century and India is no exception. High growth in population, increasing urbanization, rising incomes and standards of living of the people of India are expected to result in high and steady consumption growth in petroleum products and natural gas in the country.

In India, natural gas has 9% share in the primary energy consumption.

Current gas supply profile

| | |
|---|---|
| Oil and Natural Gas Corporation Ltd (ONGC) | 82% |
| Oil India Limited (OIL) | 6% |
| Joint Ventures/Private Companies | 12% |

Sector-wise end use pattern

| Sr. No | Sector | Percentage Utilization |
|---|---|---|
| 1. | Fertilizer Sector | 38 |
| 2. | Power Sector | 36 |
| 3. | Sponge Iron | 6 |
| 4. | Shrinkage & Internal Consumption | 7 |
| 5. | Others | 13 |

## India Hydrocarbon Vision 2025

Government of India released the report "India Hydro Carbon Vision 2025" in April 2000. Comprehensive long-term framework was announced for the development of oil and gas sector in India under globally competitive scenarios.

Strategy for development of

- Timely and continuous review of gas demand & supply to facilitate policy interventions.
- Import of LNG
- Trans-national gas pipelines
- Rationalize duties and taxes
- Coal bed methane
- Gas hydrates
- Regulatory framework

## Gas Sector Policies

- Import of gas, either as LNG or through Trans-national gas pipeline, is under OGL
- Formulated the New Exploration Licensing Policy (NELP) as well as Coal Bed Methane Policy.
- Process of decontrolling the Domestic Gas prices was initiated by introduction of market related price determination mechanism linked to prices of Fuel Oils in the international market.
- Regulatory Framework for the Gas Sector and appointment of Regulator.

## Natural Gas Demand in the country:

Assumptions:

- Gas demand assumes that the Gas supply shall remain at current level till 2004 first quarter.
- Gas demand projection is the incremental gas demand which includes the following:
    - i) Unmet demand
    - ii) Switching over demand – Gas demand from consumers who are currently using alternate liquid fuels.
    - iii) Planned expansion
    - iv) Greenfield projects demand
- Incremental Gas demand from power sector assumes 25% of power deficit to be met by gas based power plants and currently planned projects in the region would come up as per plan.

Incremental Gas Demand – All India

Figs. In MMSCMD

| Sector | Year 2005 | Year 2010 | Year 2015 |
|---|---|---|---|
| Power | 25.0 | 29.8 | 40.1 |
| Fertilizer | 36.0 | 65.1 | 112.6 |
| Other Sectors | 10.4 | 14.4 | 16.7 |
| Total | 71.5 | 109.3 | 169.3 |

Source- CRISIL, India

## What Fuel Price Indian Power Sector is looking for - A survey of the market

Before looking for the answer let us look into certain issues in this sector as detailed below:

- Dispatch order is linked to the demand and cheapest fuel cost.
- If the power producer is not getting the dispatch order due to higher variable cost, he is taking risk towards 'Take or Pay' obligation despite he is not getting the variable cost.
- Power sector value chain comprises of Fuel supplier, Power Producer, Distributor & End Users. In this chain, Power production is regulated. Their pricing formulation and other variables are open for public scrutiny. The distributor is also regulated. Their pricing is also discussed in public forums before finally being decided by a regulator.
- Off-peak and peak differences in power demand.

So when the power producer will decide the fuel he will look for transparency in pricing mechanism since around 60 percent of the power price is the fuel price.

## Power sector reforms in India

The new Electricity Act, 2003 along with other recent initiatives resulted in a new liberal and competitive framework for the development of the power sector in the country. The Government of India have also taken several initiatives as mentioned below:

- Accelerated Power Development and Reforms Program
- Accelerated Generation and Supply Scheme
- Accelerated Rural Electrification Program
- Hydro Electric Power generation initiative to the tune of 50000 MW
- Creation of National Grid

## What is the benchmark fuel price Indian power sector is looking for?

According to major power producers in India, benchmark will always be the comparative cost with coal. According to their opinion, the fuel cost works out around \$2.06/MMBTU considering coal-based project near the coast. If the coal is moved 800 km the cost is approximately \$2.36/MMBTU. Today delivered price of indigenous gas is around \$2.5/MMBTU. The coal power generated at the pithead and moved 1500 kms by HVDC system will cost almost the same what the power generation cost with indigenous gas. Prudent operator will make a balancing between transport of power and transport of fuel.

## Fertilizer Sector

At present there are 63 large size fertilizer units in the country, manufacturing wide range of nitrogenous and phosphatic/complex fertilizers. Of these, 38 units produce urea whereas 9 units produce ammonium sulphate as a by-product. Besides, there are about 79 small and medium scale units producing single super phosphate. 48% capacity is based on Gas and balance are based on Naphtha/Fuel Oil and mixed feed.

With effect from 1.4.2003 Government of India approved new pricing policy for urea units, which will replace the existing retention price scheme. The primary consideration and goal of the new pricing policy is to encourage efficiency parameters of international standards based on the usage of the most efficient feedstock, state-of-art technology and also ensure viable rate of return to the units. Gas is a preferred feedstock for urea manufacturing plants due to ease of operation, lower energy consumption, environment friendly, lower capital cost.

With the change in fertilizer policy demand of gas in this sector is going to increase which has been estimated in previous para under Natural Gas Demand in the country.

## How to meet the incremental Gas Demand and at what price?

To meet the incremental Gas Demand, the gas supplies are expected to materialize from number of possible sources

- Supplies from presently developed fields and possible upside from these due to better reservoir management practices and improvement in recovery factors.

- Supplies from proved but not yet developed gas reserves.
- Potential supplies from successful exploration in NELP blocks.
- Potential supplies from coal bed Methane exploration.
- LNG imports
- Gas imports – Trans-national gas pipelines.

Recently significant gas discoveries have been reported in Krishna Godavari Basin, Cambay Offshore and on land blocks. Commercial supply and development of related pipeline infrastructure will take some time. Piped gas import may be feasible, subject to overall techno-economic considerations and also subject to satisfactory resolution of other complex aspects of cross border gas trade and transit issues.

Hence import of LNG is the viable option to meet the immediate demand. Question is at what price? Let us look at the prevailing natural gas price in India.

## Natural Gas Price in India

Historically, Natural Gas pricing in India was based on Government Awards, opportunity costs by the National Oil Companies and cost of alternative fuel to consumers. Gas pricing came under the administrative control of the Government of India in 1986 and this was predominantly based on a cost plus approach for providing adequate returns to the producer and transmitter of Gas. The consumer prices were constant over a period of time and also uniform irrespective of the End Use. From January 1992, a variable consumer-pricing concept was introduced with an annual escalation to take the consumer price close to price parity with fuel oils.

With effect from 1.10.1997 India's consumer price of natural gas has been linked to the price of basket of LS/HS fuel oils are as follows:

| Year | General Price | Concessional Price for North Eastern States |
|---|---|---|
| **1997-98** | **55%** | **30%** |
| **1998-99** | **65%** | **40%** |
| **1999-2000** | **75%** | **45%** |

However, these prices are due for revision w.e.f. 1.4.2000 so as to achieve 100% fuel oil parity.

## Gas Pricing from JV Exploration – JV Gas

With the economic liberalization policy of Government of India, private participation was allowed in the petroleum exploration. Exploration contracts were awarded under Petroleum Exploration Licence & Mining Lease. Pricing formula were based on Low Sulphur Fuel Oil (LSFO) and in some cases based on High Sulphur Fuel Oil (HSFO) prices with floor and ceiling.

During 1997-98 Government of India formulated New Exploration Licensing Policy (NELP) to provide level playing field in which all parties can compete on equal terms for the award of exploration acreage. In three rounds of bidding, NELP-1 (1999), NELP-II (2000) &

NELP-III (2002), PSCs have been signed for 70 blocks. Under NELP-IV, 24 blocks have been offered for exploration.

The floor and ceiling price at which the gas was sourced, from JV under pre-NELP is higher than the price of gas under APM. Therefore, pooling the gas price from the indigenously available gas sources and selling the same at the uniform price to all the consumers met the differential cost. However, from March 2003, Government of India started withdrawing pooling support and pre-NELP JV Gas is expected to be sold at MDP.

## Arrival of LNG in India

From April, 2004, commercial supply of re-gasified LNG will start from 5 MMTPA capacity Dahej LNG Terminal of M/s Petronet LNG Limited (PLL). Promoters of PLL are Bharat Petroleum Corporation Ltd. (BPCL), Indian Oil Corporation Ltd. (IOCL), Oil & Natural Gas Corporation Ltd. (ONGC) and GAIL (India) Ltd. Gaz de France (GDF) has taken 10% equity and RasGas, Qatar is in the process of taking 10% equity. PLL has signed SPA with RasGas for importing 5 MMTPA LNG at Dahej and 2.5 MMTPA at Kochi.
Commercial operation will begin from last quarter of 2004 from Shell's Hazira LNG terminal. Upon commissioning the terminal will have a capacity of 2.5 MMTPA, rising to 5 MMTPA.

## Proposed LNG Projects in India

Concurrently, many private parties also prepared plan to import LNG at various port locations in the country. Details are:

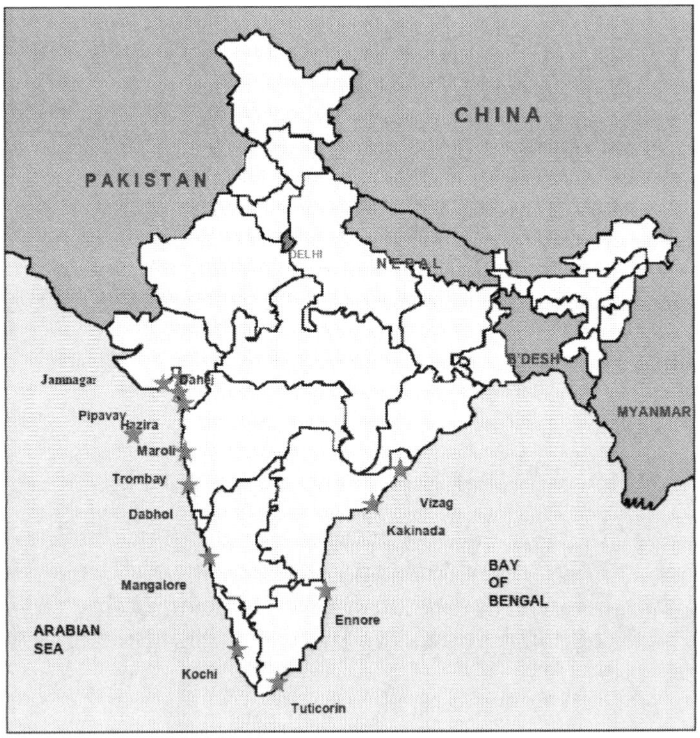

| WEST COAST | | | |
|---|---|---|---|
| Location | Company | Capacity (MMTPA) | Commissioning by |
| Dahej | Petronet | 5.0 | April 2004 |
| Pipavav | BG | 2.5 | - |
| Jamnagar | Reliance | 10.0 | Proposed |
| Hazira | Shell | 2.5 | 2004 |
| Maroli | Unocal | 2.5 | - |
| Mumbai | Tata/Total | 2.5 | - |
| Dhabol | Enron | 5.0 | Major work comp. |
| Mangalore | Finolex | 2.5 | - |
| Kochi | Petronet | 2.5 | Proposed |

| EAST COAST | | | |
|---|---|---|---|
| Location | Company | Capacity (MMTPA) | Project Status |
| Ennore | TIDCO | 5.0 | - |
| Tuticorin | Indian Gas | 2.5 | - |
| Kakinada | IOC | 2.5 | Proposed |
| Vizag | Total/HPC | 2.5 | - |

**<u>Petronet LNG Limited</u>**

- **Commercial supply from April, 2004 from Dahej LNG terminal**
- **Head start over competition will result in first mover advantage.**
- **SWOT Analysis**

| Strengths | Opportunities |
|---|---|
| ➢ Strong promoters profile.<br>➢ Location advantages<br>➢ Agreed supply & off take contracts in place<br>➢ Head start advantage<br>➢ Project implementation is on schedule | ➢ Consumption of gas as a proportion of total energy consumption at very low levels<br>➢ Large unmet demand<br>➢ Cross country trunk pipeline in place and work in progress for further augmentation.<br>➢ CNG and PNG demand. |
| ➢ Price comparison of Re-gasified LNG and domestic gas | ➢ No. of LNG projects in the pipeline at various stages of implementation<br>➢ Possible increase in domestic gas production with award of new E&P contracts under NELP |
| **Weakness** | **Threats** |

• Creating value in the entire chain

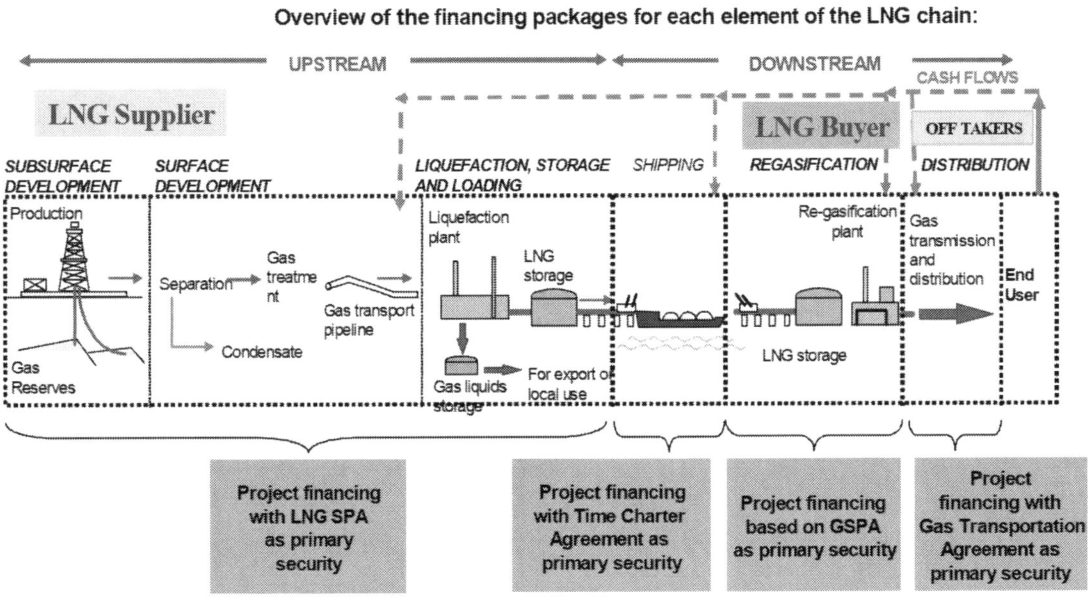


## LNG price and contractual terms in Indian market

Globally utility markets are open to competition. It will no longer be possible to sell LNG to monopolies. The old compact, whereby the Seller effectively took most of the price risk and the Buyer took the volume risk is breaking down. Importers are facing uncertain market conditions and possible loss of market share. This situation will force importers to seek more flexible volume terms and/or more market responsive pricing. Under this context the following developments are taking place in the LNG chain:

- Capital cost intensity is declining.
- Gas resource base is not constrained
- Oil price linkage is falling away – Different geographical markets will continue to develop their own tailored solutions whether these be benchmark gas spot prices, power netbacks, or just a redefinition of the oil price linkage with, for example, Floors and Ceilings or a new Oil/Gas Btu price relationship.
- Long term and rigid trading conditions are being reviewed.
- Spot trading of LNG, which is one of the yardsticks of flexibility, is increasing at a rapid rate.

Indian market is looking for LNG from suppliers with softer terms and conditions some of which are mentioned below:

- **Shorter duration contracts**
- **Pricing formula**
  - Fixed price
  - Quasi fixed Price by setting a formula
  - Price with Floor and Ceiling
  - Negotiated ratio of fixed elements while lowering the ratio of the crude oil-linked portion.

- **Flexible Terms and conditions**
  - Making the contract two-tired, with a flexible delivery portion (to meet seasonal demand) and a fixed delivery portion based on base load.
  - Varying TOP coverage
  - Price linked to net Calorific Value

## Conclusion

India is a rapidly growing energy market. This market may evolve New Risk Distribution Model.
The importers not only will demand that the suppliers bear some of the risks they have taken, but also that they may seek new profits in return for the additional risks they are going to assume by enrolling themselves into entire LNG chain.

New LNG re-gasification terminals along with pipeline infrastructure are expected to be commissioned for providing terminalling, storage, re-gasification and transportation services nearer the demand centers. Thereby utility owners shall have direct access towards ownership of the gas under long, medium, short-term contracts and access of gas under spot purchase. The major utility owners are negotiating new generation contracts as "Terminalling, Re-gasification and Transportation Agreements (TRTA)".

Indian importers shall endeavor to create an optimal portfolio of LNG contracts by negotiating different pricing formula with different contractual terms. This will enable them to match demand patterns, customer needs and lower generating costs.

# Selecting the Optimal Process for C2+ Extraction from LNG

*Stanley Huang*
*David Coyle*
*Joseph Cho*
*Charles Durr*
*Kellogg Brown & Root, Inc.*
*601 Jefferson Avenue*
*Houston, TX 77002*

*Prepared for Presentation at AIChE Spring National Meeting*
*4th Topical Conference on Natural Gas Utilization*
*New Orleans, LA, April 2004*

## ABSTRACT

Kellogg Brown & Root, Inc. (KBR) expects LNG imports to the U.S. to increase significantly in the near future. While securing LNG supply through long-term contract will continue to play an important role, acquiring spot LNG from scores of foreign producers will also be a popular practice. Also, gas specifications of Pacific Rim countries are significantly richer than the U.S counterparts. Although there may be changes toward U.S. specifications if sufficient long term contracts are in place, it will take years. As a result, a receiving terminal in the U.S. can expect imported LNG to have wide compositional variations.

It will be the responsibility of a receiving terminal owner to assure the re-gasified LNG meets U.S. pipeline specifications. Toward this goal, certain compositional adjustments on batches of LNG are anticipated by the industry. There is a good chance that gas received into the U.S. will need to have the heating value lowered either by nitrogen injection or heavy component ($C_2$+) extraction. $C_2$+ removal is more versatile than nitrogen injection, and can generate additional revenue if suitable demand for higher value extracted components exists.

This paper classifies the $C_2$+ extraction processes of receiving terminals into three categories based on the pressure-boosting mechanisms for the lean gas, also called residue. Each category is illustrated by several examples and process advantages and limitations are discussed. Additionally, from our extensive experiences in plant design, attention is paid to plant operability and flexibility with respect to inlet variations. Finally, this paper provides indexed capital costs of the different process schemes as a means of comparing economic merits. The conclusions of this work will serve as a useful guideline for selecting the optimal $C_2$+ extraction process for receiving terminals.

## Introduction

Currently, the imported LNG accounts for roughly 1 % of the total gas demand in the U.S. With declining domestic gas production, the LNG import is projected to increase to 10 % by 2010, or around 2 to 2.5 Tscf annually. As a result, the issue of "gas interchangeability" has become important [Rogers, 2003a and b]. The intention is to assure the re-gasified LNG is acceptable to all gas users after injecting into existing pipeline grid.

LNG specifications are subjected to supply-demand negotiations, similar to other commodities. Since Japan imports about 50 % of globally traded LNG, the Japanese gas specification of relatively high $C_2$+ content will probably continue to be the basis of many future LNG supplies. Consequently, most LNG suppliers produce LNG richer than U.S. pipeline specifications. With LNG imports to U.S. projected to increase, some LNG producers may be willing to produce lean LNG suitable for the U.S. market. Nonetheless, the spot market will remain popular for opportunistic purchasers in the foreseeable future. The capability of a receiving terminal to assure gas interchangeability would enhance business opportunities.

$C_2$+ extraction is one way to achieve gas interchangeability. Other methods include injections of air, N2, or inert gases. Aside from being versatile (able to change gas properties over a wide range), one advantage of $C_2$+ extraction over others is that the $C_2$+ products may have higher market values. Several $C_2$+ extraction schemes have been proposed in recent years, and some are protected by patents. Without exception, all $C_2$+ extraction schemes rely on volatility differences of components in large-scale operations. The problem with this approach is that the operating pressures of fractionation towers generally fall below the pipeline delivery pressures. Consequently, boosting the pressure of residue gas after the $C_2$+ extraction is required.

This paper classifies existing schemes into three categories based on their handling of the residue gas after $C_2$+ extractions. Advantages and disadvantages of each category are discussed. An order-of-magnitude cost index is also provided as a general guide. Finally, factors affecting the selection of an optimal scheme are addressed.

## C2+ Extraction Schemes

The heart of a $C_2$+ extraction scheme is a demethanizer(DeC1). Other associated facilities for product handling may also be provided. The upper limit in operating pressure for a typical DeC1 is about 46 bar, the critical pressure of methane. Lowering operating pressures not only reduces column cost, but also facilitates component separation which, in turn, also reduces reboiler duties.

Pipeline pressure specifications are region-specific, and may run as high as 100 bar for long-distance transportation. Pressure boosting of the residue gas from the DeC1 is required in order to meet the high pressure specifications.

Existing $C_2$+ extraction schemes are classified into three categories based on their handling of the residue gas. The three categories are Residue Compression, Residue Compression and Condensing, and Residue Condensing. The categories are denoted A, B, and C, respectively. Table 1 and Figures 1 to 8 present an overview of Categories A to C, which follow a general trend of descending requirements of mechanical (compression and pumping) horsepower. The references are listed in Section 5. Table 2 lists the simulation basis used in this work.

The Residue Compression scheme as shown in Figure 1 is straightforward, has good process flexibility (no theoretical lower limit in operating pressure) and operability (insensitive to inlet LNG temperature and DeC1 heat input). However, this scheme can be costly in both installation and operations because of the high horsepower requirement. On the other hand, Residue Condensing schemes, Figures 5 to 8, require absorbers or exchangers to re-condense the residue gas. The recondensed LNG is then pumped for pressure boosting and vaporized. The process is more complex because two-phase changes are involved in the process. When properly designed, Residue Condensing is less expensive to install and operate than Residue Compression.

Several schemes, Figures 2 to 4, are combinations of the two previously described schemes. They require compression, but not to the same extent as Residue Compression. These schemes achieve significant savings in compression horsepower and have improved process flexibility and operability.

Figure 4 of Category B was originally proposed for LNG purification at production sites, it is presented here for completeness. However, cost analysis for Figure 4 is not included in the next section because it is not competitive in LNG vaporization applications. The additional refrigeration recovery exchanger at the DeC1 top not only adds significant capital cost, but also increases compression horsepower. The purpose is to reject the heat input from the compressor and DeC1 reboiler. However, the heat input actually helps unload the downstream LNG Vaporizer in this application.

Exergy is a measure of the maximum amount of work potentially extractable from a given thermal (cold) source. By the second law of thermodynamics, the greater the irreversibility of a process, the greater the exergy loss. Minimizing the exergy loss is of interest if the cold energy of LNG would be utilized to generate power for export. Figures 9 and 10 show typical heating-cooling curves of the LNG pre-heater/condensers in Categories B and C, respectively. For schemes in Category B (Figure 9), there is a close match between heating and cooling sides. This close match indicates small exergy loss, although the advantage is gained at the expense of large exchanger areas. Figure 10 shows schemes in Category C and these factors are simply reversed.

The improved flexibility and operability of Category B over C can be explained by Figures 9 and 10 as well. The prerequisite of pumping LNG to boost pressure is the total condensation of the residue gas in a Residue Condensing scheme. To meet this requirement, sufficient cold energy in the inlet LNG must be available to condense the residue gas without incurring temperature cross-over in the pre-heater/condenser. Category C (Figure 10) achieves this by maintaining the inlet LNG at subcooled condition (through raising the pressure to a sufficiently high level) and returning the residue gas at a relatively high temperature to avoid the cross-over. However, the operating pressure of the DeC1 is limited by the methane criticality. As a result, there is a limited pressure range where the residue condensing scheme works, and limited capability to handle varying LNG inlet temperatures, extraction levels, and compositions.

Category B (Figure 9) accepts some increase in equipment cost and compression energy by adding vapor compression as a design variable. This vapor compression effectively eliminates the possibility of temperature cross-over. Thus it allows a wider range of DeC1 operating pressures (flexibility) and also enables the facility to handle larger variations in LNG inlet temperature and compositions (operability).

There are methods to improve the flexibility and operability of Category C. Figure 11 demonstrates the impact of adding a residue gas heater in Category C, as shown in Figure 8. This added exchanger raises the residue gas to a higher temperature in order to avoid the temperature cross. The same amount of heat, if goes through the DeC1 reboiler, would significantly reduce the C2 recovery level.

## Cost Optimization for Receiving Terminals

In a typical receiving terminal, the three major areas of capital expenditure are: marine facilities (including seawater intake facilities if applicable), LNG storage tanks, and process equipment including $C_2$+ extraction. The inclusion of $C_2$+ extraction facilities may impact the selection of plant heat source which in turn impacts environmental aspects. As such, the optimization of $C_2$+ extraction facilities should be an integral part of the overall plant design.

Using the equipment cost as a starting point, Figure 12 provides an indexed comparison for different schemes of $C_2$+ extraction facilities. Only key equipment items which are presented in the figures are included to highlight the differences. The heat source for operating the fraction column (furnaces and heating medium circuit) and $C_2$+ product handling facilities are excluded.

The capital cost can be correlated loosely by the total mechanical (compression and pumping) horsepower. As a rule of thumb, the compression horsepower should be minimized whenever possible because the contribution by compressors is dominant in cost evaluations. For example, between the two schemes presented in Figures 2 and 3 under Category B, the one in Figure 3 significantly reduces the compression requirement by installing an intermediate vapor separator

.

Without major compression requirements, such as schemes in Category C, no specific design distinguishes itself from others from equipment cost viewpoint. The optimization of capital cost should consider other factors, such as heat sources for reboilers, C2 recovery level required, clients' preferences, etc.

The $C_2$+ extraction facilities may also impact the operating cost, because reboilers of fractionation columns demand relatively high temperatures (above ambient) which are obtained by combustion or waste heat. For a receiving terminal using seawater as vaporization medium, the impact on the fuel gas consumption estimate can be significant. In this case cost optimization may steer the process design toward reducing reboiler duties by lowering the column operating pressure, process heat integration, or exploring other means to help achieve gas interchangeability.

On the other hand, if a receiving terminal is designed to use combustion as the vaporization source, the impact of $C_2$+ extraction facilities on the fuel gas consumption would be relatively minor. Therefore, the impact of adding $C_2$+ extraction facilities would be mainly on the capital spending, but not on fuel gas estimate (the fuel shifts from the vaporizer service to the reboiler). There will be efficiency differences between Submerged Combustion Vaporizer (SCV) and conventional furnaces to consider in the fuel gas estimate and economic evaluation.

KBR routinely performs Life Cycle Cost analysis for cost optimization in order to capture the impacts of both one-time capital expenditure (CAPEX) and longer-term operation expenditure (OPEX). In recent studies, increases in domestic natural gas cost have influenced the analysis results. Also, environmental regulations may also significantly impact the plant design. Of direct relevance to $C_2$+ extraction facilities would be the emission limit on the NOx level of combustion burners.

## Conclusions

The conclusions of this paper are listed as follows:

- Facilities for $C_2$+ Extraction from LNG are grouped into three categories based on their handling of residue gas from the DeC1 overhead: Residue Compression, Residue Compression and Condensing, and Residue Condensing.
- Residue Compression has a simple design and is easy to operate, but can be expensive in capital and operational costs.
- Residue Condensing, followed by LNG pumping and vaporization, requires more complex process designs and controls. This approach is more economical in capital and operational costs.
- Schemes of Residue Compression and Condensing require compression, but not to the same extent as Residue Compression. These schemes achieve significant

savings in compression horsepower and have improved process flexibility and operability

- Residue Condensing schemes are fairly similar in capital costs because no compressors are needed. The compressor contribution is dominant in cost evaluations.
- The capital cost of $C_2$+ extraction facilities represent a small portion of overall plant cost. However, providing and operating such facilities can have far-reaching effects on the selection of plant heat sources and air emissions.

## References


McCartney; D. G. "Method for Vaporizing and Recovery of Natural Gas Liquids from Liquefied Natural Gas", U.S. Patent 6,564,579 to Black & Veatch Pritchard Inc. (2003).

Rambo, C. I., Wilkinson, J. D., and Hudson, H. M., "Liquefied Natural Gas Processing", U.S. Patent 5,114,451 to Elcor Corporation (1992).

Rhoades, G. D., Weber, L. R. C., and Engdahl, G. E., "Process for Purifying Liquid Natural Gas", U.S. Patent 5,359,856 to Liquid Carbonic Corporation (1994).

Rogers, D. R., "Gas Interchangeability and Its Effects on U.S. Import Plans", Pipeline & Gas Journal, September (2003a).

Rogers, D. R., "Long-term Solution Needed to Embrace Imports with Pipeline Gas", Pipeline & Gas Journal, September (2003b).

Yang, C. C., Kaplan, A., and Huang, Z., "Cost-effective Design Reduces C2 and C3 at LNG Receiving Terminal", Oil & Gas Journal, Issue of May 26 (2003).

## 6. List of Tables

**Table 1** Three Categories of $C_2$+ Extraction Schemes Based on Their Handling of Residue Gas

| Figures | Description | Advantages | Disadvantages |
|---|---|---|---|
| 1 | **Category A:** Residue Compression | Straightforward process, no phase changes; Good flexibility: no theoretical lower limit in DeC1 operating pressure; Good operability: insensitive to inlet LNG temperature and DeC1 heat input; | High capital and operating costs for compressors; |
| 2, 3, 4 | **Category B:** Residue Compression and Condensing | Between Categories A and C; Small exergy loss; | Between Categories A and C; Relatively large exchanger areas for LNG re-condenser; |
| 5, 6, 7, 8 | **Category C:** Residue Condensing | Low capital and operating costs; Relatively small exchanger areas for LNG re-condenser; | More complex process, phase changes twice; Reduced flexibility: relatively high DeC1 operating pressure; Reduced operability: sensitive to LNG inlet temperature and DeC1 heat input; Large exergy loss; |
| Note: Exergy is a measure of the maximum amount of work potentially extractable from a given thermal (cold) source. By the second law of thermodynamics, the greater the irreversibility of a process, the greater the exergy loss. | | | |

**Table 2** Simulation Basis of this Work

| **LNG inlet** | |
|---|---|
| Inlet Temperature | -160 °C |
| Inlet Pressure to $C_2$+ Facilities | 10 bar |
| Components | Mole Fraction |
| Nitrogen | 0.00127 |
| CO2 | 0.00008 |
| Methane | 0.90692 |
| Ethane | 0.08617 |
| Propane | 0.00552 |
| i-Butane | 0.00003 |
| n-Butane | 0.00001 |
| i-Pentane | 0.00000 |
| TOTAL | 1.00000 |
| HHV, MJ/Sm3 | 40.3 |
| Btu/SCF | 1,082 |
| Plant Capacity, inlet LNG | 500 MMscfd |
| Send-out Gas Pressure | 100 bar |
| C1 Composition in DeC1 Bottom Product | 1% |

**7. List of Figures**

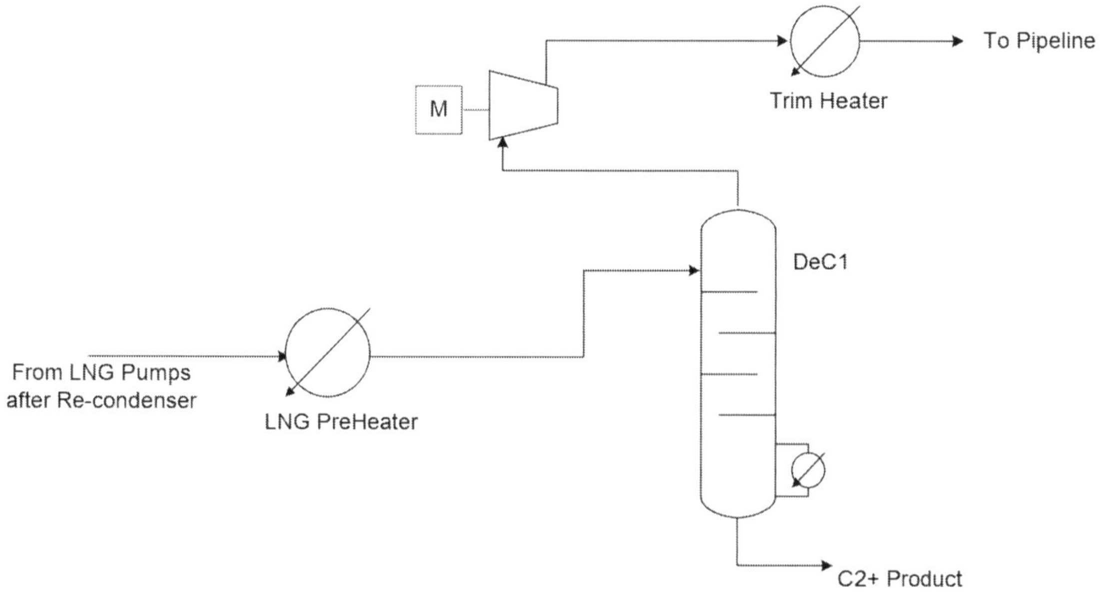


Figure 1 Residue Compression [Rambo et al., 1992]

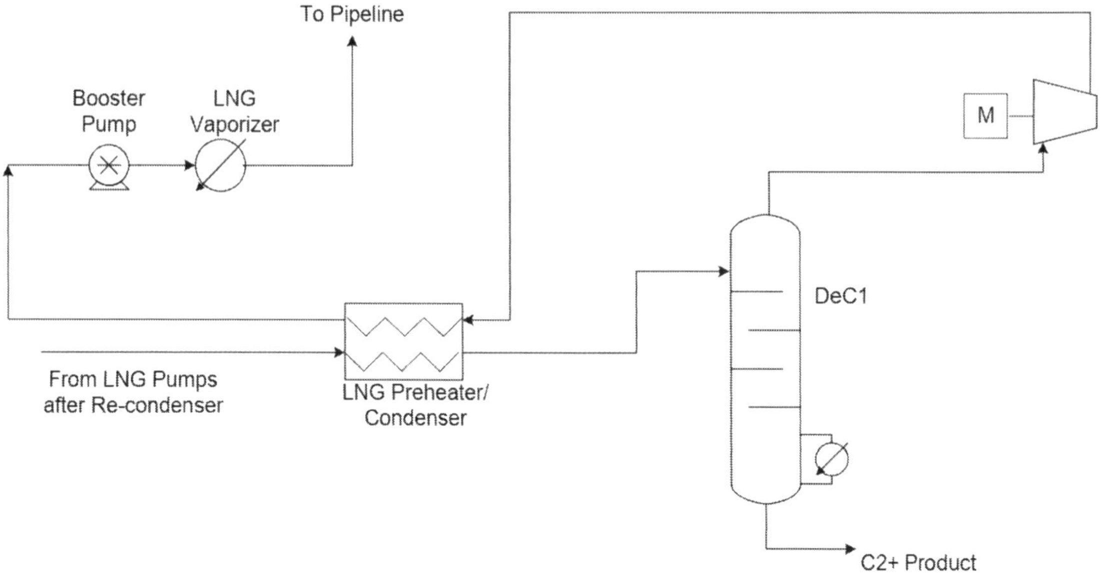


Figure 2 Residue Compression and Condensing, Configuration 1 [McCartney, 2003]

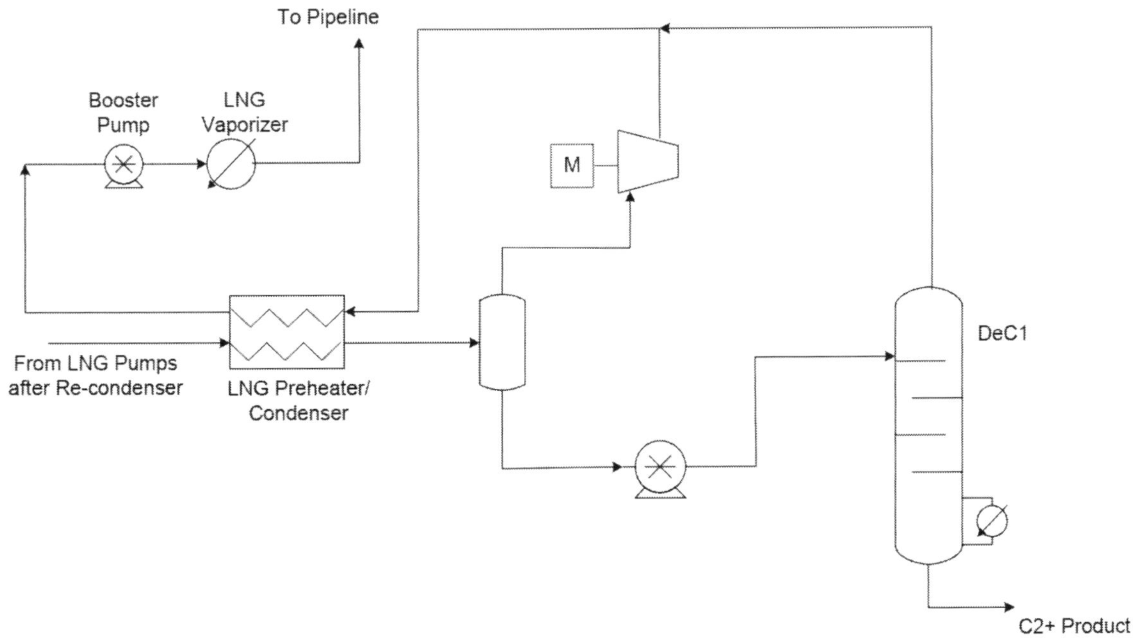


Figure 3 Residue Compression and Condensing, Configuration 2 [McCartney, 2003]

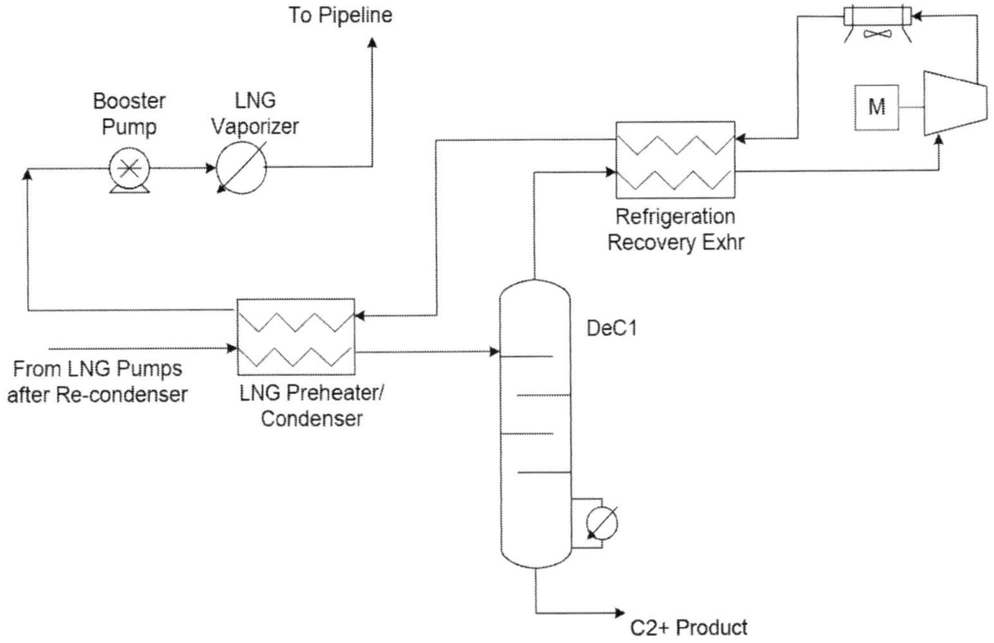


Figure 4 Residue Compression and Condensing, Configuration 3 [Rhoades et al., 1994]

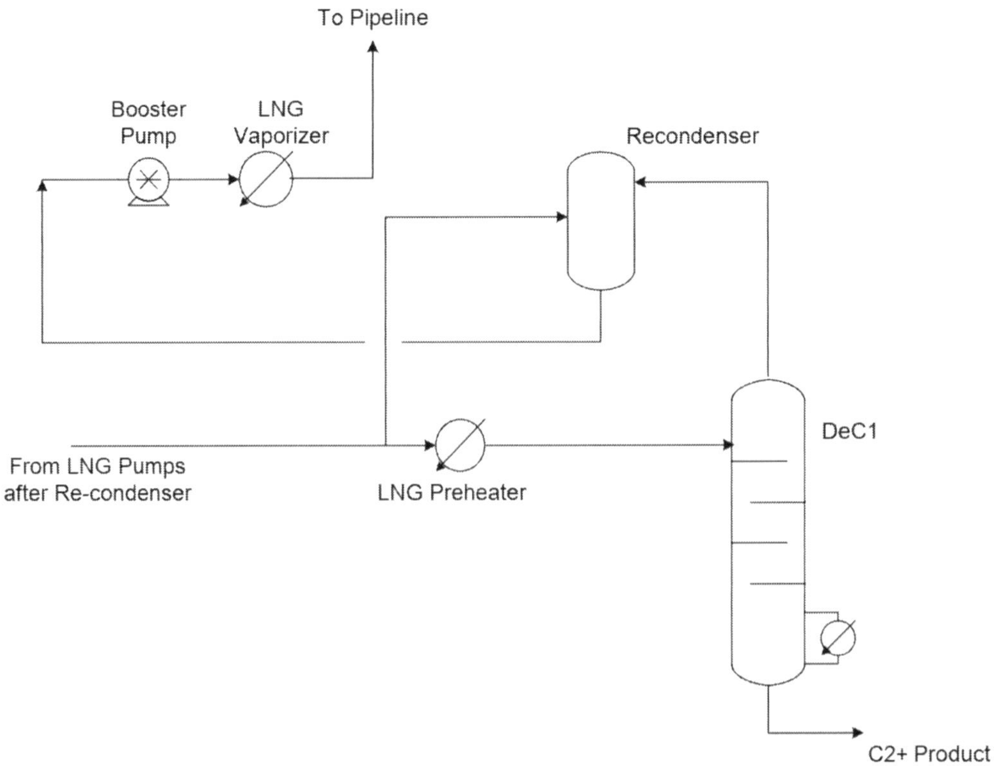


Figure 5 Residue Condensing - Liquid Absorption [Yang et al., 2003]

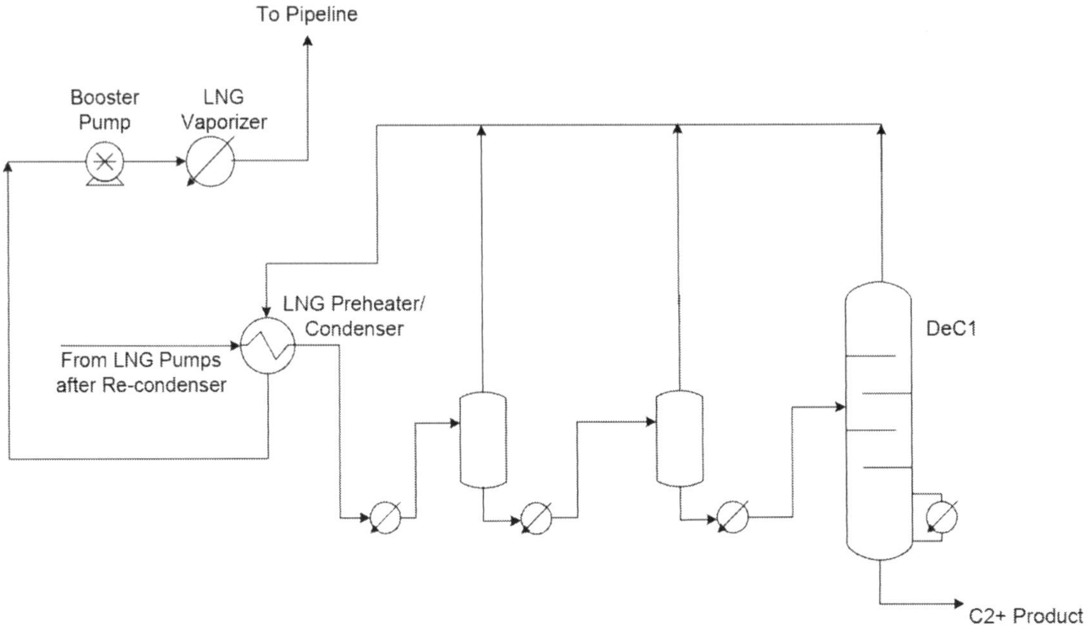


Figure 6 Residue Condensing - Exchanger, Configuration 1 [KBR Design for Cove Point Receiving Terminal, 1972]

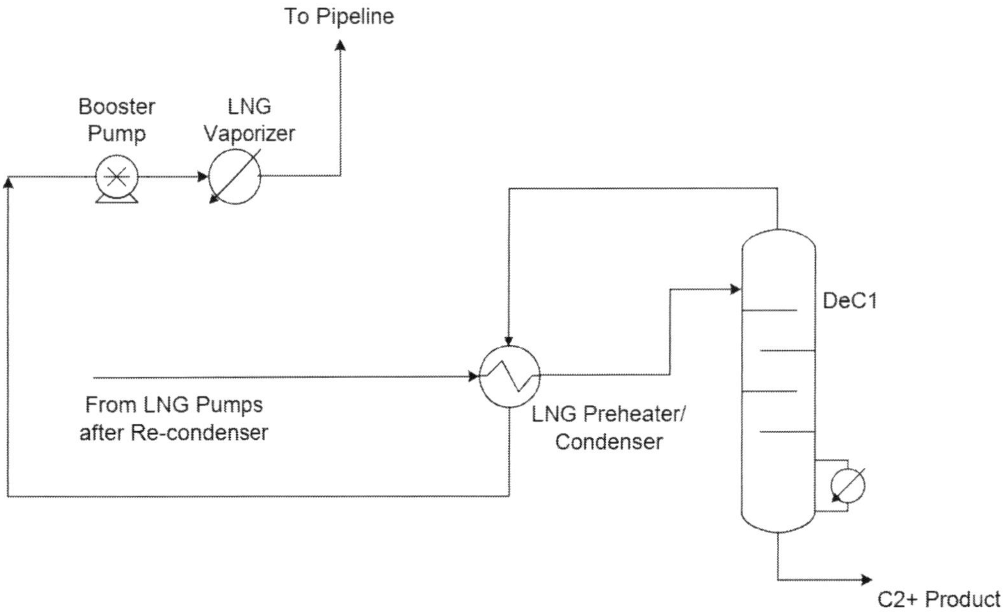


# Figure 7 Residue Condensing - Exchanger, Configuration 2 [KBR Design]

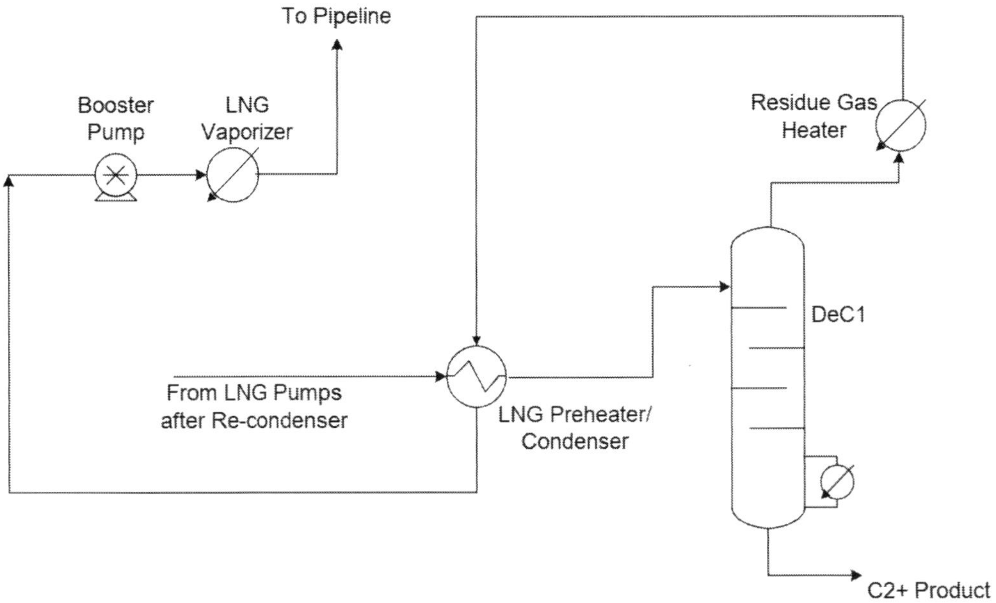


# Figure 8 Residue Condensing - Exchanger, Configuration 3 [KBR Design]

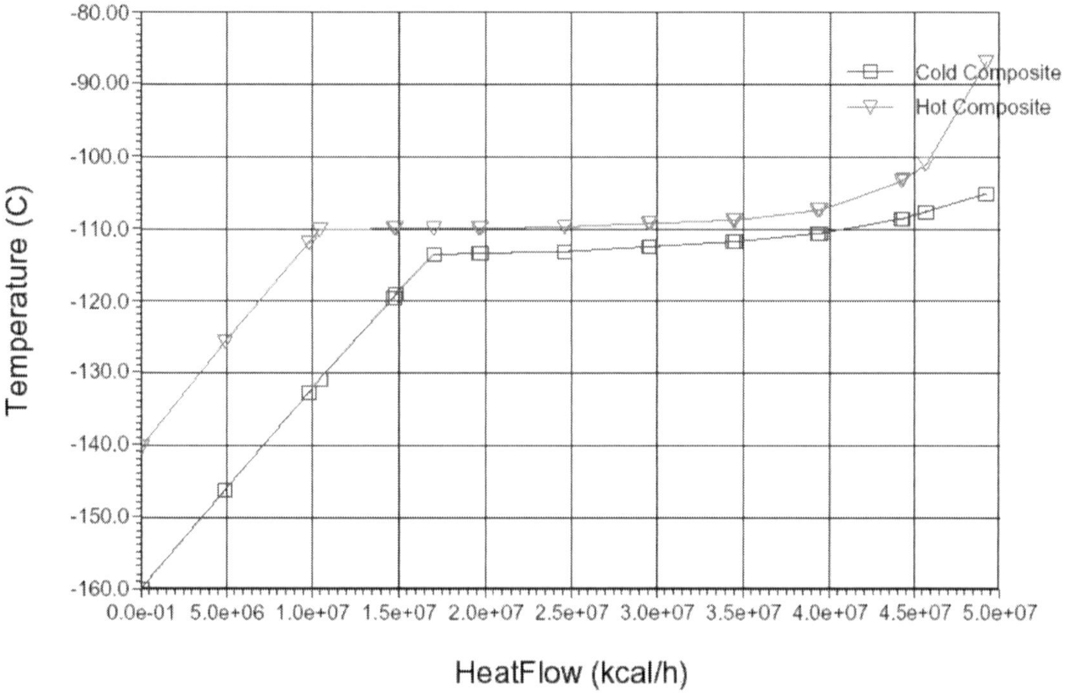


Figure 9 Typical heating-cooling curves for Residue Compression & Condensing Schemes

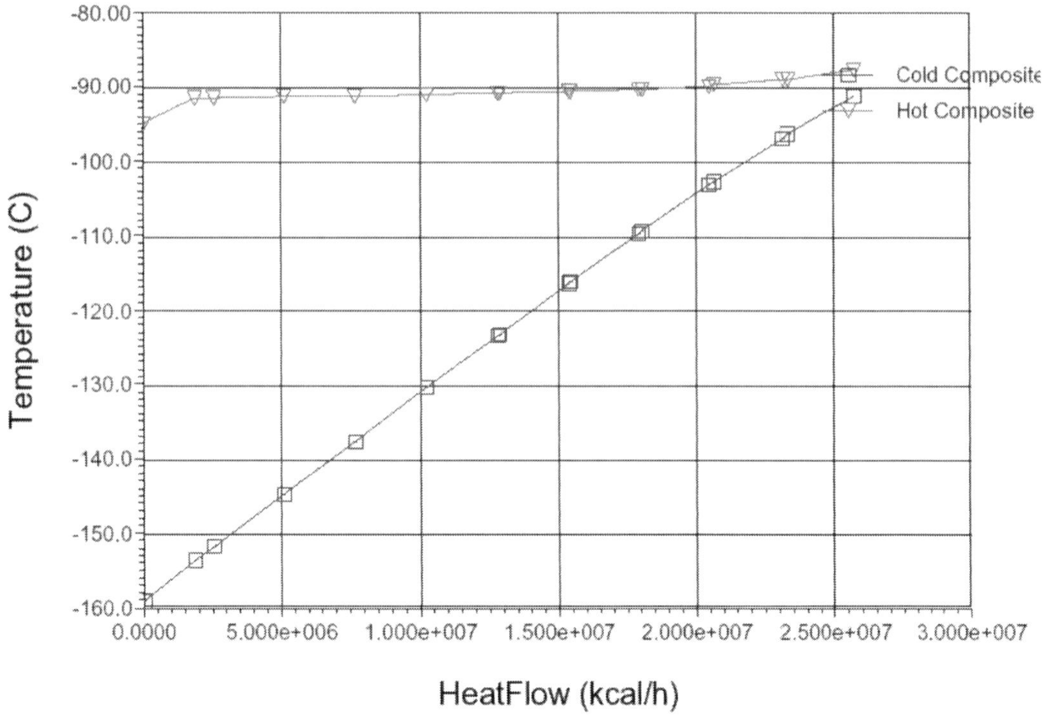


Figure 10 Typical heating-cooling curves for Residue Condensing Schemes

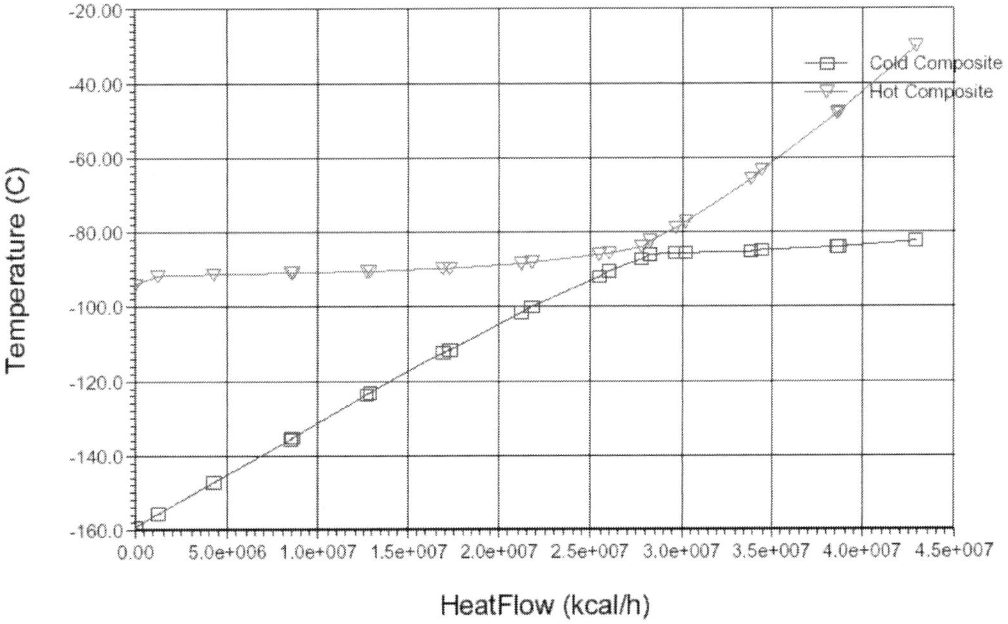


Figure 11 Residue Gas Heater in Figure 8 improves flexibility and operability of Residue Condensing Schemes.

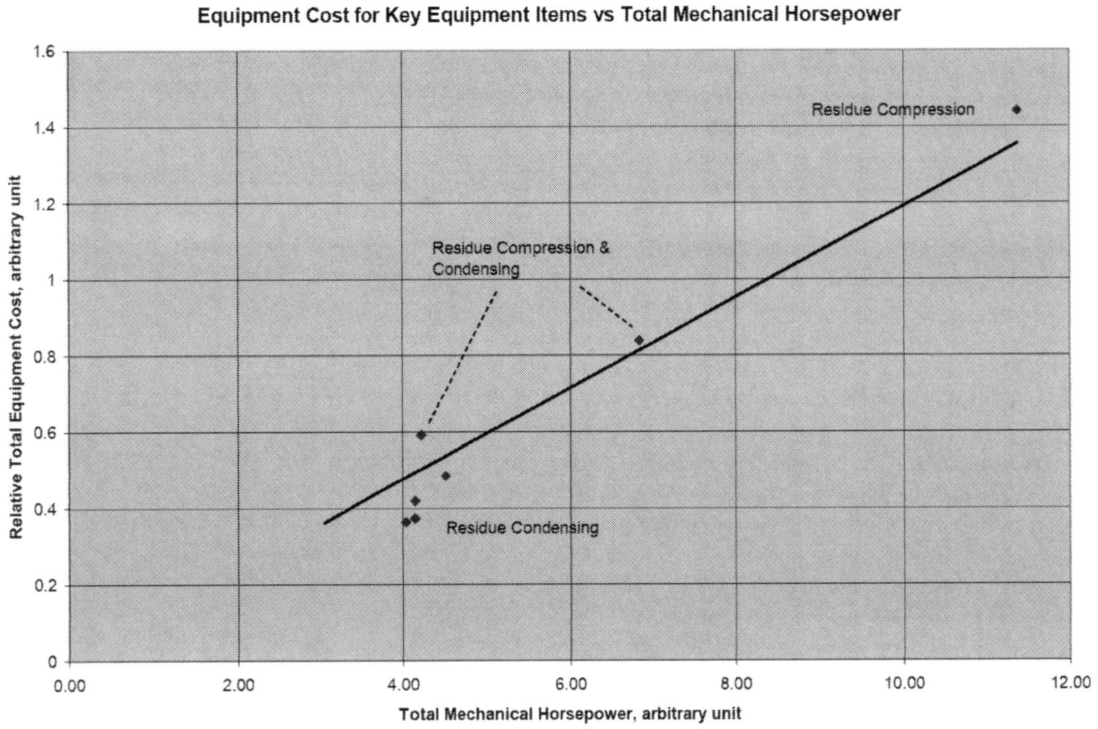


Figure 12 Equipment cost for key equipment items versus total mechanical horsepower. The compressor contribution is dominant in cost evaluations. Data of Figure 4 is not included.

**ZERO EMISSION LNG VAPORIZATION:**

**AN ENVIRONMENTAL FRIENDLY AND COST EFFECTIVE DESIGN**

**FOR LNG RECEIVING TERMINALS**

*C.C. Yang* and Z. Huang*

*Foster Wheeler USA Corporation*

*2020 Dairy Ashford Road*

*Houston, Texas 77077*

*A. Kaplan*

*Marathon Oil Company*

Keyword: LNG, Natural Gas, Vaporizer, Power Plant, Cold Utilization

## Abstract

Open Rack Vaporizer (ORV) and Submerged Combustion Vaporizers (SCV) are normally the equipment of choice at LNG regasification terminals. However, there are concerns about the environmental impacts of both these types of equipment. For ORV vaporizers, there are concerns regarding the large seawater intake and return system, the colder returned seawater temperature and the residual chlorine (as seawater is treated with sodium hypo-chlorite) in the returned seawater. For SCV vaporizers, there are concerns regarding burning expensive fuel gas and NOx and $CO_2$ emissions.

This paper describes an innovation using the waste heat from either a power plant or industrial facilities for LNG vaporization. This will eliminate fuel requirements, while reducing or eliminating air or water emissions and improves the LNG terminal and plant facilities thermal efficiency.

** Author to whom correspondence should be addressed*

## 1. Introduction

The selection of LNG vaporizers for the LNG receiving terminals has recently been more critically evaluated for the impact of the discharges of the flue gas, and/or seawater on the environment compared to the traditional LNG vaporizers.

This paper describes an innovation using the waste heat from either a power plant or industrial facilities for LNG vaporization. This will eliminate fuel requirements, while reducing or eliminating air or water emissions and improves the LNG terminal and plant facilities thermal efficiency.

LNG cold has been previously utilized in power plants for gas turbine inlet air chilling. An additional water/glycol or methanol water loop can be provided and an air-cooling system for chilling air added. To take advantage of the cooling water in a combined cycle power plant or in an industrial facility if available in the vicinity of the LNG receiving terminal, the cooling water can be utilized for the LNG vaporizers to further improve the power plant or industrial facility performance with minimum additional new equipment.

With the recent development of various new types of vaporizers as described below, the cooling water can be effectively used.

There are several types of LNG vaporizers commonly used in LNG receiving terminals. The following five types have either been used or demonstrated in LNG receiving terminals:

A. Open Rack Vaporizers (ORV)
B. Submerged Combustion Vaporizers (SCV)
C. Combined Heat and Power unit with Submerged Combustion Vaporizer (CHP-SCV)
D. Shell & Tube Vaporizers and RELI-VAP vaporizer
E. Other type of Vaporizers - Ambient Air-Heated Vaporizers,

Key issues to be considered in the evaluation and selection of vaporizer type are:

- Availability and quality of seawater
- Cost of fuel
- Environmental issues such as air and water emissions

LNG receiving terminals commonly use one of two types of LNG vaporizers. These are the Seawater Open Rack Vaporizer (ORV) and the Submerged Combustion Vaporizer (SCV). In general, the ORV system uses seawater as the heat medium. It has a lower operating cost than the SCV, but normally a higher capital cost because of the vaporizer equipment cost, the added seawater intake/outfall system, the large diameter seawater pipes, the seawater pumping system, and the seawater treating system. The SCV requires fuel for the LNG vaporization, and the fuel consumption is about 1.5% of the send-out gas. Thus, it has a higher operating cost than the ORV as the fuel has a significant economic value at the LNG terminals. Shell and Tube type vaporizers have been used in the application of utilizing LNG cold for pre-cooling the air in power plant applications. Intermediate Fluid Type LNG Vaporizers (IFP) or Shell and Tube type LNG Vaporizers have been used in the LNG receiving terminal by rejecting cold to the seawater or a heat transfer medium.

## 2. Comparison and Operational Issues of LNG Vaporizers

### 2.1 ORV

Open Rack Vaporizers (ORVs) use ambient seawater as their source of heat in an open, falling film type arrangement to vaporize LNG passing through the tubes. ORV's are widely used in Japan and Europe, and are well proven in baseload LNG terminal service. There are two main manufacturers of ORV's, both located in Japan (Kobe Steel and Sumitomo). In general, the seawater temperature is preferably always above 8 °C for using ORV.

<u>General Description</u>

The ORV is made of an aluminum alloy for good mechanical characteristics at low temperatures, excellent workability, and high thermal conductivity. The ORV system consists of an aluminum alloy header and a heat conductor panel comprising a large number of heat exchanger tubes in a row like a curtain. The ORV is made up of separate panels, with several of these panels forming independent panel groups. The entire unit is configured from several of these panel groups. The panels are coated externally with zinc alloy, providing corrosion resistance against seawater. ORV's require regular (usually annual) maintenance to keep the finned tube surface clean.

Seawater is fed from an overhead distributor, falls over aluminum panels and is collected in a trough below where it is routed and discharged back to the sea. The seawater flows over the outer surface of long finned tube panels, vaporizing the LNG flowing inside, while cooling the seawater.

The seawater is chlorinated to protect the surface of the tube panel against bio-fouling and to prevent marine growth inside the piping. Fluctuations in product gas demand; gas outlet temperature and seawater temperature will be handled by turndown of the unit, which can be over 90%.

<u>Water Quality</u>

Water quality is a critical requirement for successful operation of an ORV system. Key requirements are:

- ORV's require significant amounts of water. This will lead to a requirement to carefully evaluate and assess the amount of underwater fish and plant life that are ingested by the intake system. Often, significant design requirements will be involved to minimize this.
- Chlorination for water treatment is desirable, however residual chlorine content can have a negative impact on the marine environment by killing significant marine life. This can be minimized by "shock" treatment of chlorine.

- The water should not contain solids exceeding 2 mm in diameter in order to assure uniform water flow without jamming of the solids between the water trough and the top of the tube panel.
- Water containing heavy metal ions, $Cu^{++}$, $Hg^{++}$ must not be used in the water lines, as these ions will shorten the lifetime of zinc-aluminum spray coating on the tubes. $Cu^{++} < 10$ ppb and $Hg^{++} < 2$ ppb are required.
- Sand and sludge deposits contained in sea or fresh water must be negligible. Suspended solids shall not exceed 80 ppm (standard spec). In Japan, they are limited to 10 ppm.
- pH of the seawater must be between 7.5 and 8.5.
- Isolated chloride ions (active $Cl^-$) must be less than 0.05 ppm at the connecting point between ORV and water supplying piping.

## 2.2 SCV

Submerged Combustion Vaporizers (SCV's) burn natural gas as their heat source and require electric power to run the combustion air blower. The SCV vaporizes LNG contained inside stainless steel tubes in a submerged water bath with a combustion burner. In the baseload terminal SCV, the fuel gas is burned in a large single burner rather than multiple smaller burners. The single large burner is used because it is more economical and it achieves low NOx and CO levels. The hot flue gases are sparged into a bath of water where the LNG vaporization coils are located.

The SCV's are designed to utilize the low-pressure fuel gas derived from the boil-off gases of the facility and the let-down gas from the send-out gas. They may also use an extracted heavier fuel gas ($C_2$ plus) from the LNG at the LNG terminal[(1)].

For the SCV operation, since the thermal capacity of the water bath is high, it is possible to maintain a stable operation even for sudden start-ups/shutdowns and rapid load fluctuations. Thus, they provide great flexibility for quick start-up after shutdowns and the ability to quickly respond to changing demand requirements.

The SCV unit combusts fuel gas to supply heat to the vaporizers. Due to the larger amount of flue gas, there is a concern for NOx and $CO_2$ emissions for the operation, though a low NOx emission of less than 46 ppm is attainable. The NOx level may be reduced to 15 ppm by using a newer combustion technology and may be further reduced by using Selective Catalytic Reactor units (SCR) to 5 ppm, but it adds significant cost to the SCV unit, almost doubling the cost of the system.

The bath water becomes acidic as the combustion products are absorbed in it. Alkaline chemicals (e.g. dilute caustic, sodium carbonate and sodium bicarbonate) must be added to the bath water to control pH, and resulting excess combustion water must be neutralized before discharge.

## 2.3 Combined Heat and Power unit with Submerged Combustion Vaporizer (CHP–SCV)

In order to decrease the gas auto-consumption of these type of submerged combustion vaporizers, as well as to increase the efficiency and economics of the entire regasification process, the receiving terminal can be modified to use a cogeneration concept that offers energy saving and environmental advantages. This has only been implemented at the Zeebrugge LNG Terminal Cogeneration Project to date[2].

Following is summary of CHP performance at Zeebrugge:

1. The gas turbine produces 40 MW of power with an electric efficiency of 38% (we estimated 35%).
2. Turbine fuel consumption can be calculated as 40 MW/0.38 = 105.3 MW.
3. 65.3 MW of heat in the exhaust gases is available to the terminal.
4. The terminal can recover 62 MW in addition to the 10.5 MW originating from the condensation of water vapor resulting from the combustion of natural gas in the turbine, therefore giving a total heat recovery of 72.5 MW.
5. When a record of rational energy application is obtained by taking into account the condensation heat recovery, then the energy conversion of the whole installation amounts to 106.8%.
6. The energy saving compared to a separate production of these energies is 27.8% considering a 50% efficiency for a modern combined cycle plant and 110% for submerged burners of the LNG vaporizers.
7. In addition, a reduction in CO emission of 27.8 % from the flue gas of power generation unit is achieved.
8. NOx emissions are reduced by approximately 65% when compared to the reference of separate production of electricity and heat.

The CHP plant changed the operational characteristics of the vaporization process significantly. After the modifications made to the vaporizers, they have become "hybrid", i.e., they can now be operated in three heating modes. Firstly, with the submerged combustion as per original design; secondly, with the closed circuit of warm water from the CHP plant with the submerged burner off; and thirdly, with both heat sources on. With the gas turbine operating at baseload (about 41 MW) and without submerged combustion, the natural gas sendout rate of the terminal is 500,000 $Nm^3/h$ (424 MMSCFD). Normally, the load of the gas turbine follows the heat demand. The load variation is limited to the range where the turbine is still working at low NOx emissions; i.e., the practical minimum lies at 60% of nominal load. If terminal sendout rate and thus heat demand falls lower, part of the exhaust gases can be bypassed through the diverter and bypass stack to atmosphere or the gas turbine can be stopped. At higher sendout rates above 500,000 $Nm^3/h$, the submerged burners are started one by one for additional heat supply.

There are some disadvantages for the CHP units:

- The CHP unit will lower the power plant efficiency. A typical Combined cycle power plant has a fuel efficiency of 54% vs. a simple cycle power plant having a fuel efficiency of 35%. Also, a smaller power plant has a higher cost, about $425 per kw vs. $300 per kw for a larger unit (Frame 9 based power plant).

- The buoyancy of the flue gas will be poor as the flue temperature is lowered to 16°C. Flue gas tends to stay at ground level as the flue temperature is lowered. The flue temperature will have to keep above the ambient temperature to ensure the flue gas buoyancy.

This CHP system would require 100 % sparing with SCV's to ensure that in the event that the power plant was not operational for maintenance or demand requirements, the ability to send-out gas would not be impaired.

This option will have high-energy efficiency. However, there will be some potential efficiency loss to ensure air emissions do not remain at ground level by heating the exhaust gases.

## 2.4 Shell and Tube Vaporizers (STV)

The Shell and Tube Exchanger and Intermediate Fluid Shell and Tube Exchanger type Vaporizers are general smaller in size and cost competitive to an ORV or SCV system. Heat is usually supplied to the LNG vaporizer by a closed circuit with a suitable heat transfer medium. They are mainly used when a heat source is available. Design of these type vaporizer systems requires a stable LNG flow at design and turndown conditions with provisions to prevent the potential for freeze-up within the vaporizer. These have had only limited application to date. They are manufactured by Kobe Steel and Chicago Power.

The design of Double Tube Bundle Shell & Tube Vaporizer incorporates both a lower and an upper set of tube bundles, and uses an intermediate heat transfer fluid (e.g. propane, isobutane, freon, $NH_3$) between the LNG (upper tubes) and the seawater or glycol water (lower tubes) inside a single shell. A small shell and tube superheater is required to heat the vapor to 5 °C.

The Cove Point design uses glycol-water in the lower tube bundle instead of seawater as the heat source. The cooled glycol-water is pumped, circulated and heated by the gas turbine exhaust (waste heat recovery) in another exchanger. The intermediate fluid is isobutane.

In the recent development of Shell and Tube Vaporizers using the ambient seawater as the heating medium, they exchange heat directly with seawater. They are similar to ORV's, but instead pump seawater through the shell and tube exchangers.

Similar to the issues addressed for ORV units, Shell and Tube Vaporizers have additional concerns in corrosion and erosion inside the exchanger when the seawater is used as the heating medium. Other issues on the lowered seawater return temperature and hypo-chlorite in the returned seawater are identical to the ORV system.

The turn-down of seawater flow in the exchanger will be limited in consideration of the possibility of icing when the seawater flow rate is reduced in the heat exchanger.

## 2.5 Other Types - Ambient Air-Heated Vaporizers

Ambient air heated vaporizers utilize ambient air in either a natural draft mode or a forced draft mode to vaporize LNG. Such vaporizers are in service for warm climate and with plot space available. They are manufactured by a conventional air cooler manufactures and have been used at the Petronet LNG Terminal in which the air coolers are manufactured by GEI Hamon Industries.

## 3. Environmental Impact Comparison

### 3.1 Open Rack Vaporizer (ORV)

The ORV has a lower energy usage than the SCV. For a terminal capacity of 7.5 million tonnes per annum LNG sendout, as an example, at least two very large diameter lines, one for inflow and one for outflow, will be required. These lines will be 5 - 7 feet diameter and likely to cause significant impact, during construction, to the marine environment. This is in addition to the impact on the marine environment due to ongoing operations, which will cause large volumes of marine plankton to be withdrawn, and of cold water to be discharged. In addition, the visual impact on the marine and landsite, and impact on terrestrial ecology, may be significant.

The primary benefit of an ORV is that no fossil fuels are burnt and thus there are no carbon dioxide and NOx emissions from the plant. However, the global impact must be considered against the local impact. For example, the abstraction of a large amount of seawater and return of much cooler water must be assessed for impact to the local marine environment.

The environmental impact summary of the ORV system (in comparison to the SCV system) is shown as follows:

<u>Advantages</u>

- Does not contribute to global warming resulting from fossil fuel combustion.
- Uses renewable resources.

<u>Disadvantages</u>

- Abstraction of a large amount of cooled seawater back to the marine environment has a significant impact on sensitive marine environment. The return seawater temperature is 5 to 12°C lower than the intake seawater temperature.
- The returned seawater may contain 0.2 to 2 ppm hypo-chlorite.
- Significant environmental impacts from laying of pipes through sensitive / protected marine habitats.
- Large open rack system would require more topside space and could pose visual impact of the rack itself as well as visual impact of increased localized mist, fog, etc.

### 3.2 Submerged Combustion Vaporizer (SCV)

This option will burn fuel to supply the heat of LNG vaporization and will have a lower energy efficiency than an ORV system in which seawater is pumped to vaporize LNG.

Air emissions ($CO_2$, NOx & CO) are from fuel combustion.

Environmental impact summary of the SCV system (in comparison to ORV system) is shown as follows:

Advantages

- Requires no seawater and eliminates the issues associated with the returned seawater in temperature and hypo-chlorite.

Disadvantages

- Contribution to global warming from burning gas to fire burners.
- Condensate discharge may have an impact on the marine environment depending on the composition of the discharge. Degree of impact would depend on location of outfall point and may also result in minor habitat loss from laying of discharge pipes.
- Emissions of $CO_2$, CO, NOx, etc.
- High content of water vapor in stack emissions will likely produce white vapor plume on cool days.

3.3 Waste Heat Utilization

When waste heats are utilized for vaporizing LNG, either by the STV, air cooler type or CHP option, the concerns of the discharges of the flue gas, and/or seawater on the environment are eliminated.

## 4. Zero Emission LNG Vaporization Process

In the Combined cycle power plant or industrial plants, cooling water is often used to reject heat from the facilities. Cooling water is then cooled down in a cooling tower.

This paper describes an innovation using the waste heat from either a power plant or industrial facilities for LNG vaporization. This will eliminate fuel requirements, while reducing or eliminating air or water emissions and improves the LNG terminal and plant facilities thermal efficiency.

When a LNG Terminal operation is integrated with a Combined Cycle Power Generation Unit by utilizing the cooling water, which used for condensing the power plant steam turbine exhaust steam for LNG vaporization, the LNG vaporization process will achieve zero emissions.

As an example (refer to the following Flow Diagram), about 536 MMBTU/Hr cooling water duty at about 39°C steam temperature level is required for the GE 9FA Unit Combined Cycle system with 390.8 MW power export. This heat duty can be used to vaporize about 950 MMSCFD (or 6.9 million tonnes per annum) LNG. For 6.9 million tonnes per annum LNG vaporization, about 30,000 GPM cooling water is needed to pump around the system between the power plant and the LNG terminal based on a 20°C water temperature drop utilized in the LNG vaporizers.

In this zero emission process design, the power plant efficiency is also improved as the steam turbine exhaust steam to be condensed at a lower pressure because of using colder cooling water. For example, if the power plant steam turbine exhaust steam-condensing temperature lowered by 10°C, the steam turbine power export will be increased by 1.3 MW.

A Process Schematic Diagram showing the cooling water flow between the Power Plant and the LNG Receiving Terminal has been developed and is shown below. This proposed process scheme will reduce the power plant cooling tower duty. It also increases the overall temperature difference ΔT, the temperature driving force and therefore, improves the heat transfer for the steam turbine steam condenser. Hence, it provides an opportunity to lower the power plant steam condensing temperature by 4 to 7 deg. C. Consequently, the steam condensing pressure will be lowered by 0.02 bar. Power plant output will therefore increases when the heat is used for vaporizing the LNG.

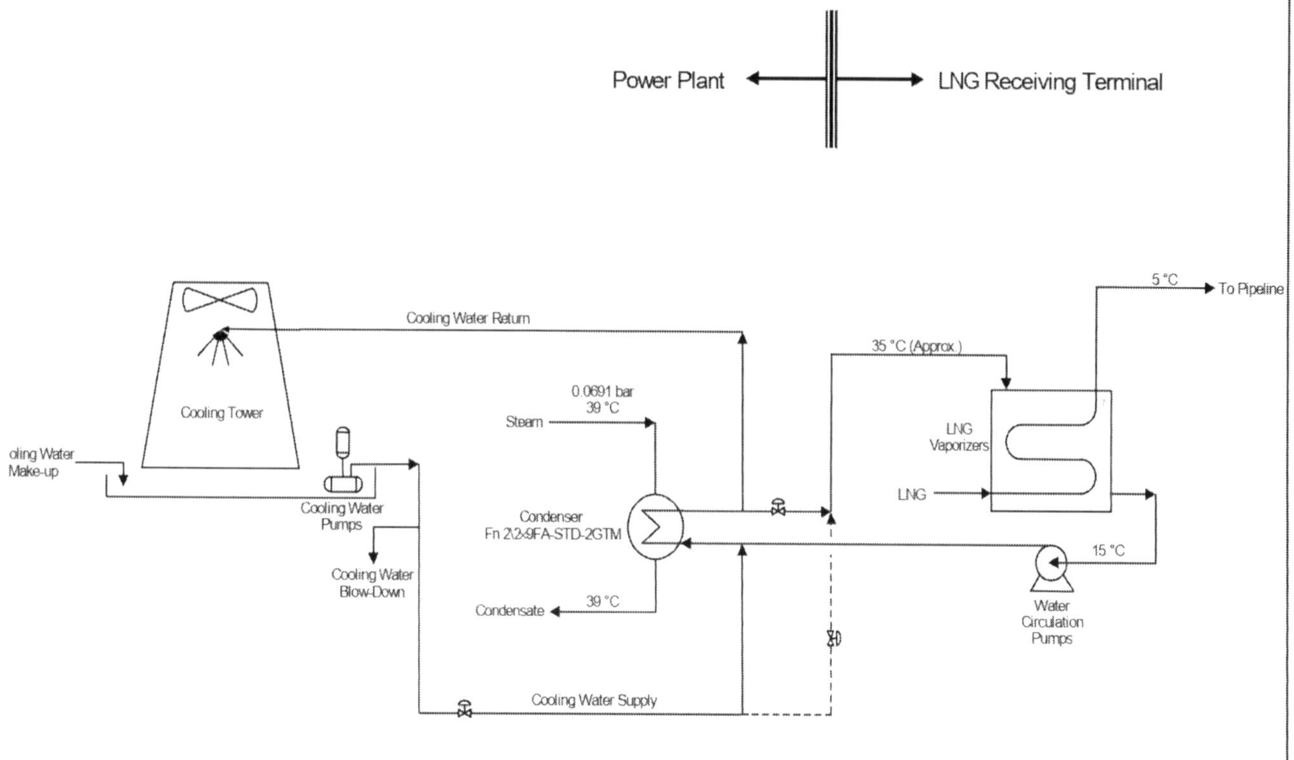


**Schematic Process Flow Diagram of Water Circulation for LNG Vaporizers**

**Figure 1 – Schematic Process Flow Diagram of Water Circulation for LNG Vaporizers**

Various types of the heat transfer equipment may be used for vaporizing the LNG by utilizing the cooling water:

i. The LNG vaporizers designed by Chicago Power & Process (CPP) or Kobe Steel (Intermediate Fluid Type LNG Vaporizer) can be used for this zero emission LNG vaporization process. The CPP LNG VAPORIZER is a special, patented, shell and tube heat exchanger design, based on established heat transfer theory, experience with low temperature liquids and proven, unique flow path design. A typical CPP vaporizer sketch and the detailed description are shown in Appendix 2.

ii. The Reli-Vap vaporizer designed by Wheaton Process Systems offers a modified shell and tube heat exchanger design, which can be used to vaporize and heat LNG. The vaporizer consists of tube side heat transfer circuits positioned within a trough-like heating medium containment structure. The detailed descriptions are shown in Appendix 2.

iii. A modified SCV system similar to the Combined Heat and Power Unit described earlier (CHP-SCV) operation. Instead of uses the scrubbed warm water from the gas turbine exhaust gas stack, uses the cooling water from condensing the steam turbine exhaust steam as the heat medium.

## 5. Conclusions

Both ORV and SCV units produce some impacts on the environment. The proposed water circulation by utilizing the cooling tower water for LNG vaporization also furnishes synergy between Power Plant or industrial facilities and LNG Receiving Terminal. The process scheme described above produces no emissions, and therefore significantly reduces the environmental impact of the facility.

## Appendix 1 Vaporizer Vendors

Submerged Combustion Vaporizers (SCV)

1. T-Thermal, USA
2. Kaldair LTD in Langley, Slough, UK
3. Sumitomo Precision, Japan
4. Reli-Vap Cross Flow LNG Vaporizer

Open Rack Vaporizers (ORV)

1. Kobe Steel Kobelco
2. Sumitomo

Shell and Tube Type Vaporizers and Intermediate Fluid Vaporizer (IFV, using Propane, Butane, Freon as an intermediate Fluid)

1. Kobe Steel
2. Chicago Power & Process
3. Wheaton Process System

Combined Heat and Power Unit (CHP)

1. Tractebel Group

## Appendix 2 Vaporizer Descriptions and Sketches

### Open Rack Vaporizers (ORV)

The ORV uses water as the heating source for vaporizing or heating low-temperature fluids into gases at ambient temperature. The heat conductor is called a panel. A panel comprises a large number of heat transfer tube in a raw like a curtain. The ORV is made of an aluminum alloy for good mechanical characteristics at low temperature, excellent workability, and high thermal conductivity. To protect the ORV against corrosion by seawater, an aluminum zinc alloy is thermal-sprayed on panels and upper and lower headers exposed to seawater.

Recently, Kobe Steel, Ltd. and Osaka Gas Co., Ltd. have jointly developed a new type of ORV called SUPERORV with the vaporization capacity improved by approximately 3 times compared to conventional ORV. The volume of seawater required for vaporization has also been reduced by 15% and the area required for installing SUPERORV is 40 to 50% smaller. The operation cost can also be reduced by 15%. SUPERORV is under commercial operation (base load) in Osaka Gas Himeji LNG Terminal since autumn, 1998 and had delivered to Spain, Fukuoka and Portugal in 2001-2002.

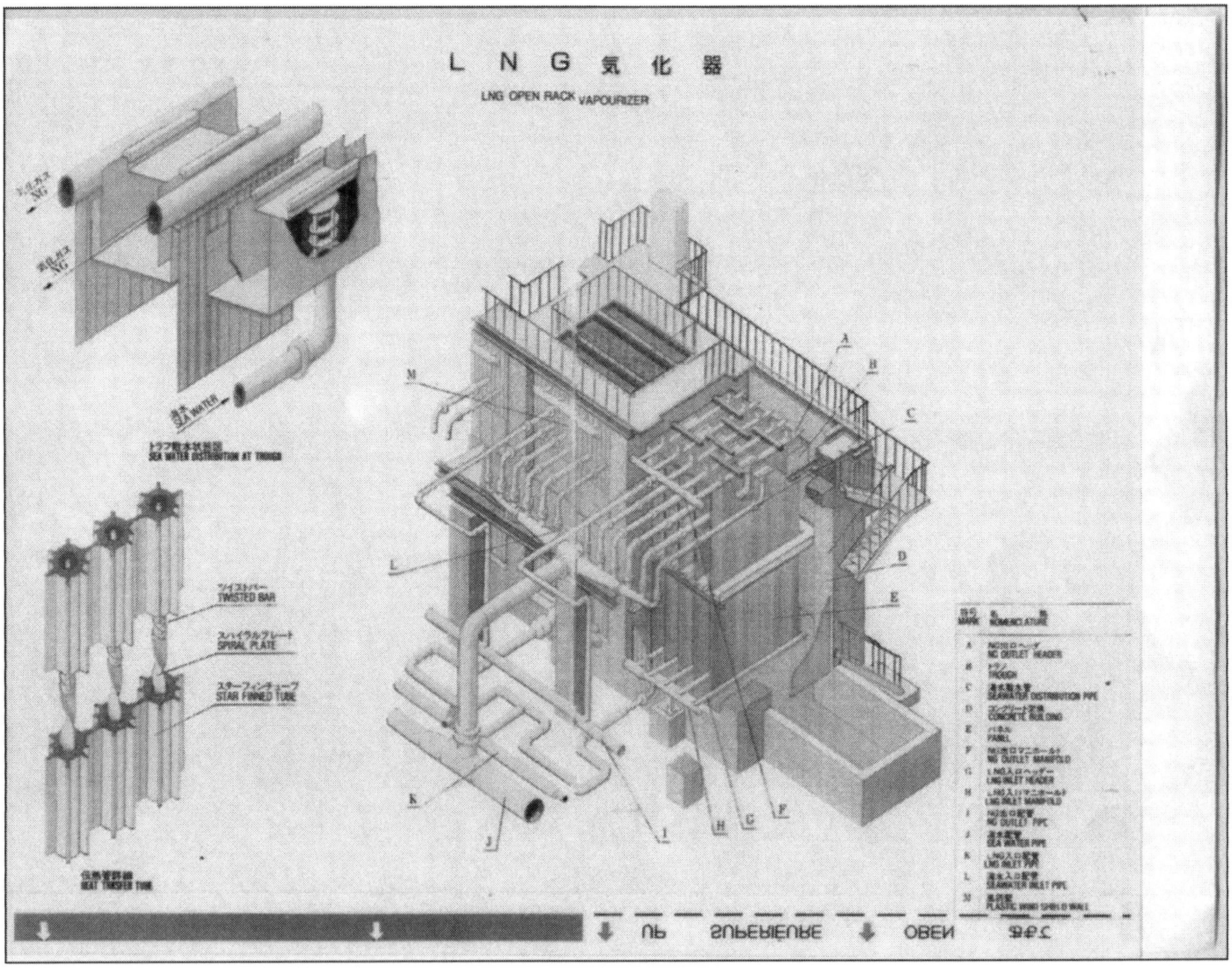


**Figure 2 – LNG Open Rack Vaporizer (Courtesy by Sumitomo Precision Products, Co.)**

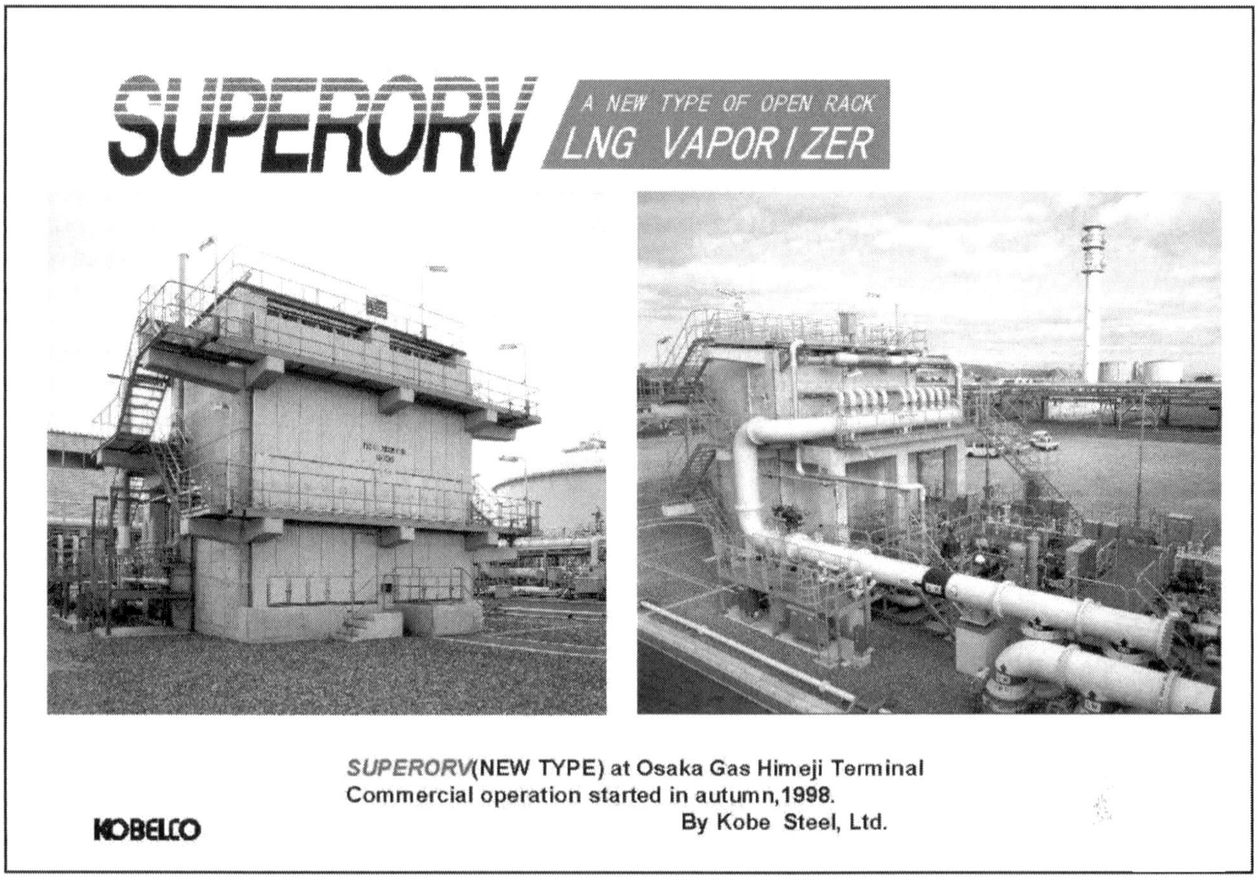


**Figure 3 – A New Type of Open Rack LNG Vaporizer (Courtesy by Kobe Steel Kobelco)**

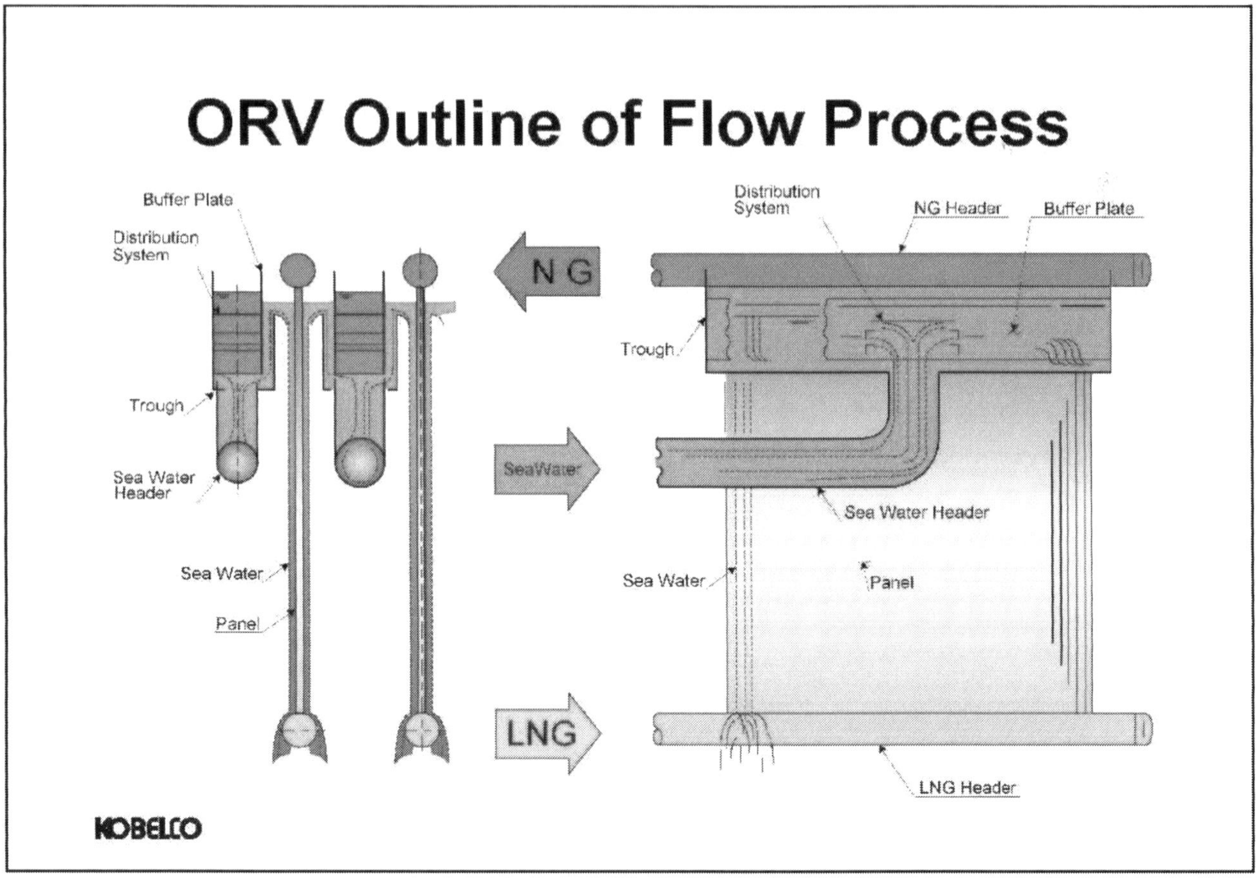


**Figure 4 – ORV Outline of Flow Process (Courtesy by Kobe Steel Kobelco)**

## Submerged Combustion Vaporizers (SCV)

Submerged Combustion Vaporizers (SCV's) burn mainly natural gas as their heat source. The SCV vaporizes LNG contained inside stainless steel tubes in a submerged water bath with a combustion burner. In the baseload terminal SCV, the fuel gas is burned in a large single burner rather than multiple smaller burners. The single large burner is used because it is more economical and it achieves low NOx and CO levels. The hot flue gases are sparged into a bath of water where the LNG vaporization coils are located.

The SCV's are designed to utilize the low-pressure fuel gas derived from the boil-off gases of the facility and the letdown gas from the send-out gas. They may also use an extracted heavier fuel gas ($C_2$ plus) from the LNG at the LNG terminal.

For the SCV operation, since the thermal capacity of the water bath is high, it is possible to maintain a stable operation even for sudden start-ups/shutdowns and rapid load fluctuations. Thus, they provide great flexibility for quick start-up after shutdowns and the ability to quickly respond to changing demand requirements.

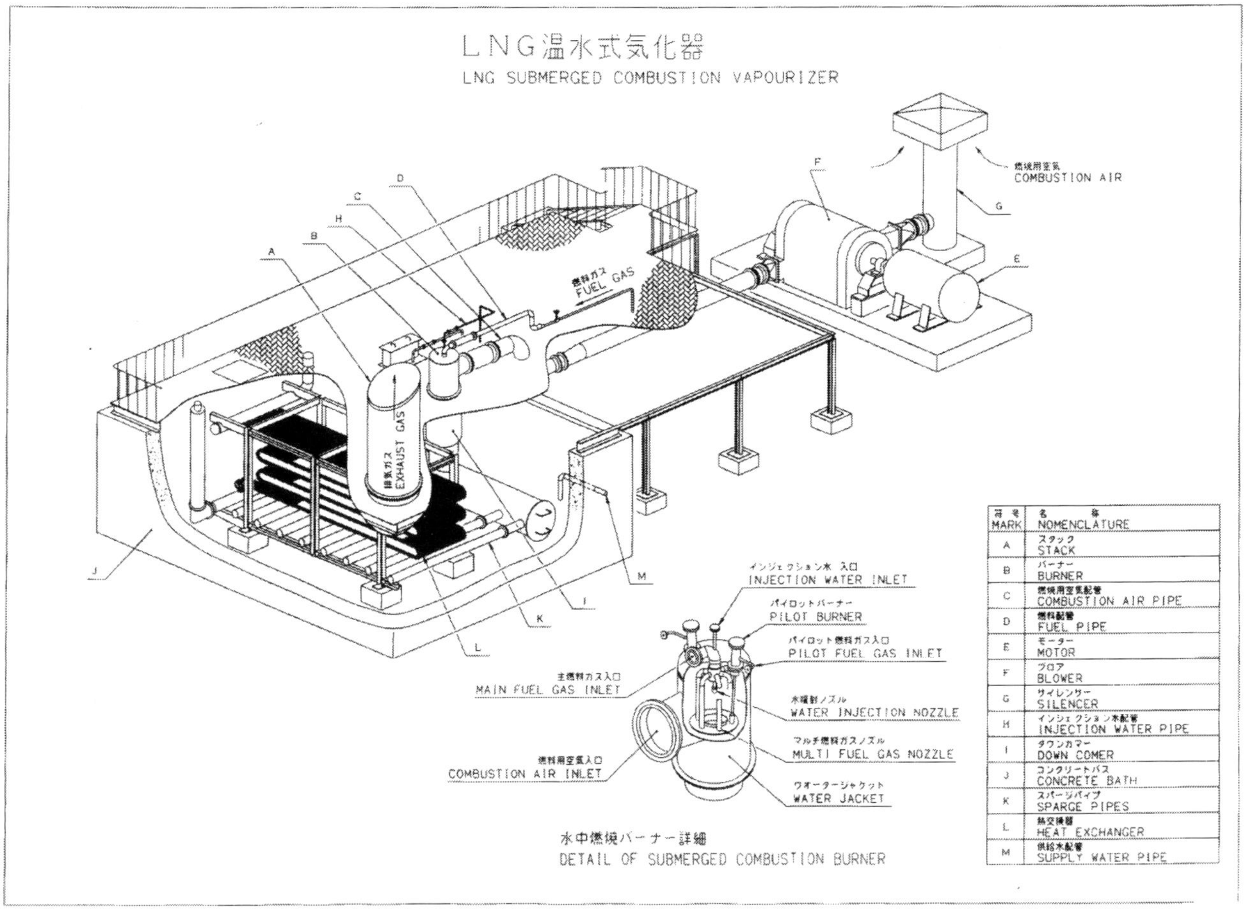


**Figure 5 – LNG Submerged Combustion Vaporizer (Courtesy by Sumitomo Precision Products, Co)**

## Combined Heat and Power Unit (CHP-SCV)

The heart of the CHP facility is a gas turbine type LM6000 that generates 40 MW of electrical power. The hot exhaust gases from the turbine pass through a heat recovery tower and transfer their heat to raise the temperature of a closed hot water circuit. This hot water will then be circulated and injected in the water bath of the vaporizers and transfer its heat to regasify the LNG.

The Sketch of Combined Heat and Power Unit (CHP) is shown in the following internet address: http://www.fluxyslng.net/Infra_CHP.htm

## Shell and Tube Type Vaporizers

1. The Wheaton Process Systems - RELI-VAP Vaporizers

   The RELI-VAP is a modified shell and tube (MST) heat exchanger. It has features similar to shell and tube heat exchangers. The heating medium now flows through the MST in a completely cross flow serpentine arrangement. Plenums are used to provide the cross flow heating medium passages. The full cross flow design provides higher and more uniform heat transfer rates. The RELI-VAP provides an economical solution to both open loop and closed loop vaporizer applications.

   For open loop applications, the RELI-VAP vaporizer design provides less susceptibility to fouling. The high turbulence in the completely cross flow heat transfer circuits helps to reduce material settlement within the heat transfer circuits and reduce the fouling of the exchanger surface area. Material would have more of a tendency to settle in the vaporizer plenum areas. The material can also be removed from the plenum areas during system operation.

   The RELI-VAP vaporizer can utilize heat sources in several arrangements:

   i. Using once through seawater or river water in direct contact with the MST heat transfer area.
   ii. Using boilers or burners to heat water or water glycol circulating in a closed loop through the MST.
   iii. Using seawater or river water to heat a circulating loop fluid. The RELI-VAP would be heated using the circulating loop fluid.
   iv. Using cooling tower water and other waste heat mediums circulating through the vaporizer.
   v. Heat from combustion gas turbine inlet air-cooling.
   vi. Combinations of the above.

   Using the RELI-VAP vaporizer in seawater service compared to the Open Rack (ORV) type vaporizer has several advantages. They include the following:

   i. Very considerable installed cost savings.
   ii. Can be constructed of austenitic stainless steel or other special stainless steels compared to aluminum for the ORV for more rugged seawater and other corrosive water service.
   iii. Can be constructed of austenitic stainless steel or other special stainless steels compared to aluminum for the ORV in high-pressure pipeline service.
   iv. Trouble-free design
   v. More compact to save installation space and foundation requirement cost.
   vi. Compact and rugged design particularly important for offshore applications.
   vii. The seawater is not aerated in the vaporizer. There is less potential for foaming at the vaporizer seawater discharge outfall and the environmental impact of foaming.

   The RELI-VAP vaporizer in open loop and closed loop service has several advantages compared to conventional shell and tube LNG vaporizers including:

i. Heat transfer tube thermal expansion is accommodated by the serpentine heat transfer circuit arrangement for increased vaporizer reliability.
ii. The vaporizer plenum areas are accessible for cleaning. The heating medium containment covers can be removed for easy access. Other cleaning means are also available.
iii. Individual heat transfer tubes and circuits can cool down or warm up at different rates without high thermal stresses.
iv. Shell expansion joints are not required.
v. The RELI-VAP can be used to vaporize LNG without intermediate fluids and several sets of heat exchange surface area.
vi. In many applications using fired heaters, warm water can be circulated as the heating medium rather than a water glycol solution providing many advantages.

The following shows RELI-VAP vaporizer sketch.

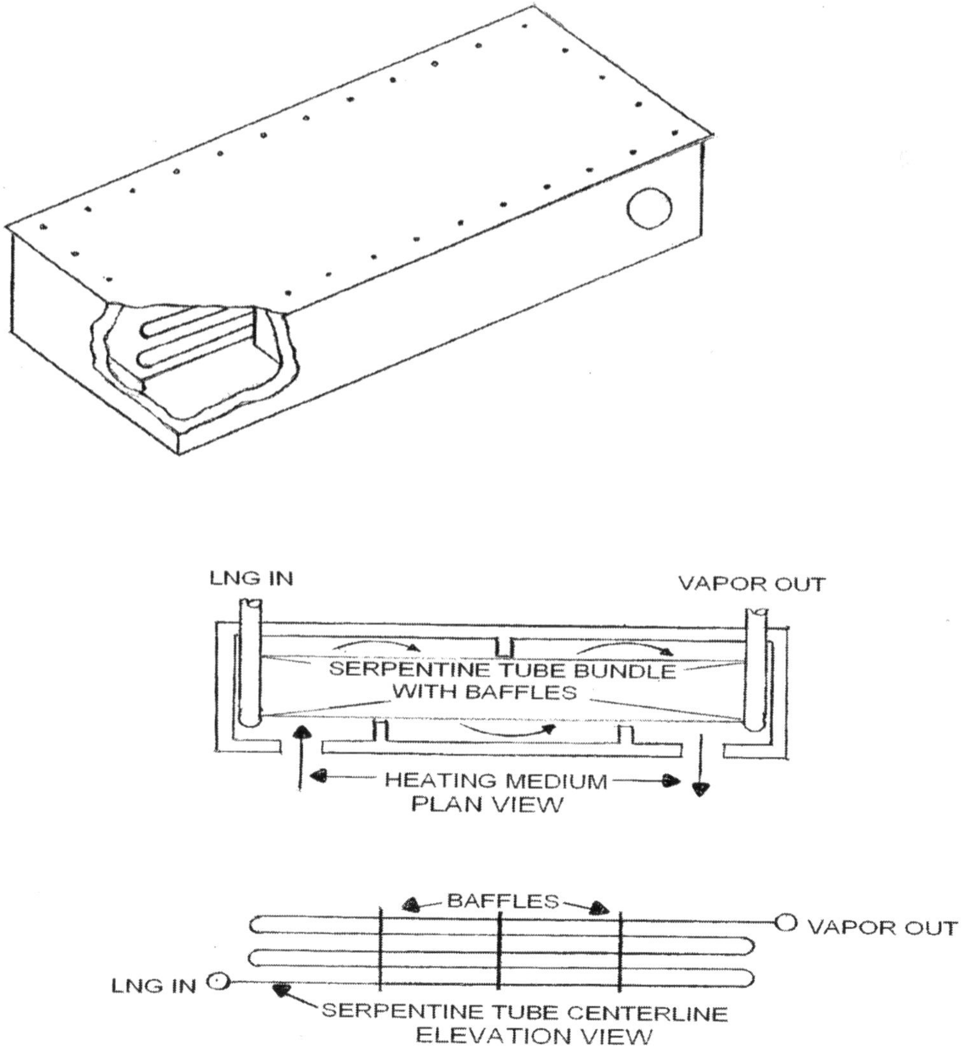


**Figure 6 – RELI – VAP Vaporizer Sketch (Courtesy by Wheaton)**

2. CPP Vaporizers

The Chicago Power & Process (CPP) LNG VAPORIZER is a special, patented, shell and tube heat exchanger design, based on established heat transfer theory, experience with low temperature liquids and proven, unique flow path design. A warm fluid source, commonly water, water/glycol mix or sea water, is required for the heat input and will provide natural gas vapor at common distribution temperatures, with no liquid carry-over. Rapid start-up and shut-down with the capability of operating down to 10% load or less, makes the CPP LNG VAPORIZER a versatile performer for base load or intermittent service. Performance has been proven with over 75 domestic and international installations during the past 25 years. Capacities range from 4 to 300 MMSCFD per vaporizer with unlimited output in multiple unit installations. Gas pressure designs extend from 75 to 3500 psig.

The CPP LNG Vaporizer is a TEMA Type NEN or NJN welded up design for vertical mounting (or horizontal mounting with approval), single pass, with gas in the tubes and a co-current or split flow heating medium in the shell. The shell of the exchanger has an integral bellows type expansion joint to accommodate differential expansion. The tubes are arranged on a triangular layout, designed to insure a specific gap between each. Tubing is welded T304 SS or T316 ELC stainless steel in fresh water/ glycol systems and AL-6XN alloy (N08367) or 254 SMO alloy (S31254) for seawater service. The tubes are rolled into double grooved tube holes, seal welded and the tube-to-tubesheet joints are helium leak tested. The heat exchanger is designed to TEMA R and stamped per ASME, Section VIII Division 1, Code.

Each exchanger is fitted with 2 or 4 chair type mounting support lugs and lifting lugs. Suitable vents and drains are provided on the shell and tube sides. The shell side can be provided with an extra nozzle to mount a relief valve or rupture disk assembly.

The unique, patented inlet system, evenly distributes the LNG into the tubes insuring equal tube loading and accommodating any LNG vapor, formed within the inlet bonnet. Tube geometry minimizes the effect of pressure surges and prevents liquid carry-over in the superheated discharge vapor. The gas exits through the outlet bonnet which is usually fitted with a 90° elbow and flange suitable for the vaporizer design pressure.

The heating medium mixture enters the lower shell side nozzle with parallel flow or in both the bottom and top shell side connections with split flow. The shell side nozzles normally will be one or two pipe diameters larger than the piping system to achieve good tube bundle flow distribution. All external piping must be designed to prevent transferring any significant loads to the heat exchanger nozzles. The shell side is designed to minimize bypass across the baffles and spiral liquid flow patterns between baffle spaces. Baffles are specifically selected to accommodate very large shell side flow rates, however the unsupported tube lengths are always less than TEMA standards. Shell side nozzles will either be on the same or 180° opposite side, depending on the heating fluid flow rate and heat exchanger tube length. LNG inlet and vapor outlet nozzle arrangements will not affect the shell side nozzle locations. On split flow shell designs, the outlet nozzle will straddle a full baffle to insure fluid mixing occurs externally.

# CPP VAPORIZER

## TYPICAL ELEVATION

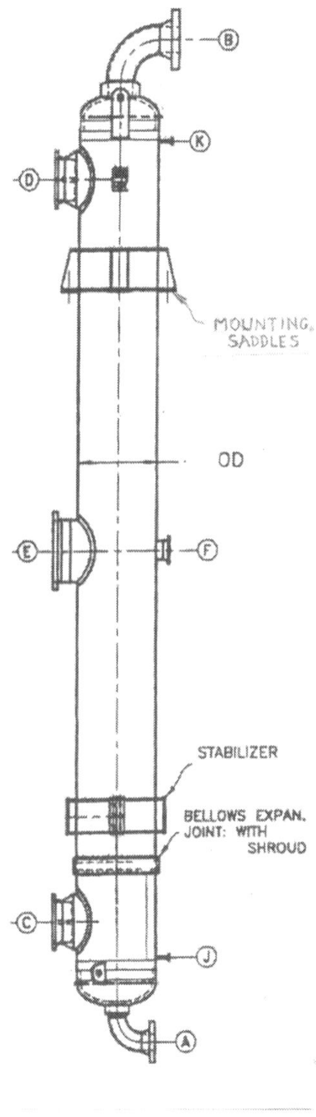


VERTICAL DESIGN
WITH MOUNTING SADDLES
AND STABILIZER

CONNECTIONS:

A. LNG INLET

B. VAPOR OUTLET

C. HEATING FLUID INLET

D. HEATING FLUID OUTLET
(PARALLEL FLOW)

D. HEATING FLUID INLET
(SPLIT FLOW)

E. HEATING FLUID OUTLET
(SPLIT FLOW ONLY)

F. RUPTURE DISC NOZZLE
(OPTIONAL)

J. SHELL DRAIN

K. SHELL VENT

**Figure 7 – CPP Vaporizer / Typical Elevation (Courtesy by Chicago Power & Process)**

## References


1. C.C. Yang, A. Kaplan, Z. Huang, Cost-effective Design Reduces C2 and C3 at LNG Receiving Terminals, Oil & Gas Journal, May 26, 2003, 50

2. J. Pairon, et al, Zeebrugge LNG Terminal Cogeneration Project: Results after one Year of Operation, LNG 12, May 4, 1998, E.3-1

**Continuously Transient Operation of Two-Phase LNG Expanders**

*Christopher D. Finley*
*Ebara International Corp.*
*350 Salomon Circle*
*Sparks, NV 89434*

*Prepared for Presentation at AIChE Spring National Meeting*
*4th Topical Conference on Natural Gas Utilization*
*New Orleans, LA, April 2004*

ABSTRACT

Variable speed LNG expanders are able to control the mass flow of LNG plants by changing the rotational speed. Some LNG plants, particularly small Nitrogen rejection plants have continuously changing mass flows, which the expanders control with continuously changing rotational speed. The expanders operate therefore in a continuously transient mode. The paper presents long-term field experience of two-phase LNG expanders operating with continuously changing rotational speed.

INTRODUCTION

A significant quantity of natural gas is not produced economically because its quality is too low to enter the pipeline transportation system without some type of processing to remove the undesired gas fraction. Such low-quality natural gas contains significant concentrations or quantities of gas other than methane. These non-hydrocarbons are predominantly nitrogen, carbon dioxide, and hydrogen sulfide, but may also include other gaseous components. The most reliable and widely used process for nitrogen rejection from natural gas consists of liquefying the feed stream using temperatures in the order of -300°F and separating the nitrogen via fractionation. In order to reduce the gas temperature to this level, the gas is compressed, cooled by heat exchangers, and expanded to a low pressure over a Joule-Thompson valve.

The turbulent friction losses in a pressure drop across a Joule-Thomson valve are very high as is the energy wasted in frictional heating of the fluid. To keep frictional heating in the process to a minimum, two-phase liquid expanders replace Joule-Thompson valves. This is due to the near isentropic expansion process achieved while generating shaft power from previously wasted two-phase energy.

Some nitrogen rejection plants, especially smaller ones, experience a continually changing mass flow rate. To control this, two-phase expanders utilize variable speed drives to adjust the mass flow rate through the expander by continually adjusting its rotational speed. Two of these type of expanders were built and installed at the existing nitrogen rejection plant and have been in continuous operation for over one year. Regular inspections of these expanders show no failures in bearings or materials and the vibration levels are less than 20% of the API 610 allowable levels [10].

VARIABLE SPEED EXPANDER DESIGN

The design of the two-phase expander used to replace the Joule-Thompson valve in nitrogen rejection plants is based on existing turbine and expander technology. A hydraulic nozzle transforms the hydraulic energy of high-pressure fluid into kinetic energy. A runner and exducer attached to the shaft transforms the kinetic energy into mechanical shaft power. Finally, an electric power generator converts shaft power to electrical energy [5].

There are two basic methods used for power generation in cryogenic technology. One method utilizes an externally cooled generator coupled to the expander and mounted outside the cryogenic liquid. The second method uses an induction generator mounted integrally with the expander shaft. Cryogenic liquid completely submerges and cools this configuration [6].

Figure 1 shows the cross section of the two-phase expander installed in the nitrogen rejection plant. This particular design utilizes a completely submerged hydraulic expander and induction generator mounted to the same shaft. The hydraulic expander consists of a radial inflow nozzle ring that generates rotational fluid flow, a reaction turbine runner and two-phase jet exducer that transform the kinetic energy of the fluid to shaft power.

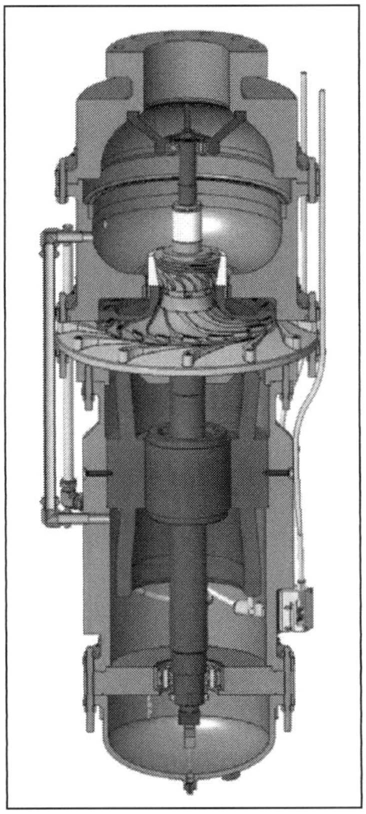

Figure 1: Cross Section of Two-Phase Expander

Figure 2 shows an enlarged view of the two-phase runner assembly. The flow direction of this two-phase expander is upward to take advantage of the buoyant forces of the vapor bubbles and to minimize flow induced vibrations. The hydraulic assembly is designed for continuously decreasing pressure to avoid any cavitation along the two-phase flow passage [1].

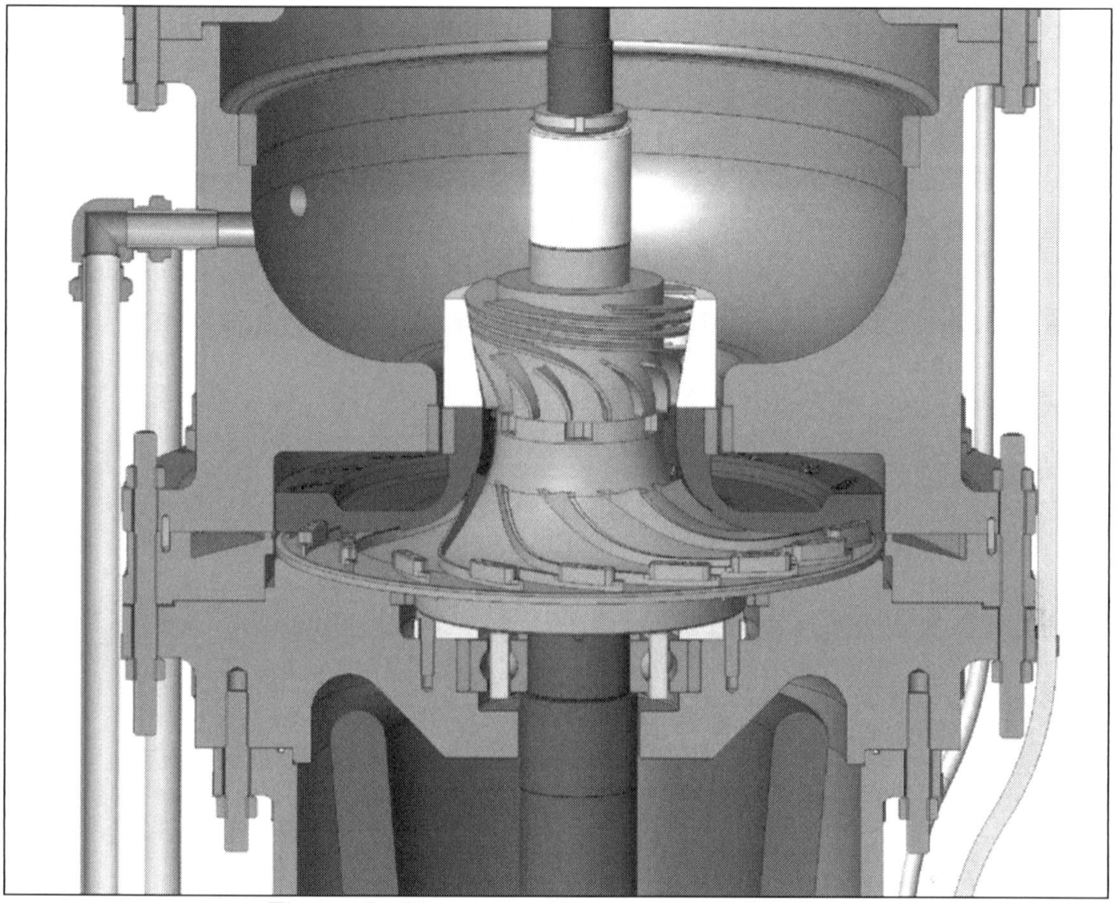

Figure 2: Close-up of Expander Hydraulics

To effectively replace the Joule-Thompson valve in a nitrogen rejection plant, two-phase expanders must be able to adjust to accommodate continually changing mass flow rates. To accomplish this, a variable speed constant frequency drive (VSCF) is used. A VSCF drive allows the generator to operate at any speed desirable and still supply constant frequency power to the power grid [6].

## TEM DESCRIPTION

One problem observed in non-thrust balanced cryogenic turbine expanders is heavy axial loads on the rotating assembly and bearings causing a decrease in efficiency due to friction losses. To alleviate this problem, some cryogenic turbine expanders are equipped with an axial thrust-balancing device to create a no axial load condition. Common thrust balancing devices use the hydraulic pressure of the fluid to equalize the axial thrust. Originally developed for centrifugal pumps, the design of the Thrust Equalizing Mechanism (TEM) has permitted them to operate maintenance free in the field for several decades. The mechanism is entirely and unconditionally applicable to two-phase expanders. The benefits of installing this Thrust Equalizing Mechanism in hydraulic expanders are; extended life, increased reliability, higher efficiency and reduced operational costs [2].

This two-phase expander uses the self-adjusting mechanism design because it performs equally well over a wide range of differential heads, flow rates and variations in turbine speed. It operates independently of the direction of the rotational axis, which can be either vertical or horizontal. It will also operate if the turbine outlet is located above or below the inlet.

Situated on the backside of the first stage turbine runner, the runner implements the thrust equalizing mechanism into its design. It consists of a two-orifice system, one fixed orifice and one variable orifice. The design of the turbine shaft allows for slight movement in the axial direction thus opening or closing the variable orifice. On the left side of Figure 3, the turbine shaft is in its maximum upward position closing the variable orifice. This creates a high-pressure condition in the inner ring chamber below the runner effectively creating down thrust. On the right side of Figure 3, the shaft is in its' lowest axial position opening the variable orifice to its maximum position. This causes a lower pressure condition in the inner ring chamber creating an up thrust condition. Between these two limiting axial shaft positions, there is one position at which the sum of all the axial forces due to upstream and downstream thrust is equal to zero. In this position, the equilibrium of axial forces on the shaft causes it to run in a balanced position [3].

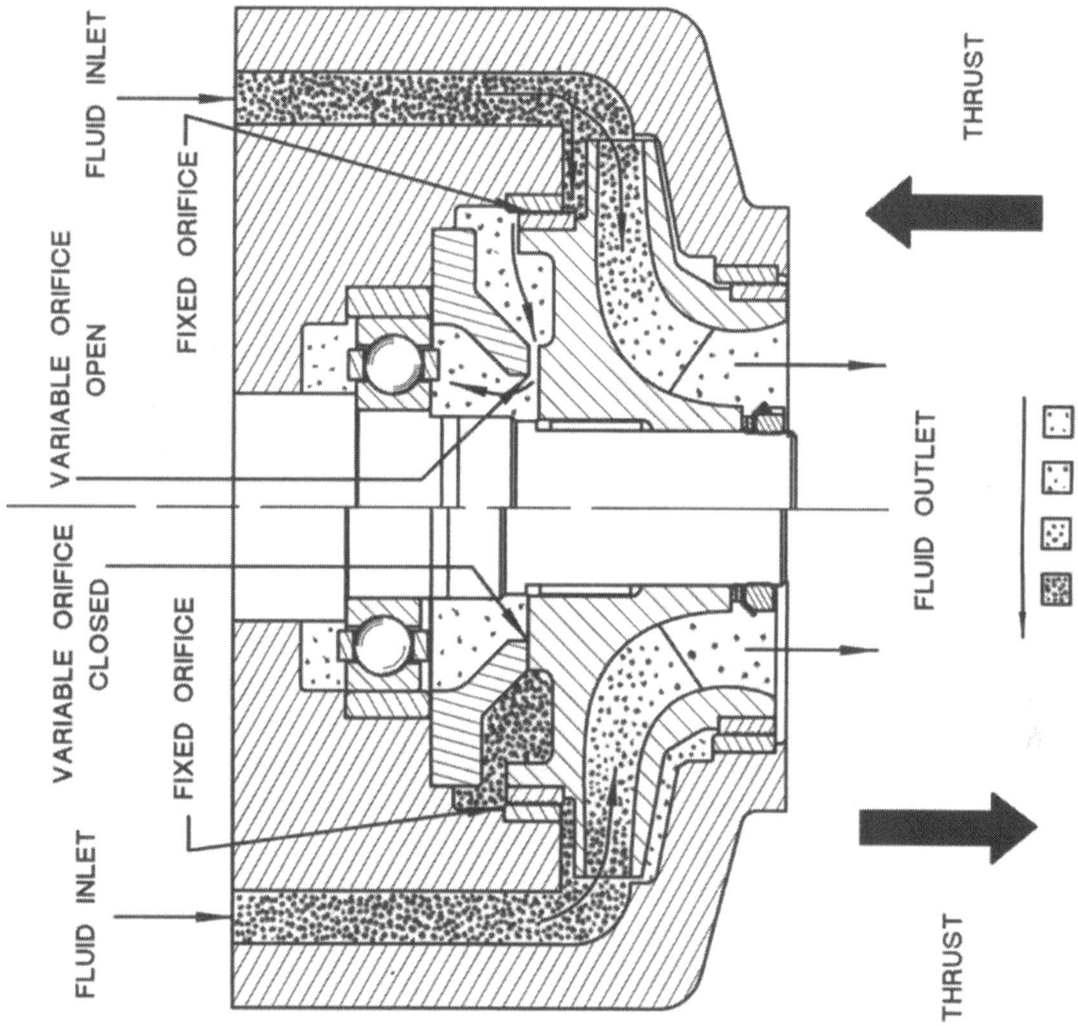


Figure 3: Thrust Equalizing Mechanism

OPERATION OF TWO-PHASE EXPANDER

During the operation of the two-phase expander, the mass flow rate required to flow through the expander is continuously changing. To cope with this, the rotational speed of the expander shaft must continually adjust to match the required flow rate. Figure 4 shows the typical relationship between expander speed, differential pressure and flow rate. At a constant pressure drop across the expander, represented by the dotted line, the mass flow rate through the expander increases as the rotational speed decreases and decreases as the rotational speed increases.

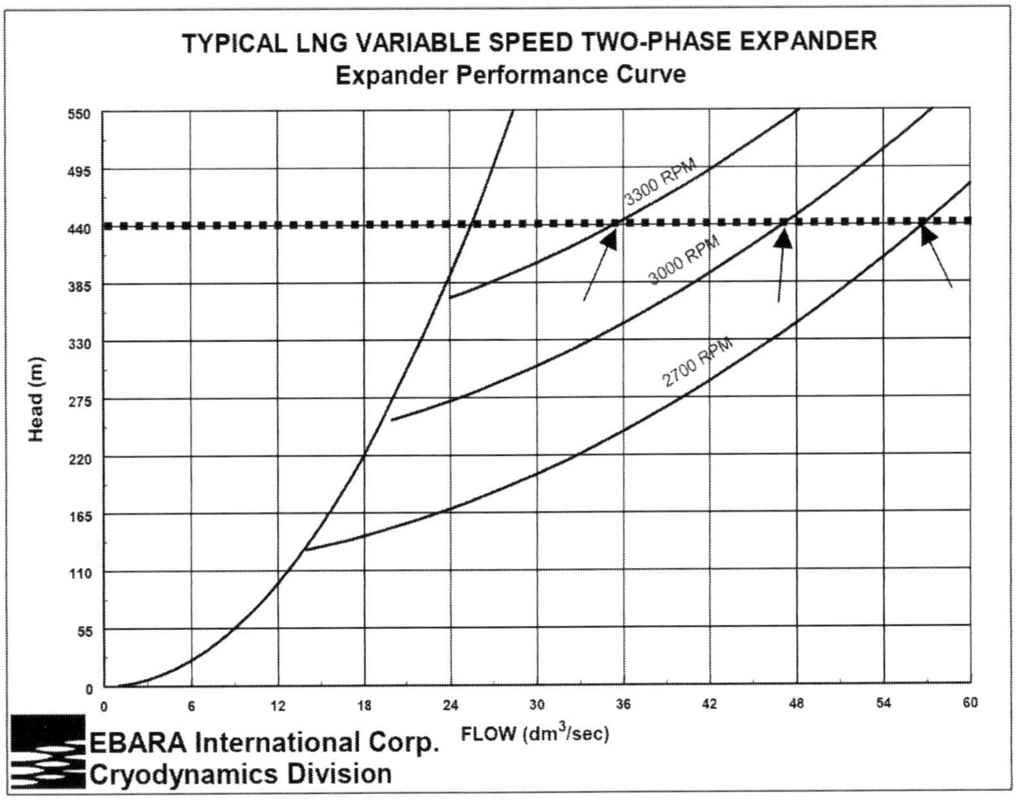


Figure 4: Typical Two-Phase Expander Performance

Adjustment of the rotational speed is accomplished by adjusting the VSCF controller using a PID control system. Figure 5 shows a simple schematic of the two-expander system. In this system, plant throughput is measured by the change in the liquid level in the column. When the liquid level in the column begins to increase, the mass flow rate into the system is increasing. This causes the PID controller to decrease the rotational speed of the turbine allowing a higher mass flow rate through the expander. When the liquid in the column begins to decrease, the PID controller increases the rotational speed of the expander effectively restricting the mass flow rate through the expander.

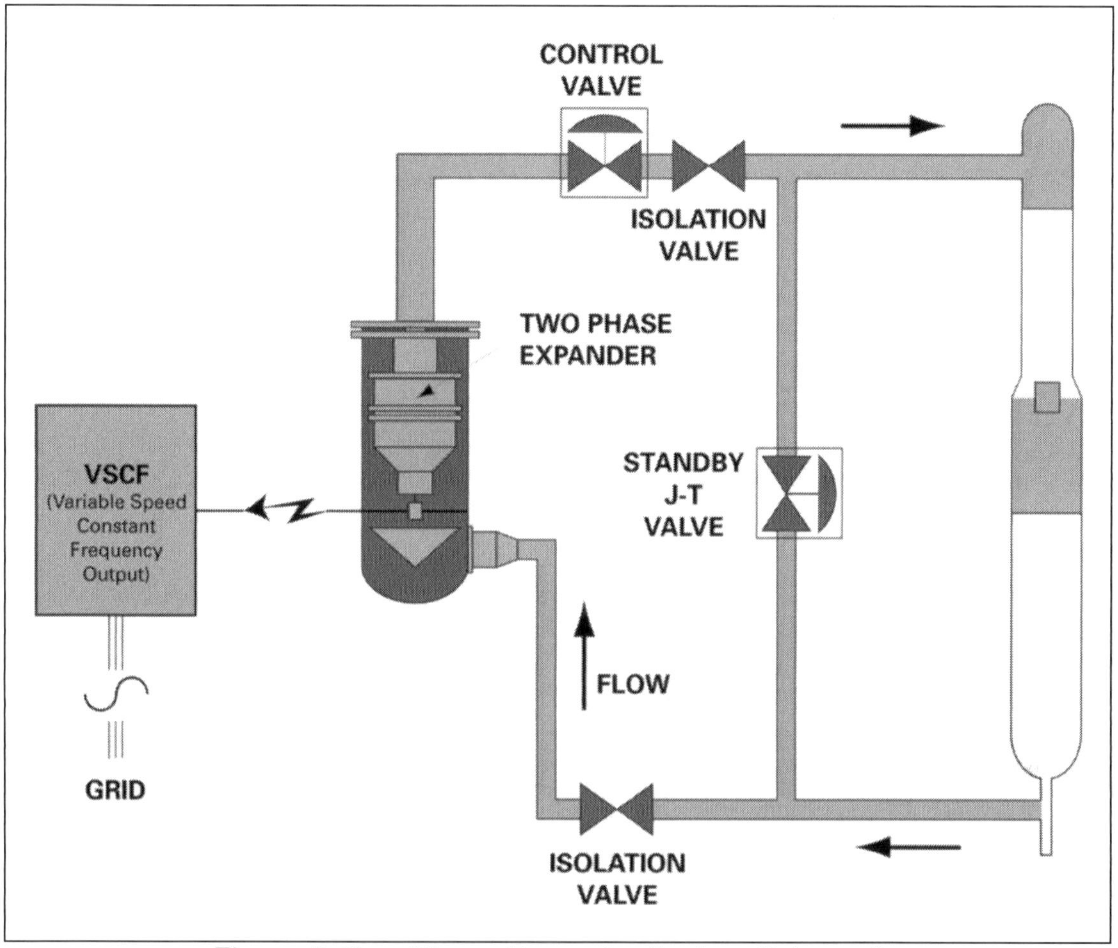

Figure 5: Two-Phase Expander System Schematic

OPERATIONAL TESTS

An on-site operational test schedule was implemented to verify the performance and long-term reliability of this novel two-phase expander. Two turbines were installed on site and subjected to normal operating conditions for a period exceeding 9000 hours. During this period of operation, the expanders operated at a wide variety of constantly changing mass flow rates, differential pressures and rotational speeds. Several different hydraulic component designs were mounted to the two-phase expander during the long-term operation and testing to verify that two-phase expanders designed for a wide range of duties would operate reliably, These different hydraulic components included several different nozzle rings of varying exit angles, two different runners with different blade heights and exducers with varying blade lengths.

A data acquisition method was developed to accurately measure and record all parameters of expander operation. This was accomplished by accessing data automatically stored in the nitrogen rejection facility's DCS. This enabled measurements of differential pressure, mass flow, temperature drop, vibration and power to be continually monitored and recorded without disrupting the facility's day-to-day operation. An Eddy probe system was mounted to the end of the expander shaft to monitor the expander shaft's axial position and rotational speed. The voltage signal from the Eddy probe driver was then routed to a high-speed data acquisition system that monitored and recorded axial position and rotational speed.

These measurements of all relevant operational data were recorded and analyzed throughout the test to observe the long-term effects of continually transient operation.

## RESULTS

Before long-term operational tests of the expanders could begin, it was necessary to confirm that the expanders would operate within their design specifications. This was done by measuring a wide range of expander hydraulic performance. Figure 6 shows the hydraulic performance during this initial test phase. The graph shows a wide variety of mass flow rates, corresponding differential pressures and expander efficiency for the expander operating at three different speeds. The data presented in this figure verifies that the expanders were operating at their design specifications.

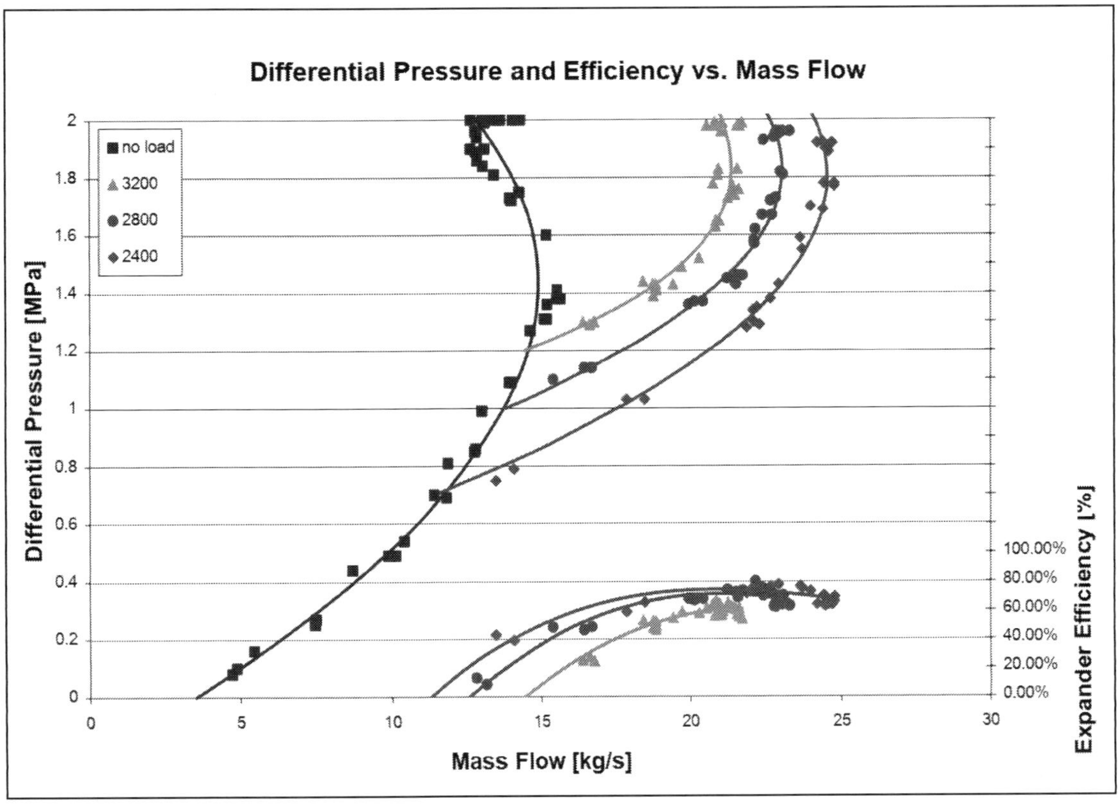


**Figure 6:** Tested Hydraulic Performance Of Two-Phase Expander

The next step in the operational test process was to show that the thrust-balancing device was working properly. To do this, axial shaft position was measured and recorded for expander start-up, shutdown and automatic operation. Figure 7 shows the axial shaft position during expander start-up. As the expander control valve was opened, the expander shaft raised from its' lower axial limit to its' upper axial limit. Within 25 seconds, the axial thrust on the turbine shaft was completely balanced with the axial shaft position between its' upper and lower limits.

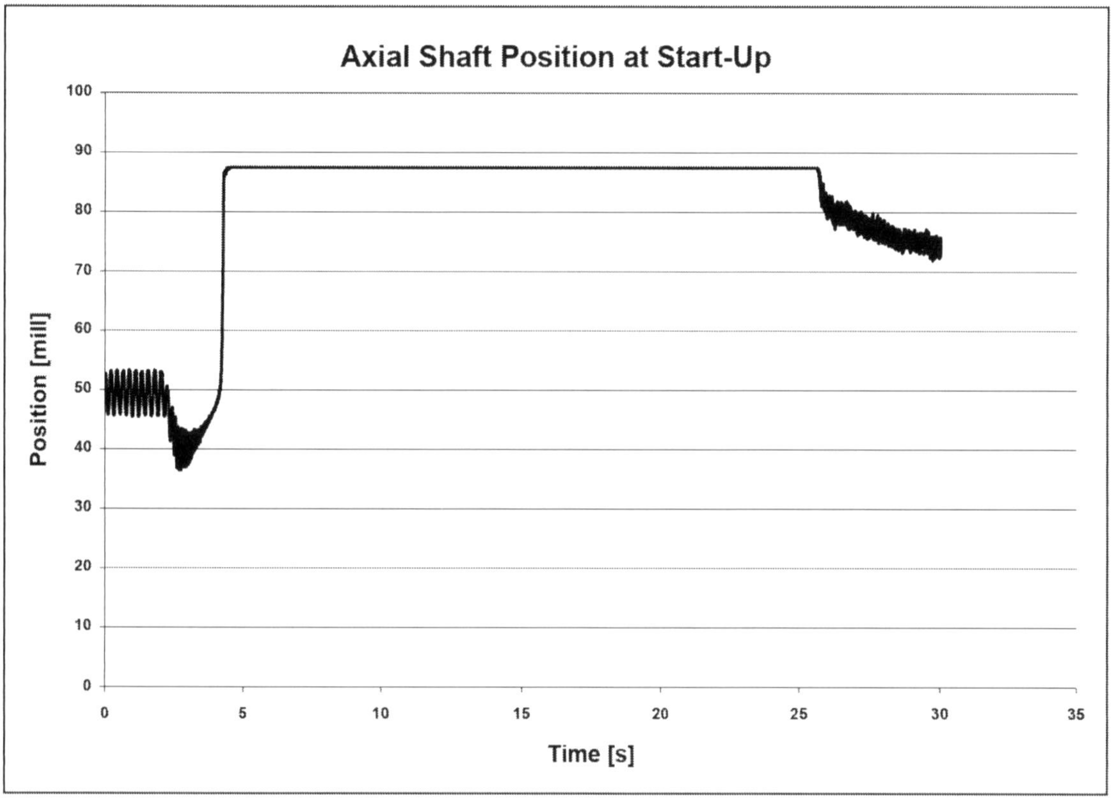


Figure 7: Shaft axial position during start-up

With the expander running at normal capacity under automatic control, the shaft axial position was measured and recorded. The axial shaft position during automatic operation is important in determining that the thrust-balancing device will operate at a wide variety of flow rates. Figure 8 shows the expander shaft axial position during this automatic operation. The axial position of the shaft can be seen to remain between its' upper and lower axial limits throughout the operation. Figure 9 shows the axial shaft position during expander shutdown. When the flow through the expander is reduced to zero by the control valve, the shaft returns to its' lower position. This process ensures that the thrust-balancing device was not damaged during start-up or normal operation.

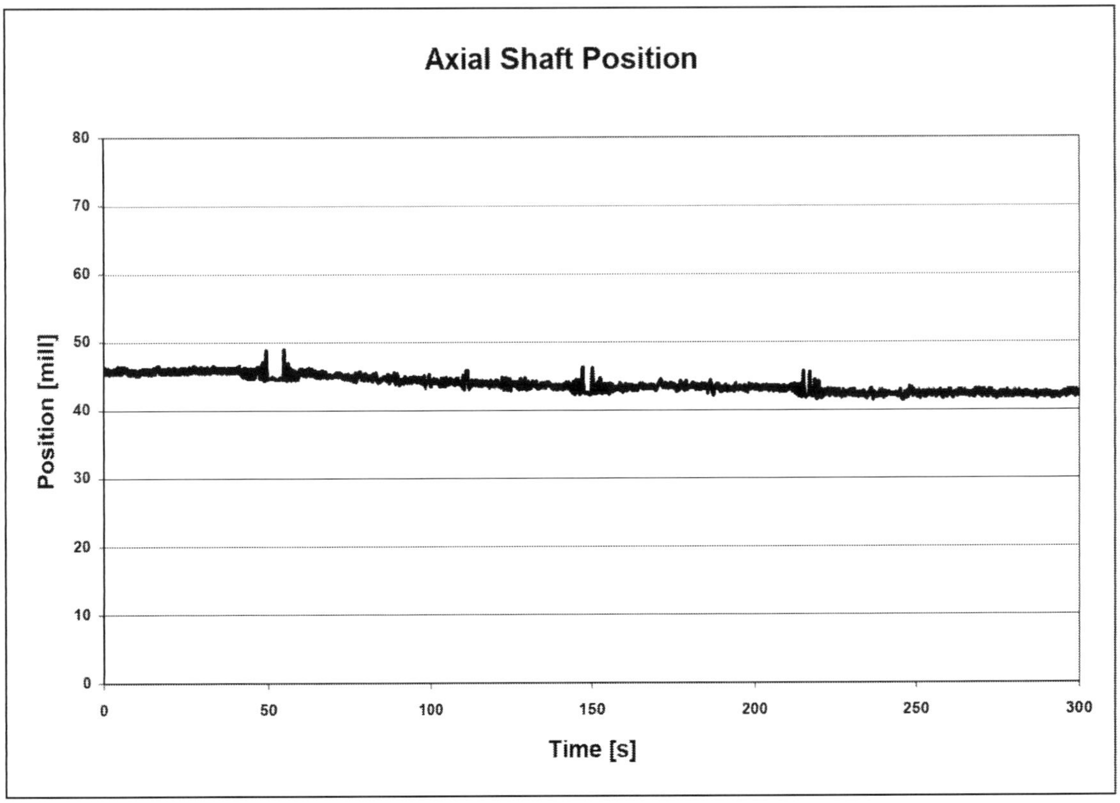


Figure 8: Shaft axial position during normal operation

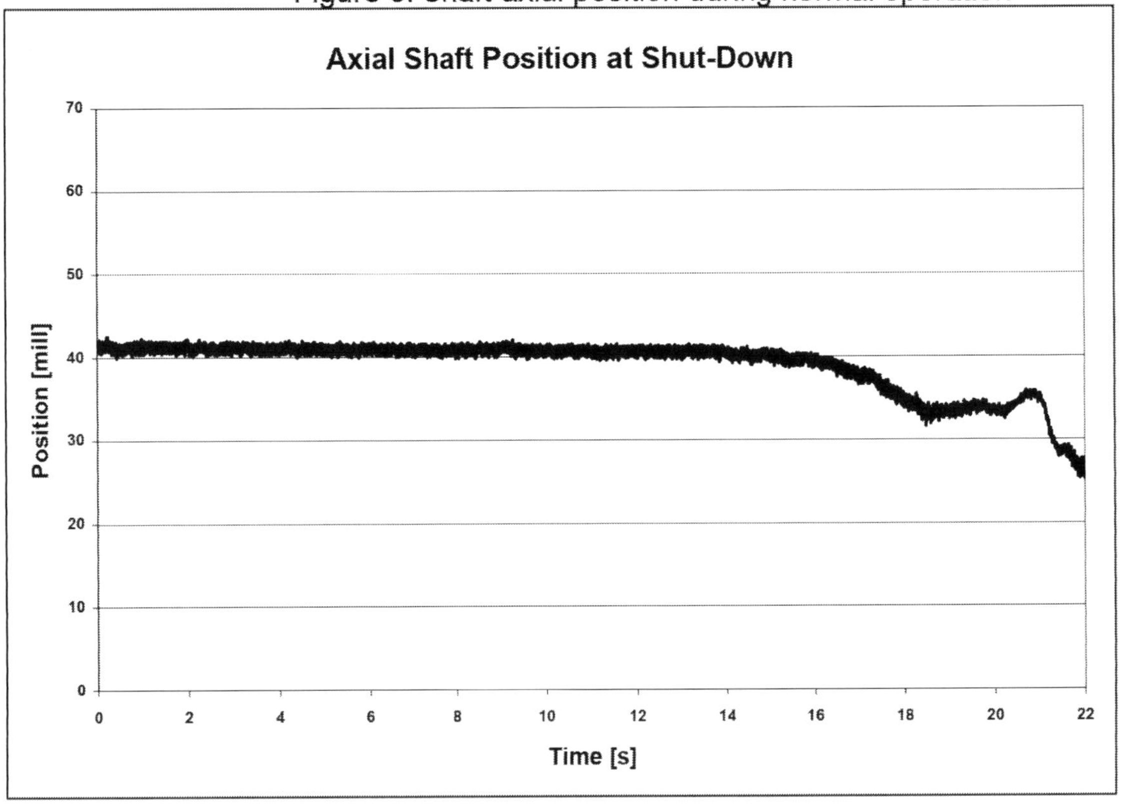


Figure 9: Shaft axial position during shut-down

The next step in the operational test process was to verify the expander's ability to automatically adjust to the continually changing mass flow rate through the plant. Figure 10 shows the flow rate and expander speed over 30 minutes. The dashed line represents the volumetric flow rate measured at the inlet of the turbine. The solid line shows the expander speed measured by the expander VSCF's monitoring system. These measurements show that as the flow rate through the expander decreases, the shaft speed increases and as the flow rate increases, the speed decreases.

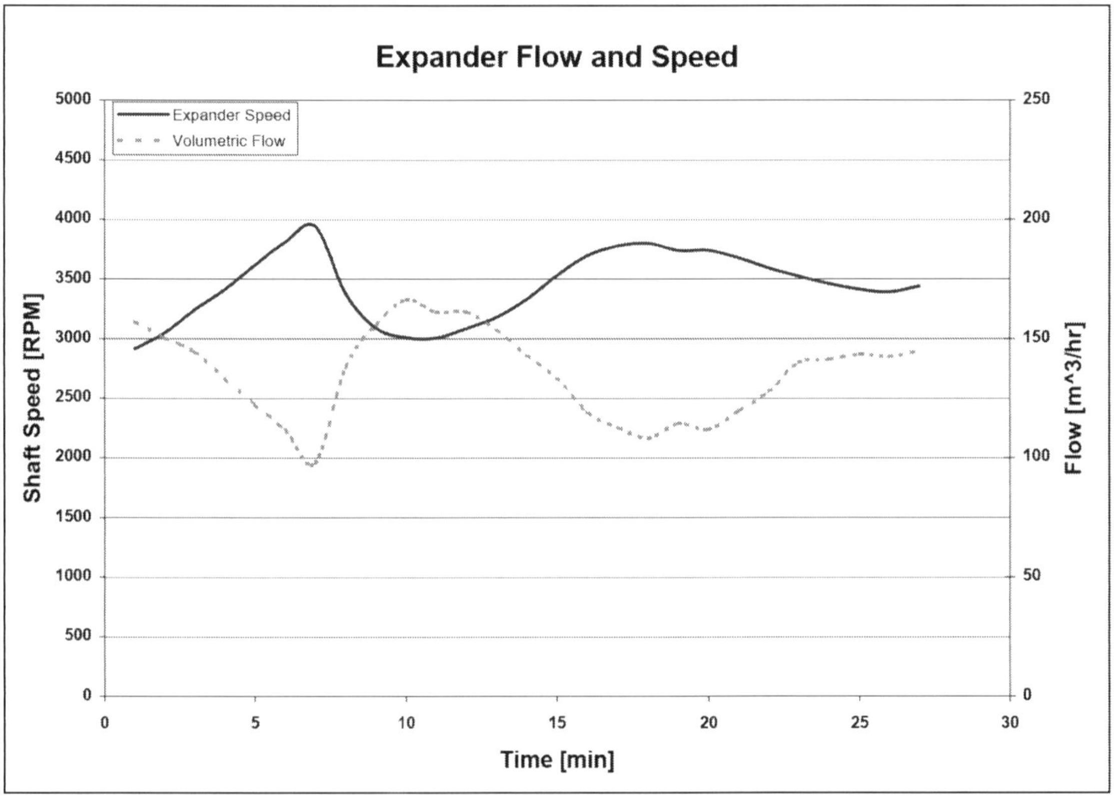


Figure 10: Expander flow and speed during normal operation

In figure 11, the expander flow is plotted against the expander shaft speed over the same period as figure 10. The linear relationship between flow and speed proves the accuracy of the measurements taken. The expander follows the rotating machinery affinity laws that are actually corollaries of the Buckingham Pi theorem with the dimensionless flow coefficient. The first affinity law gives a linear relationship between rotational speed and flow [7].

Due to the automatic balancing of the TEM system, continually transient flow conditions will result in continually transient shaft axial position. To further verify this type of operation of the expander, analysis of the shafts axial position over a relatively short period of time was conducted. Figure 12 shows the shaft axial position over a period of .2 seconds. This figure shows that the shaft is continuously moving in the axial direction to balance the changing thrust acting on the shaft.

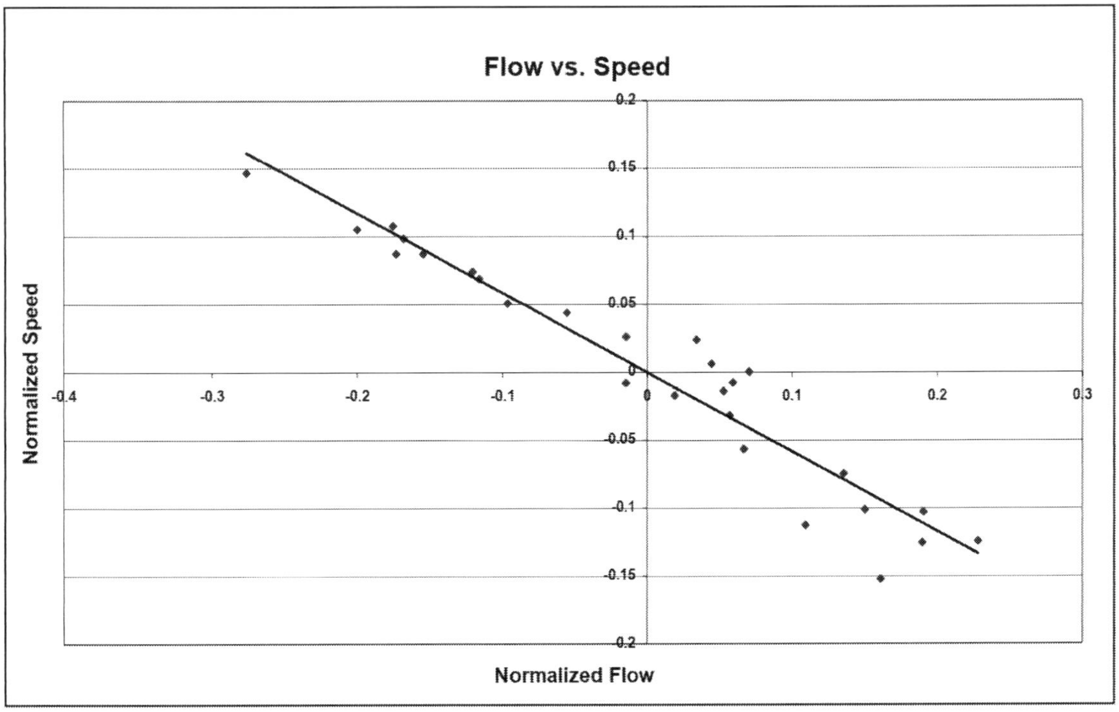


Figure 11: Normalized Flow vs. Speed

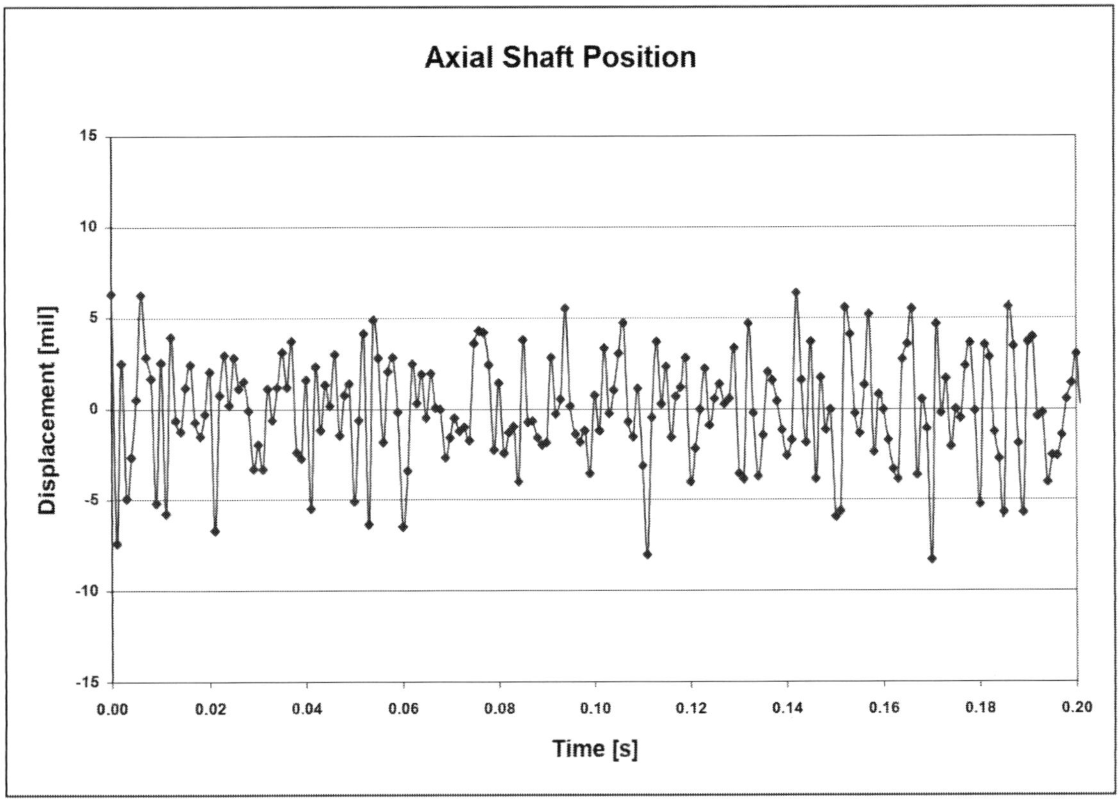


**Figure 12:** Axial Shaft Position

## CONCLUSION

Long-term operational tests conducted at a nitrogen rejection plant in have verified the use of two-phase variable speed expanders to control mass flow rate through the plant. These tests, conducted for over one year, have shown that these expanders can operate consistently and reliably for a wide range of flow conditions. Post test inspections of critical components on both expanders show no signs of damage or excessive wear. Due to the complex nature of two-phase systems, the development of new methods of predicting performance and monitoring the operation of two-phase variable speed expanders in nitrogen rejection plants will be necessary to ensure an optimized design

## REFRENCES


1. Cengel, Y.A., Kimmel, H.E., Cullen, C.M.; “Optimization of Liquid Expanders”, LNG Journal, July/August 1998

2. Elliott, D.G.; Weinberg, E; "Acceleration of Liquids in Two-Phase Nozzles", Technical Report no.32-987, Jet Propulsion Laboratory, California Institute of Technology, Pasadena, California, USA, 1968

3. Evrensel, C.A., Finley, C.D., 2002, “Verification of Axial Rotor Oscillations in Cryogenic Turbine Generators”, The 9th International Symposium on Transport Phenomena and Dynamics of Rotating Machinery, Honolulu, Hawaii.

4. Evrensel, C.E., Kimmel, H.E. and Cullen, D.M., 1999, “Axial Rotor Oscillations in Cryogenic Fluid Machinery”, Proceedings of the 3rd ASME/JSME Joint Fluids Engineering Conference, San Francisco, California

5. Filina, N.N.; Weisend II, J.G.; "Cryogenic Two-Phase Flow: Applications to large-scale systems", Cambridge University Press, 1996, ISBN 0-521-48192-9

6. Finley, C.D., Evrensel, C.A., 2002, “Observation of Jump Phenomena in Non-Linear Axial Rotor Oscillations”, The 9th International Symposium on Transport Phenomena and Dynamics of Rotating Machinery, Honolulu, Hawaii.

7. Japikse, D., Marscher, W.D. and Furst, R.B., 1997, Centrifugal Pump Design and Performance. Concepts ETI, Inc., Wilder, Vermont, USA

8. Kimmel H.E.; “Variable Speed Turbine Generators in LNG Liquefaction Plants”, Proceedings of GASTECH ‘96, Vienna, Austria, December 1996.

9. Kimmel, H.E., 1998, “Power Generation Using Thrust Balanced Hydraulic Turbines”, Proceedings of the American Power Conference, Chicago, Illinois.

10. Perlmutter, M.J.; Kimmel, H.E.; Chiu, C.H. and Paradowski H.: "Economic and Environmental Benefits of Two-Phase Expanders" Proceedings LNG 14, Doha, Qatar, March 21-24, 2004

# Technical and Economic Outlook for GTL Projects

Bruce Burke, Y. Larry Song
Nexant, Inc.
44 South Broadway
White Plaines, NY 10601-4425
bburke@nexant.com
ylsong@nexant.com

and

Sheldon J. Kramer
Nexant, Inc.
3000 Post Oak Boulevard
Houston, TX 77056-6503
sjkramer@nexant.com

Prepared for presentation at the 2004 AIChE Spring National Meeting
New Orleans, April 25-29, 2004



## Abstract
The volatile environment in world energy markets over the past few years has changed the ease of implementation for GTL projects. In a positive development, recent high crude oil prices have significantly improved the economics of GTL projects making it easier to justify developing these projects. However, the uncertain world situation, including recent global economic weakness, slowing energy demand and political uncertainty in the Middle East, has increased the perceived risks associated with GTL projects.

On the other hand, high natural gas prices in key consuming markets such as the United States and environmental drivers to reduce carbon dioxide emissions have promoted a dramatic increase in development and implementation of LNG facilities. Not only do these LNG plants compete for the same natural gas reserves and limited corporate resources, they have the advantage of a simpler and more proven technology over GTL projects that produce liquid transportation fuels.

This paper reviews the current environment and provides an independent outlook for the future. It also reviews GTL technology trends and the fundamentals of GTL economics including the sensitivity to gas and crude oil prices.

## Macroeconomic Outlook
The **world economic** recovery accelerated during 2003, building on modest growth achieved in 2002. Global GDP growth is estimated to have averaged 2.4 %/year in 2003, up from 1.9 %/year in 2002. Recovery to trendline growth of 3.2 %/year will be achieved in 2004 and thereafter. This level of future growth is consistent

with long-term historical performance of about 3.0 %/year growth over the 1970 to 2003 period (all rates reflect growth expressed on an exchange rate basis; growth would be 0.5 to 1.0 % higher per year on a Purchasing Parity basis). As shown in Figure 1, above average growth is forecast for developing regions, such as Asia (ex. Japan), while developed regions (United States, Western Europe and Japan) will grow in the 2.5-2.7 %/year range. Recovery in Japan is expected to take a number of additional years to achieve required structural changes in the economy. Since economic growth is the driver for energy use, recovery to trendline growth levels is important since this can be expected to result in energy growth in line with historical levels of 2-2.5 %/year. Such growth is needed to support development of new sources of energy supply, such as GTL fuels.

**Crude oil prices** have been highly volatile during the last few years, and there is no reason to expect that this will change. However, for planning purposes, future crude oil pricing scenarios must be assumed. Nexant addresses future pricing uncertainty by projecting an envelope of prices to capture expected future volatility. Our current pricing envelope for project analysis is presented graphically in Figure 2, and reflects roughly a 13 $/bbl range for Brent, from roughly 15-28 $/bbl.

## Fischer-Tropsch Technology

Although other processes are under development, the only commercially proven processes for converting natural gas into liquid transportation fuels are Fischer-Tropsch (FT) based processes. These processes use chemistry that was developed by Franz Fischer and Hans Tropsch in the mid 1920's. In the first step of the process, natural gas is reacted to produce synthesis gas or syngas that is composed primarily of CO and $H_2$. In the second step, the syngas then is catalytically converted into a synthetic crude via the following basic FT reaction:

$$CO + 2\,H_2 \rightarrow -[CH_2]- + H_2O$$

The $-[CH_2]-$ groups are the building blocks that combine to form larger straight-chain paraffin and 1-olefin molecules by a chain growth process. The result of these reactions is a mixture of molecules containing from 1 to over 50 carbon atoms. Generally gasoline contains molecules which have from 5 to 11 carbon atoms, while diesel fuel contains molecules which have from 12 to 20 carbon atoms. Paraffin molecules containing more than 20 carbon atoms are classified as wax.

The third step of the process then involves the recovery and catalytic conversion of the FT products into a synthetic crude oil that can be converted into either transportation fuels or specialty products, such as high quality lubricating oils, solvents or paraffin waxes. Although these specialty products have a high value, they have limited markets. Thus, the emphasis has been on the conversion into a synthetic crude oil by hydroprocessing to produce a shippable product that can be upgraded in conventional petroleum refineries to fungible products. Upgrading into commercial fuels at the GTL site may or may not be practical depending upon the specific situation. All FT products essentially are sulfur and nitrogen free.

## Technology Challenges

The challenges of improving the FT process lie in three areas; cost reduction, product quality, and efficiency improvement. Syngas generation is the most expensive part of the plant, and significant research is being directed in this area. Another method to minimize the cost per barrel of product is to build large plants to take advantage of the economy of scale. Figure 3 shows the effect of plant capacity on the GTL capital cost. Large plants require large reserves of inexpensive gas. Many of these locations generally are in remote areas away from population centers where little or no infrastructure exists. If they were near population centers, the gas already would be pipelined to users or converted to electricity, and hence, it would not be inexpensive. Thus, the need to import workers and to build infrastructure (such as port facilities, roads, and towns) to support the plant significantly increases the project cost.

Modifying the product distribution from the FT reactor to selectively produce material in the naphtha and diesel boiling ranges would reduce the plant cost by minimizing the amount of downstream processing that is required to produce fungible products. Efficiency improvements such as increasing the amount of useable products from a given amount of gas would reduce feedstock costs and simultaneously reduce the amount of wasted energy.

## Competition to Fischer-Tropsch GTL

There are other processes for monetizing remote natural gas that are competitors to the FT GTL processes, such as the production of liquefied natural gas (LNG), methanol, dimethyl ether (DME), ammonia and fertilizer.

LNG is a physical process in which natural gas is liquefied by cooling to about -161°C. It is shipped in this condition from the liquefaction plant site in special tankers to regasification plants where it is vaporized and pipelined to users. The market for LNG is limited, but growing rapidly. The need to limit $CO_2$ emissions is pushing many power plants to use natural gas, rather than coal or fuel oil, because of its lower carbon content. The current rush to LNG will slow down the development of FT GTL projects because of the limited capital available for development of such costly projects.

Currently, there are over twenty LNG projects in various stages of development with a projected annual capacity of over 100 million tons/year. LNG train sizes also are growing with some new plants expected to have annual capacities of 7 to 8 million tons/year. As a result, the cost of LNG plants on a $/tpy basis has significantly decreased.

Methanol is another liquid product that is produced from natural gas via syngas. It's primary use is for chemicals production. However, it also can be used as a fuel. The market is limited. In the 1980s, a methanol to gasoline plant was built in New Zealand, but it was not economic and since has been converted to a methanol plant.

DME is another product that also can be produced from natural gas via syngas. It essentially is the condensation product of two methanol molecules after a water molecule is split off. To increase its limited market potential, DME is being proposed as another fuel for power generation.

## Market Potential for GTL Products

The yield of FT products differs dramatically from a conventional refinery. As shown in Figure 4, GTL facilities primarily produce middle distillates (diesel/jet) and naphtha, with flexibility to produce higher value products such as waxes and base oils with additional investment.

**FT diesel** generally is superior to conventional ultra-low sulfur (ULS) petroleum-based diesel because of its virtually zero sulfur content, high cetane number, and lower emission characteristics. With the necessary additives, particularly for lubricity, FT diesel can be blended with other refinery streams to make finished products. When used as a blendstock, the primary impact on refining operations is to reduce the required desulfurization severity, resulting in lower hydrogen usage and lower catalyst and utility costs. By 2015, about 50 percent of global diesel demand is expected to have a sulfur content of 15 ppm or less as shown in Figure 5.

Globally, the distillate fuel market is expected to increase about 2.3 %/year, or about 5 million barrels per day by 2015. Approximately 60 percent will be consumed as on-road diesel fuel. A typical worldscale GTL facility is expected to produce about 50,000 barrels per day of FT diesel. Thus, to achieve a 10 percent market share of new global demand by 2015, a total of about 60 new GTL plants would be required. Realistic projections for GTL plant construction call for at most 20-25 projects to be built by 2015. It appears that the global diesel fuel market is sufficiently large and robust to absorb the industry's ability to build GTL facilities over the next 15 years.

**FT naphtha is** an ideal feedstock for producing ethylene and other basic chemicals with ethylene production expected to be the primary market. Other potential uses, which are not considered viable for high volume consumption, include refinery gasoline blendstocks, specialty products (e.g. white oils, hydrocarbon solvents, light lubricant base stocks), and fuels for fuel cells.

Global demand for petrochemical-grade naphtha is forecast to increase by 3.6 %/year, or about 2.3 million barrels per day by 2015. A typical worldscale GTL facility is expected to produce about 18,000 barrels per day of FT naphtha. To achieve a 10 percent market share of new global demand by 2015, a total of about 130 new GTL plants would be required. Therefore, similar to the conclusion for FT diesel, the global market for naphtha as an ethylene feedstock is sufficiently large and robust to absorb the industry's ability to build GTL facilities over the next 15 years. The most likely target market for FT naphtha produced in a Mideast GTL plant is the Asian ethylene industry due to the region's high growth and its position as a major naphtha importer.

Due to the very high quality specifications for conventional diesel fuels in 2010, limited, if any, price premium is expected for FT diesel. FT naphtha is expected to command a modest price premium, due to its very high paraffin content, which has no counterpart in conventional naphtha production.

### Current Status of GTH projects worldwide

Current GTL technology providers include:

- BP
- Conoco
- ExxonMobil
- Rentech
- Sasol
- Shell
- Statoil/PetroSA
- Syntroleum

There are numerous projects that are being pursued, each with its own issues and challenges. Due to issues of confidentiality, specifics by project cannot be provided here. However, leader projects share a number of important characteristics, including:

- a strong competitive technology position
- have (or are negotiating for) access to gas resources at favorable prices
- a willingness to undertake significant financial/commercial/offtake risk
- a commitment to developing natural gas as a clean fuel of the future
- a strategy for marketing the products.

The primary regions that are pursuing projects are presented in Figure 6, which indicates the dominance of the Middle East due to extensive gas reserves and programs for development.

### GTL Economics

The cost of a GTL plant can vary greatly depending upon location as well as how much infrastructure has to be built. For example, construction costs in Alaska and on the northwest shelf of Australia are very high relative to more developed locations, such as the Middle East or the U.S. Gulf Coast.

Other factors that can markedly influence the economics of a GTL plant include natural gas cost, product selling price, and the logistics of delivery to target markets. Figure 7 shows the impact of gas price on the production cost of GTL liquids. For planning purposes, a reasonable cost for gas is about 1.00 $/MMBtu. As shown, at this price level, the cash cost of production equals about 14 $/bbl of liquid product. Coverage of capital investment (depreciation and ROI) results in a required revenue of about 26 $/bbl (netted back to the plant gate). Thus, at today's high crude oil prices, such a plant would be financially attractive. A shift in natural gas price of 0.50 $/MMBtu changes costs by 4-5 $/bbl

### Future Developments

Looking to the future, the primary driver for project implementation is anticipated economic returns, which are primarily dependent upon:

- the spread between natural gas feedstock cost and the selling price of the GTL products
- capital cost of the GTL plant

Capital costs are declining, but with little technology partnering between the developers/licensors, the pace of capital cost decline is expected to slow, as shown in Figure 8. The "next investment wave", to be built within the next 5-8 years, is anticipated to have capital costs of around 20,000-25,000 $/bpd of capacity.

Nexant considers that capital investment levels of the "next generation" plants (built post 2010) could realistically be around 15,000-18,000 $/bpd of capacity. This relatively slow decline in cost primarily is because a GTL process is the combination of existing technologies that are themselves relatively mature, though catalysis offers the key for potential improvement. Improvements in process efficiency, measured in terms of gas consumed per barrel of liquid product, and thermal and carbon efficiencies will take place over this period.

In conclusion, global fuel and petrochemical markets have the potential to adsorb all of the GTL fuel products that will be produced. Under the current environment of high crude oil prices, a worldscale GTL plant consuming gas at 1.00 \$/MMBtu would be financially attractive. However, due to uncertainty in future oil prices, plus slow progress in obtaining acceptable pricing for gas feedstocks, it has proven difficult to finance and develop projects. In addition to financial uncertainty, other challenges include practical limits on global construction and capital resources, as well as competition with LNG plants, which have less technical and commercial risk. As a result, Nexant expects that only a few GTL plants will be built by 2010. Over the next decade GTL technology advances will continue to reduce capital and operating costs. Those companies that can overcome these challenges and build the initial plants will gain advantage by proving their processes sooner, thus gaining the know-how to make required improvements to drive down operating and capital costs.

**Figure 1 Global Economic Outlook**

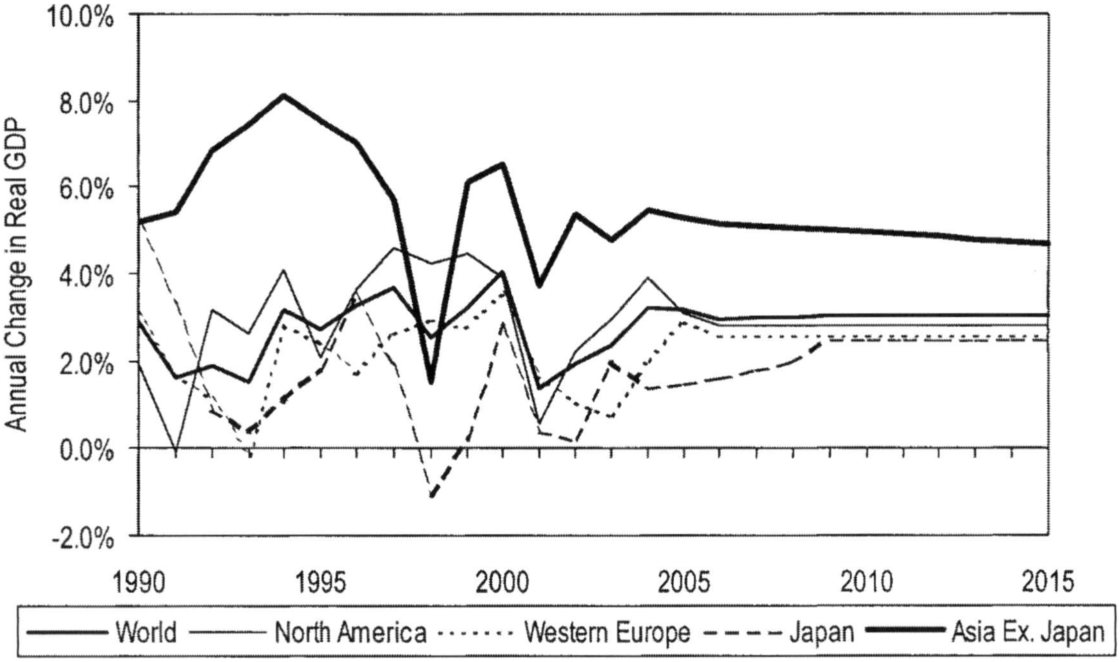


Q403_71440.003.01_Charts.xls

## Figure 2 Outlook for Crude Oil Prices (BRENT)

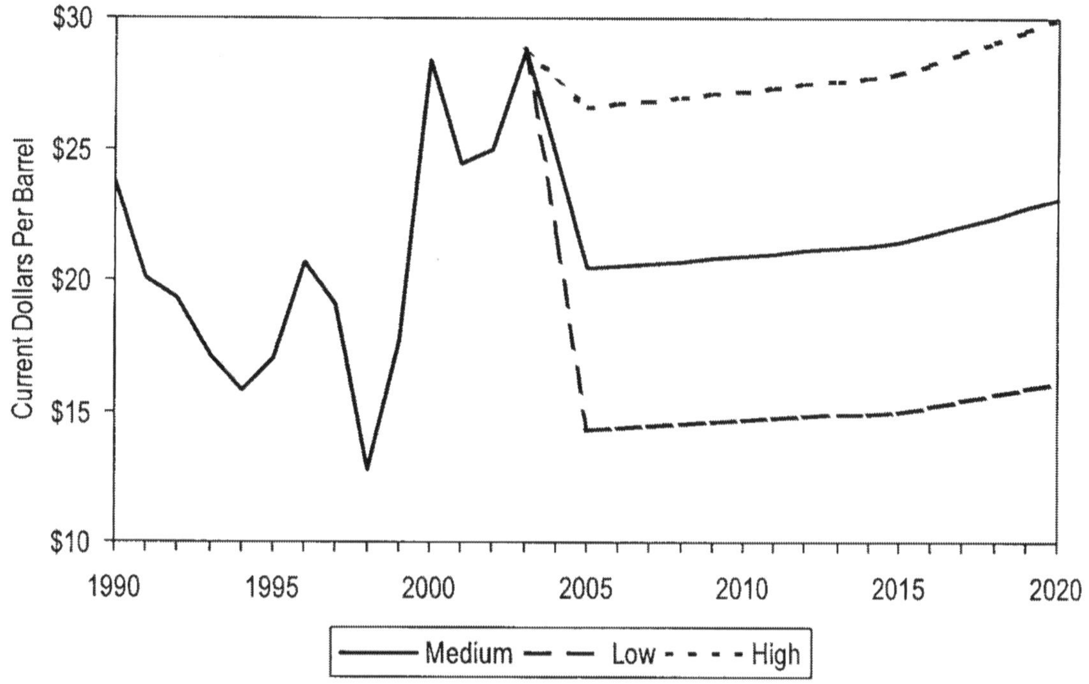


U:\Q403_71440.003.01_Charts.XLS\Sheet2

## Figure 3 GTL Plant Economies of Scale (Middle East, 2005)

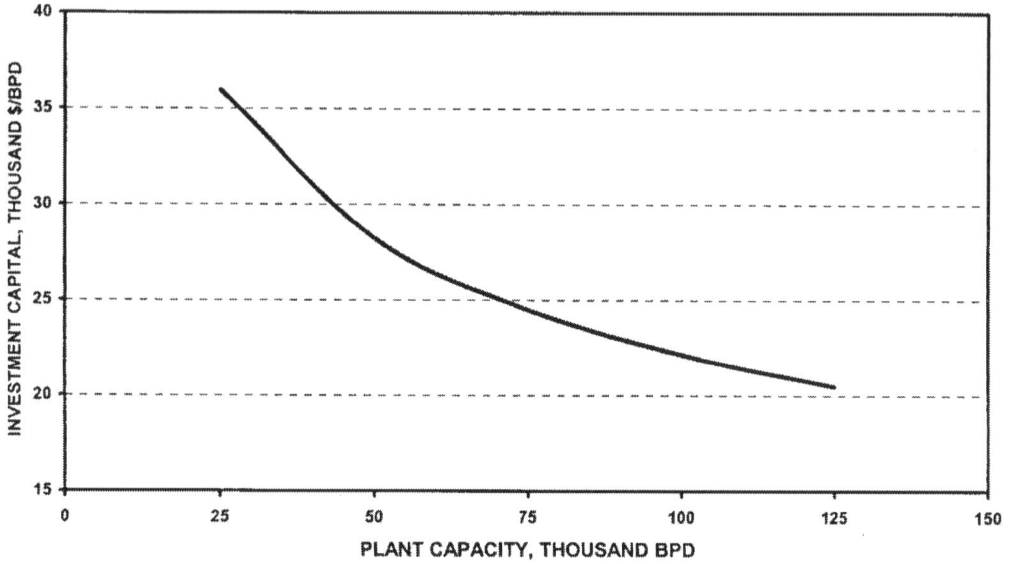

Figure 4 Refinery/GTL Product Profiles

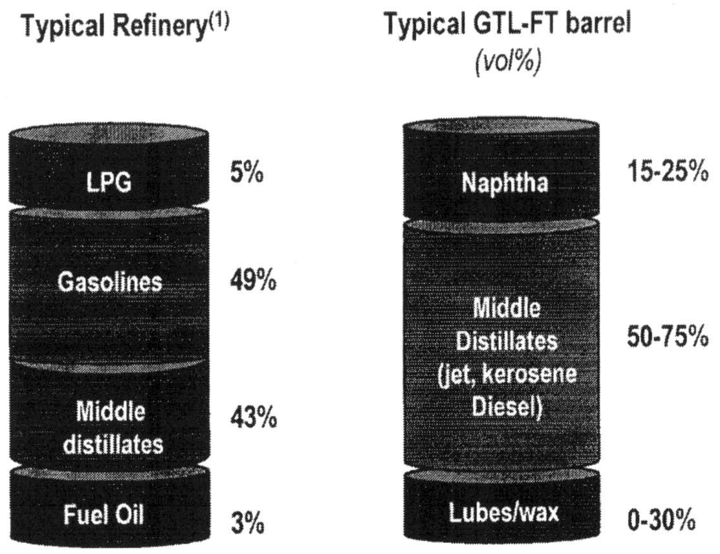


(1) USGC coking refinery, 50/50 Arabian light/heavy crude oil

Figure 5 Diesel Demand by Sulfur Level

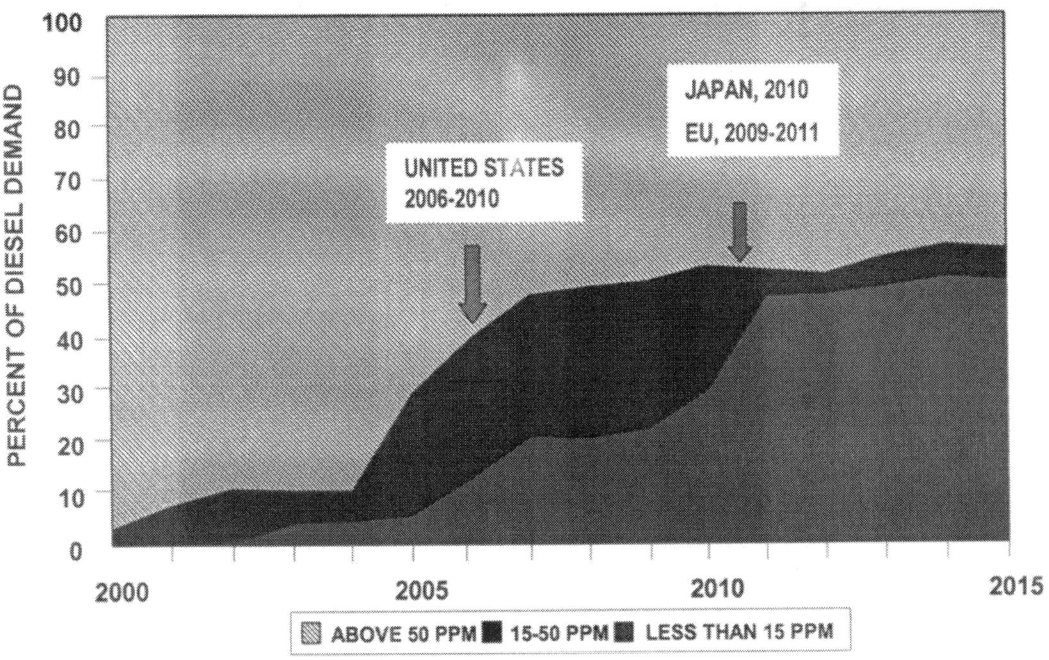

Figure 6 Global Commercial GTL Project Activity

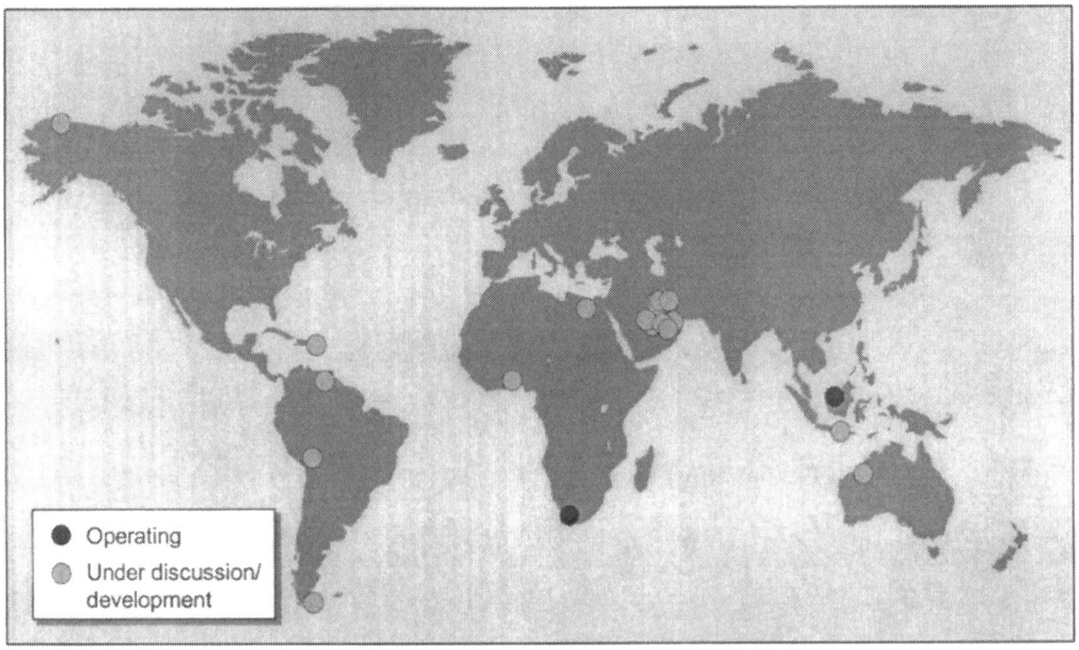


Figure 7 Manufacturing Economics of GTL

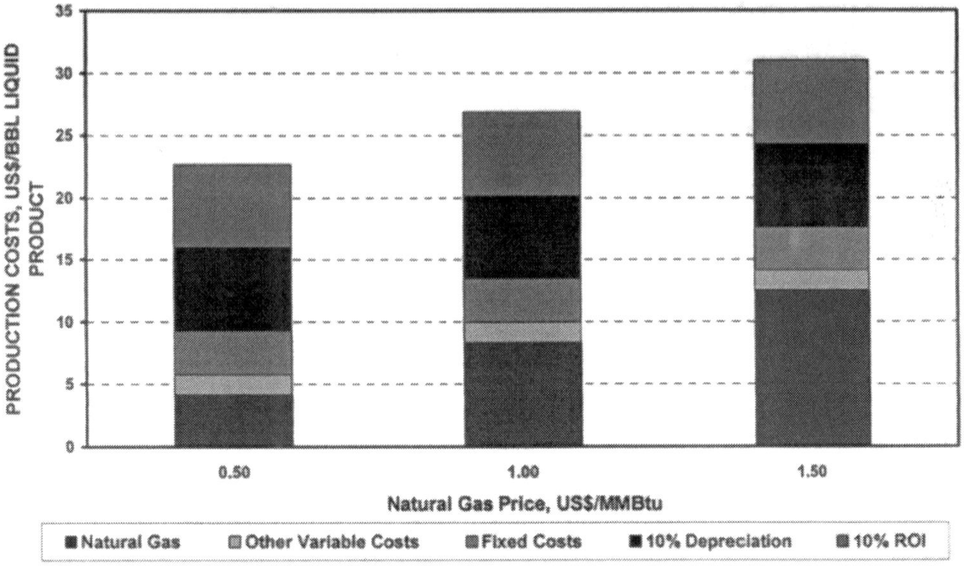

**Figure 8 Conceptual Outlook for GTL Plant Costs**

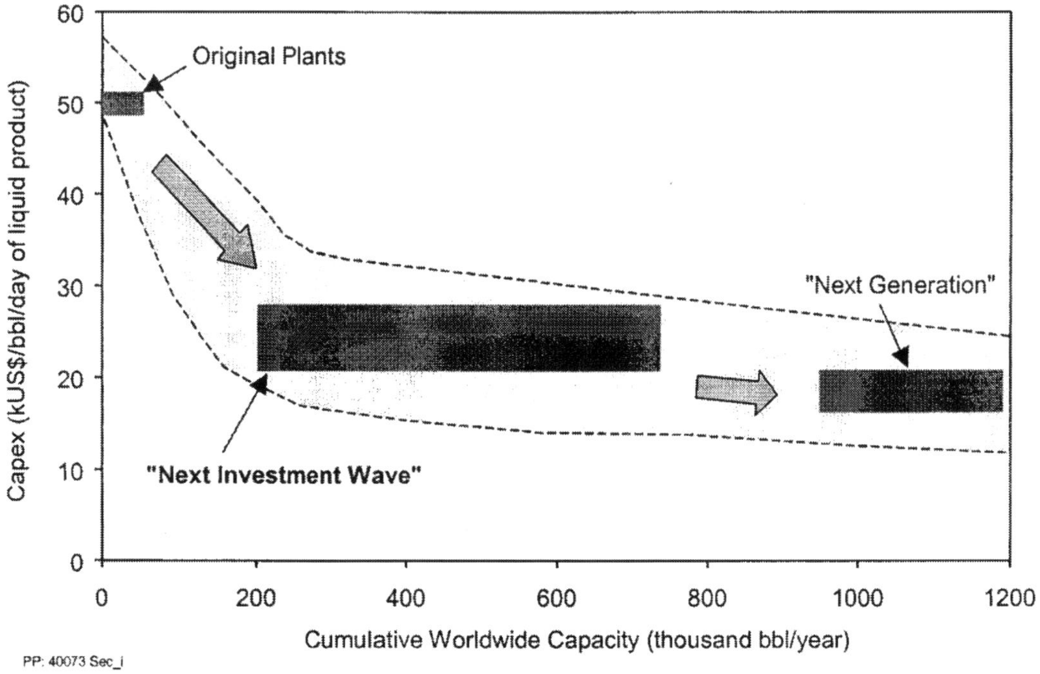


PP: 40073 Sec_i

**Pricing, Output, and Investment Decisions Under Uncertainty: A Game Theoretic Approach for the Study of Chemical Process Industry Economic Behavior**

*Arthur S Gow*
*University of New Haven*
*300 Orange Avenue*
*West Haven, CT 06516-1999*
*Phone: (203) 932-7173*
Fax: (203) 931-6077
Email: agow@newhaven.edu

**Abstract:**

Chemical firms make production and investment decisions based on several endogenous (firm-level) and exogenous (competitor-driven) factors (Bell and Campa, 1997; Lieberman, 1987). Internally, the firm has relatively certain knowledge of its immediate (short-run) marginal cost function, budgetary allocation for new capital outlays, and other quantitative and qualitative characteristics (e.g., labor unions, environmental regulations, and foreign trade policies). At the other extreme, the firm is uncertain (cannot assign a riskiness or probability) about the actions or reactions by its rivals within the industry. At the intermediate level, the firm has vague knowledge (can assign a probability to) about such factors as product demand and the near-term prices of production input factors (i.e., capital equipment, labor, raw materials and energy). Furthermore, the firms decisions may be characterized as either short-run (pricing or output) or long-run (capacity expansion) moves.

Game theory (Haskel and Scaramozzino, 1997; Brock and Scheinkman, 1985) is a powerful tool with which to study organizational behavior in a setting such as described above. Firms will react in a rational manner to the actions or anticipated actions of competitors. Both cooperative and noncooperative games may be played within this (oligopolistic) framework. This paper exploits the main features of the game theoretic approach applied to the global ethylene industry. Thermal cracking of ethane, propane, and light naptha feedstocks is a major technology for producing ethylene and propylene intermediates for polymers production. Several sample games are presented in which the initial composition of the ethylene industry (i.e., number of firms, plants per firm and national dispersion), market position (leader or niche player), the plant technology (embodied in the short-run marginal cost curve), the initial factor (i.e., capital, labor, raw materials and energy) endowment, infrastructural factors (interest and tax rates), and rules of the game (i.e., cooperation versus noncooperation), are first defined. Moreover, a game is subdivided into rounds in which players (firms) complete moves with respect to product price and output (short-run games) or capacity expansion (long-run games). Not surprisingly, the overall results of many games show that industry leader firms are more willing to expand capacity in large lumps, whereas small niche firms infrequently expand incrementally. Conversely, firm pricing decisions appear to be made in a manner, which is independent of the firm's position within the industry. Finally, game results are compared with actual ethlyene industry economic data.

REFERENCES

Bell, G. K. and J. M. Campa, 1997. Irreversible investments and volatile markets: a study of the chemical processing industry. Review of Economics and Statistics, 79: 79-87.

Brock, W. and J. Scheinkman, 1985. Price setting supergames with capacity constraints. Review of Economic Studies, 52: 371-382.

Haskel, J. and P Scaramozzino, 1997. Do other firms matter in oligopolies? Journal of Industrial Economics, 45: 27-45.

Lieberman, M. B., 1987. Post-entry investment and market structure in the chemical processing industries. RAND Journal of Economics, 18: 533-549.

## Effects of Price Uncertainties on GTL Project Valuation

*Richard G Mallinson*
*University of Oklahoma*
*100 E. Boyd Street, Room T335*
*School of Chemical Engineering*
*Norman, OK 73019*
*Phone: 405-325-4378*
*Fax: 405-325-5813*
*mallinson@ou.edu*

**Abstract**

A Model has been developed for GTL project valuation that shows the effects of Uncertainty on the Probability distribution for valuation of GTL projects. Such an approach may be used to value all types of gas utilization projects and also the value of the resource to the overall project.

# Economic Analysis

***The most influent factors on the project valuation***:

- Value of the final products
- Cost of gas feedstock
- Capital expenditure for the GTL plant

} Uncertain ⇩ Scenario Approach

- Operating cost for the GTL plant
- technology risk (yields, catalyst life, etc.)
- Product market demands = Production
- Weighted average cost of capital:
- Interest rate = 8% per annum, Type of finance = 100% Debt
- Income Tax = 35%
- NPV Discount Factor = Desired Rate of Return = 15%
- Product yields = 67.5% Diesel - 26.5% Naphtha - 6.0 % Lubes
- Plant Lifetime = 25 years

# Crude Oil Price

***The price of Diesel, Naphtha and Lubes is Determined mostly by the price of Crude Oil***

From EIA International Energy Outlook:

Lognormal Random Variable Lognormal Random Variable with
$\mu = 3.00\text{-}3.11$
$\sigma = 0.2$

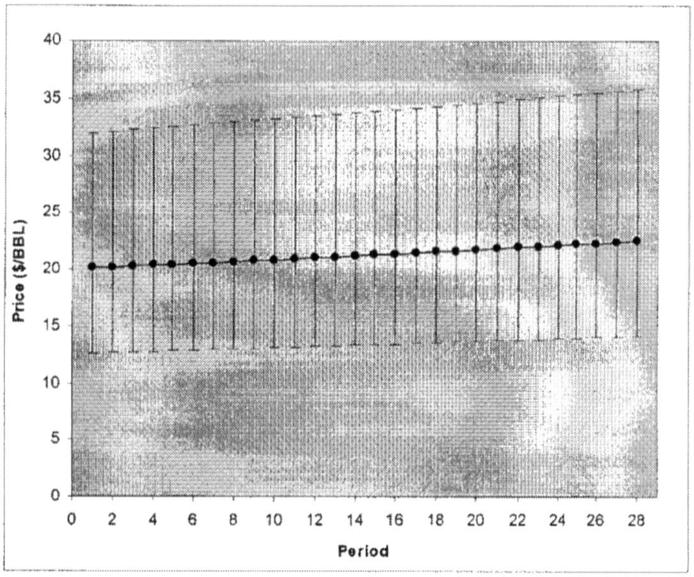

# GTL Diesel Price

***Price Premium for GTL Diesel is highly uncertain. Some analysts believe a premium of 2-2.5$/BBL above refined Diesel is possible if sold as a low sulfur fuel. Others consider that no premium is viable because it will be used as a blendstock with refined diesel.***

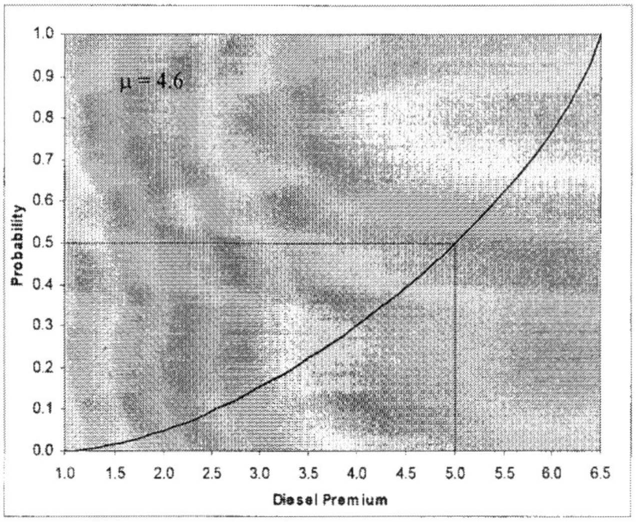

# GTL Naphtha Price

***No price Premium for GTL Naphtha was considered compared to refined naphtha. From historical data, the premium for naphtha varies between 0 and 2.5 $/BBL***

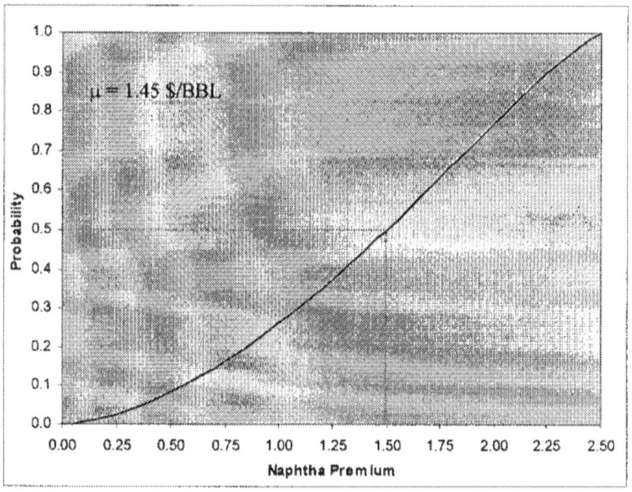


# GTL Lubes Price

***Price is highly dependant on the lubricant grade. Price premium above crude oil price ranges from 20 to 100 $/BBL. mean premium of 50 $/BBL with the following distribution:***

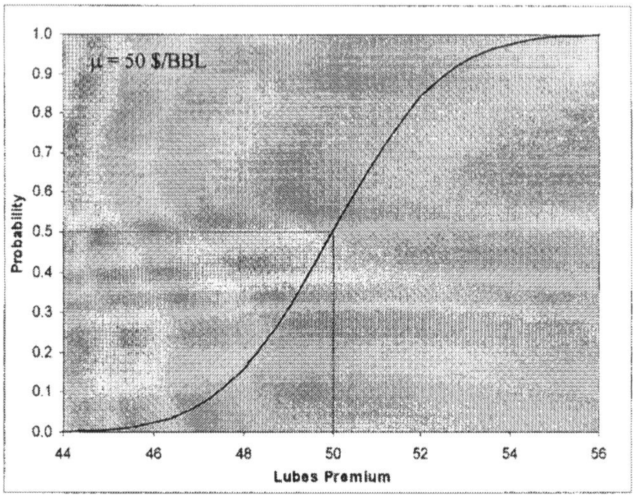

# Natural Gas Cost

***A mean natural gas price of 1 $/MSCF, with small variations around this value is assumed.***

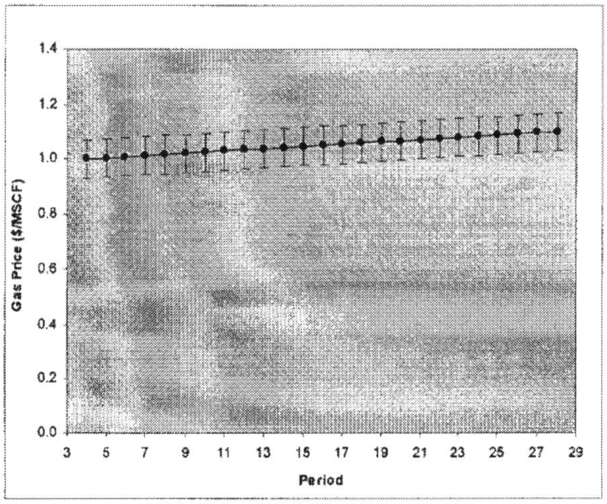


# GTL Cost Breakdown

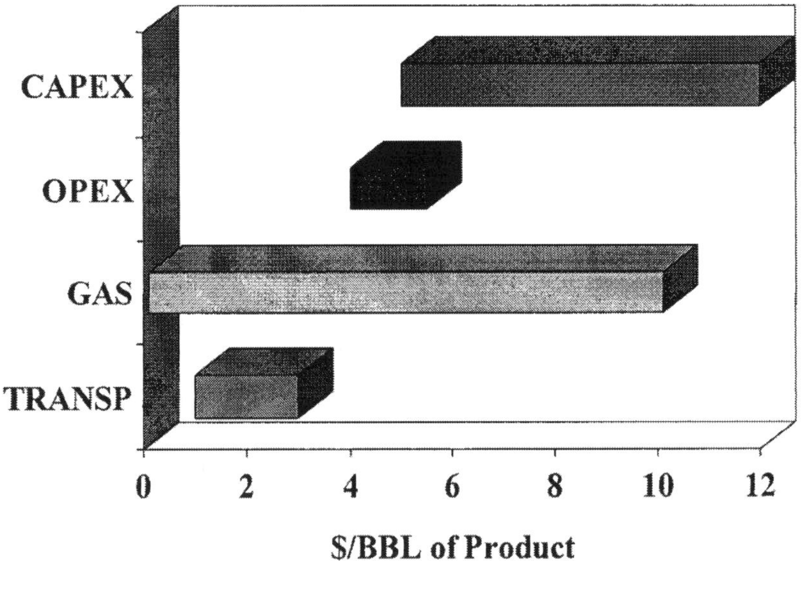

# Plant Capital Cost

- Total Capital Cost: 660 MM$ (22000 $/BBLD)

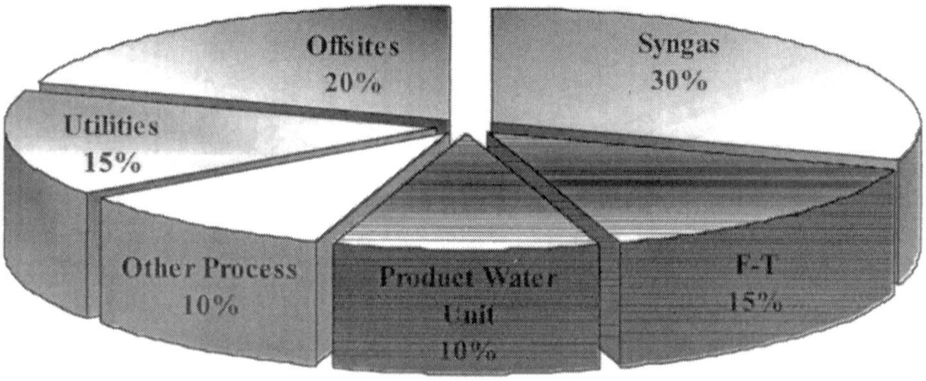


# Operating Cost

- Normally Distributed Random Variable with
  Mean = 5 $/BBL    Standard Deviation = 1 $/BBL

# Transportation Cost

- Normally Distributed Random Variable with
  Mean = 1 $/BBL    Standard Deviation = 0.1 $/BBL

# Net Present Value – Scenario Approach

- **Net Present Value on each Scenario:**

$$NPV_s = \sum_t d_t (Revenue_{ts} - GasExp_{ts} - OpExp_{ts} - TraExp_{ts} - Inv_t)$$

- **Expected Net Present Value:**

$$E[NPV] = \sum_s p_s NPV_s$$

- **Risk of not achieving a certain value of NPV:**

$$Risk(NPV) = \sum_s p_s z_s(NPV) \qquad z_s(NPV) = \begin{cases} 1 & \text{If } NPV_s < NPV \\ 0 & \text{Else} \end{cases}$$

# Project Risk Profile

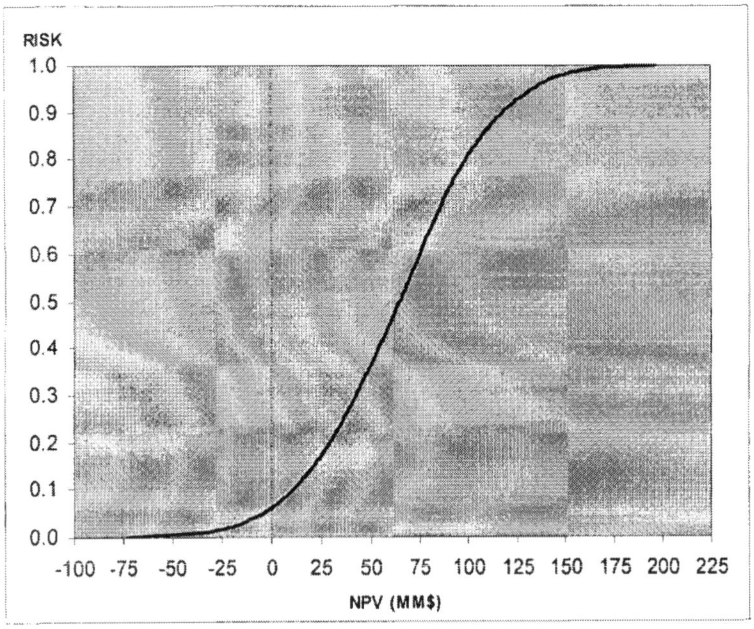


**E[NPV] = 64.4 M$**

**Risk[NPV=0] = 0.06**

(6% chance to have a NPV less than zero)

# Project Cash Flow

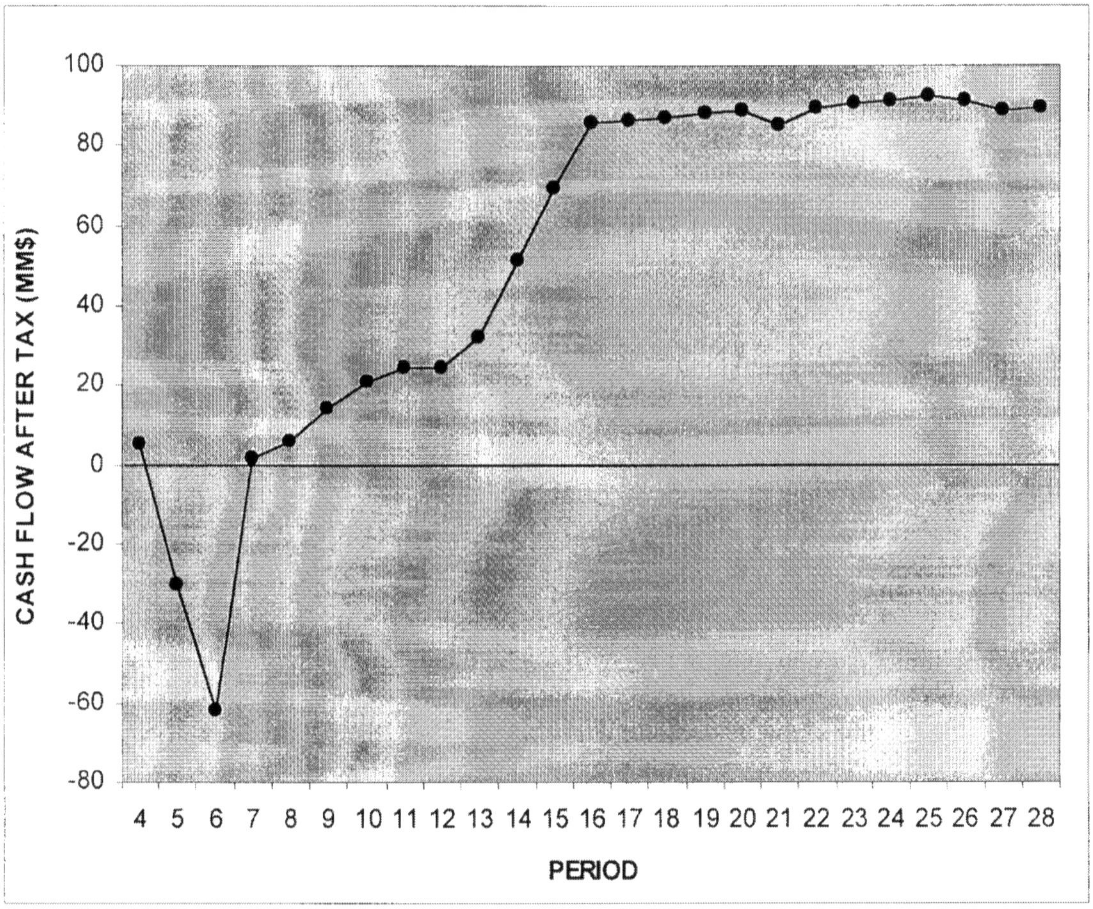

# Reducing the Project Risk

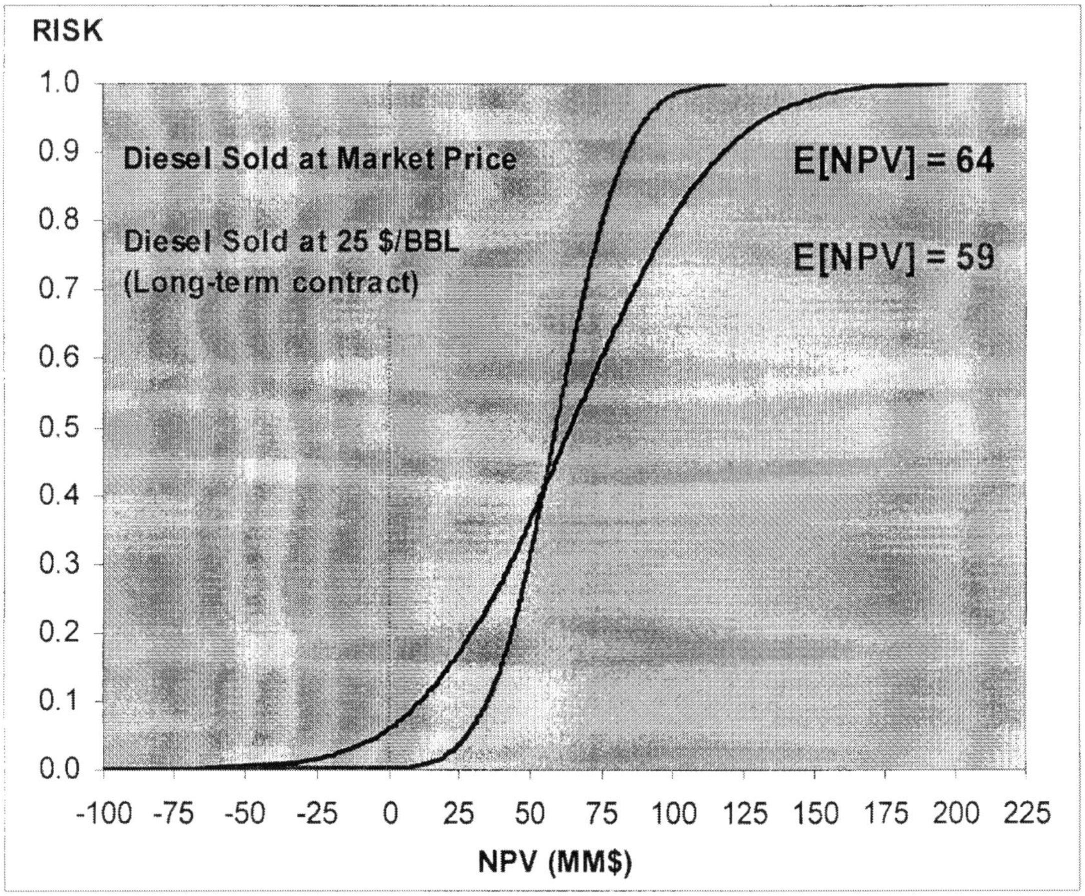

# Project Cash Flow

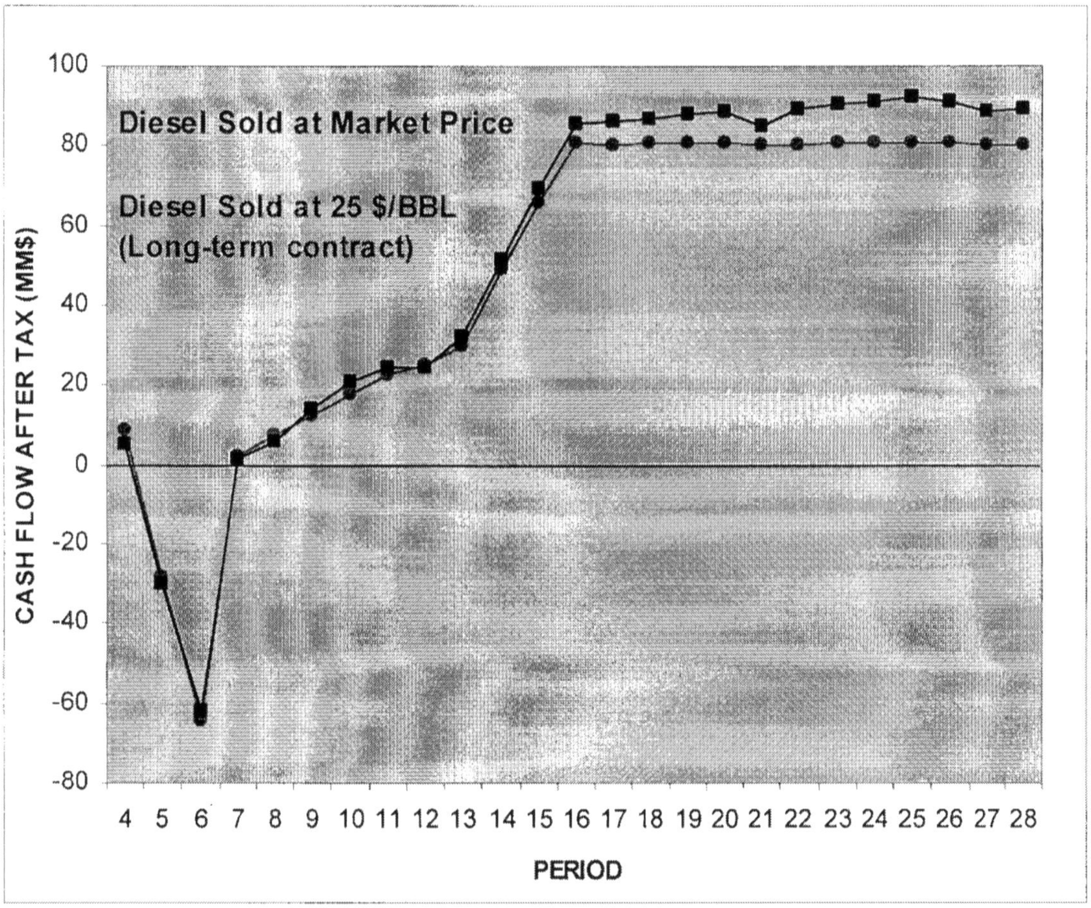

# Full Value Chain Analysis

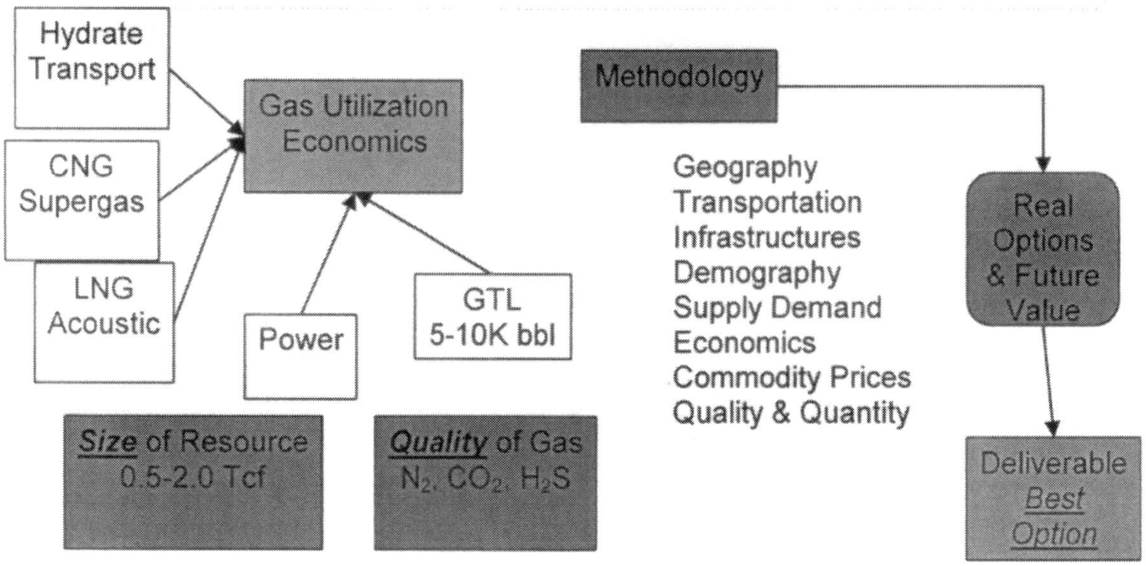


Acknowledgements

Helpful discussions with Dr. Jonathan Kwan and Dr. Miguel Bagajewicz
Results from a study by Andres Barbaro and Bojan Mitrovic for ChE 5643, Fall 2002.

Paper No.

**Financing Large Gas Stranded Assets Development**

Claude H. Devillers
Merzbach Group, LLC

Prepared for Presentation at AiChE meeting on April 25-29, 2004, session

Abstract:

The Challenge of developing gas stranded reserves is both global and regional:

- Global:
  - Supply Growing Energy Demand
  - Replace Aging Facilities (e.g. nuclear, coal)
  - Optimize Use of Natural Resources
  - Reduce Impact on Environment
- Regional
  - Compete for Resources
  - Apply Proven and Efficient Technology
  - Organize Energy Flows
  - Optimize Market Offtake

Such projects are highly capital intensive. We will use four examples to illustrate the approach to financing ventures such as:

- LNG, CNG, Hydrates:
  - Means of transportation of natural gas.
  - Sponsors have a direct interest to sell large quantities of gas from their upstream (>10TCF for LNG).
  - Integrated companies have capacity to organize the marketing from their upstream division.
- GTL:
  - Transformation of natural gas into a slate of products that blend with or substitute for products traditionally obtained from the refining process.
  - Minimum Gas reserves are large (7TCF),
  - Decision process involves upstream and downstream departments of integrated companies.



1- LNG

1-1 Why Finance LNG project with limited recourse to their sponsors?

Capital needs until 2010 for LNG projects is estimated at $144 billion, consisting of:

- Expansion, planned & proposed liquefaction units: USD92bn
- Ships: USD26bn
- Terminals: USD26bn

The industry funding capacity stood at $125 billion, based on the cumulative cash-flows of the major LNG sponsors in 2000, a good year for oil and gas companies.

Therefore, among the international sponsors of LNG projects, one good year generates sufficient cash to cover nearly 90% of the capital needed for new development until the end of the decade, so that lack of cash flow is not a driver for these sponsors.

Furthermore, in the LNG arena, most sponsors have access to cheaper capital than the banks, based on their credit ratings in 2003 (Standard & Poors' data 04/03):

Cost of capital:
Relative long-term ratings of LNG sponsors and LNG expert banks (Standard & Poors' datas 04/03)

| | | | |
|---|---|---|---|
| ExxonMobil | AAA | | |
| Shell | AAA | | |
| BP | AA+ | | |
| ChevronTexaco | AA | Barclays | AA |
| Total | AA | CAI | AA |
| ENI | AA | RBS | AA |
| | | Citibank | AA- |
| | | Deutsche Bank | AA- |
| | | HSBC | AA- |
| | | BNP Paribas | AA- |
| | | ABN Amro | AA- |
| | | RBC | AA- |
| | | SG | A+ |
| Statoil | A | | |
| Norsk Hydro | | | |
| CLP | A- | | |
| Repsol | BBB | | |
| Marathon | | | |

The banking world has undergone a dramatic restructuring during the past few years, implementing drastic consolidations. In addition, banks have weathered major crisis in the emerging markets (Asia 97, Russia) and the emerged markets (US energy deregulation crisis), yielding:

- Fewer project finance teams,
- Fewer banks with large balance sheets and LNG Expertise,
- Less emerging markets appetite,
- Higher risk profile of banks involved,resulting in an increasing need for underwritten Club Deals among few Arrangers of better quality.

Other financing solutions are not available or adapted:

- Project Bonds' track record for LNG Projects in Emerging Markets has been inconclusive at best (RasGas I), and the current appetite is not demonstrated.
- Alternative Financings:
  - Multilateral (some ADB, IFC/World Bank Group) is limited,
  - Carbon Financing is not mature yet, and
  - It is difficult to find an upside sharing proposition with Private Equity and Mezzanine Debt.

ould they be on their own, the main international sponsors of LNG projects would most unlikely seek direct financial leverage on a project. However, LNG projects have been finance on a limited recourse basis for over 20 years.1-2 Drivers to using the financial leverage:

Why use the financial leverage? There is an array of reasons for the use of available financial leverage despite its weaknesses and inconvenient:

- Find common financing ground: such projects have multiple sponsors with different relative strengths and cost of funding,
- Sourcing Capital for the Local Partner(s) Equity Share: most of the local partners have sub-investment ratings (NNPC, Sonangol, QP, Pertamina, Sonatrach, PDVSA), although not all (Trinidad, Woodside).
- Political risk:
  - Country Limits can be full e.g. China

Such reasons have driven the use of limited recourse financing, mainly with bank debt since the 1980's for LNG projects.

1-3 Fundamental conditions to obtain limited recourse financing of LNG projects:

- Allocate Risks to Those Best Suited to Evaluate and Manage Such Risks
- Predictability
- Enforceability
- Accountability

1-4 Risk sharing proposition: views on risks and their mitigation:

- Gas reserves:
  - Legal Title and Enforceability
  - Dedicated volume (ring fenced)
  - Gas Quality (e.g. sulfur, butane, propane, CO, CO2)

- Proven technology
  - Proven track record. LNG track record before 1st financing: 20+ years
  - Assessment of new developments e.g. floating facilities, 8MMTPA+ trains
  - Safety (e.g. H2S)

–

- Proven capacity to develop, build and operate
  - Develop and Build: Timing to book access to markets becomes increasingly important in deregulating markets
  - The right operating team and design can yield a significant competitive edge re.QatarGas versus RasGas.
- Proven market(s)
  - Access to markets
    - Delivery capacity
      - Ships: cyclical access to capacity
      - Reserved capacity with existing and projected receiving terminal
    - Regulatory Environment
      - US
        - FERC Section 3 versus Open Access
        - Coast Guards
      - Europe:
        - Unbundling of gas transmission on July 1st ,2007
        - Japanese and Korean gas markets under revisionLong term contracts versus market risk:
    - Trend: shorter contracts, re-pricing periods (re-openers)
    - Volume
      - Multiple regions marketing strategy
        - US East Coast-Europe / Europe-Asia / Asia-US West Coast
    - Price risk
      - Gas based formulae versus traditional oil replacement product one: Sustainability of gas prices in target region, availability of hedging instruments
    - Strength of the marketing entity to support market risk
        - Ability to book contracts and terminal capacity
        - Credit quality of the off-take
    - Arbitrage options:
      - Futures markets: back to bought deals
      - Volumetric management: limited (e.g. gas storage)
- Competitiveness:
  - Gas prices:
    - stability of agreement with reserve owner
    - Predictability of tax regime with country of operation?
  - Break-even Cost:
    - Assets costs down 30% and falling:
      - Project position on the cost curve
      - Assessment of future cost curve
    - Train capacity x3 (1.5mm TPA to 8mm TPA). Timing for new volumetric production capacity increase

- Greenfield versus Brownfield
- By-products revenues (liquids, sulfur, ER)
- Synergies with other projects (ammonia, methanol, GTL)

– Location:

- Miles to market(s)
- Distance of market entry to consumers

Conclusion: LNG Financing

- Opposing trends:
  - Most international sponsors are reluctant to project finance
  - Some project sponsors have a need for capital
  - Fewer lenders and less lending capacity are available
  - Other Capital Markets solutions are not available
- Compromise with banks on the Risk Sharing Proposition is possible using traditional clauses - DSCR (high for LNG projects), reserve accounts, volumetric guaranteed off-take, minimum equity in the marketing company, etc-,
- Overall Terms of Exchange between Banks and Sponsors likely to be less favorable to Sponsors compared to past financing: bank debt supply is constrained by the elements described above, while risks have increased with the on-going market deregulation in all consumers' countries.

Sample of recent and prospective LNG project financings

| Project | Fin ($mm) | Sponsors | Status | # Int. Arrangers |
|---|---|---|---|---|
| NLNG+ (Bonny trains 4&5) | 1,970 | Shell, NNPC, ENI, Total | Signed-12/02 | 5 |
| LNG tanker | 125 | TEPCO, Mitsubishi, NYK | Signed-03/03 | 1 |
| Qatar LNG Vessel | 179.5 | Mitsui, Qatar Shipping, NYCo | S-03/03 | 3 |
| Bayu Undan | 3,000 | ConocoPhillips (lead) | In Progress | - |
| Brass LNG | 2,250 | ConocoPhillips, NNPC, ENI | In Progress | - |
| RasGas II | 1,400 | ExxonMobil, QP | In Finance | NA |
| Qatargas 2 | NA | ExxonMobil, QP | In Progress | NA |
| Qatargas 3 | 5,000 | ConocoPhillips,QP | In Progress | |
| Idku LNG | 1,311 | BG, EGPC,GdF Petronas, EGAS, Bechtel | In Finance | 12 |
| Angola LNG | NA | ChevronTexaco, | In Progress | NA |

| | | | | |
|---|---|---|---|---|
| | | ExxonMobil, BP, Total, NorskHydro, Sonangol | | |

- Magnitude of the investment: comparable to LNG
- Top gas capacity tailgates LNG bottom one (500MMcfd)
- Sponsor mix: likely to reflect the ship cost component, the largest one, thus less the upstream
- Optimum use in a radius where pipeline and LNG don't overlap. However this area is progressively disappearing
- Fewer fixed infrastructures onshore, leading to a more versatile solution for smaller fields
- Important cost savings (harbor, regasification)
- All issues discussed for LNG are valid for CNG.
- CNG specific issues are:
  - CNG has no track record
    - CNG completion and operation risks are not yet demonstrated
  - Cost:
    - Not yet on the starting blocks on the cost curve
    - Competitiveness needs constant comparison with falling costs of pipelines and LNG projects
    - Potential sharp gains paralleling steep ship building cost curve
    - Gas: incremental benefits versus massive ones for LNG or pipelines
  - Sponsor strength:
    - Majors are already invested in competing solutions
    - Improbable competitive edge versus projects that fill up gas development schedules for majors' fields development:
      - development time comparable at best to GTL, Methanol, or Ammonia
    - Likely to attract higher risk sponsors

2-3 CNG Conclusion: CNG sponsors are likely to seek financial leverage. Their success is predicated to offering sufficient risk mitigation to mitigate the lack of visibility on technology and markets of the early movers.

- Chemical specifications still being investigated at R&D level
- Too early to consider using financial leverage
- Could be screened by Venture Capitalist

of which:

- USD20bn for mass products (e.g. clean diesel)
- USD6bn for niche products (e.g. waxes, lubricants)

Major GTL sponsors long-term ratings (Standard & Poors' data 04/03):

| Mass products | Rating | Niche products | |
|---|---|---|---|
| ExxonMobil | AAA | | |
| Shell | AAA | | |
| BP | AA+ | | |
| ChevronTexaco | AA | Barclays | AA |

| | | | | | |
|---|---|---|---|---|---|
| ENI | AA | | | CAI | AA |
| | | | | RBS | AA |
| | | | | Citibank | AA- |
| | | | | Deutsche Bank | AA- |
| | | | | HSBC | AA- |
| ConocoPhillips | A- | | | | |
| Sasol | BBB | | | | |
| | | Syntroleum | Sub-Investment grade | | |
| | | Rentech | Sub-Investment grade | | |

Like LNG, major sponsors do not have a financial incentive to use limited recourse project financing as explained above.

Drivers to use financial leverage are identical to the ones described for LNG i.e.

- Harmonization of funding between partners
- Local sponsor funding requirement
- Political risk allocation

4-2 Views on CNG Project financing risk mitigation:

- Track record:
  - Technology (Fischer-Tropsch) known for 70 years
  - Industrial application Coal to Liquids (Sasol, Mossol)
  - Existing plant with 1/2 capacity of existing projects (Bintulu, Shell)

Therefore, Lenders are likely to request an Extended Completion Guarantee into starting production period to insure the scaling up of the production to 30,000 bbl/day or more.

- Upstream drivers:
  - Regulations to reduce flaring / re-injection of gas with incentives / penalty schemes sponsored by producing countries (e.g. Nigeria)
  - Optimize production of gas in between LNG projects

While GTL projects are complementary to other gas utilization projects, ring fencing of gas reserves and legal enforceability of title will be important to Lenders.

- Market drivers:Environmental regulations (sulfur, aromatics) implemented through tax policies:
  - Incentive to introduce new clean diesel - requests sufficient tax level on fuels e.g. Europe, Japan, Korea-
  - Reduction of sulfur level in the U.S. refining industry through GTL blending
  - Product sold through Sponsor distribution franchise

Control of the marketing by the some of the Sponsors distribution channels will trigger request by the Lenders for marketing support, e.g. volumetric take-or-pay.Price risk: price of fuels are volatile

A mitigation could be sought in a minimum price guarantee.

- Competitiveness:Break even cost:
    - Current project based on USD25,000 investment to produce 1 bbl. Pilot plants have announced costs of plants down to USD[15-20],000/bbl, creating a potential first mover competitive disadvantage

Lenders will request robust DSCR and cash sweeps to mitigate such risks.

Conclusion:

- Equity participation remains the driving source of capital for large gas monetization projects such as LNG or GTL,
- Bank financing constitutes the bulk of the available third party capital for such projects in the current environment and is relevant to the financing of LNG or GTL projects: however, such intermediation of capital is scarce, and more difficult to access with few specialists left in the market,
- Other significant capital sources are not available, not adapted or not yet mature,
- There is a stronger drive of capital towards project quality,
- The terms of the project value sharing proposition with third party capital has changed: In the present environment of supply and demand of capital, value of debt, when successful, has increased, but only a small group of banks with relevant expertise has the capacity to benefit from this value.

# GTL Fuel Economics and Strategy – A System Look

**Jonathan T Kwan** *(speaker)*
University of Oklahoma
Mewburne School of Petroleum Engineering
Norman, OK 73019
Phone: 281-350-6636
Fax:
Email: jtkwan@yahoo.com

**Michael Matejka**
Shell Chemical Company
100 E Boyd St
Norman, OK 73019
Phone: 4053252921
Fax:
Email:

**Abstract**

Automobile industry must embrace low sulfur-fuels to meet future fuel economy and emission standards. With their own profits under pressure auto companies are adamant to maintain and even expand sales of the popular Sport Utility Vehicles (SUVs). These vehicles, however, have and will be unable to comply with a variety of regulations. The inability to comply with the stiff Corporate Average Fuel Economy (CAFÉ) standards have set manufacturers up for huge fines and the introduction of new emission standards by California will virtually outlaw these vehicles come 2004.

In order to comply with both standards the auto manufactures need wide availability of ultra-low-sulfur fuels fast. It will enable them to substitute currently utilized gasoline SUV engines with new compression ignition engine designs, thereby gaining a fuel efficiency improvement of up to 40%, which is inherent to that type of power plant, and they can achieve the reduction through advanced emissions control technology enabled by the clean fuels.

This paper is to take a system look at the strategic issues and economics of the diesel fuel and their potential substitutes.

**Decision Focused Assessment: Effective Use of Probabilistic Methods in LNG**

*Dave Charlesworth and William J Haskett*
*Decision Strategies, Inc.*
*2028 Quenby*
*Houston, TX 77005*
Phone: 713-526-2410Email: dsi_bill@yahoo.com

Large gas projects (especially LNG) require broad based business assessment methods in order to establish reserve confidence, appropriately size facilities and feed rates, and optimize marketing/commercialization efforts. Simple mean-value calculations, even with sensitivities, are insufficient to ensure appropriate and timely business decisions in the presence of downside risk and mitigation efforts. Decision focused stochastic evaluation methods provide a valid assessment of the range of potential outcomes as well as the materiality of various risk mitigation investments across the entire value chain, from exploration and development through processing, shipping, and marketing.

Decisions regarding tactical and strategic supply security are of paramount importance to the profitability of the overall project. Stochastic evaluation enables the assessment of the probability of shortfall, the materiality of the shortfall should it occur, and highlights the causative shortfall factors. This enables an allocation of time, capital, and personnel to deal with the downside potential and return the project to tolerable levels of risk.

## *Probabilistic Methodology Applied to LNG Value Chain*

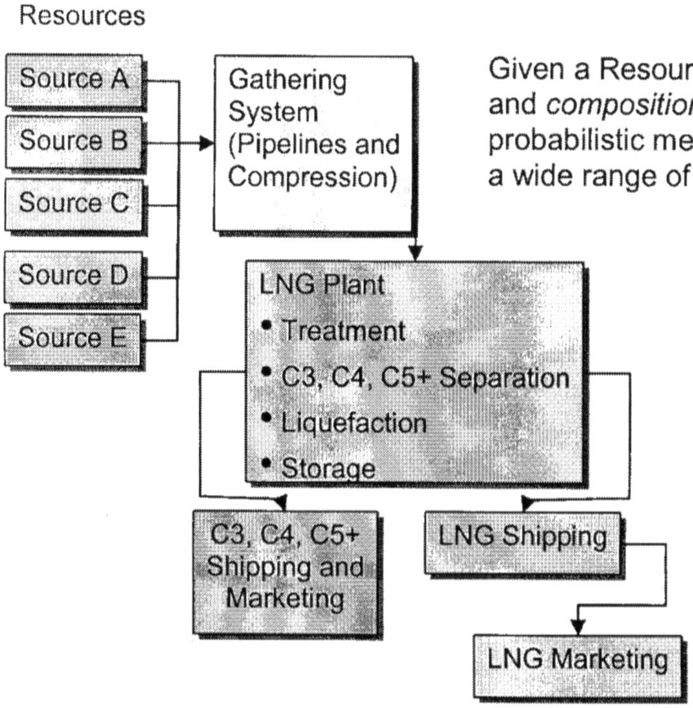


Given a Resource Portfolio with uncertain *size* and *composition* associated with each source, probabilistic methodology can give insight into a wide range of issues:

- Plant Sizing
- Plant Design Specifications
- Product Storage Requirements
- Marketing Contract Duration

## *Definitions*

- **Deterministic**
  "Single point" calculations carried out to show results given a specific set of certain inputs.

- **Probabilistic**
  Encompasses a range of methods which, to some degree, incorporate uncertainties within the sysem.

  - **Simple Probabilistic** – Sampling of $10^{th}$, $50^{th}$, and $90^{th}$ percentiles comprehensively from a limited set of uncertain inputs, or randomly within a simulation

  - **Stochastic** – Fully sampled uncertainty distributions, including dependent and/or correlated variables within a simulation tool.

## *Resource and Facility Situation Map*

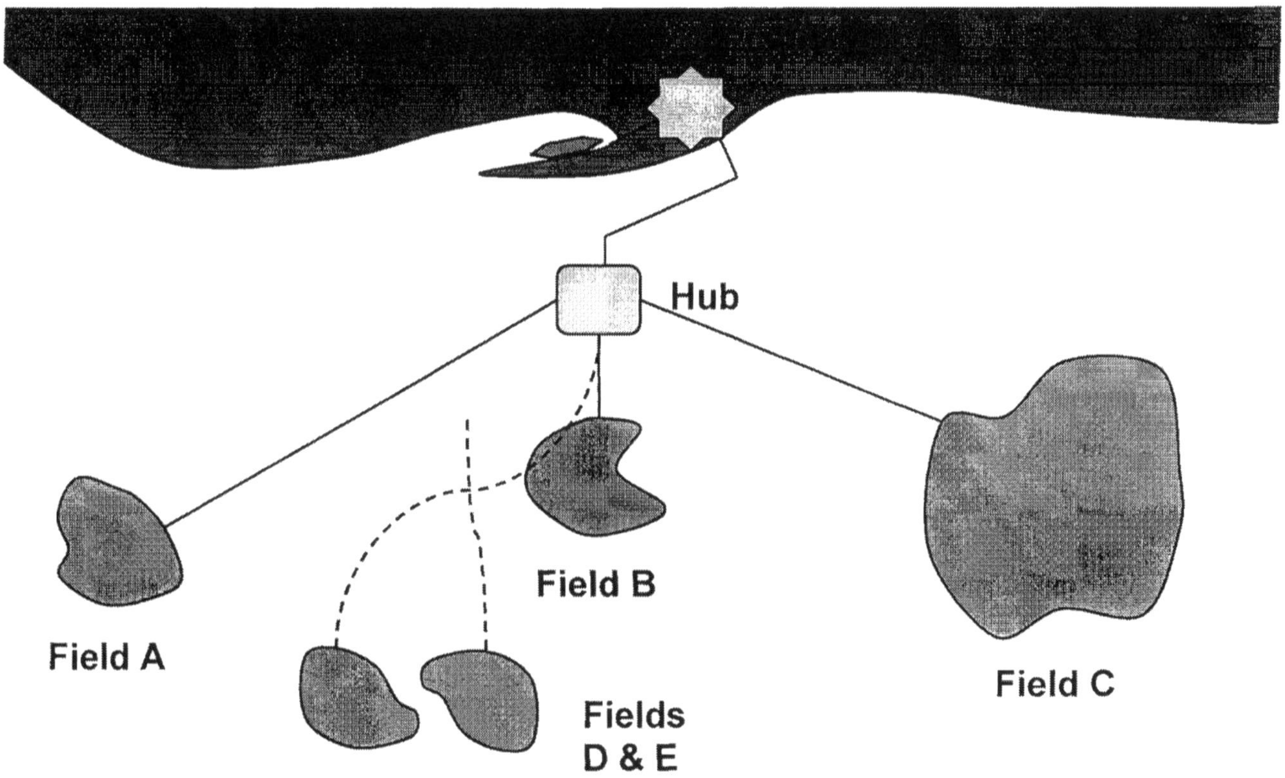

## *Decision Analysis Process and Tool Sets*

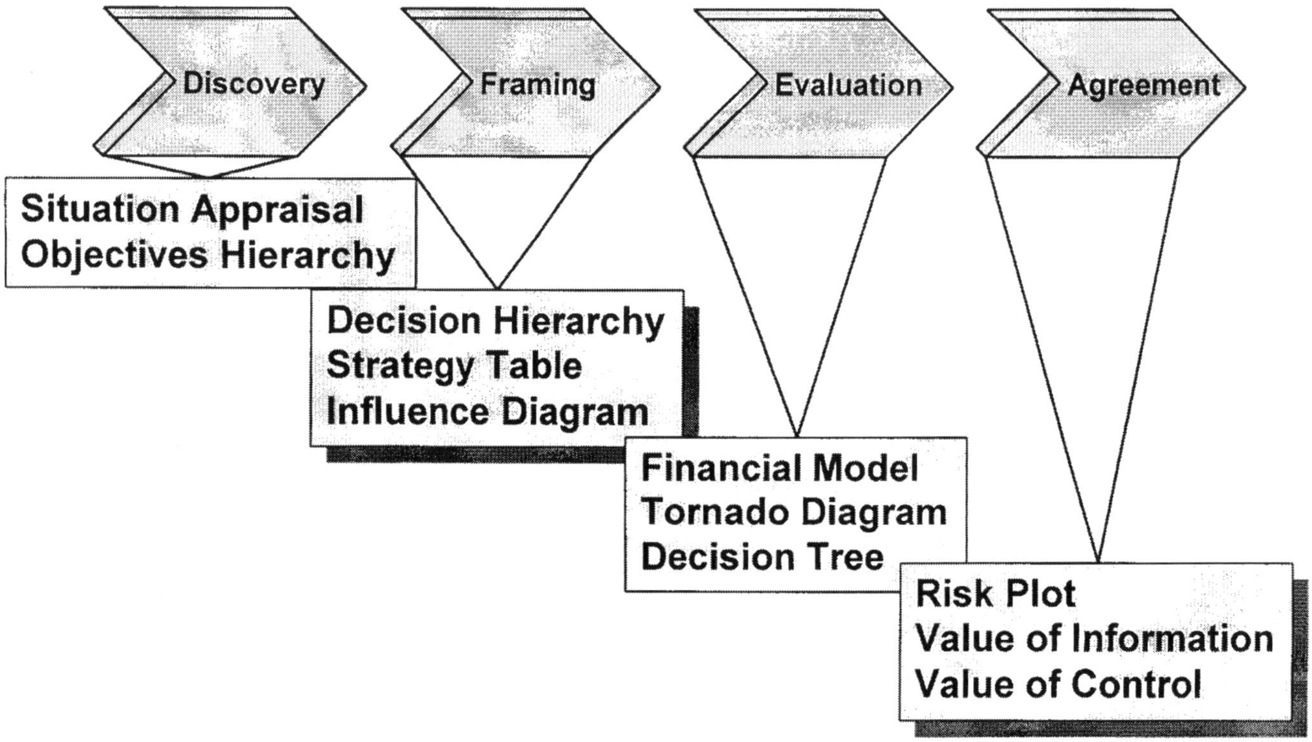

## *Resource Presence – Distribution Output/Input*

Using base reservoir characteristics and geometry uncertainty

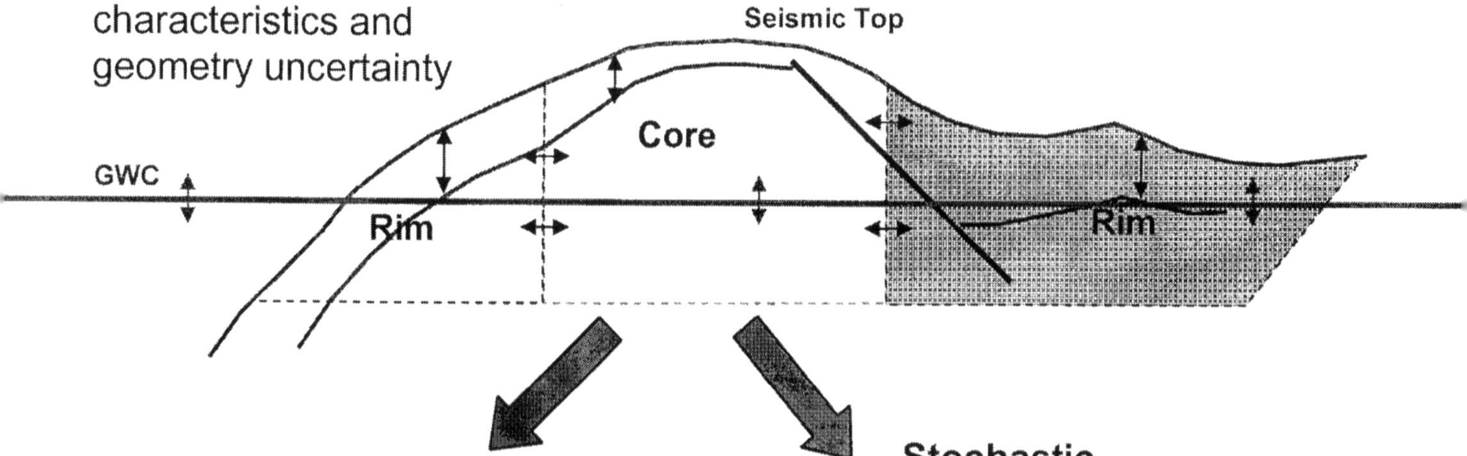


**Simple Probabilistic**

| P10 | P50 | P90 |
|---|---|---|
| X | X' | X'' |

**Stochastic**

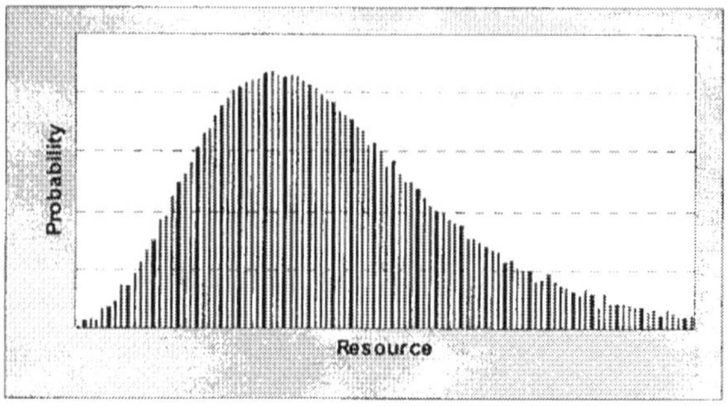

# Project Sensitivity to Resource Risk and Uncertainty

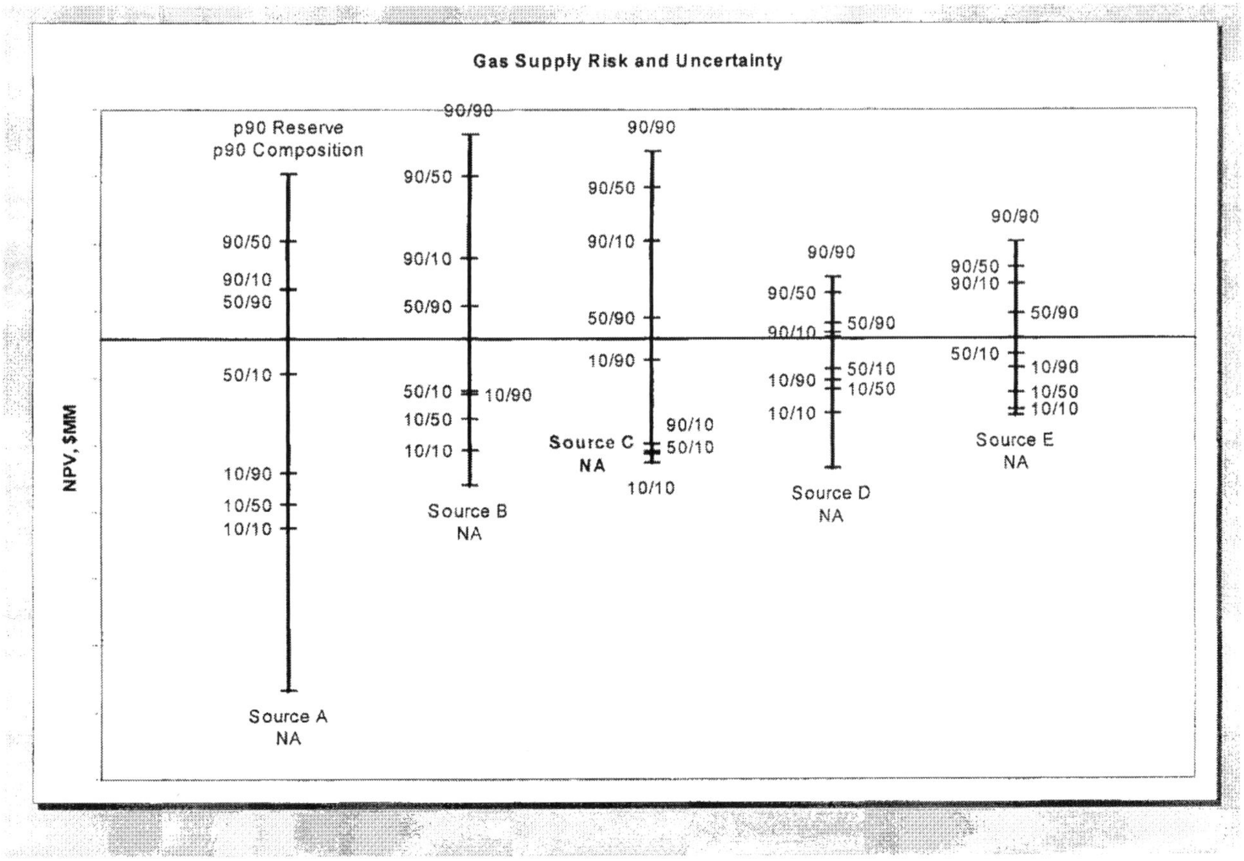

# *Production Profile Variability – Field C*

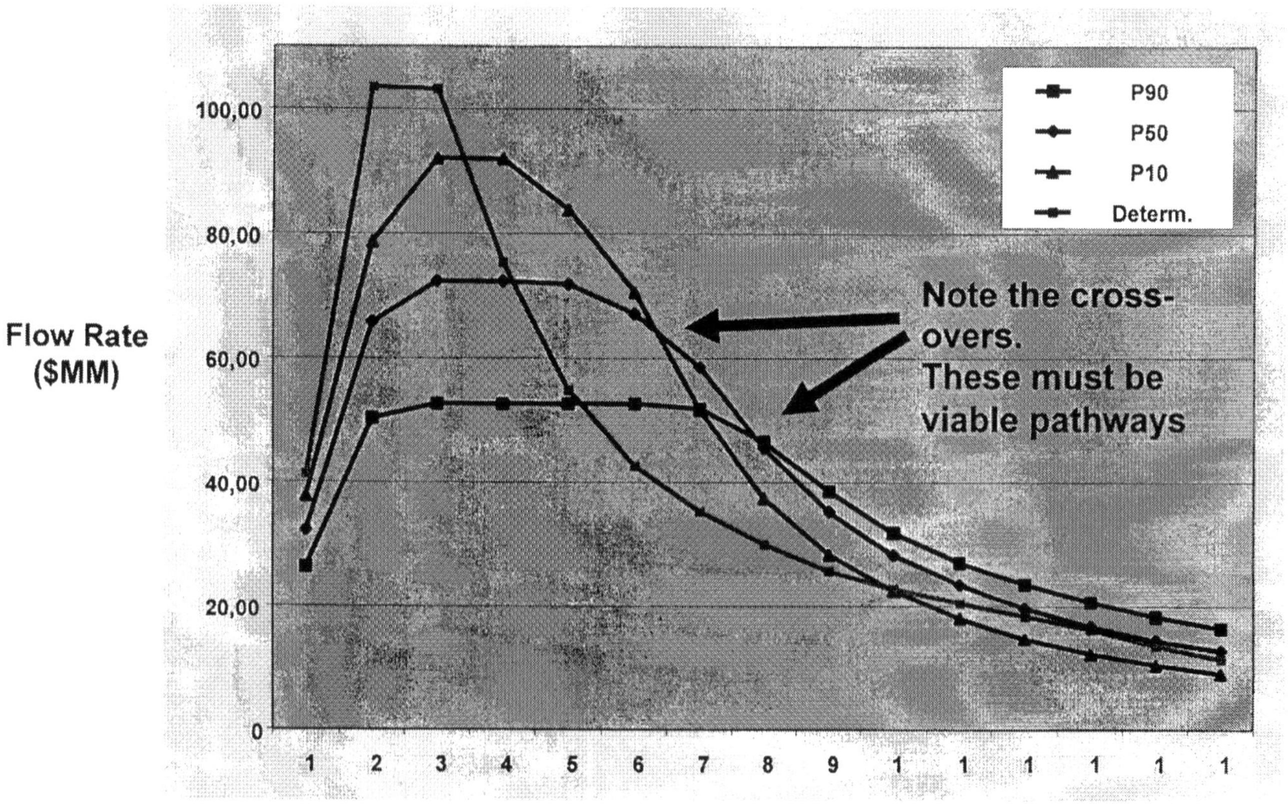

# Facility Capex Uncertainty

CAPEX ($MM)

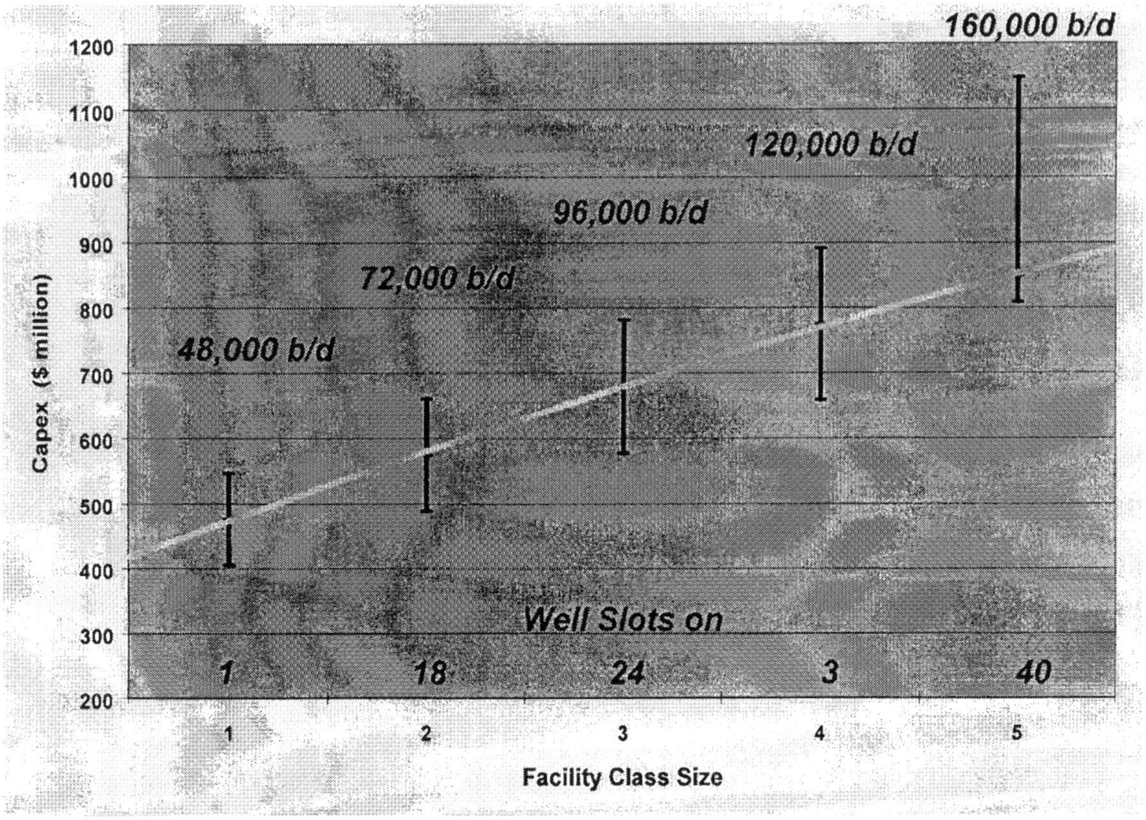

# *Facility Sizing Decisions - Capex vs NPV*

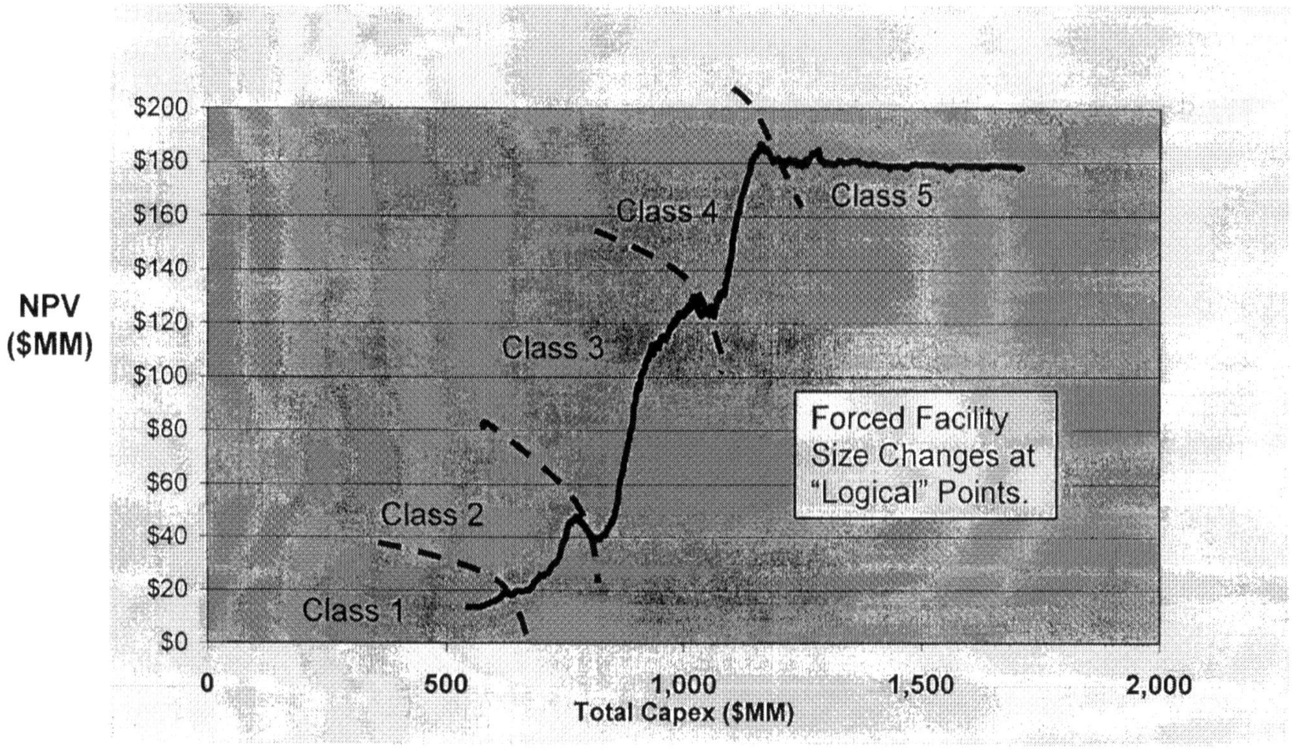

## *Facility Size - Reserves vs NPV*

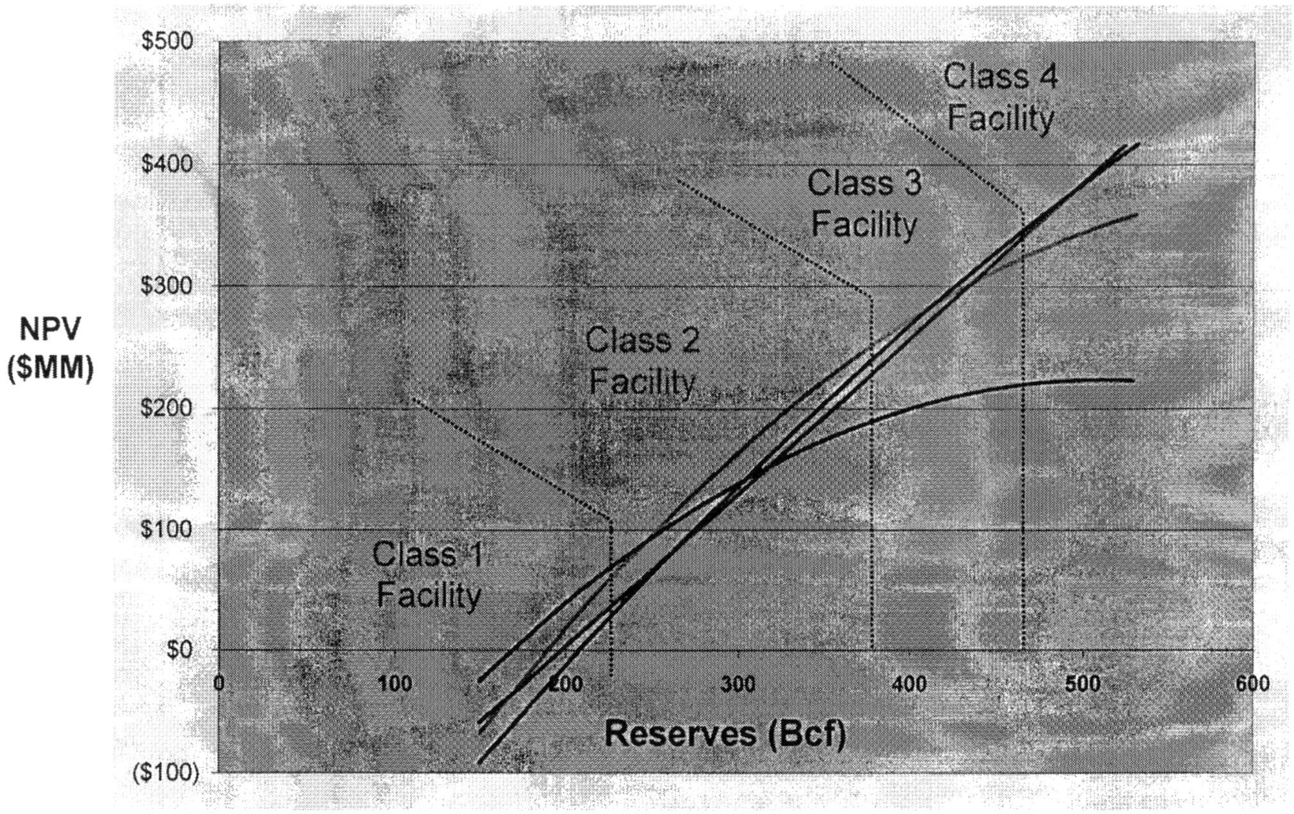

## *Better Evaluation Leads to Better Decisions*

NPV
($MM)

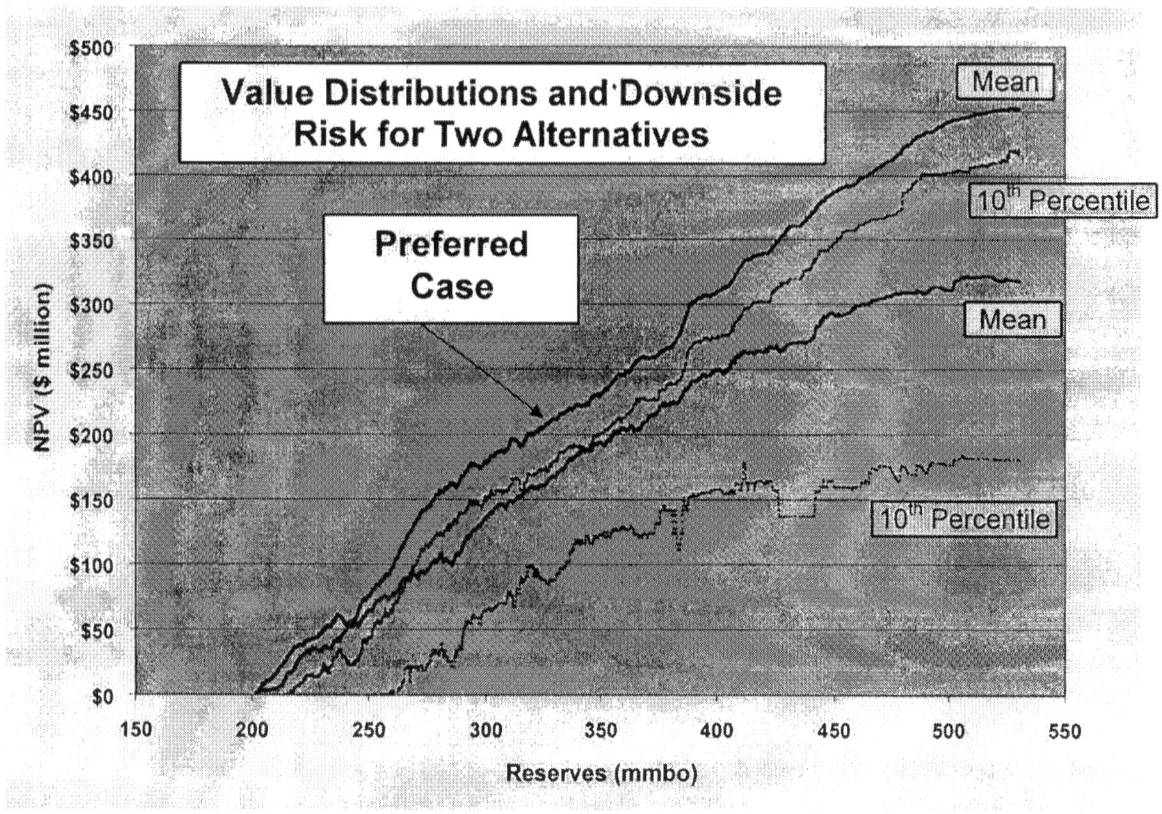

## *Composition Uncertainty and Variance Through Time*

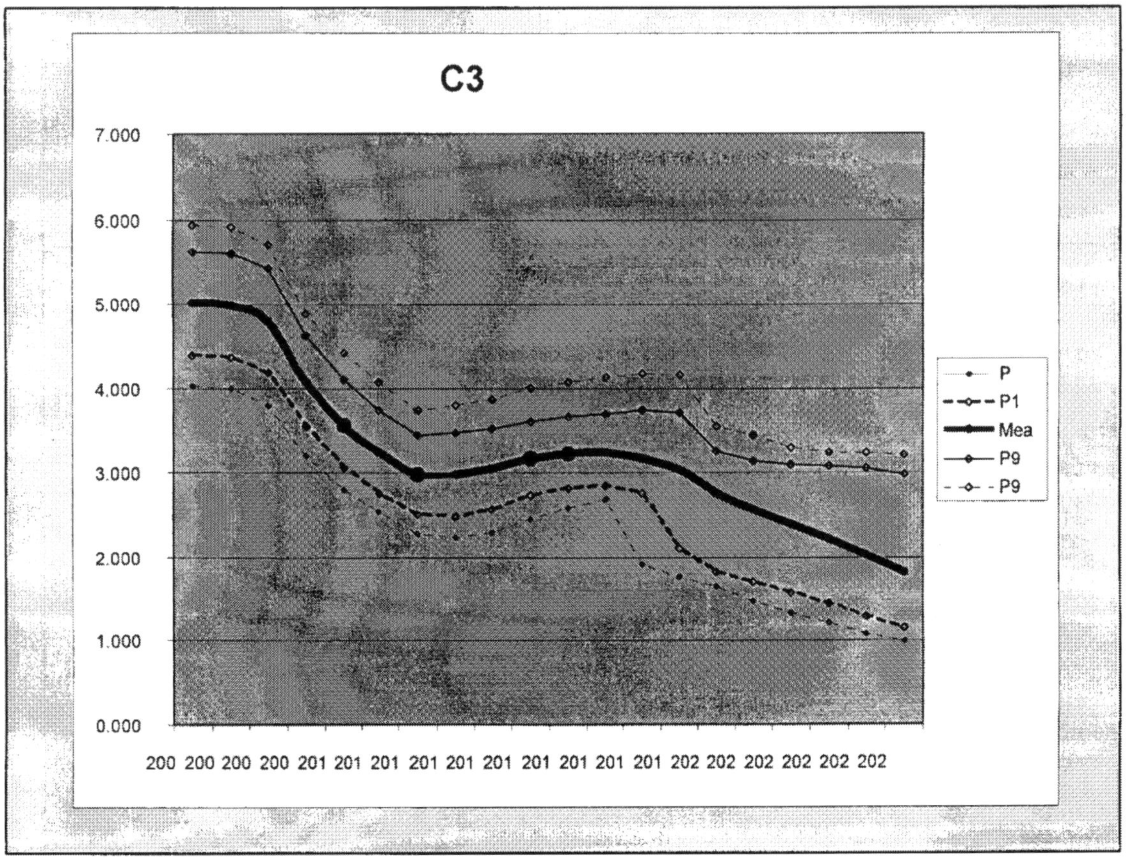

# Propane Removal Design Specification

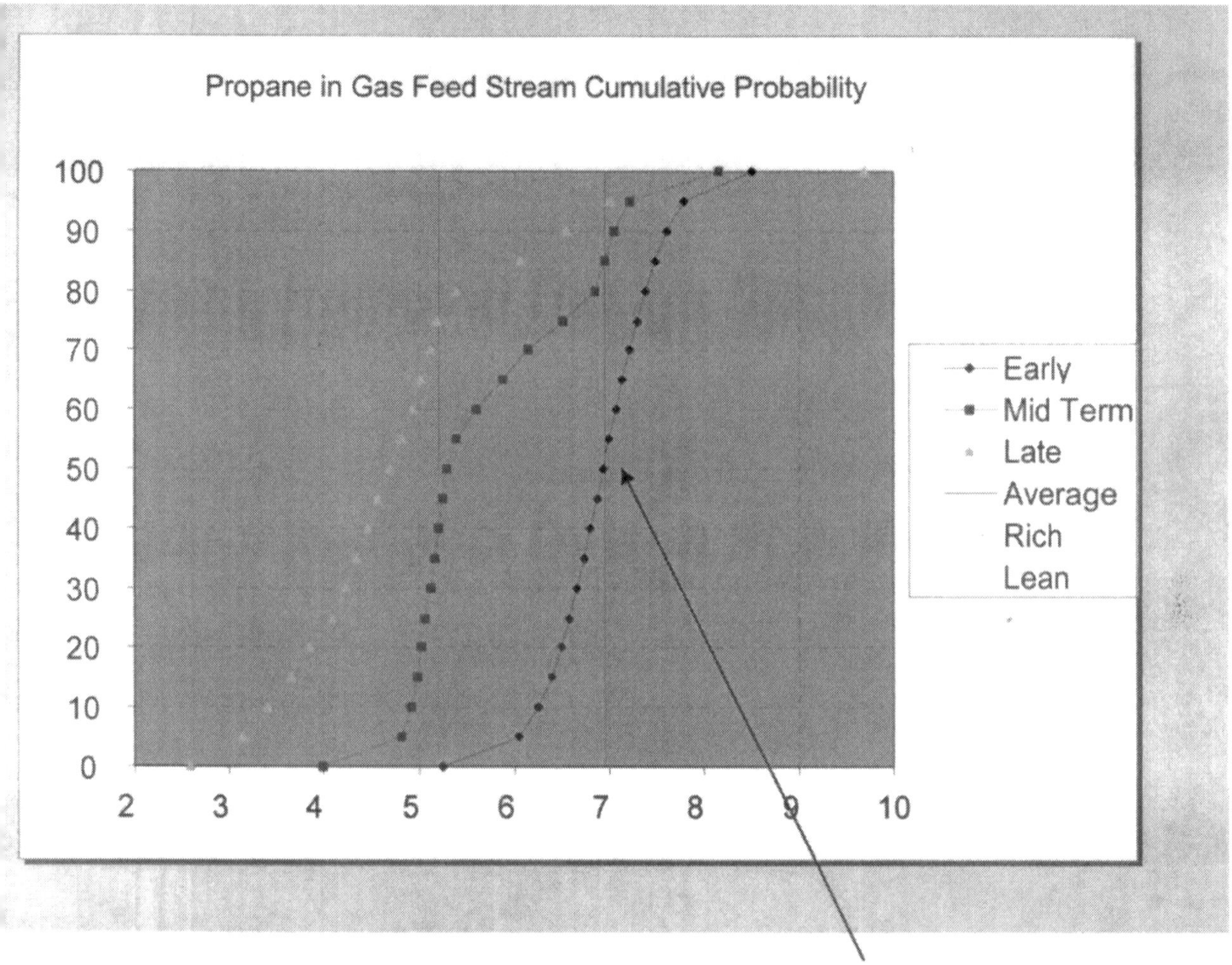


The selected “average” C3 removal design specification resulted in high confidence in the Plant’s ability to separate C3 relative to the uncertain gas supply.

# *Shipping and Storage Trade-offs*

CAPEX:
Add Storage Capacity

- *CAPEX for each additional storage tank*

CAPEX and OPEX:
Add Shipping Capacity

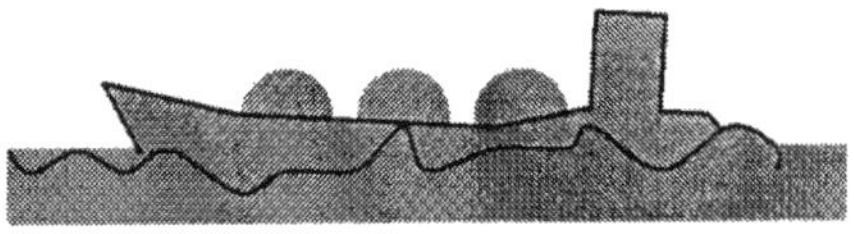

- *If an incremental ship is required: $160 MM CAPEX*
- *~$65K per day per ship*

<u>Full Tank Event</u>:
Idle Plant while Waiting for Ship

- *> $1MM per day incremental revenue*

<u>Empty Tank Event</u>:
Pay Demurrage/OPEX for Idle Ship(s)

- *~$65K per day per ship*
- *Could have adverse effects with customers, especially if sales are FOB*

# Example Shipping and Storage Probabilistic Results

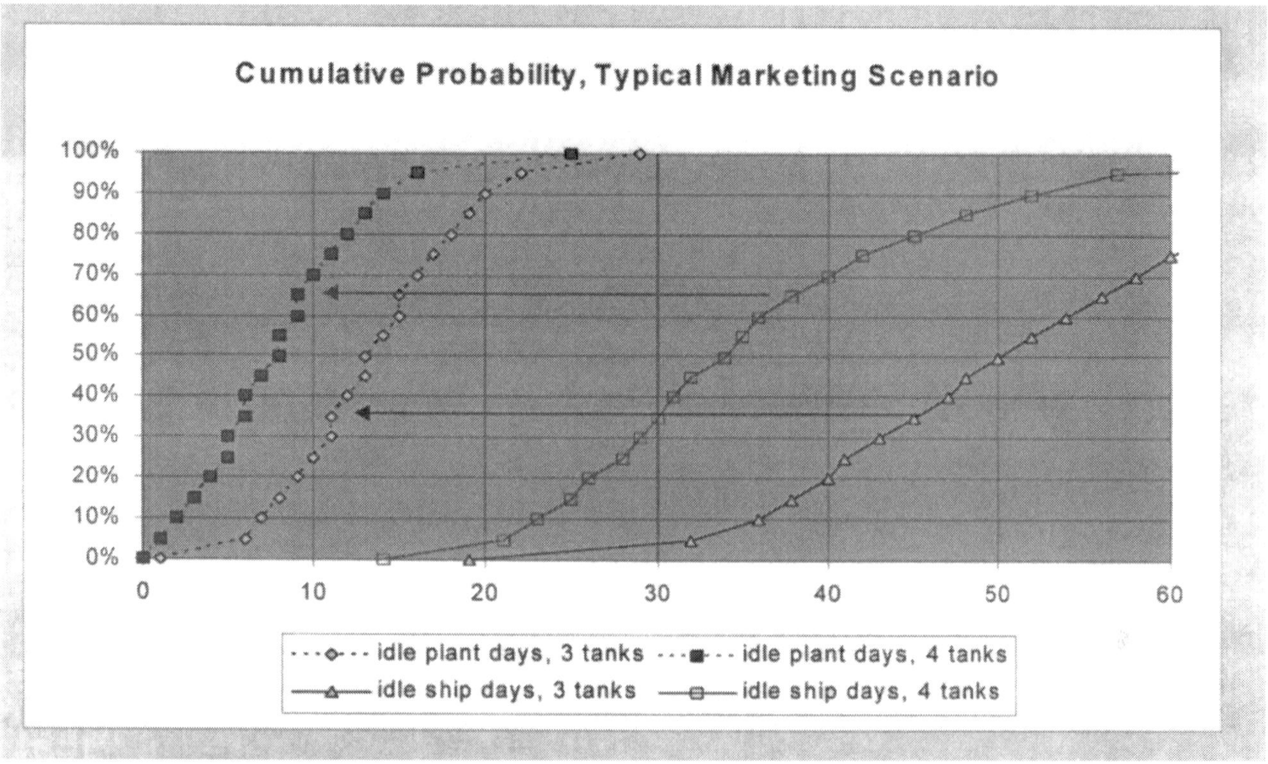


Increasing storage capacity from 3 to 4 tanks can significantly reduce the risk of full tank events and empty tank events...

## *Example Market Confidence Level and Contract Duration*

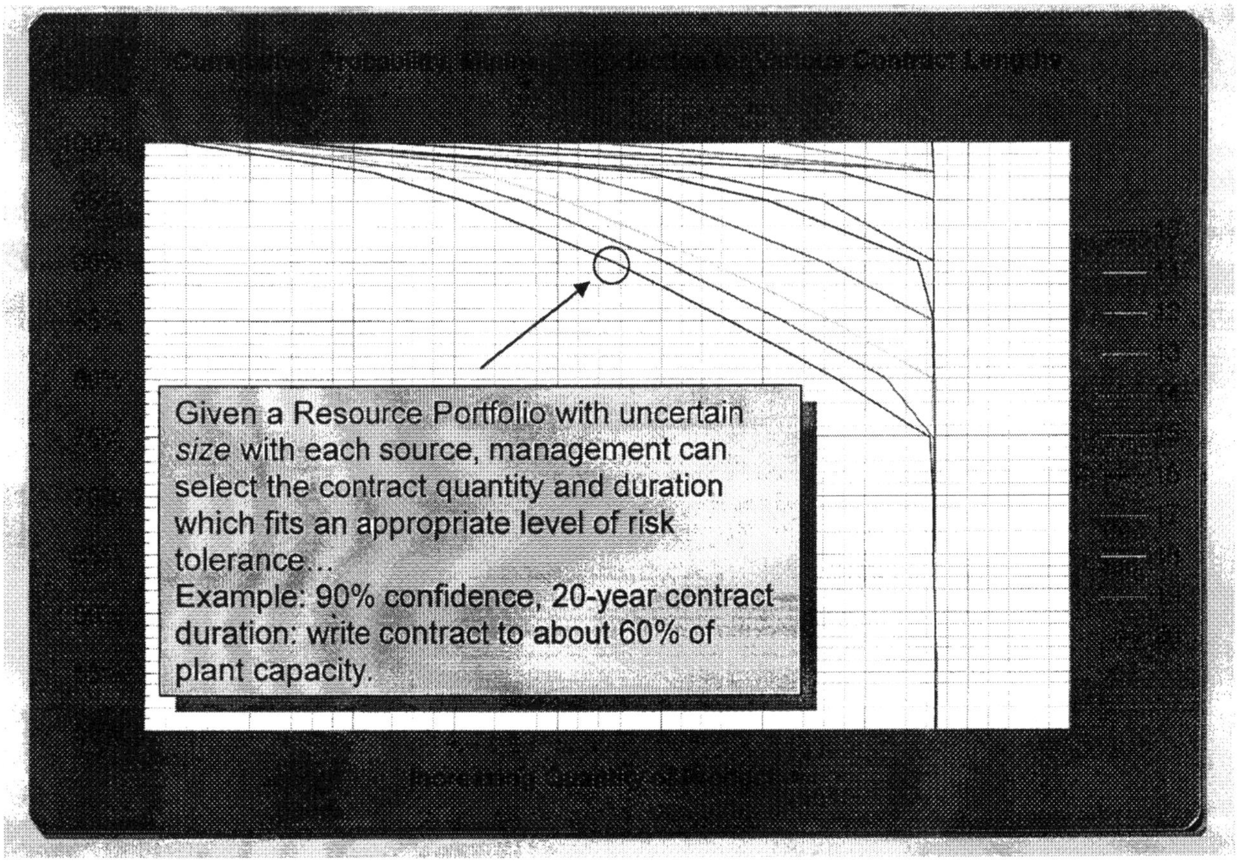

## *Summary*

- Probabilistic methods contribute value throughout the evaluation and planning process
- Decision alternatives should be selected on the basis of the expected result <u>and</u> variance.
- Assess the downside results to provide downside risk and mitigation potential
- Allocate resources to those uncertainties that may alter decisions (carry out Value of Information and Value of Control assessments)
- A fully probabilistic assessment and planning process is best at showing bottlenecks and potential roadblocks to timely / profitable project completion

## MODELING LNG SPILLS ON WATER

*John B. Cornwell and David W. Johnson*
*Quest Consultants Inc.®*
*908 26th Avenue, N.W.*
*Norman, Oklahoma 73069*

The size of a spreading pool following a release of LNG from an LNG tank ship has been the subject of numerous papers and studies dating back to the mid-1970s. Several papers, not specific to any one location, have presented idealized views of how the LNG would be released and spread across a water surface. There is a considerable amount of publicly available material describing these idealized releases, but little discussion concerning the effect the assumptions (which are implicit in the idealized simulations) will have on the final result of the study—the size of the potential hazard. This paper presents an approach for modeling the spreading of LNG on water that accounts for the presence of a potential obstacle that the liquid must overcome—local wave action.

### Spreading Model

A spill of LNG on water will spread radially from the release point until the LNG reaches an obstacle or reaches some minimum thickness that diminishes the liquid's ability to spread due to hydrodynamic head. The speed at which the LNG pool spreads is a function of the spill rate, LNG vaporization rate, physical properties of the LNG, radius of the LNG pool, and the nature of the spill surface.

By assuming that the viscosity and surface tension of the LNG have a negligible effect on spreading, the rate of spread becomes a function of spill rate (or initial volume spilled in the case of an instantaneous spill), vaporization rate, and pool radius.

At any instant in time, the rate of LNG spreading can be found by use of a simple hydrodynamic model:

$$\frac{dr}{dt} = \alpha\sqrt{h}$$

$$r = \alpha\int_o^t h^{1/2}dt \qquad (1)$$

where: $r$ = pool radius at time $t$, m

$t$ = time, s

$\alpha$ = constant = $\left[2g\left(\rho_w - \rho_{lng}\right)/\rho_w\right]^{1/2}$

$h$ = pool height, m

$g$ = gravitational acceleration, 9.8 m/s$^2$
$\rho_w$ = density of water, kg/m$^3$
$\rho_{lng}$ = density of LNG, kg/m$^3$

The term $(\rho_w - \rho_{lng})/\rho_w$ accounts for the LNG displacement of water, thus lowering the effective driving force, $h$, for spreading.

The height, $h$, when the LNG pool has spread to a given radius, can be found from:

$$h = V_{lng}/\pi r^2 \tag{2}$$

The total volume of LNG released is:

$$V_{lng} = \int_o^t \frac{dV_{lng}}{dt}\,dt \tag{3}$$

and

$$h = \left[\int_o^t \frac{dV_{lng}}{dt}\,dt\right]/\pi r^2 \tag{4}$$

where: $V_{lng}$ = volume of LNG in pool at any time $t$, m$^3$

Equations (2) through (4) assume a flat, circular shaped pool. At a given instant in time, the rate of vaporization of the LNG pool can be found using:

$$\frac{dm_e}{dt} = \pi r^2 \dot{e} \tag{5}$$

where: $m_e$ = mass of LNG vaporizing, kg
$\dot{e}$ = vaporization flux for LNG on water, kg/(m$^2$·s)

The vaporization flux for LNG on water can be estimated using a film boiling approach:

$$\dot{e} = h_f(T_w - T_{lng})/(\Delta H_v) \tag{6}$$

where: $h_f$ = LNG film boiling coefficient, J/(m$^2$·K·s)
$T_w$ = temperature of water, K
$T_{lng}$ = temperature of LNG, K
$\Delta H_v$ = LNG heat of vaporization, J/kg

Combining equations (5) and (6) gives:

$$\frac{dm_e}{dt} = \pi r^2\left[h_f\left(T_w - T_{lng}\right)/\left(\Delta H_v\right)\right] = \frac{\pi h_f\left(T_w - T_{lng}\right)}{\left(\Delta H_v\right)} r^2 \tag{7}$$

The total mass of LNG that has vaporized since the start of the release is:

$$m_{tot} = \beta\int_o^t r^2 dt \tag{8}$$

where: $\beta = \dfrac{\pi h_f\left(T_w - T_{lng}\right)}{\Delta H_v}$, kg/(m$^2$ s)

This results in the following working equations:

$$V_{lng}^e(t) = \left(\frac{\beta}{\rho_{lng}}\right)\int_o^t r^2 dt \tag{9}$$

$$V_{lng} = V_{lng}^s - V_{lng}^e \tag{10}$$

$$r(t) = \alpha\int_o^t \mathrm{h}^{1/2} dt \tag{11}$$

$$h(t) = V_{lng} / \pi r^2 \tag{12}$$

where: $V_{lng}^e$ = volume of LNG vaporized, m$^3$

$V_{lng}^s$ = volume of LNG spilled, m$^3$

The derivation of equations (9), (11), and (12) has assumed that the following parameters remained constant: $e$, $\rho_w$, $\rho_{lng}$, $T_{lng}$, and $T_w$. This assumption is likely true for the majority of the time of spreading, but changes in these parameters may result if significant changes in the LNG pool composition occur due to differential vaporization (methane preferentially vaporizes in relation to heavier components of the LNG).

Equations (9), (11), and (12) can be solved by a number of numerical methods. The equations, when placed in their differential form, are:

$$V_{lng} = \sum_{i=1}^{n}\left[V_i^s - \left(\frac{\beta}{\rho_{lng}}\right) r_i^2\right]\Delta t_i \tag{13}$$

$$r = r_o + \alpha \sum_{i=1}^{n} h_i^{1/2} \Delta t_i \quad (14)$$

$$h_i = V_{lng_i} / \pi r_i^2 \quad (15)$$

Based on the experimental work of Sciance, Colver, and Sliepcevich [1967], the value of $h_f$ is expected to be about 500 J/(m$^2$ K s) for methane. The calculated physical properties presented below are based on the LNG molar composition presented in Table 1, and an LNG storage pressure of 100 mbarg. Table 2 summarizes the physical parameters used in the simulation.

**Table 1**
**Typical LNG Composition**

| Component | Mole% |
|---|---|
| Nitrogen | 0.0112 |
| Methane | 96.9668 |
| Ethane | 2.6173 |
| Propane | 0.3164 |
| Isobutane | 0.0364 |
| Normal butane | 0.0274 |
| Isopentane | 0.0078 |
| Normal pentane | 0.0044 |
| C6+ | 0.0122 |

**Table 2**
**Summary of Physical Parameters Used in the Pool Spreading Simulation**

| Parameter | Value | Units |
|---|---|---|
| LNG temperature | 112.6 | K |
| Water temperature | 297 | K |
| LNG density | 432.6 | kg/m$^3$ |
| Seawater density | 1,025 | kg/m$^3$ |
| LNG heat of vaporization | 504,270 | J/kg |
| Film boiling coefficient | 500 | J/(m$^2$ K s) |

To verify the validity of the method, calculations were made and compared to published results which were based on an instantaneous 10,000 $m^3$ spill. Table 3 compares the model results with published model results for spills on flat surfaces [Shaw and Briscoe, 1978; Otterman, 1975].

**Table 3**
**Comparison of LNG Spreading Models**
**for an Instantaneous 10,000 $m^3$ Spill on a Flat Water Surface**

| Model | Maximum Pool Radius (m) |
|---|---|
| Quest‡ | 253 |
| SPILL [Shaw and Briscoe, 1978] | 256 |
| Fay* [Otterman, 1975] | 293 |
| Hoult** [Otterman, 1975] | 271 |
| Raj/Kalelkar [Otterman, 1975] | 271 |
| Otterman [Otterman, 1975] | 279 |

‡ quiescent surface assumed
* assumes ice formation and ignores vaporization
** ignores vaporization

## LNG Spreading Calculations

For the purposes of this exercise, LNG spreading calculations were made for LNG being released through a five-meter-diameter hole in the side of one cargo tank in an LNG tank ship. The hole is assumed to occur at the waterline. Each cargo tank in the tank ship is assumed to hold 25,000 $m^3$ of LNG. Approximately 12,500 $m^3$ of LNG are rapidly released onto the water (see Figure 1). Approximately two minutes are required to empty the ship cargo tank to the waterline. The LNG was allowed to spread until one of two boundary conditions was satisfied. Either all of the released LNG had vaporized, or some minimum thickness was achieved by the expanding pool.

## Effect of Atmospheric Conditions

Much of the earlier work on evaluating worst-case spills of LNG combines a large release of LNG with the atmospheric conditions that result in the largest hazard. The assumption that is

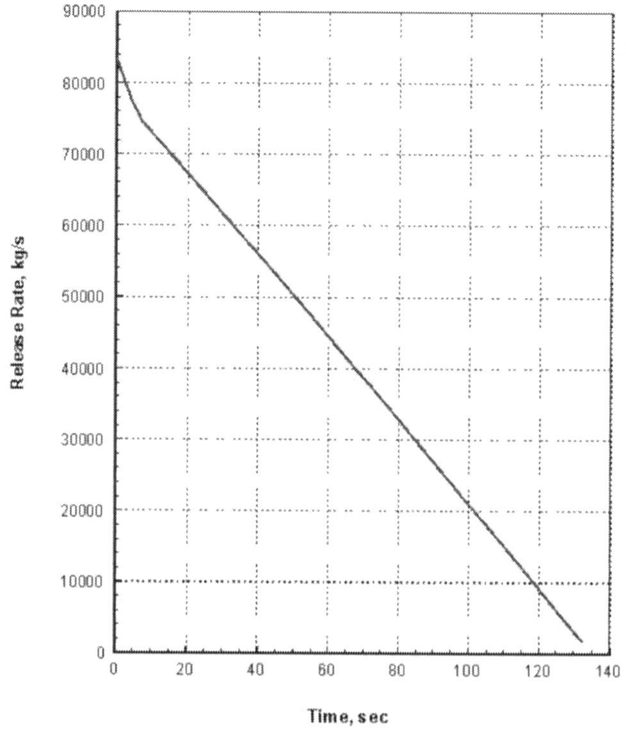


**Figure 1**
**LNG Mass Release Rate from Cargo Tank**

routinely made is that low winds and stable atmospheric conditions result in the largest potential vapor cloud travel distances. This is often true. However, the worst-case atmospheric conditions for vapor cloud travel generally do not produce the largest radiant hazard zones should the flammable cloud above the LNG pool ignite shortly following the release.

For the purposes of this analysis, three atmospheric conditions were evaluated.

Low winds (1.5 m/s) and stable atmospheric conditions (Pasquill-Gifford F)
Moderate winds (5.0 m/s) and neutral atmospheric conditions (Pasquill-Gifford D)
High winds (9.0 m/s) and neutral atmospheric conditions (Pasquill-Gifford D)

Previous model development describing LNG spreading on water surfaces assumed a quiescent surface. This simple assumption forces the use of a minimum thickness argument to be used as a boundary condition. Shaw and Briscoe [1978] recognized this limitation when reviewing the analytic models proposed by Burgess, Lind, Raj/Kalekar, Fay, and Shaw/Briscoe, and the numerical solutions provided by the SPILLS code by stating;

> The empirical formulae are based on scattered experimental data which require an enormous extrapolation in scale for the prediction of pool spread rates for real spills, whereas the theoretical pool spread rates are derived using very simple assumptions. For example, the neglect of viscous forces and surface tension

forces has already been noted. In addition, the action of the wind and waves during real spills may influence the pool spread rate. [Shaw and Briscoe, 1978]

Also, although the vaporization rates per unit area are nearly the same for all the compared models, the heat transfer conditions of real spills may not be represented adequately. For example, the experimental values quoted were measured in quiet conditions during small-scale tests, whereas the action of waves during real spills may increase the surface area of contact between LNG and water. The contact surface area may also be influenced by the frothing phenomenon... [Shaw and Briscoe, 1978]

In order to approximate the behavior of waves on the spreading of an LNG pool, the following approach was undertaken.

### Wave Characterization

A review of the National Oceanic and Atmospheric Administration's buoy data for the 2003 calendar year (one record per hour for a total of 8,784 readings in the year) provides the information on wave data for the wind speed ranges shown in Table 4.

**Table 4**
**Representative Wave Data**

| Location (Buoy Station) | Wind Speed Range (m/s) | Number of Records | Average Wind Speed (m/s) | Average Wave Height (m) |
|---|---|---|---|---|
| Boston, Massachusetts (44013) [2003 data] | 0 to 3 | 1,902 | 1.5 | 0.587 |
| | 4 to 6 | 2,727 | 5.0 | 0.715 |
| | 7 to 11 | 1,950 | 9.0 | 1.42 |
| Eureka, California (46022) [2003 data] | 0 to 3 | 2,520 | 1.5 | 1.99 |
| | 4 to 6 | 2,482 | 5.0 | 2.20 |
| | 7 to 11 | 1,702 | 9.0 | 2.79 |
| Galveston, Texas (42035) [2003 data] | 0 to 3 | 2,507 | 1.5 | 0.583 |
| | 4 to 6 | 3,751 | 5.0 | 0.804 |
| | 7 to 11 | 1,706 | 9.0 | 1.33 |

There is not a direct correlation between wind speed and wave height, although, in general, higher winds result in taller waves. Other factors, such as tidal currents, shoals, etc., influence wave height as well.

The waves are modeled using a simple cycloid. The parametric representation of a cycloid is given by:

$$x(t) = a \cdot (t - \sin(t))$$
$$y(t) = -a \cdot (1 - \cos(t))$$

where: $a$ = one half of the height of the wave from trough to crest
$t$ = parameter that varies from 0 to $2\pi$ for each cycle

Three cycles of a cycloid are shown in Figure 2, corresponding to $a = 0.5$ and $t$ varying from 0 to $6\pi$.

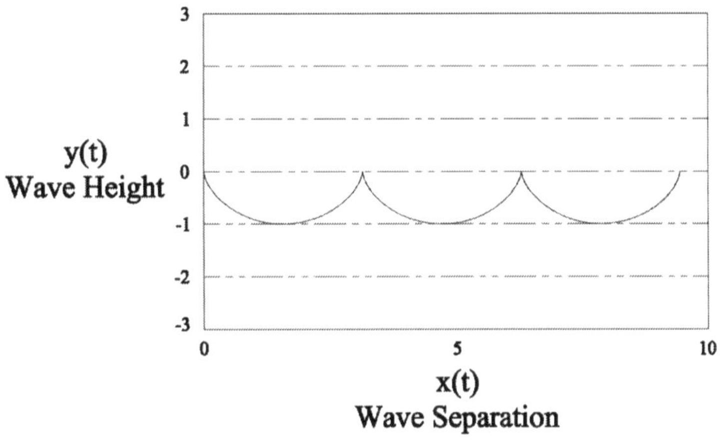


**Figure 2**
**Three Cycles of a Cycloid**

The area of the trough of the cycloid (which corresponds directly with the volume when the trough is moved though a third dimension to represent a wave) has the following characteristics.

- The trough is 50% full when the height from the base of the trough is 60% of the wave height.
- The trough is 68% full when the height from the base of the trough is 75% of the wave height.
- The trough is 75% full when the height from the base of the trough is 80% of the wave height.

As described in the spreading equations above, the ability of the LNG layer to rapidly spread is defined by the LNG's hydrodynamic head. While this formulation works well for flat water surfaces, the effect of waves on the spread of LNG must be accounted for during an actual spill. One method to approximate the interaction between the waves and the spreading LNG pool is to set a boundary condition based on the ability of the LNG hydrodynamic head to overcome the waves. The following condition was assumed.

> The LNG layer will continue to spread over the surface of the water according to equation (14) until the volume of LNG in the wave trough drops below 50% of the trough's volumetric capacity to hold the LNG. At this point, the model assumes the LNG layer is not tall enough to overcome the wave and thus ceases to spread radially.

There are two ramifications to this assumption.

1. Once the height of the LNG above the water level drops below 60% of the wave height, the LNG pool stops spreading radially. For the low wind (1.5 m/s) conditions in Boston in 2003, this results in an LNG layer thickness (above the waterline) of 0.6□0.587 m, or 0.352 m. Once the LNG pool spreads and vaporizes to the point when the LNG layer cannot overrun the wave, it stops spreading radially. Similarly, for the low wind conditions in Euerka and Galveston, the thickness of the LNG above the water is 1.19 m and 0.35 m, respectively.

2. The increased surface area of the water, due to the cycloid wave form, provides a larger effective heat transfer surface to the LNG layer. This results in a net increase in the actual vaporization flux (per radial linear unit) of 27%. (Length of wave surface from wave tip to wave tip is $8a$. Straight line distance from wave tip to wave tip is $2\pi a$. Thus, the ratio of wave path length to radial length is $8a/2\pi a$, or $4/\pi$, or 1.27.)

It should be noted that this assumption is based upon the driving force being the hydrodynamic head of spilled LNG. If the LNG acted like a lower volatility refined product (e.g., fuel oil), the dispersion of the low volatility product would slowly lead to a larger pool radius. A dispersing pool requires hours to spread significantly beyond the wave-imposed spreading barrier. Since the LNG pool is expected to vaporize rapidly, the time frame necessary for dispersion of the LNG pool simply is not available.

Incorporating the wave height boundary condition into the LNG spreading model developed above yields the maximum pool dimensions shown in Table 5 for the previously described example case in which 10,000 $m^3$ of LNG is released instantaneously. As can be seen by the results in Table 5, assuming a perfectly flat water surface under all wind conditions produces identical results that are consistent with the quiescent surface models listed in Table 3. When wave behavior is considered as a restriction, smaller diameter pools result.

Using the wave data for Boston, Eureka, and Galveston in the spreading model following a release from a five-meter-diameter hole in an LNG tank ship results in the maximum LNG pool radius and vaporization times listed in Table 6.

Table 5
Instantaneous Release of 10,000 $m^3$ onto Water (Non-Burning Pool) - Boston Data

| Wind Speed (m/s) | Wave Height (m) | Maximum LNG Pool Radius (m) | Total Time (min) to Vaporize Spill |
|---|---|---|---|
| 15 | 0.0 | 253 | 3.6 |
| | 0.587 | 74 | 19.0 |
| 5.0 | 0.0 | 253 | 3.6 |
| | 0.715 | 67 | 23.0 |
| 9.0 | 0.0 | 253 | 3.6 |
| | 1.42 | 48 | 45.7 |

Table 6
Rapid Loss of 12,500 $m^3$ LNG from a 25,000 $m^3$ LNG Cargo Tank

| Location (Buoy) | Wind Speed (m/s) | Wave Height (m) | Maximum LNG Pool Radius (m) | Total Time (min) to Vaporize Spill |
|---|---|---|---|---|
| Boston, Massachusetts (44013) [2003 data] | 1.5 | 0.587 | 79 | 20.7 |
| | 5.0 | 0.715 | 72 | 24.7 |
| | 9.0 | 1.42 | 52 | 47.2 |
| Eureka, California (46022) [2003 data] | 1.5 | 1.99 | 45 | 63.0 |
| | 5.0 | 2.20 | 42 | 72.4 |
| | 9.0 | 2.79 | 38 | 85.8 |
| Galveston, Texas (42035) [2003 data] | 1.5 | 0.583 | 80 | 20.5 |
| | 5.0 | 0.804 | 68 | 27.5 |
| | 9.0 | 1.33 | 54 | 43.4 |

**Table 7**
**Rapid Loss of 12,500 $m^3$ LNG from a 25,000 $m^3$ LNG Cargo Tank with Immediate Ignition of Vapor, Resulting in a Spreading, Burning LNG Pool**

| Location (Buoy) | Wind Speed (m/s) | Wave Height (m) | Maximum LNG Pool Radius (m) | Total Time (min) to Burn Spill |
|---|---|---|---|---|
| Boston, Massachusetts (44013) [2003 data] | 1.5 | 0.587 | 78 | 14.6 |
| | 5.0 | 0.715 | 71 | 17.1 |
| | 9.0 | 1.42 | 52 | 31.8 |
| Eureka, California (46022) [2003 data] | 1.5 | 1.99 | 43 | 44.6 |
| | 5.0 | 2.20 | 42 | 48.6 |
| | 9.0 | 2.79 | 38 | 58.7 |
| Galveston, Texas (42035) [2003 data] | 1.5 | 0.583 | 78 | 14.4 |
| | 5.0 | 0.804 | 67 | 19.1 |
| | 9.0 | 1.33 | 53 | 30.3 |

**Summary**

The approach outlined in this paper leads to the following conclusions.

1. If it is assumed that the water surface is always flat under all weather conditions, there are no waves to affect the spreading LNG pool. If this erroneous assumption is made, the proposed model will predict pool sizes similar to previous models.
2. If the local wave behavior is incorporated into the analysis by assuming the waves are a barrier for the spreading liquid to overcome, then the proposed model will predict smaller pool diameters and the subsequent vapor dispersion and fire radiation hazard calculations will be affected. This approach leads to additional questions that must be reviewed by the analyst—such as;

   A. Does a vapor cloud evolving from a smaller diameter LNG pool for a longer period of time produce a larger downwind distance to the LFL than a shorter duration release from a larger diameter pool?

   B. Does a smaller diameter pool fire generated under higher wave conditions and stronger winds produce a larger fire radiation hazard zone than a larger diameter pool under low wind conditions (i.e., less flame bending)?

The liquid spreading model described in this paper allows for the evaluation of the effect of site-specific weather and wave conditions on the spreading of LNG pools on water. This is a significant improvement over the "one size fits all" approach often used in LNG spill modeling studies.

## REFERENCES


Blackmore, D. R., J. A. Eyre, and G. G. Summers (1982), "Dispersion and Combustion Behaviour of Gas Clouds Resulting from Large Spillages of LNG and LPG onto the Sea." *Transactions of the Institute of Marine Engineers (TM)*, Vol. 94, Paper 29, 1982.

Havens, J. A., and T. O. Spicer (1983), "Further Analysis of Catastrophic LNG Spill Vapor Dispersion." *Heavy Gas and Risk Assessment II*, S. Hartwig (Editor), Battelle Institute, Frankfort am Main, Germany, 1983.

Otterman, B. (1975), "Analysis of Large LNG Spills on Water – Part 1: Liquid Spread and Evaporation." *Cryogenics*, August, 1975.

Raj, P. K. (1981), "Models for Cryogenic Liquid Spill Behavior on Land and Water." *Journal of Hazardous Materials*, Vol. 5, 1981.

Raj, P. K., N. A. Moussa, K. Aravamudan, and C. D. Lind (1979), "LNG Spill Fire Tests on Water; An Overview of the Results." *1979 Operating Section Proceedings*, 79-T-50, American Gas Association, 1515 Wilson Boulevard, Arlington, Virginia 22209, 1979.

Sciance, C. T., C. P. Colver, and C. M. Sliepcevich (1967), "Pool Boiling of Methane Between Atmospheric Pressure and the Critical Pressure." *Advances in Cryogenic Engineering*, Vol. 12, 1967.

Shaw, P., and F. Briscoe (1978), *Evaporation from Spills of Hazardous Liquids on Land and Water.* Safety and Reliability Directorate (SRD), United Kingdom Atomic Energy Authority, SRD R 100, May, 1978.

# Dynamic Simulation Studies Help Ensure Safety by Confirming Operational Readiness of LNG Plant Assets

*Peter Henderson, Product Manager, Simulation, Honeywell Process Solutions*
*Ahmet Pekediz, Senior Applications Engineer, Honeywell Process Solutions*
*Henry Schindler, Principle Consultant, Honeywell Process Solutions*

*Prepared for Presentation at AIChE Spring National Meeting*
*4th Topical Conference on Natural Gas Utilization*
*New Orleans, LA, April 2004*

## ABSTRACT

The paper was prepared to present the usefulness of a dynamic simulator and simulation engineer throughout a project life cycle. Recently, Honeywell developed dynamic models that were used to evaluate the design of LNG Trains. Dynamic simulations primarily used for operator training were also applied during detailed plant design to exercise, evaluate and optimize plant assets (operators, equipment, controls and operating procedures) prior to initial operation. This allowed engineers to focus on problems off the critical path and prior to committing to capital expenditures. This methodology has helped with the management of change process, resulting in accelerated start-ups and ensuring safe plant operation.

The development of a dynamic simulator has many parallels and synergies with the work practices and project phases in new LNG plants construction from definition, design, construction, commissioning, turnover, training and support. Its diverse personalities and application to many problems easily justifies the investment in dynamic simulation.

## INTRODUCTION

Operating a facility in today's competitive business environment challenges manufacturers to think about efficiency, reliability and responsibility while they produce products with quality. Several strategies can influence these operating objectives:

*Improving Plant Operational Efficiency* by starting plants faster/sooner, operating closer to plant constraints, minimizing capital costs of equipment and operating costs of utilities while improving product quality with less material rerun/flaring

*Reducing Plant Incidents* maximizes plant availability (production) and improves plant safety for people, environment and process equipment

*Delivering on Commitments to Government and Community* by demonstrating compliance through operator certification and satisfying environmental regulations

Dramatic changes in modern LNG plant design over the past 5 years has led to significant increases in production capacity. Plant design teams are particularly interested in how new arrangements of powerful process equipment will behave during the initial operations. This has created the need to study the overall dynamic behavior to adequately evaluate the operational readiness for the initial start-up of new process configurations and control systems.

The primary refrigerant equipment in the liquefaction sections of these LNG plants account for almost 50% of the capital investment. Dynamic studies of compression equipment at an early stage of design will help companies optimize equipment design and ensure that safety is considered in the plant operation through HAZOP analysis. Consequently, a potential safety issue identified in the HAZOP analysis could be studied in a Dynamic Simulation Study.

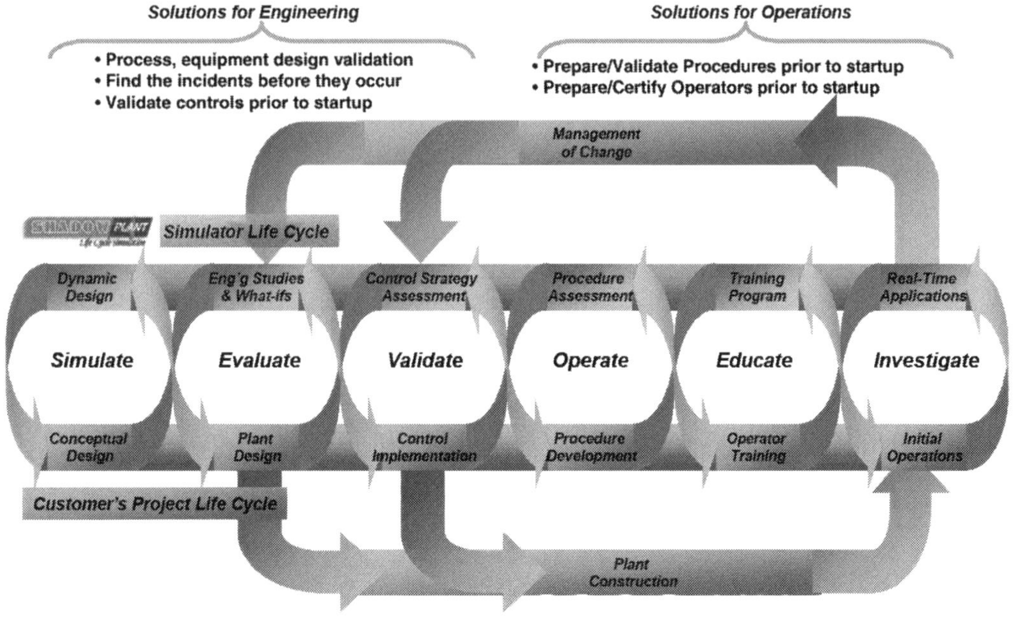


**Figure 1:** The simulator lifecycle reflects upon components of the plant lifecycle

For more than 30 years, dynamic simulation has been used as an effective tool for operator training and design verification of process automation systems (1). Improvements in dynamic simulation capabilities have created multiple uses and maximized the return on investment. As a result of such product developments, simulation tools have evolved that can be used throughout the life of a plant as depicted in Figure 1. Each cycle depicts a simulated phase of validation for each defined plant asset – process equipment, control configuration, operating procedures and operation teams.

Dynamic simulation applied throughout the project lifecycle as part of a formal management of change process helps manufacturers ensure safe and profitable production by confirming operational readiness of LNG plant assets.

*Process Assessment and Optimization* helps identify and eliminate process bottlenecks at the conceptual design phase before capital expenditures are committed and following startup for future process revamps.

*Control Validation and Optimization* of regulatory controls, safety systems, advanced applications helps identify and correct weaknesses in the automation system before commissioning and operation.

*Procedure Preparation and Optimization* ensures safe, efficient production of products.

*Operator Training and Qualification* on procedures, best practices and simulated systems ensures that human performance is improved, monitored, captured, assessed, reported and archived with computer-based tools, facilitating objective rather than subjective assessment. These methods are also applicable to operator re-certification.

The cumulative effect of ongoing cost savings and improved operation can be substantial and typically return the initial simulation investment within the first year of operation. Dynamic process simulation will also continue to contribute to profitability on an annual basis as the model is updated to match the plant changes. A Management of Change Process will help ensure that the benefits are realized into the future.

## LNG SIMULATION TECHNOLOGY

In order for projects to take advantage of lifecycle benefits made available through dynamic simulation, the models and simulation must be of a high fidelity to yield meaningful results to engineers undertaking studies. Dynamically speaking, this means the accurate prediction of operating conditions throughout the process flowsheet in consideration of the:

- dynamic modeling equations for the holdup of heat and material
- range of applicability of the modeling equations and parameters representing the process equipment, controls and logical devices within the scope of the system.

As an example, Honeywell has packaged their proprietary LNG modeling technology as dynamic components in Shadow Plant's simulation engine. Proven in comparison with operating data from more than 10 LNG plants, this capability ensures that the dynamic response of the model presented to trainees is realistic and helps prepare them for their role in plant operations.

First principles, dynamic models also allow engineers to see future behavior of their plant assets prior to operation by executing simulation scenarios. The behavior predicted by dynamic simulations is considered suitable for study of equipment performance during the detailed design phase of their projects.

Critical to the success of the LNG simulators are several fundamental dynamic process units that should be considered – specifically, centrifugal compressors, spirally wound heat exchangers, distillation columns and liquid expanders. Specialized physical property models are also important in providing an accurate dynamic behavior for gas absorption and cryogenic processes in these non-ideal operating conditions.

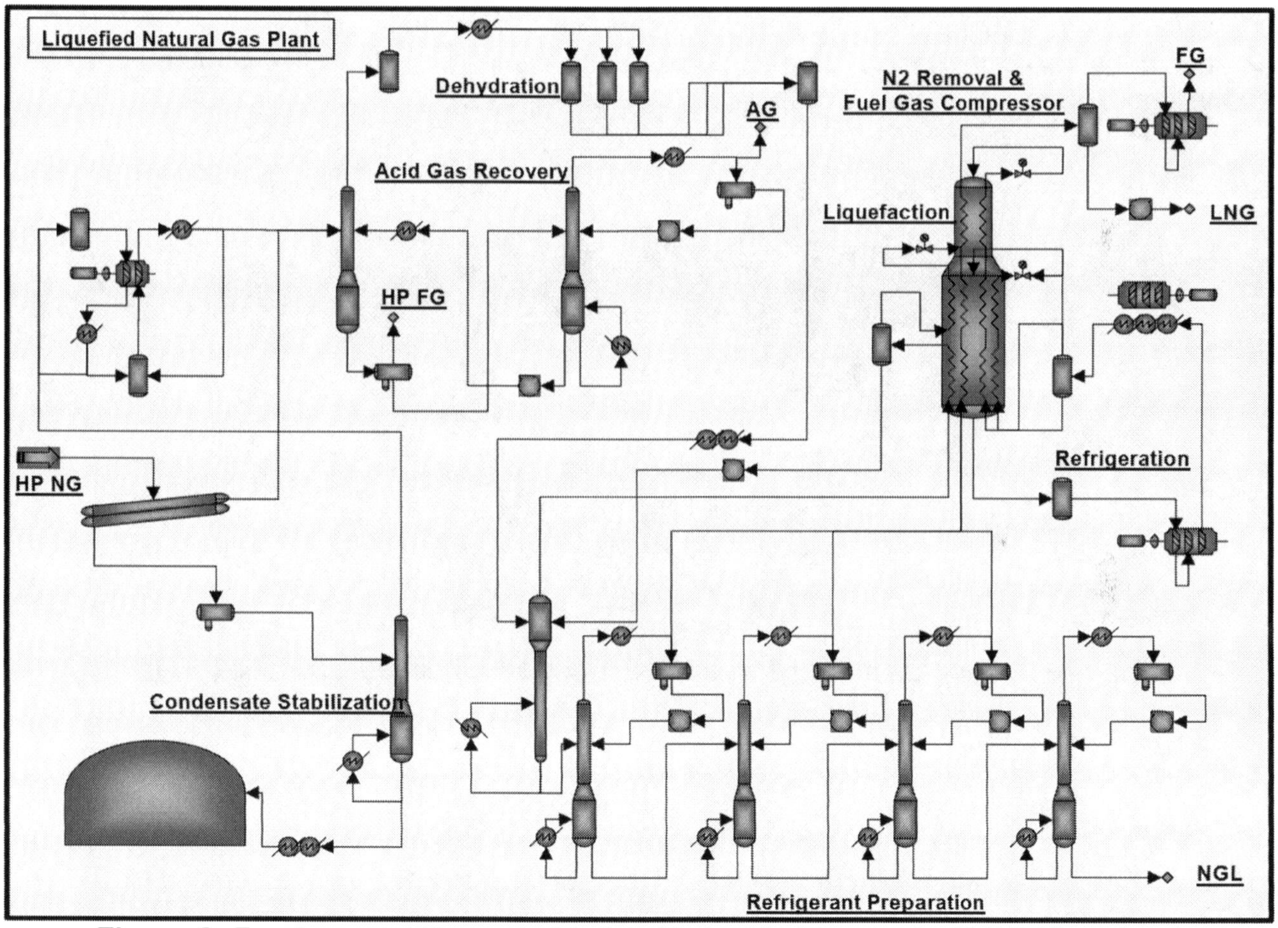


**Figure 2:** Fundamental Dynamic Models Embedded in a typical LNG Flowsheet

Another important component is the experience and ability of the process engineer preparing the simulation of the process and control models. It is often rare to find an individual with the detailed knowledge of the process design, control strategies and operating philosophies for complex LNG plants.

*Fundamentally Dynamic Process Units*

When the process model contains sufficient fidelity, a wide range of process operations can be simulated, including the initial start-up through to normal operation, shutdown and abnormal situations due to faults. Dynamic models are an order of magnitude greater in complexity than the steady state models and must be developed using a fundamentally dynamic approach from the ground up. A model that starts from a steady state basis and adds additional holdup equations for dynamics will generally fail to match a sufficient range of process operations.

A fundamentally dynamic approach facilitates high fidelity, dynamic simulation over the full range of process operations. For example, Honeywell's Distillation Column module incorporates a rate-based approach that ensures correct dynamic behavior of distillation processes for simulation applications spanning the entire engineering lifecycle. It has been shown (2) that this approach is far superior to the equilibrium-stage approach for modeling the dynamic behavior of distillation columns because the equilibrium-stage assumption does not hold for most operating modes, notably column startup (3).

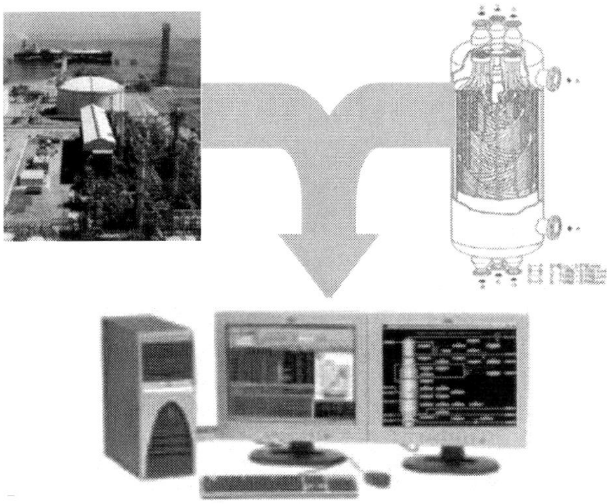

**Figure 3:** Fundamental Dynamic LNG Models Embedded in Shadow Plant

*Centrifugal Compressors*

The prime mover in a refrigeration system is the centrifugal compressor. Dynamic execution of a compressor relies upon several critical dependent and independent operating variables that influence performance in the refrigeration system. Laws of kinematic similarity known as the fan laws are used to make predictions however, they are only estimates of compressor performance. Although the specifications provided by compressor vendors are typically based on a single rated condition in terms of polytropic head and efficiency, operation of a process is not constrained to fixed inlet conditions. To be valuable in a dynamic simulation, enhancing the formulation of these fan laws permits accurate:

- regressions of the entire design performance map from surge-to-stonewall (as a function of speed or guide vane angle) and

- predictions for variable inlet conditions (composition, temperature, pressure) commonly encountered throughout normal startup operations and during plant upsets.

*Liquid Expanders*

Dynamic equations can be formulated where potential exists to generate power by converting head (pressure drop) through liquid expanders. Refrigerant flow, exit temperature, expander power and expander speed dynamically varies as a function of available head, efficiency and inertia of the expander. Liquid expanders help to close the mixed refrigerant process loop contributing to the dynamic behavior of the mixed refrigerant system. Expander performance curves from the equipment vendor are considered as the design basis for this equipment.

*Spirally Wound Heat Exchanger*

In order to accurately predict the dynamic, closed loop behavior and suction conditions for the mixed refrigeration compressors in the liquefaction process, the Spirally Wound Heat Exchanger (SWHE) should be included. A simplified heat exchanger will not produce the expected dynamic behavior throughout the initial startup as volumetric capacitance influences the refrigerant charging procedures and variable compressor suction conditions (flows, pressure, temperature, composition) are influenced by the supply of refrigerant as a function of the exchanger duty. Simplified modeling of heat exchangers also produces irrational temperature profiles with subsequent crossovers at segment boundaries and between individual tube-side and shell-side streams.

Spiral Wound Heat Exchangers can be modeled using a segmented approach. Each segment of the SWHE is modeled with one a single multi-stream exchanger module, generally taking into account the changes in flow streams. The SWHE is structured as "n" discrete process streams serving as a heat source that also represent a bundle of many more tubes with variable internal arrangements, tube lengths, wall thickness, diameter and spacing. There is only a single shell side serving as the heat sink.

Two or more SWHE segments connected in series, together with appropriate valves and piping, can be used to represent the overall heat exchanger arrangement required in the study of LNG plants. Additional modules can be used to account for reservoir drains, control valves, malfunctions (tube leaks), additional process connections and branches to external devices.

To be valuable in a dynamic simulation, the formulation of these SWHE should dynamically predict the:

- exit flow, temperature, pressure, vapor fraction and composition for each of the outlet streams arising from changes in the incoming streams and the environment

- phase change within each of the tube-side streams and the shell-side stream

- temperature of the tube and shell walls

- intermediate temperatures along the heat exchanger

- thermal profiles in the shell wall and the insulation

Furthermore, the SWHE should perform its calculations in a manner that ensures the real time operation of the process model is unaffected during strong transient operating scenarios normally encountered during startup/shutdown procedures or significant process upsets.

A thorough engineering evaluation of the heat and material balance will be necessary in order to reconcile duties in the various sections of the SWHE. In this case, general equipment information, estimates of all equipment parameters can be calculated to define exchanger performance consistently with the design basis.

*Physical Properties*

Rigorous physical property models are also required to calculate K-values, enthalpies, densities and the transport properties required by the models. With the recent increase in computer processing speeds, it is now possible to use rigorous equation-of-state models during real-time dynamic simulation. Within an operator training application, the properties must be significantly more robust compared to steady state design simulation (i.e. provide the correct answer without mathematical failure). This is because the range of temperatures, pressure and mixture compositions is so much greater than encountered in steady state simulation or dynamic simulation of perturbations about a steady state.

*Control System*

Simulators are composed of three basic parts; the model, the interface and the control processor. Including an exact replica of the control and logic algorithms is a key component of the simulator system. The engineering tools used to configure, maintain and test the controls should be useable on the simulator in a transparent fashion. Control and PLC configurations should be easily transported between the simulator and the actual system to increase project efficiencies. The replication of the control algorithms can be achieved by using the actual control devices or emulated control devices. The decision of which method to choose depends on the project constraints and objectives.

## APPLICATIONS FOR DYNAMIC SIMULATION OF LNG PLANTS

*CASE STUDY – Ras Laffan LNG – Train 3*

Refrigeration systems in LNG plants are complex processes of multi-stage compressors, exchangers, vessels and valves. They operate through a very broad range of conditions during startup. In order to design a control system that reliably and economically maximizes plant availability and protects the process from hazardous operations, the user must employ instruments and valves that react fast enough to avoid incidents. Furthermore, the occurrence of these incidents could be investigated well in advance by staging scenarios and evaluating them through dynamic simulation.

The primary objective of the Dynamic Simulation Study (DSS) was to conduct dynamic simulation tests identified by the compressor vendor and/or by the Engineering Company to confirm operational readiness of key plant assets. These studies generally commence during the detailed design stage of a construction project in collaboration with the engineering contractor and the owner/operator. Under specific review is the engineering information defined by the Basic Design Package from the Front-End Engineering Design stage of the project. Several iterations of study are required as the detailed design evolves as specific performance parameters are obtained from selected suppliers of the process equipment, instrumentation and automation products.

An additional objective of this DSS was to reuse the detailed dynamic model of the liquefaction section as an Operator Training Simulator (OTS). Interestingly, this strategy matched the execution stages of the Project Life Cycle depicted in Figure 1.

*Scope and Test Methodology of the Simulation Study*

Simulation tests focused on the response of the process and control system under many different upsets of the principle components in the propane and mixed refrigerant compressor systems. This included the anti-surge valve trip, gas turbine trip and loss of cooling to condensers (water/propane). A complete startup of the liquefaction and refrigeration system was also performed so that the sequence of operations could be optimized and reasonable guidelines for initial refrigerant charging operation could be established.

Simulation variables under investigation were pre-defined and their values were captured for post run analysis. To enable repeatable test methodology, pre-configured scenarios were prepared to consistently invoke operating events under study. These scenarios permit the execution of many events with respect to simulation time.

*Plant Design Basis (FEED Phase)*

In the initial phase of the study, the objective was to confirm that the project specifications and plant design basis was suitable for equipment selection and that the control strategy met operability and protection criteria. The dynamic model was prepared based on the released Piping and Instrumentation Drawings and theoretical equipment design data. A simplified control layer was based upon the regulatory control loops and cause and effect diagrams defined by the engineering contractor.

As the dynamic response is a closed-loop interaction between transmitters, valves, regulatory control equations, piping and process equipment, a detailed simulation is required to gain full appreciation of the overall time constant of the system. Each plant asset has dynamic characteristics that influence the behavior of the plant. From this perspective, the simulated dynamic responses of the shutdown and safety systems are evaluated against the equipment protection criteria. Attention was paid to the accuracy of the process model by distributing dynamic modeling parameters throughout the process flowsheet (thermal/hydraulic capacitance, pressure profile and equipment algorithms) in order to provide a reasonable process response.

The liquefaction process included feed dryers, feed pre-coolers, a scrub column and the main cryogenic heat exchanger. The refrigeration process included a closed loop of the mixed refrigerant and propane compression systems.

*Plant Design Basis (Detailed Design Phase)*

In the final phase of the study, the objective was to confirm operational readiness of all actual plant assets prior to construction and initial operation. The initial dynamic model was now updated with the actual performance parameters from the selected equipment vendors as the basis for repeating study while additional units were included in the model scope (Nitrogen Rejection compressor, LNG and MR cryogenic turbines). The actual process control system (PCS), emergency shutdown (ESD) logic, gas turbine startup sequences and compressor anti-surge control were now available for integration with the updated process model. The evaluation of the integrated response of the automation system was considered to be critical since this system configuration was entirely new and unique. Attention was paid to the accuracy of the control model from the transmitter time constants/signal conditioning, control device algorithms/parameters, formulation of the sequences and stroke times/flow characteristics of the control valves.

*Reporting*

Simulation results were captured, analyzed and published in a detailed report (see figure 4). Over 200 simulator variables were monitored throughout 18 test runs in the first phase of the study and 6 supplemental test runs in the second phase. Observations of simulation test results were made and presented for the contractor's review, conclusions and subsequent design activities. From the customer's perspective, identifying weaknesses in the plant design at this stage was advantageous as it allowed for design modifications before the plant assets were purchased and it helped prepare their initial operations team for startup.

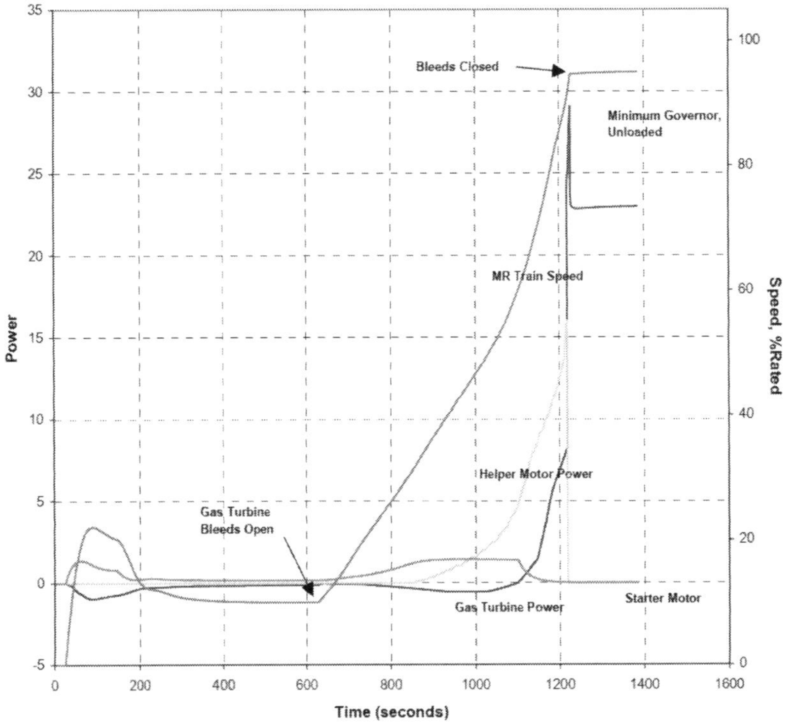


**Figure 4:** Startup Dynamics of the Mixed Refrigerant Compressor

*Simulator Migration to Operator Training*

Following the successful study of the process, equipment and control system, the focal point of the simulator migrated from engineering to operation. The objective during this phase of the project was to prepare and optimize operating procedures on the simulator while training and qualifying operators on topics of process fundamentals, control familiarization, normal operating sequences and recovery from abnormal situations.

*Conclusions of the Dynamic Simulation Study*

The participants in the study have learned many things about the dynamic behavior of the plant that will help prepare them for production:

- Sensitivity of the compressors to over-load during upset conditions with high flow-rates (anti-surge valves opened under recycle trip) was noted
- Control configuration was proven to adequately protect the equipment and personnel during upset situations
- Control system strategy and integrated components proved to be an effective automated startup sequence for the new turbo-machinery configuration in the refrigeration process
- Operating sequences developed for the study served as a basis for the actual plant startup of this new process configuration
- Guidelines for initial refrigerant charging were prepared and confirmed as a function of compressor suction pressure prior to operation
- The electric helper motors were confirmed to be adequately sized for startup of this new compressor arrangement

Possibly the greatest single result of the simulator project was the confidence achieved by methodically proving readiness for safe operation of critical plant assets throughout each phase of the project lifecycle.

*Customer Feedback*

> *We (the training department) received some compliments the other day on how well the compressor startups went and how nice it was that the panel operators knew what to do from their training on the simulator! That is a feather in all of our caps! (4)*

## CONCLUSION

The direct benefit to plant operations needs to be identified to justify creating a dynamic simulator. Many LNG facilities have achieved benefits of a dynamic simulator throughout the project life cycle by confirming operational readiness of plant assets prior to initial operations. It has also provided an objective method to train and qualify operators on a rigorous model. Operators are also able to respond to emergency conditions with confidence using proven procedures.

The accuracy of the model that is required to support engineering decision-making is completely consistent with that required to train operators in their decision-making process. Consequently, the operator training model can generate benefits as early as the FEED / Detailed Design phase of the project and later when studying the future plant modifications.

As a formal engineering work practice, the dynamic simulator can be used to evaluate the impact of future design changes prior to plant modifications. Implementation on the simulator prior to the actual plant has the additional consequence of ensuring that the simulator is perpetually and inherently updated sustaining long-term simulator benefits for engineering studies and operator training. The cumulative effect of ongoing cost savings and improved operating rates contribute to profitability on an annual basis.

## REFERENCES


1. McMeekin, Russel, "Cost reduction through simulation", Hydrocarbon Asia, March 1996, pp 70-76.

2. Shah, J., Stephenson, G.R., "Dynamic Simulation of Distillation Columns - Insights From More Than Ten Years Of Experience", Prepared for presentation at the AIChE 2001 Annual Meeting, Reno NV, Nov 4- 9, 2001.

3. Sowell, R., "Why a simulation system doesn't match the plant – Understanding the reasons will result in more successful projects", Hydrocarbon Processing, March 1998, pp102-107.

4. James Dawson, "Feedback from Operations Following Startup – Ras Laffan LNG Training Team", January 2004

**Educating the North American Public on the Safety and Security of LNG Facilities**

*Delano, Fisoye, Institute for Energy, Law & Enterprise, University of Houston*
*Foss, Michelle Michot, Institute for Energy, Law & Enterprise, University of Houston*

## 1. INTRODUCTION

Although importation of liquefied natural gas from around the world was an important component of North America's plans to restore energy security in the 1970s, there has not been any new LNG import facilities developed on the continent for the past 20 years. The apparent disillusion with LNG arises primarily from public perception regarding the safety and security of LNG transport and regasification operations, which have given rise to NIMBY (Not In My Back Yard) issues. In the face of local opposition some LNG proposals in California and in the New England states have been withdrawn.

Some of the key LNG safety and security concerns reported in the media include:

- The adequacy of mitigation measures to protect local communities in the event of a catastrophe at the facility such as, a fire caused by an LNG leak resulting from inadequate maintenance or from a natural disaster such as an earthquake. As a result there is a clamor to locate LNG terminals well away from land, and especially away from populated areas.[1] The January 2004 explosion at an LNG facility in Algeria that killed 27 people was given as an example.
- The possibility of a terrorist attack on LNG facilities, capable of causing tremendously destructive fires.
- The costs of LNG security. Growths in these costs are bound to increase as more LNG import facilities are built.[2]
- Decreased residential property values. Since the LNG facilities will be located along the shorelines, residents with beach property near the facility and within the line of sight of the facility fear that industrialization will lower property values, especially for vacation or resort property.

All of these issues need to be addressed by industry and public officials. In many aspects of daily life, risks of certain events are often perceived differently from reality. Potentially dangerous activities can sometimes become so commonplace and acceptable, that the risks associated with those activities are taken for granted, for example, driving a car.

In other cases, the focus on worst-case scenarios overshadows the relatively low probability that such events will ever occur. In many such instances, the substantial precautions taken to prevent worst case scenarios are not taken into consideration. A high

---

[1] In the senate hearing on the Pipeline Safety Act Amendments of 1979 the Government Accounting Office (GAO) stated: "We believe remote siting is the primary factor in safety" (for LNG and LPG terminals). The agency recommended and the law requires Secretary of Transportation to "prescribe minimum safety standards for deciding on the location of a new LNG facility". The law lists six factors that the Secretary must consider in setting these minimum standards. These factors include "the need to encourage remote siting". This factor is not included in the implementing regulations according to U.S. Rep. Ed Markey, D-Mass., author of the 1979 House bill outlining minimum safety standards for such facilities. *Source: Mobile Register article, Nov. 16, 2003*

[2] Paul W. Parfomak, *Liquefied Natural Gas (LNG) Infrastructure Security: Background and Issues for Congress*, Congressional Research Service (CRS) Report for Congress. Sept. 2003.

consequence event with a low probability of occurrence may be similar on a "risk basis" to a low consequence event with a high-probability of occurrence.

The extent of potential damage and injuries from an LNG incident would depend on precipitator events, the volume and location of LNG release, the release rate, the wind direction and speed among other factors.

In order to understand the total picture on LNG, the public needs precise information and the industry, and regulators need to engage in an interactive communication with the public, and have constructive exchange of information and listen to public opinions and concern about LNG. Once the public has a better understanding of the true LNG risks, in order for them to agree to a local facility they will need to see some benefit to offset the actual, though minimal, risk. Potential benefits could include an overall lower purchase price for natural gas, having the facility take on a burden of local tax requirements or requiring some community service or capital project.

In 2002, the University of Houston's Institute for Energy, Law and Enterprise organized a consortium of industry players to support studies evaluating the prospective role of LNG in North American energy markets. The most relevant of these studies addresses public concerns regarding safety and security[3]. To prepare the study report, we examined information on the physical properties of LNG, the safety record of LNG facilities and ships, the impact of the LNG operations on the environment and regulations and agencies concerned with safety and environmental protection in the LNG industry. Members of our team visited LNG facilities in the U.S. and Japan.

Based on a thorough review of the technology and operations procedures used over the past 40 years, this study concluded that LNG has an excellent safety record worldwide, particularly in the United States and that LNG can continue to be transported, stored and used safely and securely, as long as safety and security standards and protocols developed by the industry are maintained and implemented with regulatory supervision. It is in the best interest of the industry, regulators and the general public that this goal be achieved so that the benefits of natural gas can be realized for consumers. This study was completed before the tragic incident in January 2004 at the Sonatrach's Skikda LNG export plant in Algeria. This accident was at an LNG export facility and not an import facility like the ones that are being proposed to be built in the U.S.. We cannot analyze the incident because we do not yet have all the facts on the cause but we believe that the recommendations of this study would still hold in any event.

The Institute is making the results of its analysis, including accounts of specific incidents and descriptions of specific safety technology and procedures widely accessible to the public through the Institute's website, at http://www.energy.uh.edu/lng.

Using this study as a foundation, this paper presents the technologies available to address the safety and security concerns regarding LNG import facilities.

---

[3] Foss, Michelle; Delano, Fisoye; Gülen, Gürcan; *LNG Safety and Security, Institute* for Energy, Law & Enterprise, University of Houston, Oct. 2003.

## 2. U.S LNG FACILITIES

The U.S. has the largest number of LNG facilities in the world. U.S. gas utilities use LNG extensively for "peak-shaving" with facilities located near population centers. There are at least 113 active LNG facilities in the United States, Figure 1, including import terminals, storage facilities, and operations involved in niche markets such as vehicular fuel. Most of these facilities were constructed between 1965 and 1975, and were dedicated to meeting the storage needs of local utilities. Approximately 55 local utilities own and operate LNG plants as part of their distribution networks.[4] The LNG import terminals are: Elba Island, Georgia (El Paso Corporation); Everett, Massachusetts (Tractebel); Cove Point, Maryland (Dominion Energy); and Lake Charles, Louisiana (Panhandle Energy, a Southern Union company).

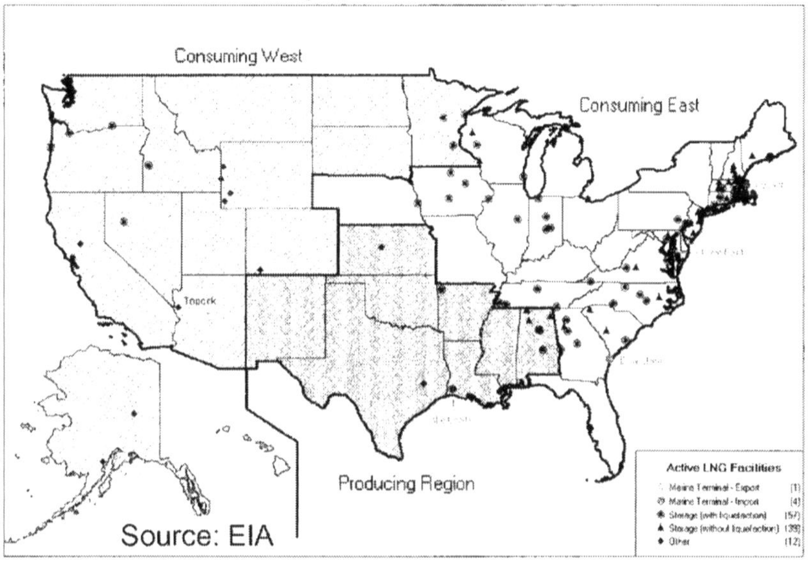


**Figure 1 - U.S Liquefied Natural Gas Facilities**

Japan is one of the world's largest users of LNG, consuming about 2.6 trillion cubic feet (tcf) of LNG or 48 percent of global LNG supply in 2002. This is more that ten times the amount of LNG import into the U.S. during the same period[5]. Japan has 23 LNG import terminals and receives 96 percent of its natural gas demand as LNG. Table 1 compares the types of LNG facilities in the U.S. and Japan.

Table 1. LNG Facilities in the U.S. and Japan

| | U.S. | Japan |
|---|---|---|
| Liquefaction Terminals | 1 | |
| Import Terminals | 4 | 23 |
| Peakshaving Facilities | 57 | |
| Satellite Storage Facilities (without liquefaction) | 39 | 26 |
| Others | 12 | |
| Total | 113 | 49 |
| *Sources: EIA, Japan Gas Association* | | |

[4] U.S. EIA: *U.S. LNG Markets and Uses*, January 2003.
[5] EIA - http://www.eia.doe.gov

The strong safety record of the LNG industry over the past 40 years is a result of several factors. First, the industry has technically and operationally evolved to ensure safe and secure operations. Technical and operational advances include everything from the engineering that underlies LNG facilities to operational procedures and technical competency of personnel. Second, the physical and chemical properties of LNG are such that risks and hazards are easily defined and incorporated into technology and operations. Third, a broad set of standards, codes and regulations applies to the LNG industry to further ensure safety. These have evolved through industry experience worldwide and affect LNG facilities and operations everywhere. Regulatory compliance provides transparency and accountability. There is a continuous improvement of LNG safety, environmental and security infrastructure, Figure 2.

Figure 2. Continuous Improvement of LNG Safety, Environmental and Security Infrastructure

By converting natural gas to LNG, it can be shipped great distances across the oceans from the countries where it is produced to those where it is in demand. Natural gas is used in homes for cooking and heating, in public institutions, in agriculture, by industry and to generate electric power. Natural gas is important not only as a clean source of energy, but also as a feedstock to enable the petrochemical industry to produce plastics, fibers, fertilizers, and many other products.

## 3. HOW IS A SAFE, SECURE LNG IMPORT FACILITY ACHIEVED?

The LNG industry is subject to the same routine hazards and safety considerations that occur in any industrial activity. Risk mitigation systems must be in place to reduce the possibility of occupational hazards and to ensure the protection of surrounding communities and the natural environment. As with any industry, LNG operators must conform to all relevant federal and local regulations, standards and codes.

Beyond routine industrial hazards and safety considerations, LNG presents specific safety considerations. In the event of an accidental release of LNG, the safety zone around a facility protects neighboring communities from personal injury, property damage or fire. The one and only case of an accident that affected the public was in Cleveland, Ohio in 1944. Research stemming from the Cleveland incident has influenced safety standards used today. Indeed, during the past four decades, growth in LNG use worldwide has led to advanced safety technologies and practices that will be used in North America as the LNG industry expands.

Generally, multiple layers of protection create four critical safety conditions, all of which are integrated with a combination of industry standards and regulatory compliance, as shown in Figure 3.

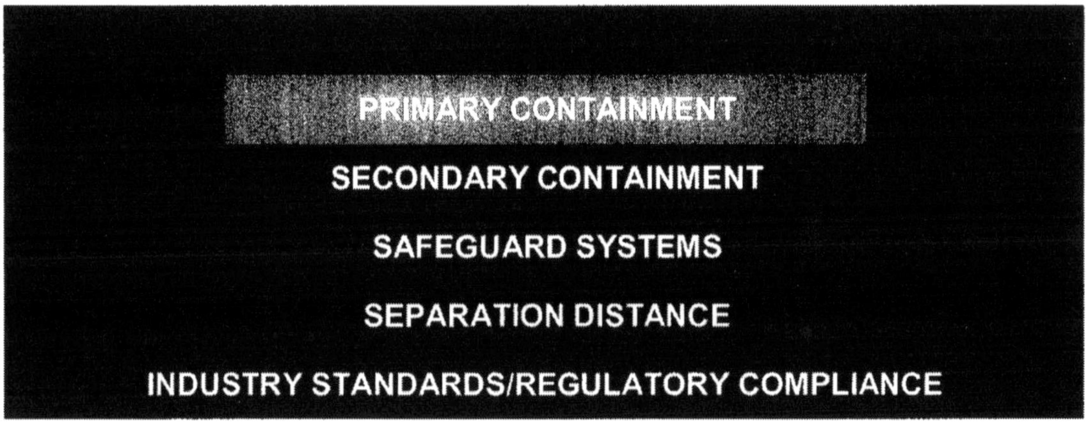


Figure 3 - **Multiple Layers of Protection for Safety and Environment**

The four requirements for safety – *primary containment,*[6] *secondary containment, safeguard systems and separation distance* – apply across the LNG value chain, from production, liquefaction and shipping, to storage and re-gasification. The discussion in this paper is limited to LNG onshore import terminals operations.

**Potential LNG Hazards**

The potential hazards of most concern to operators of LNG import facilities and surrounding communities stem from the basic properties of natural gas. LNG is odorless, non-toxic, non-corrosive and less dense than water. LNG vapors (primarily methane) are harder to ignite than other types of flammable liquid fuels. Above approximately -110°C LNG vapor is lighter than air. If LNG spills on the ground or on water and the resulting flammable mixture of vapor and air does not encounter an ignition source, it will warm, rise and dissipate into the atmosphere.

The potential hazards associated with LNG include heat from ignited LNG vapors and direct exposure of skin or equipment to a cryogenic (extremely cold) substance. LNG vapor can be an asphyxiant. This is also true of vapors of other liquid fuels stored or used in confined places without oxygen.

[6] We use the term "containment" in this document to mean safe storage and isolation of LNG.

There is a very low probability of release of LNG during normal industry operations due to the safety systems that are in place. Unexpected large releases of LNG, such as might be associated with acts of terrorism, bear special consideration although the consequences may well be similar to a catastrophic failure. In the case of a catastrophic failure, emergency fire detection and protection would be used, and the danger to the public would be reduced or eliminated by the separation distances of the facility design.

Multiple layers of protection namely primary containment, secondary containment, safeguard systems, and separation distance provide protection against potential hazards associated with LNG. Some of the potential hazards include:

**Cryogenic Liquid** - If LNG is released, direct human contact with the cryogenic liquid will freeze the point of contact. Containment systems surrounding an LNG storage tank, thus, are designed to contain up to 110 percent of the tank's contents. Containment systems also separate the tank from other equipment. Moreover, all facility personnel must wear gloves, face masks and other protective clothing when entering potentially hazardous areas to protect them from the freezing liquid. This potential hazard is restricted to the facility boundaries and does not affect neighboring communities.

**Vapor Clouds** - As LNG leaves a temperature-controlled container, it begins to warm up, returning the liquid to a gas. Initially, the gas is colder and heavier than the surrounding air. It creates a fog – a vapor cloud – above the released liquid. As the gas warms up, it mixes with the surrounding air and begins to disperse. The vapor cloud will only ignite if it encounters an ignition source while concentrated within its flammability range. Safety devices and operational procedures are intended to minimize the probability of a release and subsequent vapor cloud having an affect outside the facility boundary.

**Explosion** - An explosion happens when a substance rapidly changes its chemical state – i.e., is ignited – or is uncontrollably released from a pressurized state. For an uncontrolled release to happen, there must be a structural failure – i.e., something must puncture the container or the container must break from the inside. LNG tanks store the liquid at an extremely low temperature, about -256°F (-160°C), so no pressure is required to maintain its liquid state. Sophisticated containment systems prevent ignition sources from coming in contact with the liquid. Since LNG is stored at atmospheric pressure – i.e., not pressurized – a crack or puncture of the container will not create an immediate explosion.

**Terrorism** - There must be a general enforcement of security to protect all types of facilities and public places, including LNG operations, from acts of violence. With respect to terrorism, a system of safeguards is already in place. LNG tanks, whether on ships, on land, or offshore, require exceptionally large amounts of force to cause damage. Because the amount of energy required to breach containment is so large, in almost all cases the major hazard presented by terrorists is a fire, not an explosion. If an aircraft crashed into an LNG facility the impact would almost certainly cause a fire which would be initially fueled by the aircraft fuel. It may also ignite the LNG, causing a larger fire at the facility. Emergency fire detection and protection at the LNG facility or ship would be used in such an event. Danger to the public from this type of event would be reduced or eliminated by the separation distance of the facility.

Rigorous siting reviews coupled with separation distance requirements in U.S. safety codes[7] minimize risks to the public from land-based facilities. The U.S. Coast Guard (USCG) which by Federal law has the responsibility for implementing safety regulations that apply to LNG marine operations in the U.S. is now part of the new U.S. Department of Homeland Security. For all vessels of special interest, including LNG ships, the USCG enforces strict measures towards terror threat protection. Measures taken to prevent terrorism on LNG facilities and ships worldwide include inspections and patrols, action plans for security breach, emergency communication systems and intelligence gathering. These same measures are also used at other critical facilities, operations and ships such as passenger ships, oil tankers, container ships and so on.

**Earthquakes -** When estimating the risk of LNG projects, the companies involved in LNG facilities consider the danger of strong ground movements and failures due to seismic activity, liquefaction and landslides in the area. Regulations outline the seismic design requirements[8]. Major earthquakes can cause severe damage if the facilities are not designed to withstand such events, so the companies conduct regional and site-specific studies to see if the areas are seismically active. These factors are then taken into account during planning and design stages. The design of LNG tanks can accommodate regional seismic activity in locations of potential risk. There are no known incidences of LNG storage tank failures due to seismic activity. Japan is also one of the more seismically active areas of the world. Damage to its LNG facilities from the most severe earthquakes has been limited to that of natural gas pipelines. In fact, in 1995 during an earthquake measuring 6.8 on the Richter scale, none of the LNG storage tanks in the Kobe, Japan area were damaged.

**Maritime Incidents -** The history of the LNG industry has shown that maritime incidents with severe LNG releases are very rare. Over the industry's 60-year history of 40,000 voyages, there has never been a spill from a ship into the water from either a collision or grounding. LNG ships are well designed and well maintained, which reduces the chances and severity of incidents. Their designs prevent breaching of cargo tanks and involvement of multiple tanks in accidents. Potential hazards could come from ignition of LNG pool fires or a vapor cloud.

**Operational Incidents -** Incidents resulting from human error, equipment failures, or both can occur in any industry and at any facility. At the LNG facilities, operational incidents could happen during unloading, storage, vaporizing and pipeline transmission or other stages of production. Such errors could result in a spill or a fire. LNG facilities and ships have advanced monitoring and control systems that make an incident unlikely to occur compared to other releases. Consequences of the majority of potential incidents would be contained on site and managed before they could result in significant damage.

**Application of Multiple Layers of Protection for Safety**

The specific safety and security features embedded in the LNG import terminals and LNG ships, as they pertain to the four elements of primary containment, secondary containment, safeguard systems and separation distances, are detailed below, following our schematic in Figure 3 of the multiple layers of protection.

---

[7] 49 CFR Part 193: http://cfr.law.cornell.edu/cfr/cfr.php?title=49&type=part&value=193

[8] The National Fire Protection Association (NFPA): NFPA 59-A 2001 http://www.nfpa.org/ .

**Primary Containment** - The first and most important safety requirement for the industry is to contain LNG. This is accomplished by employing suitable materials for storage tanks and other equipment, and by appropriate engineering design.

LNG typically is stored in double-walled tanks at atmospheric pressure. High alloy steels composed of nine percent nickel and stainless steel typically are used for the inner tank and for other LNG applications. The outer container of a single containment storage tank serves primarily to retain insulation and vapor, Figure 4. It is not designed to contain LNG due to leakage from the inner tank.

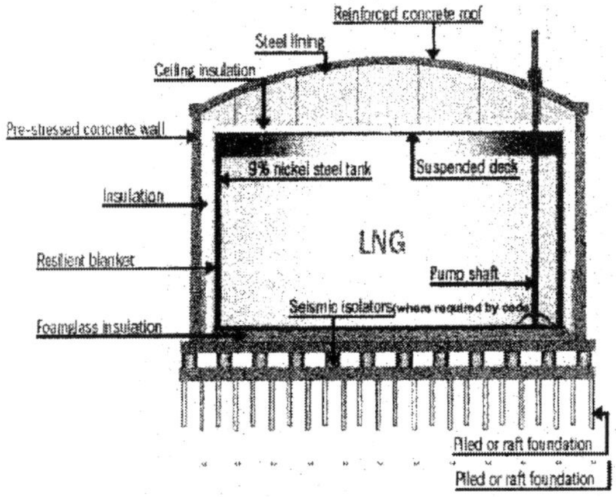


*Source: Shell*

**Figure 4 – Conceptual Design of LNG Storage Tanks**

LNG ships are double-hulled ships especially designed and insulated to prevent leakage or rupture in an accident. The primary container can be constructed of stainless steel, invar (36 percent nickel steel). The most common cargo insulation materials include polyurethane, polyvinyl chloride foam, polystyrene and perlite. Nitrogen is placed in the insulation space.

**Secondary Containment** - This second layer of protection ensures that if leaks or spills occur, the LNG can be contained and isolated. For onshore installations dikes and berms surround liquid storage tanks to capture the product in case of a spill. In some installations a reinforced concrete tank surrounds the inner tank that normally holds the LNG. Secondary containment systems are designed to exceed the volume of the storage tank. Double and full containment systems for onshore storage tanks can eliminate the need for dikes and berms. Figure 5 illustrates the different types of LNG storage systems. In communities where there are concerns about having an un-scenic view of large LNG storage tanks on the horizon, the LNG storage tanks can be buried as is the case with some inground tanks in Japan.

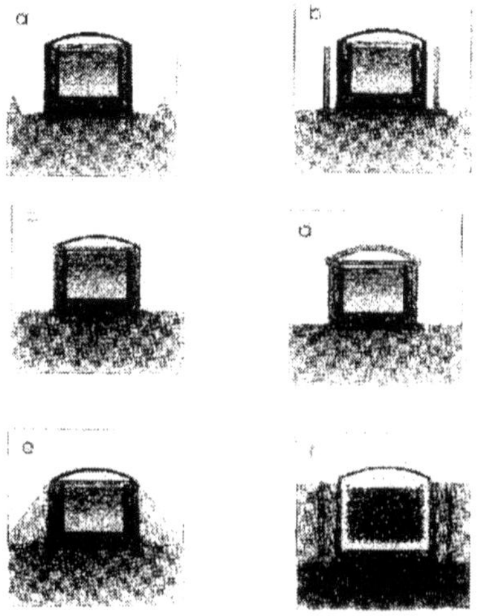

a. **Single containment** system with dyke, bund wall
b. **Double containment** system with high concrete dike
c. **Full containment** system with steel roof
d. **Full containment** system with closed monolithic prestressed concrete structure
e. **Full containment** system with steel roof and earth embankment
f. **Inground tank** (membrane system) with steel roof

Source *www.plantconst.com/indust/sub4_3.htm*

**Figure 5 - Typical LNG Storage Systems**

The secondary barrier on LNG ships depends on the type of construction of the storage tanks. Secondary containment mechanism for membrane design ships also serves as the primary barrier. For spherical and structural prismatic design systems, the secondary barrier is a splash barrier with a drip pan at the bottom from which accumulated liquid evaporates. The materials used to construct the secondary barrier include aluminum or stainless steel foil, stainless steel and invar.

**Safeguard Systems -** In the third layer of protection, the goal is to minimize any release of LNG and mitigate the effects of a release. For this level of safety protection, LNG operations use systems such as gas, liquid and fire detection to rapidly identify any breach in containment, and remote and automatic shut off systems to minimize leaks and spills in the case of failures. Operational systems - procedures, training and emergency response - also help to prevent and mitigate hazards. Regular maintenance of these systems is vital to ensure their reliability.

**Separation Distance -** Federal regulations have always required that LNG facilities be sited at a safe distance from adjacent industrial plants, residential communities and other public areas. The safe distances or exclusion zones are based on LNG vapor dispersion data, thermal radiation contours and other considerations as specified in regulations. LNG storage and other facilities have setbacks or protection distances for protecting surrounding areas should there be a release of LNG or a fire. The thermal radiation protection zone is the area beyond the impoundment area large enough so that the heat from an LNG fire does not exceed a specified limit for people and property. This area is owned or controlled by the operator and calculated specific to each location to prevent exposure to thermal radiation. The flammable vapor dispersion exclusion zone is an area large enough to encompass that part of the vapor cloud which could be flammable. It is calculated specifically for each location to prevent exposure to fire.

Safety zones are established around LNG ships while underway in U.S. waters and while moored, by the U.S. Coast Guard (USCG) and port captain. Creation of the safety zones is based on the specific risk factors at a given terminal. Two purposes for establishing the safety zones are to minimize collision while the ship is underway and at berth, and to protect surrounding property and personnel from hazards that could be associated with ignition. In some ports, the USCG may require a tug escort and specified safety zones around LNG ships. Tugs also assist in the safe docking of LNG ships. LNG dock facilities for berthing and unloading LNG ships can be located far away from shore where applicable, as in the case of Cove Point terminal in Maryland, Massachusetts. This helps minimize the disruption of harbor traffic caused by implementation of the safety zones.

To resolve some of the issues of public concern regarding the safety of LNG ships in the harbor, the Federal Energy Regulatory Commission (FERC)[9] in February 2004 hired a risk analysis firm, to investigate the hazards posed by LNG ships. The study will evaluate the flammable vapor and thermal radiation hazards created by unconfined LNG spill on water resulting from an LNG cargo release. The report was due to be submitted in March 2004. We recommend that the outcome of this study should be freely discussed with the public.

**Industry Standards and Regulatory Compliance -** No systems are complete without having appropriate operating and maintenance procedures in place and adhered to. Relevant personnel should be appropriately trained to ensure these procedures are implemented. Organizations such as the Society of International Gas Tanker and Terminal Operators (SIGTTO), Gas Processors Association (GPA) and the National Fire Protection Association (NFPA) produce guidance from industry best practices. Regulations are designed to prevent incidents from occurring, and if they do occur from human or other error to protect the public from any impact. In the U.S., several regulatory authorities govern the LNG industry. The U.S. Department of Energy–Office of Fossil Energy[10] facilitates coordination between the federal agencies maintaining regulatory and policy authorities for LNG. The FERC is responsible for permitting new onshore LNG receiving terminals in the U.S. and ensuring safety at these facilities through inspections and other forms of supervision. The USCG is responsible for assuring the safety of all marine operations at LNG receiving terminals and for LNG ships in U.S. waters.

Before LNG projects are implemented, studies must be carried out which include:

- assessments of siting requirements;
- baseline biological and land use surveys and impact analyses;
- facility process design;
- evaluation of the operational constraints and hazards associated with the terminal facilities, and shipping of LNG including earthquake tolerance;
- compatibility of LNG facilities with current and projected uses of waterways and adjacent lands;
- assessment of potential risks to the public near prospective sites;
- assessment of potential effects of facility construction and operation on terrestrial and aquatic ecosystems;
- analyses of oceanographic, navigational, and meteorological conditions to determine whether access by LNG ships is feasible and safe.

Industry organizations help to coordinate interaction between the LNG industry, and the agencies and authorities charged with creating and enforcing rules and regulations for LNG

[9] U.S. Federal Energy Regulatory Commission (FERC): http://www.ferc.fed.us/.
[10] U.S. Department of Energy – Office of Fossil Energy: http://www.fe.doe.gov/.

facilities. The *International Maritime Organization (IMO)* [11] has developed standards for the construction and operation of all ships. These standards and codes govern the design, construction and operation of specific ships, including LNG ships, and when ratified, are adopted and incorporated into the individual flag state regulations. In the U.S., the USCG has adopted the applicable IMO standards and codes in regulations covering U.S. flag ships. The USCG inspects LNG ships when in U.S. port, regardless of their flag state for compliance with these codes.

The Maritime Transportation Security Act of 2002 (MTSA) and the International Ship and Port Facility Security (ISPS) codes recommend additional security measures relating to ships and port facilities personnel and operational requirements. By July 1, 2004, as with other critical fuels and products, all LNG ships and terminals worldwide will have specific security plans in place as required by the IMO and the USCG. The LNG ship Berger Boston (which is under long-term charter to Tractebel LNG North America) was the first vessel in the world to receive the new ISPS certification, in June 2003.

**Maritime Classification Societies** provide the means by which LNG shipping operators can demonstrate that they have established clear, practical, technical standards that address the protection of life, property, and the natural environment.[12] The classification societies establish rules for the construction of LNG ships using IMO standards as a minimum. They can, on behalf of flag states, certify existing proven technologies and methods of construction and have assisted in gaining approval for the development of new technologies so that they can be tested and then built. Some of the societies that classify LNG ships include the American Bureau of Shipping (ABS), Bureau Veritas (BV), Det Norske Veritas (DNV) and Lloyd's Register of Shipping (LR).

LNG regulations and industry standards complement each other. They apply to the design, construction, and operation of LNG facilities and have been developed by using best engineering practices and incorporating many years of operating experience.

The four conditions described above for safety, along with **industry standards and regulatory compliance**, are vital to continuing the strong LNG industry safety performance. They are essential if LNG is to play an increasing role in the U.S., both for energy security and to protect the flow of economic benefits from LNG to our society as a whole.

## 4. CONCLUSION

The industry must continue building public confidence in its ability to construct and maintain safe and secure LNG facilities. Proponents and regulators need to do a better job addressing public concerns and establishing credibility and trust. Comprehensive stakeholder involvement in the development of LNG facilities could be essential to developing some degree of this credibility and trust. As illustrated by the level of public resistance to the siting of new LNG import terminals in the U.S., there is a strong perception that LNG is unsafe and would negatively impact the lives of the host communities. The public also feels insecure about the

[11] International Maritime Organization (IMO) http://www.imo.org.

[12] Sember, W.J. ABS: Development of Guidelines for Classification of Offshore LNG Terminals, GASTECH 2002, Qatar, Oct. 2002.

possibility of terrorist attack on the LNG plant. The government and industry are making some attempt to address public concern but we feel that a more coordinated effort should be made to engage the host communities and to share information with them that will resolve their concerns.

In September 2003, the U.S. Secretary of Energy Spencer Abraham announced the formation of the Department of Energy (DOE) - National Association of Regulatory Utility Commissioners (NARUC) LNG partnership to help educate critical energy decision-makers on barriers and opportunities related to the increased development of LNG resources.[13] Trade organizations such as the Interstate Natural Gas of America (INGAA)[14], the Gas Processors Association (GPA),[15] SITTGO,[16] and the International LNG Alliance (ILNGA)[17] are also engaged in public education on LNG.

To further address some of the LNG safety concerns, the DOE recently asked its Sandia National Laboratory to examine LNG safety, particularly issues related to transportation. Sandia's report to DOE is expected to be completed in Spring 2004. Thereafter public meetings involving experts representing all the stakeholders' positions should be organized.

Experts from organizations with less economic interest in the industry such as government, academic institutions or international scientific agencies should also continue to present data on LNG. Potential host communities are motivated to learn about LNG and is very focused on developments which may have direct or indirect effects on their concerns. Various advocacy groups have been formed, and they have used industry publications and the opinion of experts to support their case. In many areas the dialogue is now between competing expert opinions.

The University of Houston, Institute for Energy, Law & Enterprise (IELE)[18] is helping to disseminate accurate and timely information regarding LNG to local and regional policy-makers and to the public at large. The institute is convinced that a well-informed public is one that will ultimately be more understanding of the various energy options and choices available to them.

---

[13] NARUC: http://www.naruc.org/displayindustryarticle.cfm?articlenbr=15873

[14] Interstate Natural Gas of America (INGAA) - www.Ingaa.Org

[15] Gas Processors Association (GPA) - http://www.Gasprocessors.Com/LNG.Html

[16] Society of International Gas Tanker and Terminal Operators (SITTGO): http://www.Sigtto.Org/

[17] The International LNG Alliance (ILNGA) www.Ilnga.Org

[18] University of Houston, Institute for Energy, Law & Enterprise, http://www.Energy.Uh.Edu/LNG.

# LNG TERMINAL SITING: FEM3A VAPOR DISPERSION MODEL DEVELOPMENT

*Jerry A. Havens, Tom O. Spicer, Wendy V. Sheppard*
*Chemical Hazards Research Center*
*Department of Chemical Engineering*
*University of Arkansas*
*primary contact: jah@engr.uark.edu*

*Kent Perry*
*Gas Technology Institute*
*Des Plaines, Illinois*

## ABSTRACT

Experiments have been conducted in the Chemical Hazards Research Center (CHRC) Ultra-Low-Speed Wind Tunnel to validate computational fluid dynamics models for application to LNG terminal siting consequence analyses. Area-source ground-level releases of carbon dioxide gas in the wind tunnel provide physical simulation of near-field dispersion of cold natural gas vapors. Current emphasis is on effects upon dispersion of obstacles, such as tank and dike structures, as well as surface roughness arrays intended to simulate suburban housing. This paper reviews the current state of the development program and describes research efforts which are being initiated by the University of Arkansas and the Gas Technology Institute (GTI) sponsored by the U.S. Department of Energy.

## INTRODUCTION

The Gas Research Institute (GRI) sponsored research conducted at the CHRC to develop and verify alternative model approaches, such as the DEGADIS (Dense Gas Dispersion) and FEM3A models to account for LNG vapor buoyancy, directional diversion by topography, holdup in vapor detention systems (dikes or fences), wake turbulence, or (earth) surface-to-cloud heat transfer in the determination of vapor cloud safety exclusion zones around LNG storage and transfer facilities. (Havens et al., 1987; Havens and Spicer, 1990; Havens et al., 1994; Spicer and Havens, 1996).

Currently, GTI and CHRC are initiating a research program to provide for continued evaluation, development, and standardization of the DEGADIS and FEM3A vapor dispersion models for application to a wide range of catastrophic hazardous material release scenarios required to support siting decisions, for effective management of the hazards associated with such facilities, and to develop a national focal point for LNG safety research and technical dissemination.

## EXPERIMENTAL PROGRAM

### *CHRC Wind Tunnel*

The CHRC Wind Tunnel is an ultra-low-speed boundary layer wind tunnel capable of producing airflows that simulate the constant stress layer of the atmospheric boundary layer. It was designed and constructed specifically for the study of atmospheric dispersion of dense gases at wind speeds below 2 m/s (Havens et al., 1994). Figure 1 depicts the floor plan of the CHRC Wind Tunnel.

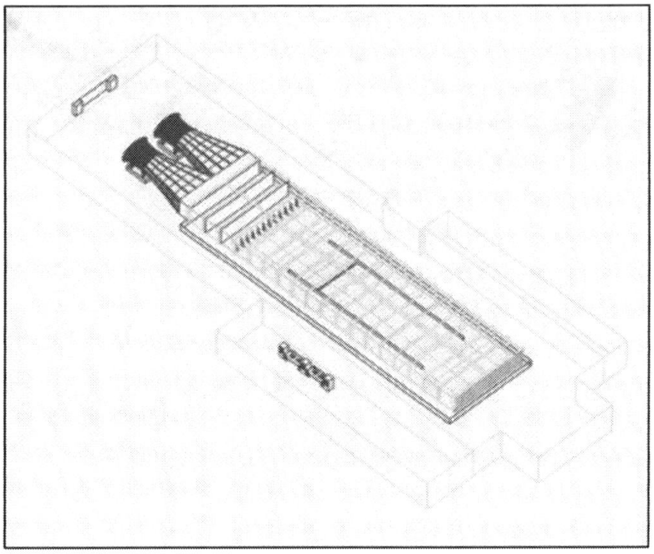

Figure 1. Floor plan of CHRC Wind Tunnel

The wind tunnel is centered in a larger room to ensure a symmetrical return space for the re-circulating air. An isolated control room situated adjacent to the wind tunnel is used to house data acquisition systems and control instrumentation. During experiments, the tunnel room is isolated to prevent extraneous effects that would disrupt the tunnel flow.

The re-circulating airflow is provided by two 75-horsepower, 72-inch diameter adjustable pitch vane-axial fans manufactured by Buffalo Forge Company. The fans, synchronized by Fenner M-Trim speed controllers, provide infinitely variable wind speed in the tunnel.

The airflow produced by the fans passes through a circular-to-rectangular transition to the working section of the tunnel with dimensions 7 ft high, 20 ft wide and 80 ft long. The working section is divided into two regions:

- *Boundary-layer generation.* A uniform flow across the cross-sectional area of the wind tunnel is generated as the air flows through a honeycomb with ½-inch cells and a series of four seamless nylon screens. Fourteen Irwin spire-shaped turbulence generators, 13.2 cm base, and 92.7 cm height, are positioned 30 cm downwind from the last seamless screen with 46.3 cm between adjacent spires. The turbulence generators produce an approximately 1 m deep turbulent boundary layer in the working section.
- *Measurement region.* The measurement region begins at six spire heights (approximately 18 ft) downwind from the last screen.

The wind tunnel floor was tiled with smooth rubber matting. A seamless screen and a back-pressure device consisting of vertical Plexiglass strips (3 inches in width and ¼ inch in thickness spaced 3-3/16 inches apart) were installed at the end of the working section of the wind tunnel.

Cut aluminum angles functioning as surface roughness elements were installed later on the floor to "roughen" the floor for experiments that studied the effects of surface roughness. These surface roughness elements were made up of a square base of 3.81 cm in length on all sides with another square of the same dimensions extended perpendicularly from one edge of the base. They were arranged in a staggered array at a distance of 30.5

cm between subsequent elements in the downwind as well as the lateral directions (a spacing of eight obstacle heights).

### *LNG Vapor Release Modeling System*

The tank and dike configurations used in the tunnel for experimental work were of 150:1 scale. The tank was 31 cm in diameter, cylindrical in shape, and had a spherical-section "dome" top. The height of the tank measured to the top of the dome was 28.3 cm. The tank was located in the center of areas enclosed by square dikes, on a platform topped with a mesh screen that was flush with the tunnel floor surface. The inner side-length dike dimensions were 0.943 m and 0.636 m, respectively, for the low-dike and high-dike designs. The dike heights were 0.02 m and 0.037 m, respectively, for the low-dike and high-dike designs.

### *Velocity Measurements*

Measurements of wind-tunnel velocity and turbulence statistics utilized constant temperature thermal anemometry, (CTA), specifically the miniature cylindrical two-sensor "X"-film probes (Model Number 1248A-10 by TSI Inc.). The X-probe simultaneously measures two components of the velocity vector. The probe measures velocities in the x-y and x-z directions by being oriented in two different perpendicular directions via a micro-electric motor.

Experiments acquiring velocity and turbulence statistics were conducted with the fans running at 90 rpm. Calibration of the CTA sensors was accomplished by centering the probe in a calibration tube (0.5 in. diameter and 30 in. long Plexiglass tube) to which breathing-grade bottled air was fed. The specific flow rates used in the calibration were designed to provide laminar flows through the tube that spanned the range of the wind velocities studied in the experiments. The centerline velocity in the calibration tube was consequently twice the value of the average velocity.

The XWA was calibrated by decoupling it as if it were two single probes, and the data were acquired using a TSI IFA 300 research anemometer. Each data point in the calibration stage was taken at 1 kilo-Hertz (kHz) and for 60 seconds. Ten voltage outputs (calibration points) were obtained corresponding to flow rates between 0.5 and 2.3 slpm with a 0.2 slpm increment between two subsequent flow rates. The yaw calibration was achieved by pivoting the calibration tube from 30-degrees right to 30-degrees left (with increments of 6-degrees) about the centerline location of the XWA, thus yielding eleven yaw calibration points.

Following computer processing of the raw data files, the voltages with their corresponding flow rates for the velocity calibration were then entered into the software to generate two separate calibration curves, one for each single wire. Mean-square-errors for both curves were also shown as an indication of the deviation of the calibration curve from a fourth-order polynomial. The voltages from the yaw calibration were also entered into the software to calculate the yaw coefficients. A typical calibration curve generated as a result of calibration of the XWA is shown in Figure 2.

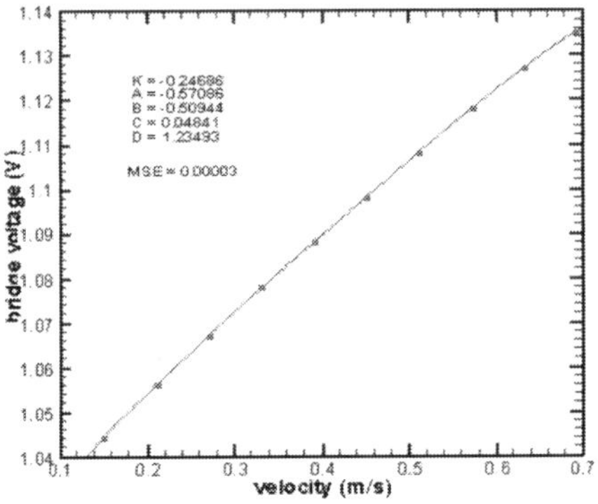


Figure 2. Typical XWA Probe Calibration.

The probe was positioned at an elevation of 0.5 (+/- 0.1) cm above the tunnel for all measurements in lateral velocity profiles in this study. The turbulence statistics were determined for all data points from data sampled at 1 kHz for 2 minutes.

### *Gas Concentration Measurements*

The gas concentration measurements were made with a Cambustion High Frequency Response (HFR) 400 Fast Flame-Ionization-Detector. Calibration of the FID entailed generating a five-point calibration curve. A calibration line "diffuser" was constructed from a 6 in. long, 1 in. diameter Plexiglass tube fitted with a cotton or porous foam filter which dampened pressure fluctuations in the calibration gas supply. Data for each calibration point, also acquired by the IFA 300 ThermoPro, were sampled at 1 kHz for 60 seconds for air-$CO_2$ mixtures (using propane tracer), corresponding to carbon dioxide concentrations of 0%, and other concentrations catering to the specific experiment. The 0% point was taken in the room air at the beginning of each experiment, so as to measure concentrations relative to the initial room gas air mixture (of air, carbon dioxide, and minor hydrocarbon contaminants). Subsequent data points were measured by inserting the sampling tube into the calibration diffuser (perpendicular to the flow in the diffuser). A typical calibration curve is shown in Figure 3.

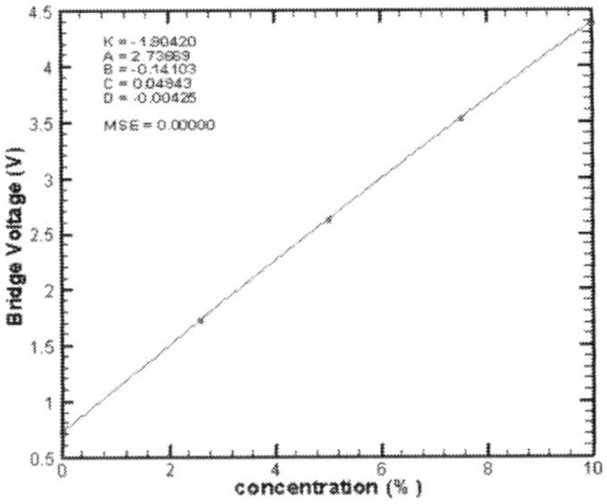


Figure 3. Typical FID Probe Calibration.

Lateral profiles of gas concentration spanning the entire gas cloud were made at pre-assigned downwind locations. All such measurements were made at 0.5 cm elevation. In selected experiments, vertical gas concentration profiles were measured on the wind tunnel centerline at 0.5 cm vertical intervals from the floor so as to span (vertically) the gas cloud down to concentrations of approximately 1%.

Five low momentum, area source $CO_2$ releases were studied in five configurations:

A. Smooth floor.
B. Low dike and tank – smooth floor.
C. High dike and tank – smooth floor.
D. Rough floor.
E. High dike and tank – rough floor.

The flow for all experiments was set at 33.4 slpm carbon dioxide traced with 0.5 slpm (~1.5%) propane. The gas concentration measurements began after the gas flow downwind of the source box reached steady state.

***Flow Visualization Tests***

Visualization tests were conducted prior to the experimental measurement phase to ensure that the gas flow was down the center of the wind tunnel, demonstrating the symmetry of the flow. This was accomplished by adding Rosco fog (theater) fluid to the carbon dioxide at the entrance to the gas source box under the tunnel. Fog fluid addition was at the same rate, as nearly as could be approximated, for all tests. A video camera was mounted near the center of the tunnel ceiling so as to give an oblique view of the gas flowing in the downwind direction. Camera position and angle of view was the same for all tests. Figures 4 - 8 show video frames from the five configurations represented by cases A - E, respectively.

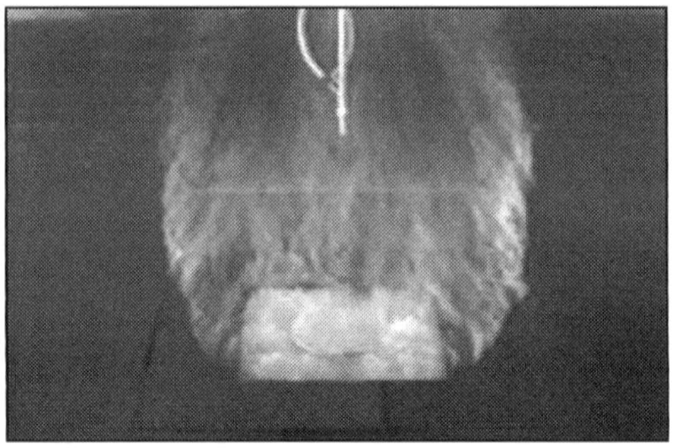

Figure 4. Case A: Low momentum area source release – smooth floor.

Figure 5. Case B: Low momentum area source release with low dike – smooth floor

Figure 6: Case C: Low momentum area source release with high dike – smooth floor

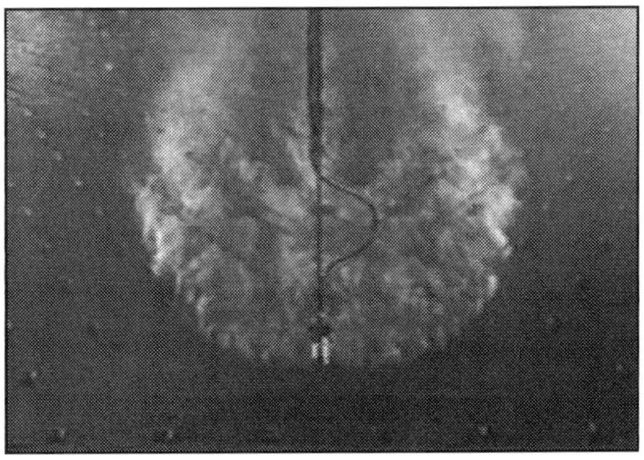

Figure 7. Case D: Low momentum area source release – rough floor.

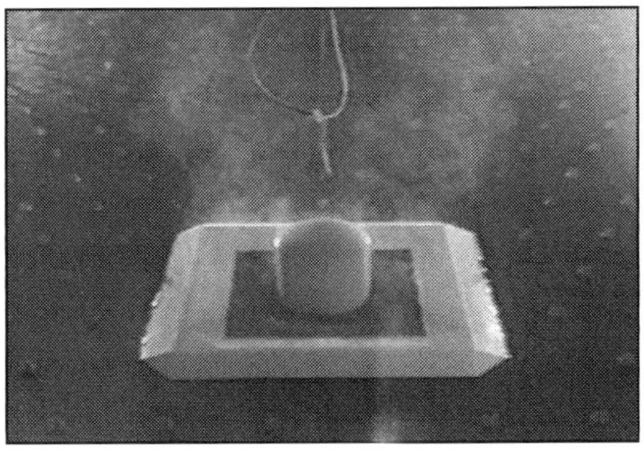

Figure 8. Case E: Low momentum area source release with high dike – rough floor

## EXPERIMENTAL RESULTS

### *Velocity and Turbulence Statistics*

The measurements made for cases A-C had been completed more than three years before any measurements were undertaken for cases D and E. In order to ensure repeatability, measurements of selected velocity and turbulence statistics for case A (which were identical for cases A-C) were performed prior to any measurements made for cases D-E.

Measurements for the vertical profiles of mean velocity and turbulence statistics were made at the centerline of the wind tunnel immediately upwind of the gas source box at the fan speed of 90 rpm for all cases. These data were used to determine the mean velocity at the (model) elevation corresponding to 10 meters of elevation at field scale

(1000 cm / 150 = 6.67 cm) as well as to determine the friction velocity.

Figure 9 shows the vertical velocity profiles over the smooth floor taken over three years apart (representing the measurement periods when cases A-C were completed and when cases D-E were completed). Ambient conditions were not identical, but were similar since the wind tunnel and its enclosure are climate-controlled. The repeatability of the velocity measurements, particularly in the range of such low tunnel operating speeds, is very good, thus indicating the exceptional control of the flow in the CHRC ULS wind tunnel.

Figure 10 shows measurements of the vertical velocity profile over the rough floor (cases D and E) with the smooth floor velocity profile (cases A - C) indicated for comparison.

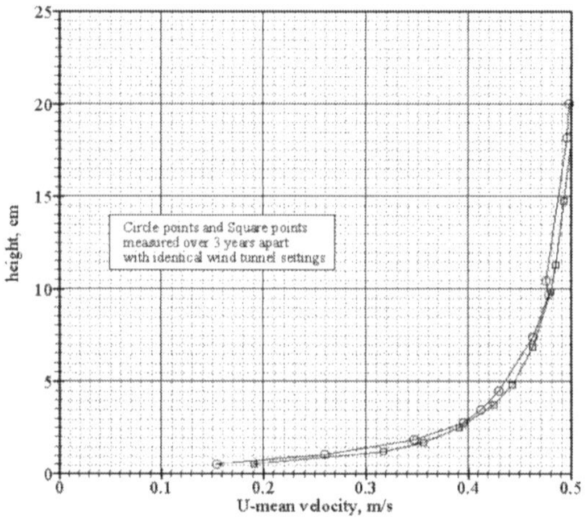


Figure 9. Verification of repeatability of velocity measurements.

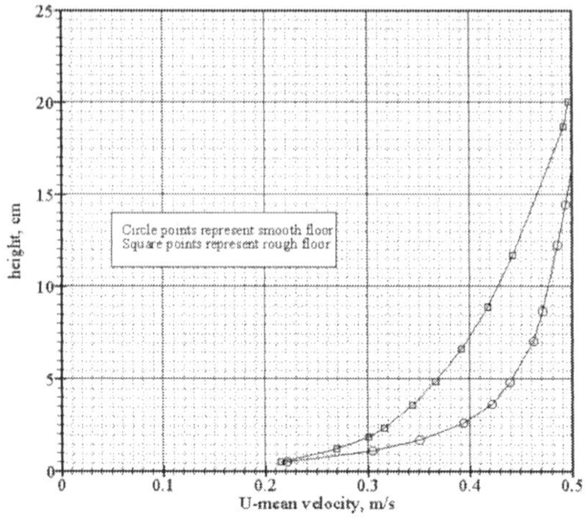


Figure 10. Smooth and rough floor velocity profiles at the same fan speed.

Table 1 indicates the values of surface roughness and friction velocity, derived from the data shown in Figure 10, for the smooth floor (cases A-C). For the experiments with roughness (cases D-E), the friction velocity and surface roughness were determined from Reynolds stress measurements (u'v'). These parameters were required for simulation of the wind tunnel experiments with FEM3A.

| Floor | Friction velocity, m/s | Surface roughness, M |
|---|---|---|
| Smooth | 0.0264 | 0.000063 |
| Rough | 0.035 | 0.00072 |

Table 1. Velocity profile characteristics for smooth and rough floors.

### *Concentration Measurements*

Due to the three year lapse between cases A-C and cases D-E, concentrations measurements were made again for case A conditions before any measurements over the rough floor began in order to ensure repeatability. Figure 11 compares the measurements made for case A (smooth floor) with recent measurements under identical (as nearly as possible) wind tunnel and gas release conditions. The repeatability was excellent.

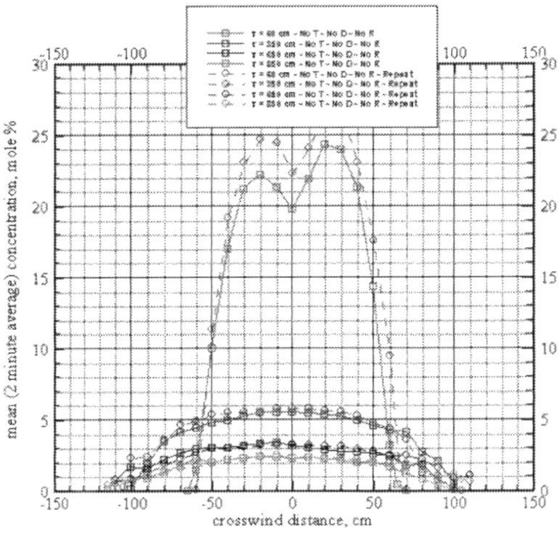


Figure 11. Verification of concentration measurement repeatability.

The lateral profiles, all at 0.5 cm elevation, were made at various designated downwind distances for each of the cases A-E in order to accurately determine the distances associated with the concentrations representing the upper flammable limit (UFL), the lower flammable limit (LFL), and one-half of the lower flammable limit (LFL/2). In some cases, the location of the measurement distances had to be varied slightly to avoid the concentration sensor touching a roughness element or the dike.

Table 2 shows the downwind locations of the lateral profiles of concentration (all at elevation 0.5 cm) for cases A-E. The maximum gas concentrations measured at that downwind distance, some of which are off-centerline, are also indicated.

Case A

| | | | | | | | | | |
|---|---|---|---|---|---|---|---|---|---|
| Downwind location (cm) | 60 | 150 | 250 | 350 | 450 | 550 | 650 | 750 | 850 |
| Maximum concentration (%) | 24.3 | 12.1 | 7.2 | 5.5 | 4.4 | 3.8 | 3.4 | 2.7 | 2.4 |

Case B

| | | | | | | | |
|---|---|---|---|---|---|---|---|
| Downwind location (cm) | 50 | 150 | 250 | 350 | 400 | 450 | 500 |
| Maximum concentration (%) | 7.6 | 3.8 | 2.1 | 1.5 | 1.2 | 1.1 | 0.92 |

Case C

| | | | | | | | |
|---|---|---|---|---|---|---|---|
| Downwind location (cm) | 60 | 150 | 250 | 300 | 350 | 400 | 450 |
| Maximum concentration (%) | 5.2 | 2.6 | 1.6 | 1.2 | 1.1 | 0.92 | 0.86 |

Case D

| | | | | | |
|---|---|---|---|---|---|
| Downwind location (cm) | 55 | 88 | 175 | 236 | 357 |
| Maximum concentration (%) | 24.0 | 18.9 | 4.9 | 2.8 | 1.5 |

Case E

| | | | | |
|---|---|---|---|---|
| Downwind location (cm) | 84 | 175 | 236 | 357 |
| Maximum concentration (%) | 4.0 | 2.2 | 1.6 | 0.85 |

Table 2. Downwind locations of lateral concentrations with observed maximum values.

Figures 12 - 16 show lateral profiles (0.5 cm elevation) of gas concentration (top) and iso-concentration contours (bottom) of carbon dioxide concentrations which correspond to methane UFL, LFL, and LFL/2 concentration levels, for Cases A-E, respectively:

| | Methane (%) | Carbon Dioxide (%) |
|---|---|---|
| UFL | 15 | 6 |
| LFL | 5 | 1.9 |
| LFL/2 | 2.5 | 0.9 |

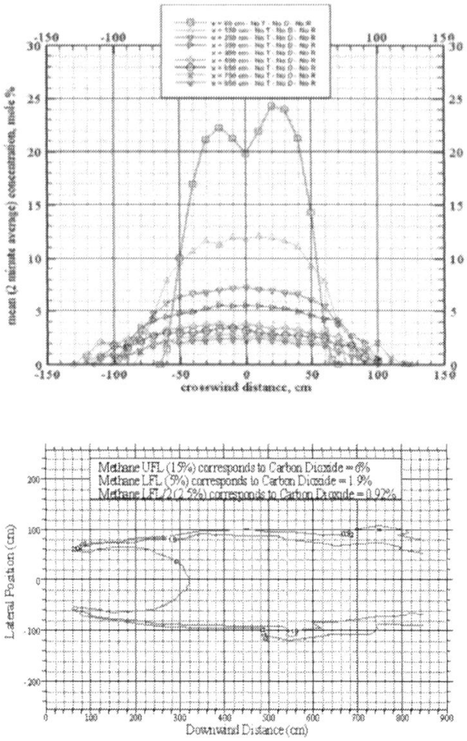


Figure 12. Case A - lateral profiles (top) and iso-contours (bottom), 0.5 cm elevation

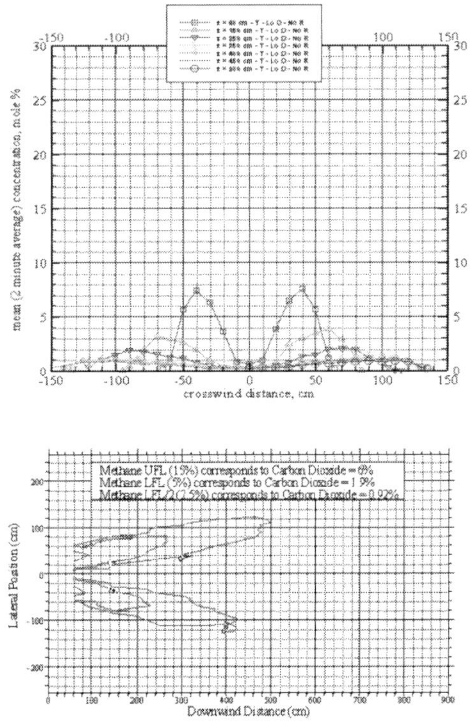


Figure 13. Case B - lateral profiles (top) and iso-countours (bottom), 0.5 cm elevation.

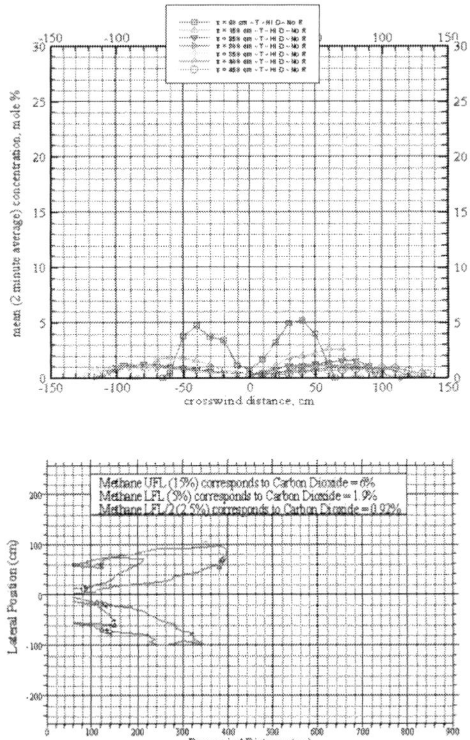


Figure 14. Case C - lateral profiles (top) and iso-contours (bottom), 0.5 cm elevation.

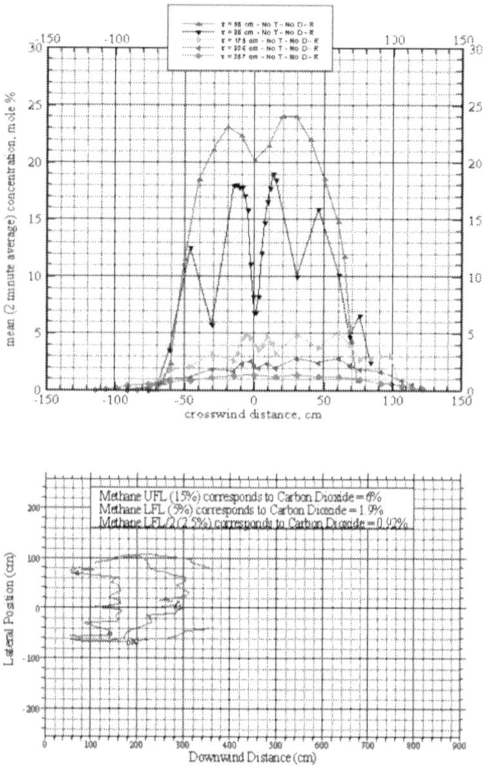


Figure 15. Case D - lateral profiles (top) and iso-contours (bottom), 0.5 cm elevation.

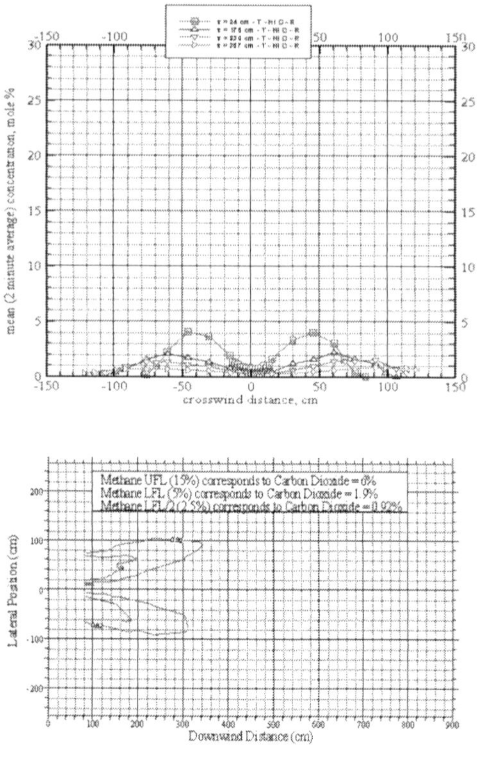


Figure 16. Case E - lateral profiles (top) and iso-contours (bottom), 0.5 cm elevation.

## CONCLUSIONS

▪ The FEM3A model is currently prescribed in 49 CFR 193 for use in siting LNG facilities. The model's acceptance by the U.S. Government for regulatory use follows a lengthy research effort, covering a decade and a half, to provide meaningful verification of the model from which the user can derive confidence in the validity of its predictions.

▪ The experimental results described here provide a large wind tunnel data collection which can be used for continuing verification of computational models, including FEM3A.

▪ Five experimental configurations have been studied:
- Cases A - C provide additional data contributing toward the verification of the models for dike designs which span current U.S. practice.
- Cases D - E were performed to study the combined effects of tank/dike obstacles and surface roughness. The roughness, approximately 10 cm at field scale (150:1) was intended to be representative of regularly arranged suburban housing.

▪ The results indicate an important reduction in downwind hazard extent due to the presence of a tank/dike system alone, surface roughness alone, and to their combination. It is noted that the surface roughness studied may be greater than that which characterizes most large LNG installations in the United States (since urban or suburban housing is not normally found immediately adjacent to the facility). Consequently, the apparent small additional reduction in hazard distance *due to the tank and dike* in the presence of the high surface roughness is probably not typical.

▪ These experiments show that further important reductions in the downwind hazard extent are possible with judicious dike/tank designs.

▪ These experiments indicate that the concentration of gas flowing over the dike appears to be determined by the dike height, thus indicating the potential for further reduction by the use of vapor fences.

▪ The data provided here exemplify the best, most cost-effective methodology available for validation of computer based dispersion models: carefully scaled physical models that can be used to directly evaluate the numerical models by comparison of experimental data with numerical simulation.

## RECOMMENDED RESEARCH PROGRAM

A joint Gas Technology Institute - University of Arkansas research program is being sponsored by the U. S Department of Energy. The research program will begin in 2004. The primary objectives of this research program are:

➢ To provide the FEM3A model with an advanced turbulence closure model (for describing the turbulent mixing with air of denser-than-air gases or aerosols) that will allow for more realistic description of dispersion problems with obstacle and terrain features of greater complexity (the real world).

- To develop the FEM3A model for evaluating hazard consequence issues for a wide range of materials, including LNG and other liquefied energy fuels, in routine industry practice as well as for consideration of terrorist impact.

## REFERENCES


Arthur D. Little, Inc., "Evaluation of LNG Vapor Control Methods", Report to the American Gas Association, October, 1974.

Havens, J. A., T. O. Spicer, and P. J. Schreurs, "Evaluation of 3-D Hydrodynamic Computer Models for Prediction of LNG Vapor Dispersion in the Atmosphere," Final Report to GRI on Contract No. 5083-252-0788, August, 1987.

Havens, Jerry and Tom Spicer, "LNG Vapor Dispersion Prediction with the DEGADIS Dense Gas Dispersion Model," Topical Report to GRI on Contract No. 5086-252-1287, September, 1990.

Havens, Jerry, Tom Spicer, and Heather Walker, "Regulatory Application of Wind Tunnel Models and Complex Mathematical Models for Simulating Atmospheric Dispersion of LNG Vapor," Gas Research Institute Report No. 92-0257, August 1994.

Havens, Jerry, Tom Spicer, and Heather Walker, "Evaluation of Mitigation Methods for Accidental LNG Releases: Volume 1/5--Wind Tunnel Experiments and Mathematical Model Simulations to Study Dispersion of a Vapor Cloud Formed following LNG Spillage into a Diked Area Surrounding a Storage Tank," Topical Report for Gas Research Institute, November 1996.

Havens, Jerry, Tom Spicer, and Heather Walker, "Evaluation of Mitigation Methods for Accidental LNG Releases: Volume 2/5--Wind Tunnel Experiments and Mathematical Model Simulations to Study Heat Transfer from a Flat Surface to a Cold Nitrogen Cloud in a Simulated Atmospheric Boundary Layer," Topical Report for Gas Research Institute, November 1996.

Havens, Jerry, Tom Spicer, and Heather Walker, "Evaluation of Mitigation Methods for Accidental LNG Releases: Volume 3/5--Wind Tunnel Experiments for Mitsubishi Heavy Industries, Ltd.," Topical Report for Gas Research Institute, November 1996.

Havens, Jerry, Tom Spicer, and Heather Walker, "Evaluation of Mitigation Methods for Accidental LNG Releases: Volume 4/5--Wind Tunnel Experiments for Osaka Gas Company," Topical Report for Gas Research Institute, November 1996.

Spicer, Tom, Jerry Havens, and Heather Walker, "Evaluation of Mitigation Methods for Accidental LNG Releases: Volume 5/5--Using FEM3A for LNG Accident Consequence Analysis," Topical Report for Gas Research Institute, December 1996.

**LNG Alarm Management**

*Harry H. West and M. Sam Mannan*
*John Sexton, Spencer Landis and E. Michael Stafford*
*Mary Kay O'Connor Process Safety Center*
*Chemical Engineering Department*
*Texas A&M University*
*hhwest@che.tamu.edu*

*Prepared for Presentation at AIChE Spring National Meeting*
*4th Topical Conference on Natural Gas Utilization*
*New Orleans, LA, April 2004*

## INTRODUCTION

Due to recent major accident events associated with alarm flood, effective alarm management has moved to the top of many plant project lists.

Many currently installed computer control systems provide only limited information regarding abnormal operation. The operator information is often not enhanced by the computer control system, but forced to search out trends and use intuition and experience to evaluate abnormal plant status.

Alarm growth is a natural outcome of the increased information load possible with the modern control system. However, if alarms are not managed in a disciplined manner, uncontrolled alarm growth can result, leading to ineffective and potentially dangerous alarm situations

A structured approach to alarm management has emerged to increase alarm effectiveness and thereby overall plant safety.

## A BRIEF HISTORY OF ALARM SYSTEMS

In the not-too-distant past, a few selected process measurements were hard-wire connected to panel board indicator light systems, which activated when the measurement exceeded a predefined limit value. These panels provided alarm annunciation to the plant operator. The panels were large but limited in capacity, and thereby tended to limit the number of configured alarms.

Modern Distributed Control Systems (DCS) and Programmable Electronic Systems (PES) provide the capability of defining limits for each field measurement. Furthermore, calculated parameters, such as rate-of-change or a combination of field measurements could also have defined limits. Therefore, computer-based control systems have the capability to vastly increase the number of configured alarms.

Alarms not only increase the amount of information provided to the operator but they often are a source of operator overload and confusion. There have been many major incidents that may have been prevented if the plant operator had recognized critical alarms amongst the flood of alarm activations[1]. Some notable examples of alarm problems include the Three Mile Island accident in 1979, where important alarms were missed in an alarm flood. The Texaco refinery explosion at Milford Haven in 1994, where, in the 10 minutes prior to the explosion, the operators had to respond to alarms that were annunciating at a rate of about three per second.

## The LNG Alarm Situation

Many older LNG facilities were built with early DCS and solid state control systems. With new process train expansions or storage tank / regasification additions, the latest DCS and SIS systems will be installed. This typically leads to an eventual control system upgrade project.

A steam or gas turbine vendor will sell a packaged unit containing systems for gas and fire detection, as well as machine monitoring, (CCC controls, GT/Comp safeguarding, Speedtronic or Woodward control systems. etc.). Most of the packaged control subsystems will provide data links to the external instrumentation including plant wide gas detection, ESD, SIS and DCS. However, the alarms present on the packaged control system may not all necessarily be available to the DCS or SIS.

On new or upgraded projects, the challenge is to exert control over the package vendors to conform to whatever plant alarm philosophy is in place overall. This may mean forcing a common PLC safeguard SIS vendor on the package supplier.

Another critical factor is to have a trip analysis function available[6]. Trip analysis (also called chain of events analysis) is a concept of determining exactly how turbine trips affect other processes. It can be considered a high level cause and effect analysis that focuses on the sequence of other alarms activated by the trip event. If alarms can be minimized by masking, what tools can be used to analyze trips and their causes. Important to have knowledge of how fast mechanical effects develop in a trip- and build up a whole trip philosophy.'

## DEFINITIONS

The most common definition of an alarm is "a signal designed to alert, inform, guide or confirm deviations from acceptable system status". Similarly, an alarm system is defined as a system for generating and processing alarms to be presented to users.

Another common definition is that an alarm system is designed to direct the operators attention toward significant aspects of the current plant status. In other words, an alarm system tries to help the operator manage his time!

Because the DCS system processes information on a tag point by tag point basis, alarms and non safety related events, such as condition monitoring alerts, equipment fault status or journal logging events for engineering or maintenance attention have all been described generically as alarms.

Hence, Alarm Management system must address events, which need operator attention and various other non-safety related events. Figure 1 provides a graphic description of the relationship between alerts and safety related alarms.

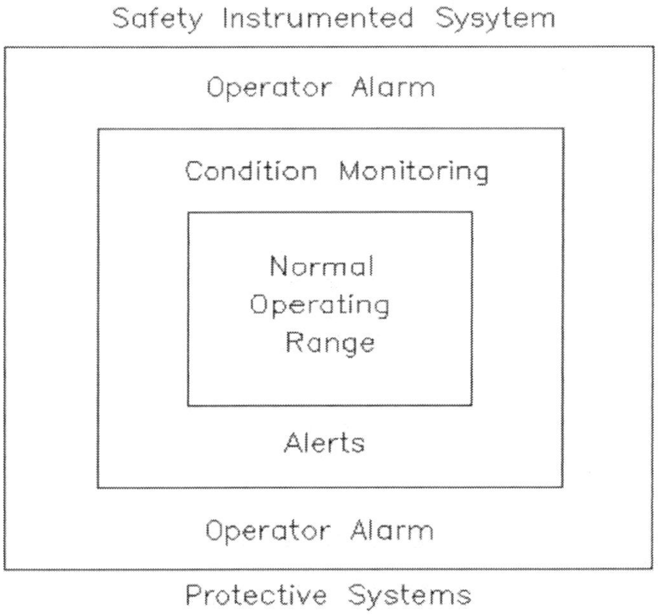


**Figure 1**

The emerging distinction appears to be:

An Alarm requires operator action

An Alert is informational and may not require specific operator action.

ALARM BASICS

The purpose of alarm systems is to alert operators to plant conditions, which deviate from normal operating limits, and thereby require timely action or assessment. Hence, the purpose of an alarm is action.

Matters that are not worthy of operator attention should not be alarmed, but rather placed into a logging journal database for the potential use by other plant staff.

An alarm is useless if the operator does not know what action is required. Similarly, if the operator knows the correct action, but cannot implement the action, the alarm is useless. Useless alarms serve no purpose.

Logging may be a suitable alternative for non-safety related discrepancy events to prevent unnecessary standing alarms. A system for assessing the significance of logged events to ensure timely intervention by maintenance personnel is absolutely required.

## RELATIONSHIP OF ALARM MANAGEMENT TO SAFETY INSTRUMENTED SYSTEMS

While alarm systems may not always have safety implications, they do have a role in enabling operators to reduce the demand on the safety related systems.

However, where a risk reduction factor of $10^{-1}$ failures on demand is claimed then the alarm system, including the operator, is a safety system. Hence, it also requires a suitable safety integrity level [3, 4]. Many Layer of Protection (LOPA) safety integrity level determination studies have incorporated alarm management as an existing layer of protection, thereby forcing the alarm system components to be designed with designated reliability.

Alarms which are not designated as safety should be carefully designed to ensure that they fulfill their role in reducing demands on safety related systems.

For all alarms, regardless of their safety designation, attention is required to ensure that under abnormal conditions, such as severe emergency situations, the alarm system remains effective within the limitations of human response.

## ALARM SETPOINTS

The type of alarm and its setpoint must be established to enable the operator to make the necessary assessment and take the required timely action.

Engineering analysis of the potential timing of events that generate alarms requires good control system engineering practice. Then alarm setpoints must be documented and controlled in accordance with the alarm system management control Timing of alarms and potential operator re-action time is critical. Figure 2 emphasizes that alerts can progress into alarms and eventually into safety shutdowns.

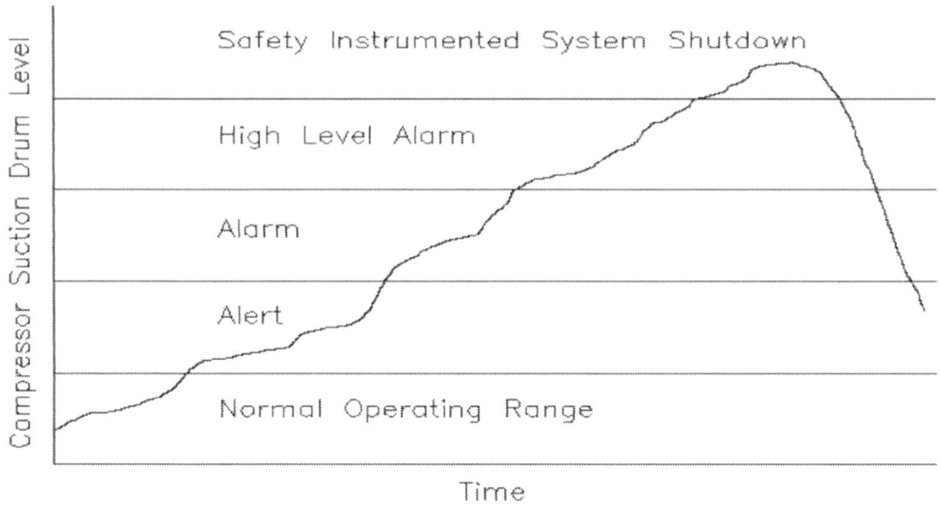


**Figure 2**

Current best practices suggest that the engineering analysis of alarm setpoint follow or co-ordinate with the plant HAZOP studies.

## ALARM PRESENTATION

The human interface must be effective[5]. Alarms may be presented on annunciation panels, individual indicators, CRT screens, or programmable display devices. Alarms lists should be carefully designed to ensure that high priority alarms are readily identified, that low priority alarms are not overlooked, and that the list remains readable even during times of high alarm activity or with repeat alarms.

Alarms should be prioritized in terms of which alarms require the most urgent operator attention. Alarms should be presented within the operator's field of view, and use consistent presentation style (color, flash rate, naming convention). The alarms should be presented in an obvious manner, distinguishable from other alarms, have the highest priority, and remain on view at all times when it is active;

The presentation of alarms should not exceed that which the operator is capable of acting upon, or alternatively the alarms should be prioritized and presented in such a way that the operator may deal with the most important alarms without distraction of the others

Each alarm should provide sufficient operator information for the alarm condition, plant affected, action required, alarm priority, time of alarm and alarm status to be readily identified.

The visual display device may be augmented by audible warnings. Where there are multiple audible warnings, they should be designed so that they are readily distinguished from each other and from emergency alarm systems. They should be designed to avoid distraction of the operator in high workload situations.

## ALARM PROCESSING

The alarms should be processed in such a manner as to avoid operator overload at all times (avoiding alarm floods). The alarm processing should ensure that fleeting or repeating alarms do not result in operator overload even under the most severe conditions. A number of alarm processing techniques include filtering, dead-band, de-bounce timers, and shelving, have been successfully used by industry. A rule of thumb for shelving alarms that has been used by some plants suggests 10 alarms repeated in 10 minutes be shelved.

Applicable alarm processing techniques include grouping and first-out alarms, suppression of lower priority alarms (e.g. suppression high alarm when the high-high activates) suppression of out of service plant alarms, suppression of selected alarms during certain operating modes, automatic alarm load shedding and shelving.

Care should be taken in the use of shelving or suppression to ensure that controls exist to ensure that alarms are returned to an active state when they are relevant to plant operation.

## SYMPTOMS OF AN ALARM MANAGEMENT PROBLEMS

1. Minor operating upsets produce a significant number of alarm activations
2. Routine operations produce a large number of alarm activations that serve no useful purpose
3. Alarm activations occur without need for operator action.
4. Major operating upsets produce an unmanageable number of alarm activations.
5. Operators do not know what to do when a particular alarm occurs
6. Occasional high alarm activation rates.
7. A large number of configured high-priority alarms.
8. Alarms that are standing for long periods of time.
9. Alarms chattering or frequently repeated transient.
10. No plant-wide philosophy for the alarm system management.
11. No guidelines for when to add / delete an alarm.
12. Operating procedures are not tied to alarm activations.
13. When the plant is within normal operating parameters, there are active alarms.
14. Lack of documentation on alarm setpoints.
15. Don't distinguish between safety, operational, environmental, and informational events
16. Erratic schedule of alarm testing.
17. A large number of defeated alarms.
18. Any operator or engineer can create an alarm or change the limits on his own authority.

## ELEMENTS OF ALARM MANAGEMENT

Some of the essential elements of alarm management are:

- An alarm management philosophy document.
- Alarm response within plant written operating procedures
- Ownership of the alarm system.
- Alarm prioritization.
- Alarm configuration records.
- Alarm presentation.
- Alarm logical processing, such as filtering and suppression.
- Informational alert management
- Operator training
- Alarm testing.
- Management of Change applied to the alarm system

Management systems should be in place to ensure that the alarm system is operated, maintained and modified in a controlled manner. Alarm response procedures should be available, and alarm parameters should be documented.

The performance of the alarms system should be assessed and monitored to ensure that it is effective during normal and abnormal plant conditions. The monitoring should include evaluation of the alarm presentation rate, operator acceptance and response times, operator workload, standing alarm count and duration, repeat or nuisance alarms, and operator views of operability of the system. Monitoring may be achieved by regular and systematic auditing.

# THE ALARM IMPROVEMENT OR RATIONALIZATION PROJECT

Alarm rationalization is a systematic procedure for the analysis and improvement of the DCS capability used to alert operators to alarm conditions. This process normally results in a reduction in the total number of configured alarms. Rationalization may occasionally identify the need for new alarms or changes in the alarm priority. Alarm rationalization is a structured procedure that involves a project team, with representatives from operations, maintenance, engineering, and safety

The rationalization procedure includes the following essential steps:

1. Assess the current alarm system performance
2. Development of a alarm management philosophy document
3. Review basis for current alarm limits
4. Review alarm history
5. Establish priority level for each alarm
6. Determine operator actions and update the written operating procedures to incorporate these actions.
7. Consider logical processing techniques to further facilitate improved alarm performance
8. Update the alarm presentation techniques
9. Implement the recommendations of the Rationalization project.
10. Benchmark the new Alarm system
11. Periodic Audit
12. Management of Change

1. Alarm System Metrics

Typical metrics needed to assess the current and then later the improved alarm system include:

| | |
|---|---|
| Rate of alarm activation | Total number of alarms |
| Pattern of alarm activations | Alarms per operator |
| Priority alarm activations | Alarm frequency by shift, |
| Number of standing alarms | Alarm rate during emergencies |
| Time before acknowledge | Fraction of unacknowledged alarms |
| Spread between trip points | Average time to return to normal |
| Time before clearing | Nuisance alarms |
| Chattering alarms | Disabled alarms |
| Correlated alarms | Bypassed alarms |
| Number of disabled alarms | Shelved alarms |

Operational metrics must also be used, such as production rate, off-quality production, number of upsets, and various other important relevant factors. Safety and environmental metrics that must be used are the number of plant shutdowns, number of incidents/near misses, releases to the atmosphere, and pressure relief activations or releases to flare.

The amount of information needed to develop these metrics is daunting, thereby requiring special software to sort the alarm journals in an efficient manner. Many DCS and

SCADA systems and vendors provide alarm management features or products. However, some of these are somewhat primitive and don't provide the necessary purpose by themselves, though they are generally improving.

There also some add-on alarm software products on the market that enhance a control system's basic alarm capabilities by providing online alarm management, advanced alarm logical processing, alarm pattern recognition, and dynamic reconfiguration of the alarm system for varying operating conditions.

2. Alarm Management Philosophy Document

A consistent, comprehensive alarm management procedure/philosophy is necessary before beginning an alarm rationalization project. The procedure typically contains, alarm type (quality, safety, environmental, maintenance information), method of prioritization of alarms, alarm logical processing methods, and testing requirements.

3. Engineering Analysis of each Alarm Setpoint

Evaluation of the alarm configuration file and the accompanying engineering reports is required to verify the basis for the alarm setpoints. If not available, then an engineering study must be done to recreate the basis for the alarm setpoints. Identifying the purpose of the alarm and it's correlation to other alarms is particularly important. Analysis of the alarm purpose includes defining the consequences of inaction to alarm notification

4. Analyze Alarm History

Dynamic analysis of several previous months alarm journal is required to get a statistically valid sample. Disadvantages include correlation with operational, process, or equipment events may not be available in the electronic database. Trip analysis may be determined at this time.

5. Prioritize Alarms

The significance of the alarm must be determined through a ranking scheme or identified during the HAZOP study. The process hazard / risk analysis must determine the importance that the operator detect and perform the expected action when the alarm occurs. Prioritization helps ensure the operator understands the importance of the alarm itself as well as the importance of the alarm in relationship to other alarms. The prioritization scheme is generally limited to the control system's capabilities along with third-party alarm management software on the system. The number of prioritization levels should be kept to a minimum to minimize operator confusion. Typical alarm priority categories include critical, high, medium, and low.

6. Incorporate Operator Actions In The Written Operating Procedure

In addition to ensuring that each alarm has a defined operator response, operator reliability must be enhanced with low workload, few false alarms, obvious display of alarm presentation, and simple operator response. Management systems must also be defined to insure a well trained operator, and include testing operator response

7. Logical Processing Techniques

In addition to grouping alarms, various suppression techniques, and artificial intelligence techniques may be piloted and then incorporated into the alarm system.

8. Alarm Presentation Techniques

The latest techniques in DCS graphic design and control system design should be considered to enhance operator effectiveness.

9. Implement the Rationalization Project

Since the rationalization may add or remove alarms; change presentation techniques, configuration parameters, it is necessary to have an implementation plan that involves the operators and operating staff and other appropriate personnel.

10. Benchmark

Once the alarm rationalization is implemented, benchmark the final system to determine the success of the rationalization effort.

For example the alarm rate should be less than 1 per 10 minutes (some plants suggest that 1 per 5 minutes is manageable). Similarly the distribution of priority alarms should be less than 5 high alarms per shift, and less than 2 alarms per hour. Journal alerts should not exceed 10 per hour.

11. Audit

A good audit process is necessary to ensure your alarm system stays manageable and in control. Online dynamic alarm monitoring systems can assist in this. If you are not following a comprehensive alarm management procedure, your alarm system may go out of control again in the future.

12. Management of Change

Adding the alarm system within the scope of the plant Management of Change program will assist in retarding uncontrolled alarm growth.

REFERENCES:


1. "Alarm Systems, a Guide to Design, Management and Procurement'", Engineering Equipment and Materials Users Assn Publication No. 191.

2. Nimmo, I. "Abnormal Situation Management," Chemical Engineering Progress, September, 1995

3. "Functional Safety of Electrical/Electronic/Programmable Electronic Safety-Related Systems," IEC 61508

4. "Functional Safety: Safety Instrumented Systems for the Process Industry Sector," IEC 61511.

5. "Better Alarm Handling", The British Health & Safety Executive, 1999

6. ISA TR91.00.02, "Criticality Classification Guideline - Annunciator Sequences and Specifications." ANSI/ISA S18.1 1992

7. Triplett, T, "Chain of Events Unlocked", Proceedings of the $56^{th}$ Instrumentation Symposium Texas A&M University, January 2004.

## The Manual for Risk Management System for Use during Plant Disasters and Specific Risk Management Methods

*Takenori Makita, Managing Director*
*Yoshimitsu Shinoda, Vice General Manager*
*Osamu Nebu, Manager*
*Haruo Sato, Chief*
*Nihonkai LNG Co. Ltd.*
*1612-32, Higashiko Seiromachi 1 Chome*
*Kita-Kanbaragun*
*Niigata 957-0195 Japan*

*Shigemi Ochiai, Ph.D, CEO*
*Jonquil Consulting Inc.*
*17-18, Honmachi 2 Chome*
*Shibuya-Ku*
*Tokyo 151-0071 Japan*

*Prepared for Presentation at AIChE Spring National Meeting*
*4th Topical Conference on Natural Gas Utilization*
*New Orleans, LA, April 2004*

## ABSTRACT

Countermeasures taken in recent years in response to accident that have occurred in high pressure gas production plants in Japan were analyzed from the perspective of risk management, revealing an increase in the number of cases where measures were delayed because of defects in disaster protection system manuals. The following are examples of such defects. (1) Disaster protection system manuals do not hypothesize evacuation flow lines according to the scale of the disaster and they are not widely distributed, so they are difficult to use in an unpredicted emergency. (2) When actual accidents occur, the manual does not act effectively, because in many cases, a training scenario is prepared and action taken in accordance with the scenario. (3) Because manuals are focused on the accident emergency liaison network system but do not clearly indicate responsibility for accidents, personnel on the scene of an accident that has occurred often contact the responsible manager and then wait for instructions concerning the action they will take. (4) Organizations including nearby organizations that will cooperate when an accident has occurred are listed in manuals, but because the specific details of the cooperation, in other words, the scope of the work each performs, are not indicated, this information is often useless.

In an effort to resolve this problem, this company that is an important energy base on the Japan Sea, has, regardless of the fact that it has never had an accident, reviewed the present disaster protection system manual from the perspective of risk management and has begun to prepare a manual for risk management system that includes an initial response system that can be applied flexibly when an accident has occurred. The specific steps taken were specifying facilities at high degree of risk of causing a disaster at the Niigata LNG Terminal and hypothesizing disasters at these facilities in order to prepare appropriate disaster protection action procedures by establishing evacuation flow lines. In particular, performing a risk evaluation of trouble records (trouble reports) at the terminal's facilities since it began operating in order to evaluate the possibility of a malfunction of each facility during present maintenance and operations in order to specify facilities that could possibly cause a disaster in order to specify facilities at high degree of risk of causing a disaster can be described as a characteristic of the preparation process. The foundations for this risk evaluation complied with the RBI standards of the ASME and API, but the data base construction method that was not included in the standards incorporated an overall risk management techniques. Our future goal is to not only enact this manual but to guarantee that it can be used for maintenance and operation management by creating this database.

We are presenting this conference with this manual for risk management system that includes an initial response system that can be applied flexibly when a disaster has occurred, assuming that it is applicable to similar terminals, and in expectation that they will use this manual and that it can contribute to disaster protection maintenance at terminals provided for public use.

## 1. Risk Management Data Base that is the foundation of the Manual for Risk Management System

### 1-1. Definition of the Risk Management Data Base

Methods of predicting plant damage caused by a disaster to include (1) an empirical rule based approach and (2) a disaster prediction evaluation approach. Examples of the first are [1] Process/System Checklist and [2] Safety Review. Examples of the second are [1] Relative ranking, [2] Preliminary Hazard Analysis, [3] "What if" analysis, [4] HAZOP, [5] FMECA, [6] FTA, [7] ETA, [8] Cause-Consequence Analysis, and [9] Human Error Analysis. With both approaches, the radiant heat, blast pressure, and gas diffusion that would be caused by a disaster affecting each type of equipment are calculated based on the degree of danger of each unit of a plant to prepare data to be used to make emergency shutdown judgments. Another is the RBM (Risk Base Maintenance) that has been stipulated by the API (American Petroleum Institute) [*1*2] and the ASME (American Society of Mechanical Engineers)[*3], etc. These methods are used as a tool to optimize a maintenance plan including inspections and repair of the constituent equipment of plants from an objective perspective. RBI (Risk Base Inspection) in the RBM defined by API-RP580 in particular was prepared by collecting the causes of losses caused by 170 serious accidents that occurred in the petroleum and petrochemical industries in the United States over three decades from 1961 to 1991. Approximately 40% of all serious losses caused by accidents were the result of mechanical damage to equipment, and approximately 80% of these 40% were the result of damage to pressure resistant parts, in other words, pipes, drums, towers, reactor vessels, tanks, pumps, heat exchangers and so on.

In order to identify equipment that is high degree of risk of causing a disaster, a working group consisting of representatives of maintenance departments, operations departments, and security departments was formed. From the above methods, the working group selected the ASME and API standards, began its work by first selecting plants that had been in operation for several decades, and then collecting and analyzing reports on trouble with all the equipment at these plants from the time they started operating till the present, as a method of supplementing ASME and API standards, because they suspected that according to the state of design and construction when they were constructed and the maintenance methods employed since that time, not only the state of gas leaks from their equipment, but the degree of deterioration and aging is likely to have varied from plant to plant. The data they obtained in this way were named the Risk Management Data. The Risk Management Data Base was prepared by supplementing the Risk Management Data with evaluation methods as such as risk categorization, degree of risk, risk avoidance, degree of risk monitor, and risk avoidance training and education, as shown in Table 1-1. The Data Base is not only applied to enact the Manual for Risk Management System, but to also perform future maintenance and operating management.

Table 1-1 The ingredients of evaluation

| R code | Evaluation items | The ingredients of evaluation | |
|---|---|---|---|
| R1 | Risk Categorization | ●Natural disaster<br>●Maintenance, deterioration, aging<br>●Interface | ●Design mistake, quality control / a guarantee of quality<br>●Human error |
| R2 | Degree of risk | ●A trouble of being simple part<br>●Influence degree to the main body machinery / facility<br>●Influence degree to machinery / facility of the main body circumference | ●Influence degree to machinery / facility of the whole plant<br>●Influence degree to regional people |
| R3 | Risk Avoidance | ●The spot restoration<br>●Part exchange<br>●A leak / a fire | ●The human body disaster<br>●Operation stop with damage |
| R4 | Degree of risk monitor | ●Nothing<br>●Once<br>●Each inspection | ●Several times<br>●Prosperity |
| R5 | Risk avoidance training and education | ●On-the-job training<br>●Education / training<br>●Study only | ●Circulation of a report<br>●Nothing |

### 1-2 Risk Management Data Base preparation step

The Working Group performed its task according to the steps shown in Table 1-2, confirmed objective empirical rules and records of various kinds regarding both maintenance and operations, and interviewed persons involved in the trouble cases to perform risk evaluations for each trouble case and overall evaluations of all types of equipment. Figure 1-1 shows the results of these risk evaluations of representative equipment.

Table 1-2 Specification of equipments at high degree of risk of causing a disaster

| STEP | Enforcement contents |
|---|---|
| STEP1 | Risk management data base depending upon trouble report |
| STEP2 | Risk evaluation on each trouble report (risk categorization, degree of risk, risk avoidance, degree of risk monitor, risk avoidance training and education) |
| STEP3 | Total risk evaluation |
| STEP4 | Risk evaluation on burst repair by leaking of LNG and NG |
| STEP5 | Risk evaluation on each equipment (possibility of trouble) |
| STEP6 | Specification of equipments at high degree of risk of causing a disaster |

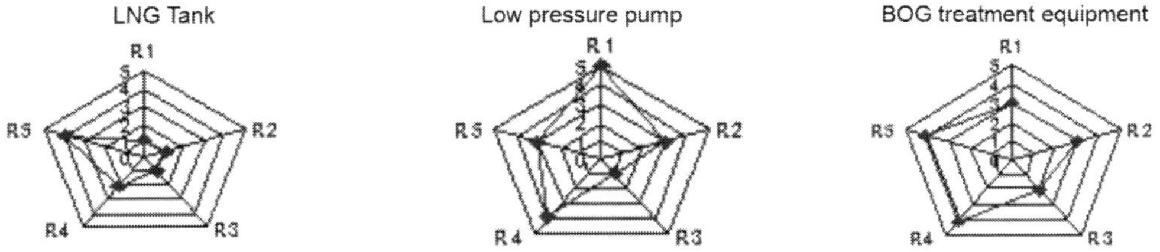


Note) R1: risk categorization R2: degree of risk R3: risk avoidance

R4: degree of risk monitor

R5: risk avoidance training and education

Figure1-1 Risk evaluation result of typical equipment

The trouble reports deal with four systems: operating systems, maintenance systems, electrical systems, and mechanical systems. Reports of operating system trouble are prepared mainly by personnel on duty at the time of the trouble. These accounted for more than half of the total of 800 trouble reports. Trouble during operation is discovered by patrols, alarms, or by checks performed during start-up and other operations. The content recorded in the trouble reports were not predicted or discovered during inspections regardless of the fact that periodic maintenance inspections were performed. The operations of maintenance systems are managed separately by the sections responsible for instrumentation systems, electrical systems, mechanical systems, and civil engineering systems etc. Instrumentation systems include instrumentation of transmitters installed in the plant and devices and computers used for control in the central control room. Many of the trouble reports deal with these instruments, because a malfunction of any one of them directly causes gas pressure fluctuations (disturbances) in gas transport systems. Because electrical systems consist of many unified standard parts, and because breaker malfunctions do not cause any loss of the reliability of electric power supply, even types of equipment that do not obstruct the transmission of gas are mentioned in many reports in relation to major equipment. Many types of equipment in mechanical systems are managed and maintained using the manufacturer's inspection records in order to replace parts or perform work according to the description of the malfunction. These trouble reports include many cases where the cause was human error. Although departments that prepare trouble reports use descriptive methods with varying degrees of detail, it can be stated that their content is adequate to evaluate the risk to all the equipment at a plant. The effective cases to be evaluated were established as shown in Table 1-3 as a result of analyzing the trouble reports to prepare the Risk Management Data Base.

Table 1-3 Rearranging result of the trouble report

| Item | The number | A summary |
|---|---|---|
| Total number | 795 | 157 number (repetition 86, facility update / removal) |
| The effective number | 637 | Risk fixed 570, risk estimate 29, unknown factor 38 |

Note) Trouble report contains a slight record as trouble memo

## 2. Risk evaluation results and the identification of highly dangerous equipment that can cause disasters

The Risk Management Data Base prepared as described in part 2 has revealed that the highly dangerous types of equipment that can cause disasters are low pressure LNG pumps and their auxiliary equipment, BOG treatment systems and their auxiliary equipment, LNG shipping systems and their auxiliary equipment, cryogenic power generation systems and their auxiliary equipment, and piping systems. Of these systems, cryogenic power generation systems and piping systems were omitted from highly dangerous equipment that can cause disasters. The former were trouble cases caused not by mechanical damage to the equipment nor for reasons related to the technical ease or difficulty of maintenance: but rather by human error. The latter were caused by exterior surface corrosion and by the way the gaskets or packing were handled, and it was determined that these could be prevented by measures based on normal inspections. Therefore, because in addition to the factors mentioned above, there are risk factors such as those in (1) and (2) below, the low pressure LNG pumps and their auxiliary equipment and BOG treatment systems and their auxiliary equipment were identified as highly dangerous systems that can cause disasters.

(1) Low pressure LNG pumps and their auxiliary equipment

a. They are inside storage tank dikes where it is difficult to see white smoke or hear the sound of leaking gas.

b. Because it is known that the air inside a storage tank dike tends to move in a vortex and cannot be uniformly analyzed, and because experiments have shown that the wind speed is strong in the winter, it is unlikely that leaks can be unfailingly detected with a gas sensor.

c. The risk evaluation has shown that its deterioration is more advanced than that of an LNG pump with the same structure, LNG gas leaks from the connections with the auxiliary equipment, and the auxiliary instrumentation is highly likely to be seriously deteriorated.

(2) BOG treatment systems and their auxiliary equipment

a. Gas leaks almost every year. There are cases where it is discovered by abnormal means such as by gas sensors used by contractors or by patrols, but not by gas detection alarms.

b. In the first floor of a BOG compressor room, the compressor pipes are complexly installed in narrow spaces and there are open spaces in the building, so it is difficult to clarify behavior during a gas leak. Because the piping outside the building is also in small spaces, it is impossible to patrol easily, with the result that it is difficult to discover leaks.

## 3. The Manual for Risk Management System and Disaster Protection Training

### 3-1 The Manual for Risk Management System

The Preparation of the Manual for Risk Management System began, as shown in Figure 3-1, with the calculation of the scale of damage accounting for the wind direction when LNG leaks from low pressure LNG pumps and their auxiliary equipment or from BOG treatment systems and their auxiliary equipment: types of equipment that are highly dangerous and can cause disasters that were identified by the Risk Management Data Base. The scale of damage caused by the diffusion of leaking LNG was calculated by obtaining the diffusion concentration based on DEGRDIS recommended by the EPA (Environmental Protection Agency, U.S.A) to evaluate whether or not damage would spread according to the presence/absence of a nearby ignition point. Two evacuation flow lines of motion were established and an emergency system command network was completed based on the results. Establishing two nearly realistic evacuation flow lines of motion permitted the in-plant disaster protection team, regional disaster protection team, and emergency rescue team to share roles and created detailed linkages between these three teams. Preliminary consultations with the regional disaster protection team and the rescue team were carried out and reflected in the manual in order to gain their understanding of the significant change in the access routes so that these teams can perform unobstructed fire extinguishing activity and rescue operations.

## 3-2 Disaster protection training

Disaster protection training is performed periodically in order that in an emergency situation, rescue activities conducted cooperatively with the region and fire extinguishing activities can be carried out smoothly. At many enterprises, scenarios are prepared in advance and fire extinguishing activities, rescue activities and guidance to safe access routes performed in flow line with these scenarios in order to provide disaster protection training that conforms to reality. This disaster protection training (on November 17, 2003) was not normal habitual disaster protection training, but rather, training performed in accordance with the Manual for Risk Management System.

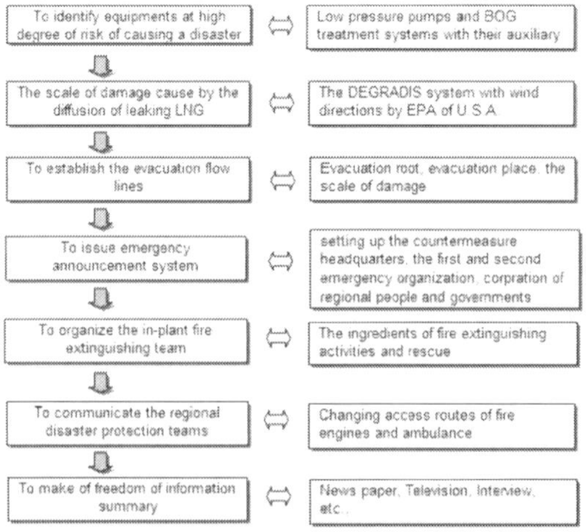


Figure 3-1 The manual for risk management system flow

The following are examples of major differences between it and conventional disaster protection training;

(1) The equipment that caused the disaster was an LNG pressurized pump.
(2) The countermeasure headquarters (consisting of the top managers) decided to issue an emergency announcement according to the wind direction at the time the disaster protection training started and the hypothetical scale of the damage.
(3) The evacuation flow lines of motion were selected according the wind direction at the time the disaster protection training started and the hypothetical scale of the damage.
(4) The location of the in-plant fire extinguishing activities team command center was determined according to the wind direction and hypothetical scale of the damage.
(5) The access routes for the regional disaster protection teams (fire engines and ambulances) were determined according to the wind direction and hypothetical scale of the damage.
(6) Divisions of roles that did not conform to work authority were prepared.
(7) Thorough rules that can be visually confirmed were created.

The switch-over from the former Disaster Protection System Manual to the Manual for Risk Management System has permitted training in conformity with actual circumstances at each plant and the reduction of the time from the start of training to its completion by about 15 minutes from the hypothetical time for conventional training. Photographs 3-1 to 3-4 show views of the disaster protection training.

Photograph 3-1 The location of in-plant fire extinguishing activities

Photograph 3-2 Aspect of fire extinguishing activities

Photograph 3-3 Aspect of rescue activities(ambulance)

Photograph 3-4 Aspect of evacuation

## 4. Conclusion

In the previous Disaster Protection System Manual, it was assumed that the highly dangerous equipment likely to cause disasters were LNG tanks or vaporizers, and the evacuation flow lines of motion were almost all planned for cases where LNG leaked from one of these types of equipment. It can be concluded that identifying the highly dangerous equipment likely to cause disasters from the Risk Management Data Base that is one of the features of the new Manual for Risk Management System has permitted calculation of a scale of damage that conforms with reality, guaranteed two evacuation flow lines of motion, and contributed to more efficient issuing of emergency announcements. And the Risk Management Data Base that has been established by carefully analyzing reports on trouble at plants from the time they started operating to perform risk evaluations has provided a large quantity of information concerning human error, the state of aging and of deterioration of equipment, and the effectiveness of education and training that are needed to guarantee the safety of plants. In the future, we want to analyze these kinds of information more quantitatively to complete the Manual for Risk Management System and to propose more effective maintenance methods.

### References

*1 American Petroleum Institute : Base resource Document On Risk-Based Inspection Preliminary Draft API Publication 581(1995.5)

*2 American Petroleum Institute : Risk-Based Inspection Base resource API Publication 581 (2000.5)

*3 ASME : Risk-Based Inspection-Development of Guidelines Vol.3 Fossil-Fired Electric Power Generating Station Application CRTD-Vol.20-3 (1994)

## Quantitative Risk Analysis of Single & Full Containment LNG Tanks

*Tor Egil Nielsen, President,*
*Scandpower Risk Management Inc, Houston, Texas, USA*
ten@scandpower.com

*Devendra Agrawal, LNG Technology Consultant*
Agrawal@houston.rr.com



*Prepared for presentation at:*
*AIChE Spring National Meeting*
*April 25 - 29, 2004, Hyatt Regency, New Orleans, Louisiana*
*Fourth Natural Gas Utilization Topical Conference*

Unpublished

## ABSTRACT

This paper presents results from a quantitative, comparative risk analysis between single containment and full containment LNG storage tanks for a generic LNG production facility in a remote area. The study focused on third party risk and equipment damage risk.

The historical experience on which to base a quantitative risk analysis for LNG tanks is scarce. The number of LNG tanks worldwide and the number of tank years of operational experience is small, compared to other types of tanks. The number of recorded accidents with release of LNG is even smaller, but an estimate of tank failure frequency (failures per tank year) can be obtained based on the experience with cryogenic storage tanks worldwide. The site specific conditions and the tank designs will determine which accident root causes will contribute more to the total frequency. For LNG tanks alone the statistical experience is too scarce to derive a valid cause distribution, but the more extensive experience with other cryogenic storage tanks offers this possibility, and this was utilized in the present study.

The quantitative results of the risk analysis in terms of risk to 3$^{rd}$ party showed that both concepts fulfill commonly applied risk acceptance criteria. The risk to employees was not included in this analysis, however, risk from the tanks will probably constitute a smaller part of the overall risk and, hence, not be as determining for the risk picture. In case of LNG plants as they are usually at remote locations - the 3$^{rd}$ party risk exposure is the same for the two tank concepts.

For equipment damage and operation interruptions, the conclusion on what is the best design option should be based on a Life Cycle Cost analysis, taking into account the costs of the established damage scenarios.

Abbreviations used in the paper:

| | |
|---|---|
| QRA | Quantitative Risk Analysis |
| HRA | Hazards Risk Analysis |

| | |
|---|---|
| LCC | Life Cycle Cost |
| 3rd party | Public outside the property line of the LNG Plant |
| LNG | Liquefied Natural Gas |
| LEL | Lower Explosion Limit; 50 000 ppm (5 by vol% in air) for methane |
| UEL | Upper Explosion Limit; 150 000 ppm (15 by vol% in air) for methane |
| PGA | Peak Ground Acceleration |
| OBE | Operating Basis Earthquake |
| SSE | Safe Shutdown Earthquake |

## INTRODUCTION

Single containment and full containment LNG storage tanks have different designs and, hence, different frequencies for releases of LNG under various scenarios. The main difference in the designs is that the single containment tank's secondary containment is a relatively wide bunding, while the full containment tank has a concrete tank outside the primary containment. This means that upon failure of the primary containment, the LNG will have a much larger area to evaporate and burn from for the single containment tank.

In single containment tanks (Fig.1), neither the LNG spilled is contained within the tank walls nor is the vapor generated from the spill. The LNG spilled from the break is contained in a low bund wall constructed around the tank. In full containment tanks (Fig.2), the LNG spill is contained within the outer concrete wall and the vapors generated from the spill are retained within the tank top. *In this study, an enhancement was made to the single containment design by replacing the bottom 8 – 10 feet of the outer carbon steel wall with 9% Ni steel.*

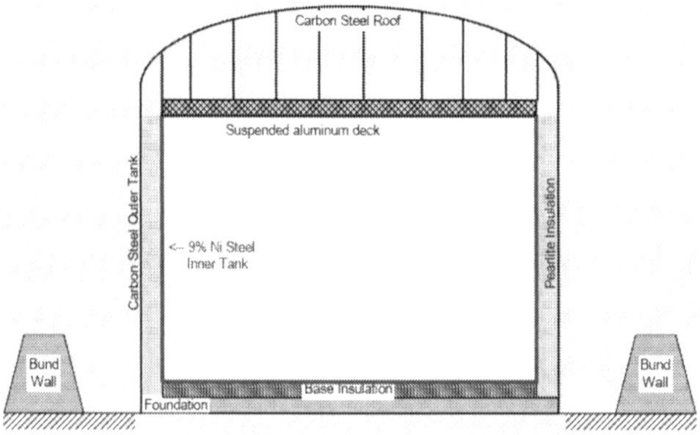


**Figure 1: Single Containment Type LNG Tank**

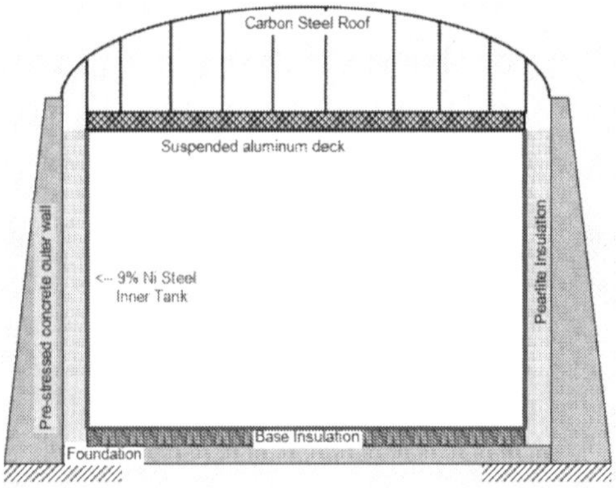


**Figure 2: Full Containment Type LNG Tank**

A Risk Analysis of single containment vs full containment LNG storage tanks has been performed to determine the risk differences to plant equipment and 3rd Party (public) outside the plant. Risk to plant equipment is defined as heat flux levels or explosion overpressures sufficient to cause permanent damage to equipment. 3rd Party Risk is defined as heat flux levels or vapor cloud explosion overpressures high enough to cause possible fatalities outside the plant property line.

## ESTIMATION OF ACCIDENT FREQUENCIES

### (a) Release from Piping and Valves

After reviewing the frequencies for releases from the piping connected to and from the tanks, it was concluded that the contribution to the risk will be marginal due to low leak frequencies, even if 1 km of piping is included. In the vicinity of the tanks (inside the bunding for the single containment tank), leaks from this piping will be collected in the spill sump as leaks of durations longer than the design spill of 10 minutes are not foreseen. In this aspect, the consequences and the frequencies are equal for the two tank concepts (leaks from the piping along the tank walls can have different consequences, however, the frequencies are very low).

**(b) Release of LNG and/ or Methane gas from Storage Tank**

Incidents of accidental release of LNG from large LNG storage tanks are extremely rare. An analysis of historical experience with low pressure, flat-bottomed, metal tanks storing refrigerated and cryogenic liquids has been carried out by the Health and Safety Executive (HSE) in the UK[1]. For LNG tanks the number of tanks included in the study was estimated as 437, with a total operating experience of 6012 tank years (13.8 years per tank, on the average) for the period 1954 - 1984. Only one "significant" release was reported. (310 $m^3$ vapor due to roll-over, at La Spezia/Italy, in 1971). This results in a failure frequency of

$$1/6012 = 1.7\times10^{-4} \text{ per tank year}$$

with a 95 % confidence interval from $1.7\times10^{-5}$ to $9.5\times10^{-4}$ per tank year.

Looking at the causes for all types of refrigerated and/or cryogenic products (LPG, NH3, LNG, ethylene, LOX, LIN) the cause distribution shown in Table 1 is obtained. It should be noted that the one incident at La Spezia resulted in vapor release from relief valves and vents, without ignition. Even though the tank was over-pressurized it did not fail and no nearby equipment was damaged, and no people were injured. To provide additional background information about LNG accidents, other information sources[2,3,4,5,6] were also studied. Incidents found are, in general, from the early days when certain problems such as low-temperature brittle fracture and roll over phenomena were not yet understood. Description of cryogenic tank failures can be found in references 2, 3, 4, 5 & 6.

**Table 1 : Cause distribution for significant releases**

| Cause | Number of Occurrences | % Occurrence |
|---|---|---|
| Over-pressurization | 7 | 41 |
| Mechanical failure | 5 | 29 |
| Overfill | 1 | 6 |
| Roll-over | 1 | 6 |
| Fire | 1 | 6 |
| Internal Explosion | 1 | 6 |

In a report[7] published in 2001, a total of 47 incidents in which secondary containment played a role were listed. None of the reported incidents involved LNG, however. For the present study, a search was conducted in the MHIDAS Data Base[8] administered by AEA (Atomic Energy Agency) in the UK. The search gave 9 results, 6 on natural gas, 1 on methane, and 2 on LNG. The two incidents with LNG were the Cleveland/Ohio accident in 1944 and an accident on Canvey Island in 1965 where a spill occurred during maintenance. The spill ignited and burned for 15 minutes, one man was badly burned. One of the 6 natural gas accidents was the Staten Island accident in 1973, which also could have been classified as an LNG accident.

Another source used by Scandpower in LNG risk analysis reports two additional incidents for storage tanks between 1965 and 1998. Two of these resulted in significant release of LNG (10,000 kg) or vapor (466,000 kg). In both cases LNG was pumped from ship to shore. The first incident was due to overfilling, the second due to roll-over. None of the releases were ignited.

| Table 2 :Tank failure frequencies from different causes | |
|---|---|
| Cause | Frequency (pr. tank year) |
| Over-pressurization | $1.2 \cdot 10^{-4}$ |
| Mechanical failure | $8.8 \cdot 10^{-5}$ |
| Overfill | $1.8 \cdot 10^{-5}$ |
| Roll-over | $1.8 \cdot 10^{-5}$ |
| Fire | $1.8 \cdot 10^{-5}$ |
| Internal Explosion | $1.8 \cdot 10^{-5}$ |

Including the two incidents, there are three incidents. The number of tank years for the period 1954 to 1998 is estimated at 10,000 (extrapolating from 6012 tank years for the 1954 to 1984 period and assuming a slightly higher number of new tanks than of tanks being taken out of service). With this one obtains an incident rate of:

$3/10000 = 3 \times 10^{-4}$ per tank year

It is assumed that this frequency is a conservative estimate for significant release from the primary containment of an LNG tank, and that the cause distribution derived for all types of cryogenic tanks and shown in Table 1 is approximately correct also for LNG tanks. This will result in the base event frequencies as shown in Table 2.

**(c) Earthquakes**

Regardless of the tank concept chosen, the tanks will be designed according to the earthquake design criteria in NFPA59A (2001). Two different criteria will be met:

(i) The container and its impounding system shall remain operable after an OBE (Operating Basis Earthquake)

(ii) No loss of containment of the primary container, safe shut down possible after an SSE (Safe shutdown Earthquake)

In the analysis it was assumed that both tank concepts fulfill NFPA59A, which means that the probability for exceeding the SSE - and losing containment - is the same, $1/4975 = 2 \times 10^{-4}$ per year. But exceeding the SSE criterion is not the same as a "significant" release. There is conservatism in the seismic analysis and load factors will be used when assessing the SSE design load. Similarly, safety factors will be applied when assessing the strength of the tank. From this, it is assumed that it will require an earthquake significantly stronger than the one with the SSE return period to cause an actual significant release. In this study, it is used that such an event has a frequency a factor 5 lower than the SSE frequency, i.e. $4 \times 10^{-5}$ per year. This event is assumed to be strong enough to give catastrophic failure of the single containment tank and the inner and outer containment of the full containment tank.

### (d) Air Traffic

It is assumed that, there is an airport in the vicinity of the LNG tank farm. The risk for an aircraft or a lost jet engine crashing into the tanks can be estimated from statistical experience. Concerning the probability of plane crash, there are various sources of information available. The Norwegian Air Traffic Control Authorities (Luftfartsverket) conducted an analysis in 1990 of the possibility of a plane crash over Oslo in connection with the establishment of new landing procedures[9].

The conditional probability to hit a target on the ground (given crash) depends on the distance from the runway, the type of aircraft, the size of the target. Taking this into account, a hit frequency of $4 \times 10^{-8}$ was calculated.

It can be questioned if the statistics used in the calculation above is applicable for all airports. However, this contribution is so small that even by increasing the frequencies by a factor 10 could not give significant contributions to the risk. From this, the frequency of $4 \times 10^{-8}$ per year is applied further.

## FAULT TREES

The different causes will most probably give releases of different magnitude and character. For the purpose of this analysis, it has been found appropriate to apply three release classes. These are:

Class 1: The release is assumed to come as gas only, i.e. no liquid LNG will be a part of the release. It is assumed that all these scenarios are released from a relief hatch or from the upper part of the tank.

Class 2: Liquid LNG may leak out from the primary containment, but will be stopped from reaching the surroundings by secondary containment. For the single containment tank, the secondary containment will be the lower, low-temperature resistant part of the outer "tank"[1], while for the full containment tank it will be the concrete walls. For the ignited scenarios, it is assumed that the roof is gone, since most of the ignitions are assumed to occur internally in the tank.

Class 3: Liquid LNG may leak out from the primary and secondary containment and will reach the surroundings.

The various causes are best visualized and top event frequencies calculated by means of fault trees. The two different tank concepts - single containment with bunding and full containment - will have an effect on the numerical values of the base events. Besides this, the two different types of secondary containment will have a marked effect on the consequences of the accidents. In the study, six different fault trees were established, one for each release class and for each tank concept. The fault tree structure is common for both tank concepts and is presented in Figure 3.

---

[1] *It is realised that the outer carbon steel tank of a single containment tank is not a containment in the stringent sense of the word. For this type of tank the secondary containment is the bunding. But for some scenarios it is believed that the low-temperature resistant lower part of the outer tank will stop LNG from reaching the surroundings. For the purpose of classifying the releases and the following analysis we will call the lower part of the outer tank secondary containment.*

Figure 3: Fault tree structure

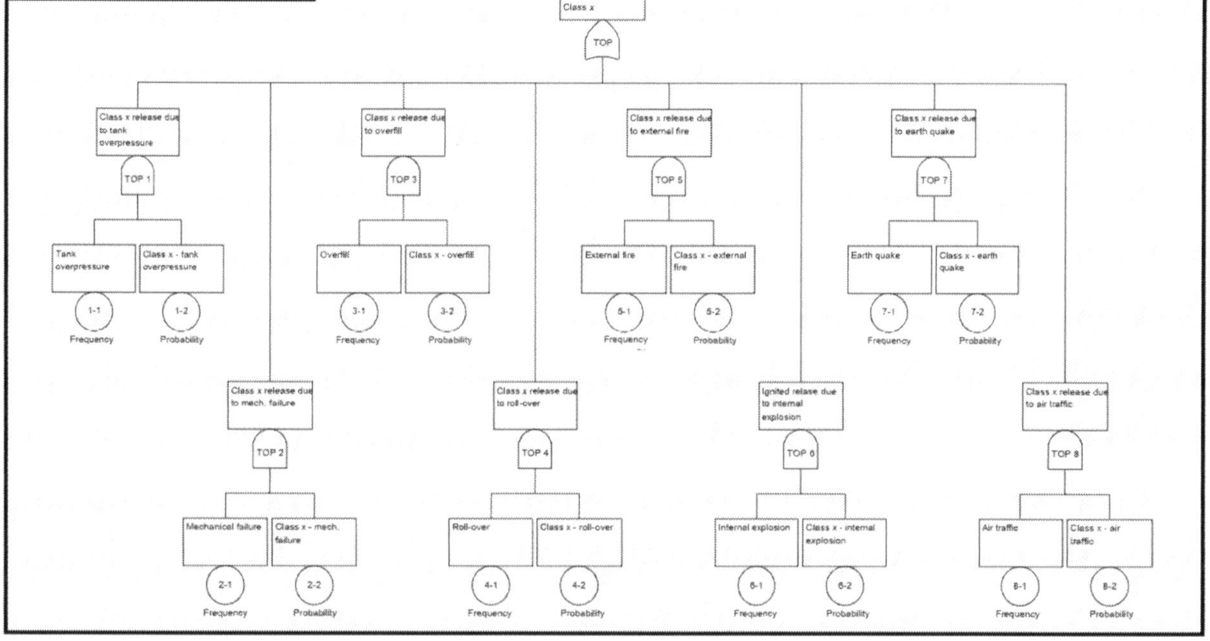


In the study, an evaluation of the different release causes was performed and conditional probabilities for Class 1, 2 or 3 were assessed based on potential failure mechanism of the tanks and effects of the secondary containment. The lower cold resistant part of the "outer layer" of the single containment tank was given a small probability of containing some of the releases from the primary containment.

Based on the evaluation of release causes, frequencies for the release classes as presented in Table 3 below were established. The rows in the table dealing with the figures not including earthquakes and air traffic should, to a large degree, match the statistical material the failure causes are based on. As it is assumed that most of the statistic is based on single containment tanks, the comparison is performed for them only.

**Table 3: Comparison with accident statistics**

| | Total Freq. (/year) | Gas leak from outer containment | | LNG leak into annulus | | LNG leak to surroundings | |
|---|---|---|---|---|---|---|---|
| | | % of total | Freq. | % of total | Freq. | % of total | Freq. |
| **Single Containment Tank** | | | | | | | |
| All Inclusive Figures | 3.1E-04 | 41.2 | 1.3E-04 | 31.7 | 9.7E-05 | 27.2 | 8.3E-05 |
| Figures When Disregarding Earthquake & Air traffic | 2.7E-04 | **47.4** | 1.3E-04 | **36.4** | 9.7E-05 | **16.2** | 4.3E-05 |
| **Full Containment Tank** | | | | | | | |
| All Inclusive Figures | 2.9E-04 | 44.3 | 1.3E-04 | 41.9 | 1.2E-04 | 13.8 | 4.0E-05 |
| Figures When Disregarding Earthquake & Air traffic | 2.5E-04 | 51.4 | 1.3E-04 | 48.6 | 1.2E-04 | **0.0** | 0.0E+00 |
| **Percent Distribution for All Cryogenic Tanks (From Ref. /1/)** | | | | | | | |
| Release Classification | | Significant Vapor Release | | Significant Liquid Release | | Catastrophic Tank Failure | |
| Reported Percentage | | **42** | | **46** | | **11.6** | |

It is important to note that the release classes as used in this study do not fully match the release classes as given in Ref. /1/, see the headings in Table 3. The most important figures in the table above are given in bold. As can be seen, the percentages of large (catastrophic) LNG releases to the surroundings match quite well (16.2% vs. 11.6%). Also, the percentages of gas/vapor releases match quite well (47.4% vs. 42%).

One problem with the figure of 46% "Significant Liquid Release" as established in Reference /1/ is to know the size of a release being classified as "Significant" since the reference does not include information about this. For the single containment tank design with the low-temperature resistant lower part, some of the "significant liquid releases" might not have reached the surroundings. Another issue is if the "significant liquid release" is large enough to fill up the lower part of the outer tank and the spill basin, which can contain 833 $m^3$ LNG. Even if it is impossible to assess the size of a significant liquid release, the figures in Table 3 do not seem unreasonable. In this connection, it should be noted that a fire from the roof of the LNG tank would, for 5 and 15 kW heat loads, have larger isopleth distances than for a fire in the spill basin. From this, it is difficult to determine exactly how the difference in defining the leak classes affects the results. When disregarding earthquakes and air traffic, the full containment tank has no LNG releases to the surroundings.

Ignition probabilities for the different release scenarios were assessed. These were based on how big the resulting gas cloud could be, the energy involved in the failure mechanism (e.g. external fires, mechanical failure) and the presence of ignition sources, i.e. the location of the release. The ignition probabilities used are estimates in line with historical experience published in the literature[10,11,12]. Table 4 summarizes the frequencies of vapor clouds and early and delayed ignitions in the vicinity of the plant area.

**Table 4 : Frequencies for accidental events near the facility**

| Tank Type | Gas leak from outer containment | | | LNG leak into annulus | | | LNG leak to surroundings | | |
|---|---|---|---|---|---|---|---|---|---|
| | $f_{vapor}$ | $f_{FireEarly}$ | $f_{FireDel}$ | $f_{vapor}$ | $f_{FireEarly}$ | $f_{FireDel}$ | $f_{vapor}$ | $f_{FireEarly}$ | $f_{FireDel}$ |
| Single Containment | 1.1E-04 | 1.9E-05 | | 8.9E-05 | 8.3E-06 | | 3.7E-05 | 3.5E-05 | 1.1E-05 |
| Full Containment | 1.2E-04 | 4.8E-06 | | 9.7E-05 | 2.5E-05 | | 2.0E-05 | 1.4E-05 | 6.0E-06 |

Figures in table 4 are based on releases close to the tanks and exposing the process facilities. An additional ignition probability was applied for the un-ignited scenarios having a long hazard range and hitting the public outside the plant area, in this case a village community approximately 3 km away. Since the village area is not "classified" with respect to ignition sources, a high ignition probability of 0.8 was applied for this scenario.

## CONSEQUENCE CALCULATIONS

For the study, the computer code TRACE from SAFER systems[13] has been used. In case of the catastrophic failure of both inner and outer tank of a full containment tank, there will be no bunding restricting the flow of the escaping liquid LNG. The flow may follow the terrain and the features near the tanks such as roads, drainage channels, slopes etc. It is uncertain how large an area may be covered by LNG. From this, calculations for two different areas were performed.

Table 5 presents the modeled scenarios and the chosen accidental load endpoints with corresponding hazard ranges. There can be differences in the consequences for releases from the piping dependent on if they are in relation to a single or full containment tank. The reason is that the full containment tank has higher resistance to impinging jet fires from holes in the piping coming down from the top of the tank.

For the single containment tanks, such jet fires could escalate to putting the tanks on fire. For the full containment tanks, the concrete walls are assumed to resist such jet fires. Even if the single containment tanks are more vulnerable to impingement from jet fires, a coarse estimate indicates that the probability for such an event is insignificant compared to tank failures due to other reasons (the yearly frequency of such a jet fire is in the order of $2\times10^{-8}$) The main reasons for this are the low failure frequency of piping only and the low ignition probability for relatively small releases along the tank walls.

## EVENT TREES

In addition to the established ignition probabilities and the hazard ranges, the wind statistics plays an important role in assessing the risk for the community. Due to a very prominent wind direction away from the community, the probability for wind towards the community, is set to 0.05.

Figure 4 represents the event tree established for this study.

**Table 5 : Hazard distances at ground level, V= Vapor, E= Explosion, F= Fire**

| ID | Tank Design | Scenario description | Hazard Distance, |
|---|---|---|---|
| V1 | Both | LNG release followed by vapor dispersion<br>- leakage: 10 minutes x 5000 m3 / hr (=632 kg/s)<br>- basin: 10m x 10m x 8.3m = 833 $m^3$ | ½ LEL: 272 m |
| V2 | Single | Sudden release of tank contents (136,000 $m^3$) into bunding (165 x 165 m) | ½ LEL: 3740 m |
| V3 | Full | Sudden release of tank contents (136,000 $m^3$) to ground, 100.000 $m^2$ | ½ LEL: 6021 m |
| V4 | Full | Sudden release of tank contents (136,000 $m^3$) to ground, 200.000 $m^2$ | ½ LEL: 7263 m |
| V5 | Full | Sudden release of tank contents (136,000 $m^3$) into outer container (86 m Ø) | ½ LEL: 817 m |
| V6 | Both | Major release through PSV or relief panel (127 kg/s) | ½ LEL: 531 m |
| V7 | Both | Methane jet dispersion from piping<br>- 285 $mm^2$ leak cross section / 3 barg internal pressure<br>- rate: 0.22 kg/s | ½ LEL: 7,4 m |
| V8 | Both | LNG jet dispersion from piping<br>- 285 $mm^2$ leak cross section / 3 barg internal pressure<br>- rate: 2.29 kg/s | ½ LEL: 18,6 m |

| Table 5 : Hazard distances at ground level, V= Vapor, E= Explosion, F= Fire | | | |
|---|---|---|---|
| ID | Tank Design | Scenario description | Hazard Distance, |
| E1 | Both | Explosion of "V1-gas cloud" (see V1)<br>- Multi-Energy | 0.01 bar: 380 m<br>0.1 bar: 51 m<br>0.3 bar: 24 m |
| F1 | Single | Pool Fire:<br>- full tank contents released into bund : 136,000 $m^3$ LNG<br>- bund: 165m x 165m x 5.5m | 5 kW: 804 m<br>15 kW: 418 m<br>32 kW: 251 m |
| F2 | Full | Pool Fire:<br>- collapse of carbon steel tank roof<br>- size of fire equal to outer tank diameter<br>- outer diameter = flame diameter: 86 m<br>- tank height = elevation of flame base = 30.5 m | 5 kW: 253 m<br>15 kW: 119 m<br>32 kW: - |
| F3 | Full | Pool fire 100,000 $m^2$ | 5 kW: 1202 m<br>15 kW: 688 m<br>32 kW: 442 m |
| F4 | Full | Pool fire 200,000 $m^2$ | 5 kW: 1478 m<br>15 kW: 880 m<br>32 kW: 588 m |
| F5 | Both | Pool fire:<br>- ignited 10-minute release into 10 x 10 m sump | 5 kW: 58 m<br>15 kW: 30 m<br>32 kW: 18 m |
| F6 | Both | Ignited major release through PSV or relief panel, see V6 (127 kg/s), heat load at roof level | 5 kW: 148 m<br>15 kW: 73 m<br>32 kW: 26 m |

| Table 5 : Hazard distances at ground level, V= Vapor, E= Explosion, F= Fire | | | |
|---|---|---|---|
| ID | Tank Design | Scenario description | Hazard Distance, |
| F7 | Both | Methane jet fire from piping (see V7) | 5 kW: 12,8 m<br>15 kW: 11,0 m<br>32 kW: 10,0 m |
| F8 | Both | LNG jet fire from piping (see V8) | 5 kW: 33,5 m<br>15 kW: 28,7 m<br>32 kW: 26,2 m |

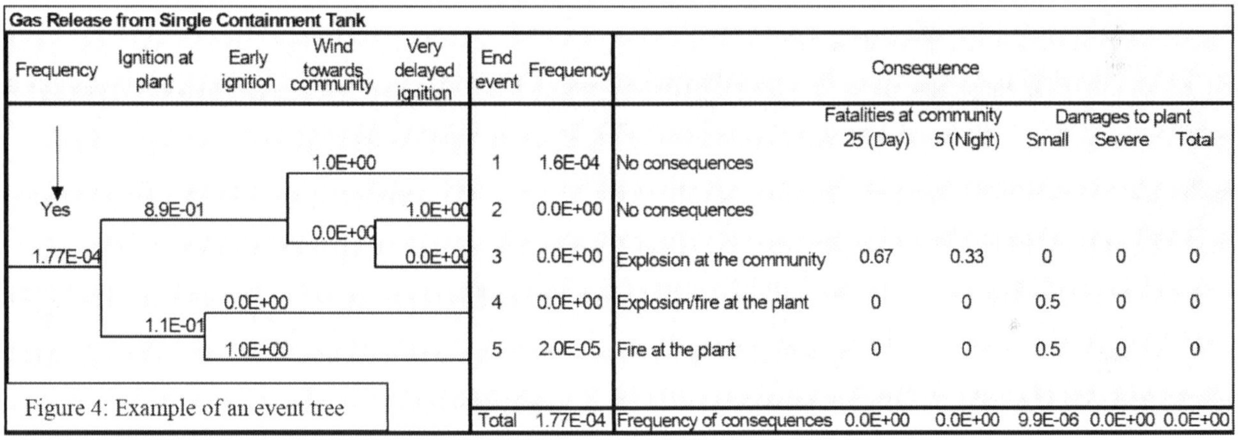

Figure 4: Example of an event tree

Third-party people in this study are considered people at an assumed village community. In this study, it is assumed that 25 people may be exposed during daytime and 5 people during the night as most people will be indoors during the night. Equipment damages has been grouped in three classes; small, severe and total damages and will be the damages coming in addition to the damages to the tank causing the accidental loads. The "small" class is assumed to be covering damages to instruments, cabling, etc., not causing structural damages to the equipment like loss of integrity and loss of support structures. The "severe" class is assumed to give structural damages to a significant part of the liquefaction plant (in the order of 5 – 60% of the plant). Anything giving structural damages to more than approximately 60% of the plant is classified as "total" damages.

| Table 6 : Conditional Probabilities for the Different Consequence Groups | | | | | |
|---|---|---|---|---|---|
| Brief scenario description | Probable Fatalities | | Equipment damage | | |
| | 25 (day) | 5 (night) | Small | Severe | Total |
| Possible Explosion at community | 1 | 1 | | | |
| Fire from gas leak | | | 1 | | |
| Fire from LNG to annulus | | | 1 | | |
| Fire from LNG to surroundings, single | | | | 0.8 | 0.2 |
| Fire from LNG to surroundings, full | | | | | 1 |
| Explosion from LNG vapor to surroundings, single | | | | 0.5 | 0.5 |
| Explosion from LNG vapor to surroundings, full | | | | | 1 |

For equipment damage, explosion overpressures and heat loads may contribute to the damages. It can in general be assumed that a fire may follow an explosion and, hence, the consequences in the explosion scenario may be equal or more severe than in the fire only scenario (from the same release scenario).

Based on the hazard ranges of the accidental loads and the physical impacts, the probabilities as presented in Table 6 were used for the different consequence classes.

## ACCEPTANCE CRITERIA, 3RD PARTY RISK

In Figure 5 below proposed acceptance criteria for the 3rd party risk are presented. Many operators worldwide apply curves equal to these. For members of the public the risk acceptance is also based on statistical experience, but not from working life but general risk exposure due to normal activities such as driving a car, taking an airplane, staying at home (home accidents), etc. Typical values from daily life are:

- car accident: 1×10-4 per year
- fire at home: 4×10-5 per year
- electrical shock: 6×10-6 per year
- struck by lightning: 5×10-7 per year

On this background a value of $1\times10^{-5}$ per year is normally used as acceptance criterion for the most exposed member of the public. Normally, the most exposed member of the public is the person (or persons) living closest to the plant, or a person passing the plant close by daily on his/her way to/from work (commuter). Obviously, all other members of the public will be exposed less than the most exposed person. So, if the criterion is fulfilled for the most exposed person, it is also fulfilled for all other members of the public.

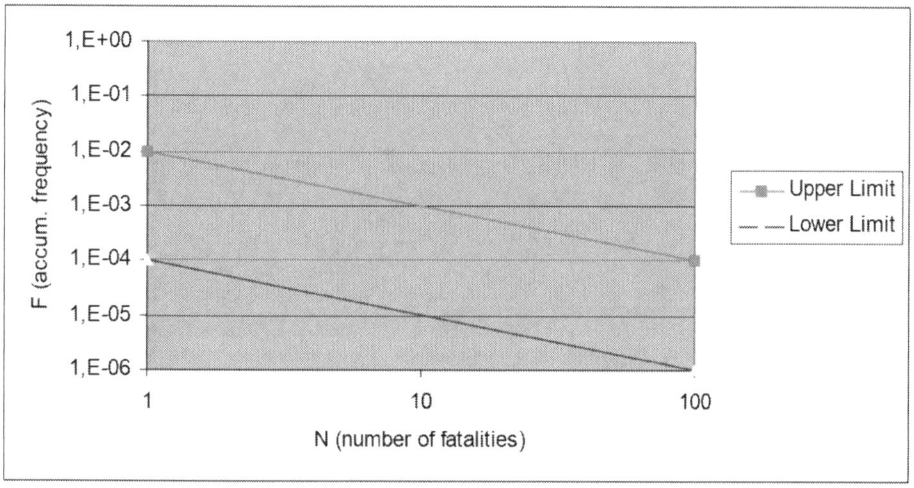


**Figure 5: Applied Acceptance Criteria for 3rd Party Risk**

The region above the upper curve is considered unacceptable, the region below the lower curve as acceptable, and the region in between the two curves is the ALARP-region (as low as reasonable possible). Note that F is the cumulative frequency.

## RESULTS

It should be noticed that the figures in Table 7 below are the results from two tanks, while all the previous tables have been dealing with figures for one tank only. This is done in order to have the correct figures to compare with the proposed acceptance criteria for 3rd party risk.

**Table 7 : Annual Frequencies Within the Different Consequence Classes**

| Scenario | Probable Fatalities at community | | Probable Damages to plant | | | |
|---|---|---|---|---|---|---|
| | Daytime (25 people) | Nighttime (5 people) | Small | Severe | Total | Sum, damages |
| | | | | | | |
| Single Containment Tank | | | | | | |
| Class 1, Gas leak | | | 1.9E-05 | | | 1.9E-05 |
| Class 2, LNG to annulus | | | 8.3E-06 | | | 8.3E-06 |
| Class 3, LNG to surroundings | 2.0E-06 | 9.9E-07 | | 6.8E-05 | 2.5E-05 | 9.3E-05 |
| **Sum** | **2.0E-06** | **9.9E-07** | **2.7E-05** | **6.8E-05** | **2.5E-05** | **1.2E-04** |
| | | | | | | |
| Full Containment Tank | | | | | | |
| Class 1, Gas leak | | | 4.8E-06 | | | 4.8E-06 |
| Class 2, LNG to annulus | | | 2.5E-05 | | | 2.5E-05 |
| Class 3, LNG to surroundings | 1.1E-06 | 5.3E-07 | | | 4.0E-05 | 4.0E-05 |
| **Sum** | **1.1E-06** | **5.3E-07** | **3.0E-05** | | **4.0E-05** | **7.0E-05** |
| | | | | | | |
| **Ratio Single/Full** | **1.8** | **1.8** | **0.9** | **n. a.** | **0.6** | **1.7** |

## CONCLUSIONS

### 3rd Party Risk

In Table 7 it can be seen that the frequencies for 3rd party risk are 1.8 times higher for the single containment tank compared to the full containment tank, however, both concepts are well below the ALARP region. For 3rd party risk, the relatively long distance between the facility and the community may give a lead time between the gas release and exposure of the community (approximately 15 minutes, dependent on the wind speed). For a catastrophic failure of the single containment tank, the wind direction that gives exposure of the community may also in most cases lead the gas through the liquefaction plant. From this, or from dedicated gas detectors or other means, it should be possible to detect this scenario very fast. As the risk at the community is based on people being outdoor during a probable explosion, emergency preparedness plans to get people indoor can be established, which may significantly reduce the 3rd party risk. In a severe earthquake scenario, this measure is not assumed to be efficient due to the chaotic situation that may occur and because of the possibility that the buildings at the community have collapsed also. An efficient emergency preparedness plan for this scenario can bring the 3rd party risk for the single containment tank close to that for the full containment tank. With a relatively simple emergency preparedness plan, the 3rd party risk for the single containment tank can be brought down close to the risk level for the full containment tank.

### Equipment Damages

The QRA shows that for equipment damage, the full containment design has higher frequencies in the "small" and "total" damage classes, while the single containment design has a higher frequency in the "severe" damage class. The optimal design will have to be chosen based on Life Cycle Cost analysis, taking into account the cost of repairs and lost production. Overall, the single containment tank design may damage the plant about 1.7 times more frequent than the full containment tank design. The reason that the single containment tank has a lower frequency in the "total" damage class is that there is a low probability that fires from the bunding may give damages in this class, while the largest fires from the full containment tank to a large degree can engulf a major part of the liquefaction plant. The full containment tank's higher frequency in the "small" damage class is because of the secondary containment, which may give smaller fire scenarios (from the roof of the outer tank) compared to the single containment tank's fires from the bunding area.

A firm conclusion regarding equipment damage should be based on LCC analysis, which was outside the scope of this study. Briefly, such an analysis should compare the overall cost for the different design options, including the initial investments and the net present value cost of possible future accidents, including the cost of the repair and the lost production. More information on LCC analysis specifically considering accidental losses can be found in NORSOK Z013[14].


**References:**

[1] J. N. Edmondson, P. D. Michell: An estimate of operating experience over the period 1954 - 1984 with low pressure, flat-bottomed, metal storage tanks storing refrigerated or cryogenic liquids and of the associated historical incident frequencies, HSE/SRD/41068/WP1, October 1988.

[2] F.P. Lees: "Loss Prevention in the Process Industries", Butterworth, 1980.

[3] US General Accounting Office: "Liquefied Energy Gases Safety", Report to the Congress, 31 July 1978.

[4] D. Reidel Publishing Company: "Risk Analysis of Six Potentially Hazardous Industrial Objects in the Rijnmond Area", a Pilot Study, 1982.

[5] Paten & Partners: "Liquefied Gas Ship Safety", Analysis of the Record 1964-1979, plus Supplement, 1980.

[6] G. Drogaris: "Major Accident Reporting System: Lessons learned from Accident notified", Commission of the European Communities, In stitute for Systems Engineering and Informatics - CDICR, 1993.

[7] WS Atkins Consultants Ltd.: Effects of secondary containment on source term modelling, contract research report 324/2001 (for the HSE).

[8] E-mail from Andrew Lelland, AEA Technology Environment, to Robert Dayton, Scandpower Houston, on 03.04.2003.

[9] Luftfartsverket: Konsekvensanalyse ILS24 Fornebu, 1990.

[10] WS Atkins Safety & Reliability, for the HSE: Ignition probability of flammable gases, Contract Research Report 146/1997.

[11] WS Atkins Consultants Ltd, for the HSE: A model for the ignition probability of flammable gases, Phase 2, Contract Research Report 203/1998.

[12] Frank P. Lees, Loss prevention in the process industries, 2nd Ed. 1996, Chpt. 16.10, Ignition Models.

[13] TRACE, SAFER Systems, USA (www.safersystem.com).

[14] NORSOK Z013, Annex E, Guidelines for Cost Benefit Analysis.

# DEVELOPMENT AND CONSTRUCTION OF A PILOT LINED CAVERN FOR LNG UNDERGROUND STORAGE

*Emmanuel Chanfreau, SN Technigaz*
*Project Engineering Manager, LNG Business*
*Energies*
*1/7 av San Fernando*
*78884 Saint Quentin en Yvelines Cedex*
*emmanuel.chanfreau@saipem-sa.com*
*Tel : 33 +1 61 37 85 98*

*Eric Amantini, Geostock*
*Head of Department Mined Caverns*
*7 rue E et A Peugeot*
*92563 RUEIL MALMAISON CEDEX*
*FRANCE*
*eam@geostock.fr*
*Tel : 33 +1 47 08 73 50*

*Key words : LNG, underground, storage, cavern, membrane, mined*



*Prepared for Presentation at AIChE Spring National Meeting*
*4th Topical Conference on Natural Gas Utilization*
*New Orleans, LA, April 2004*

*This paper is originally being presented at the 14th International Conference on LNG, Doha, QATAR, March 21-24, 2004 and is published in the proceedings of the conference. A poster session has also been held during this conference.*

## ABSTRACT

A joint development by Géostock / SN Technigaz from France has been made in the 1990's in order to provide a safe, reliable and cost-effective solution for LNG underground storage. Instead of storing pressurized methane directly in underground cavities, the concept is to store LNG at –160°C in hard rock mined cavern equipped with a liner made of membrane containment system. After a first phase dedicated to conceptual design and testing program, the validity of this design was demonstrated.

The concept of the underground storage of LNG in lined rock cavern is composed of:

- A containment system to ensure LNG containment and rock protection against thermal shock.
- A drainage system used during the first months of the storage operation, before the surrounding rock is frozen, to drain water around the cavern and prevent hydrostatic pressure acting against the containment system.

A cryogenic pilot cavern was designed and constructed in 2003 at Daejon, South Korea, in partnership with SK Engineering & Construction. The pilot cavern operation started successfully at the beginning of 2004 and the first results from this test plant shall be available by middle of 2004. The purpose of this pilot plant is to validate this new technology for underground storage at –160°C of LNG in lined, mined rock caverns.

## INTRODUCTION

Some attempts have been made to store LNG underground in unlined containment but were not successful. Underground storage failures were due to thermal stresses generating cracks in the host soil and thermal cracks contributed to induce gas leakage and to an increase in heat flux between LNG and the ground. Facilities were then decommissioned due to their excess boil-off rate.

The only way to prevent hard rock from cracking at –162°C would be to locate the unlined storage cavern deep enough for the overburden stresses to counterbalance the tensile stresses caused by the cooling. The necessary depth varies with rock type from 500 to 1000m which would make this concept very expensive.

On the other hand, in-ground insulated concrete tanks have been developed as reliable, though expensive, LNG storage techniques.

To provide a safe and cost-effective solution, Geostock and SN Technigaz both from France have worked for several years in developing a new concept storing LNG in a lined hard rock cavern. It consists in protecting the host rock against the extreme low temperature and in providing a liquid and gas tight liner. Moreover, the moderated and controlled frost development in the surrounded rock mass contributes to create a ice ring, acting as a secondary barrier.

After a first phase dedicated to theoretical analysis and to design, a validation program including several tests and experiments were performed in the late 1990s. Effects of ice in joints (measuring induced displacements), flow in joints at low temperature (to have a good knowledge of the water freezing in the rock joints according to water flow conditions) and deformability of the containment system on a full scale model were studied and demonstrated the validity of the concept.

Now, with the support of SKEC from Korea, bringing their own expertise, providing a convenient site and participating in construction and operation, the cryogenic pilot cavern project described in this paper is being undertaken (Daejon Pilot Project).

## CONCEPT OF LNG UNDERGROUND LINED MINED CAVERN

The basic concept developed is based on the combination of:

- A containment system to ensure LNG containment and rock protection against thermal shock,
- A drainage system used during the first months of the storage operation, before the surrounding rock is frozen, to drain water around the cavity and prevent hydrostatic pressure acting against the containment system. When sufficient thickness of rock has been chilled around the cavity to absorb the hydrostatic loads, drainage pumps

are stopped to allow water to seep into the cold rock and form, in a controlled manner, an impervious ring of ice, forming a double barrier concept.

**Containment system**

The containment system which is used for underground lined mined caverns is similar to the one used and improved by SN Technigaz since 1962 for the membrane type LNG storage tanks and LNG carriers. The modular structure of the containment system makes it very flexible, improving construction and adaptation to cavern geometry. The thickness of insulating panels can be adapted to increase thermal efficiency and the nearly unstressed membrane permits very large scale future projects.

The containment system is composed of several layers with, from rock to LNG (see figure 1 – All figures are shown at the end of this paper):

. A concrete structure ensuring the transition between the containment system and the rock. It is made of reinforced cast in place concrete for the Daejon LNG pilot cavern.
. Some insulating panels made of foam sandwiched between plywood sheets. They are bonded on concrete using load-bearing mastic. The insulation thickness is commonly 300 mm so that the rock temperature will not fall below –50°C after 30 years and the boil-off rate will stay acceptable.
. A 1.2 mm thick stainless steel corrugated membrane providing gas tightness at low temperature.

All surfaces (e.g. bottom, walls, chamfers and vault) are covered with insulating panels and stainless steel membrane sheets.

**Thermomechanical and hydrogeological aspects**

Combined thermomechanical and hydrogeological numerical models were adapted and used for investigating relevant mechanisms such as the propagation of the cold front, migration of water and ice formation in the host rock mass, and induced thermomechanical stress.

At storage cavern scale almost all rock masses and especially hard rock masses are fissured. These already existing natural cracks will act as relaxation structures when the temperature is lowered. It can be demonstrated through geomechanical modelling that low temperature yields opening of existing joints. Creation of new cracks is limited by controlling the cooling rate. It is considered that most rock masses can withstand temperatures as low as -80°C, as long as adequate rock support is provided (i.e. by rock bolting).

Fluid flow analysis and experiments on ice formation gradually led to the present concept: the water is temporary removed from the rock immediately surrounding the cavity during the first phase of operation. This preliminary desaturation of the host rock mass aims at preventing unacceptable hydrostatic pore pressure and ice formation behind the cavern lining.

When the cold front has advanced far enough, drainage can be stopped to allow water progressively to re-invade the rock and quickly form a thick ring of ice around the cavity so withstanding the outside water pressure. The thickness of the frozen rock ring required to sustain the outer pore pressure is about 1 to 2 metres, depending on the strength of the rock mass. Therefore, the drainage period will last from a few months to about one year depending on the hydrogeological characteristics of the site.

The drainage system is composed of a series of pumping holes drilled from the surface and underground specific galleries (water curtain systems). Reduction of high water seepage in the cavern will generally be needed. This is achieved by systematic grouting works during excavation.

**General**

The plant and equipment design is based on the concept of an LNG receiving terminal, allowing it to operate in exactly the same way as an above ground tank.

The main surface equipment adaptations to the specific requirements arising from the underground location are:

- Passage of pipelines through the storage cavern operation shaft (e.g. top of vault of cavern is about 50 metres depth),
- Length of the storage cavity (i.e. total length up to several hundred metres). The number of lines is kept as low as possible without sacrificing operational safety or reliability, and the cooling down and inerting lines are arranged to ensure that the whole length of the gallery is effectively covered.

The figure 2 shows typical overall arrangement of LNG terminal with underground lined storage cavern.

**Pipelines arrangement**

The operation shaft is typically equipped with the following process and instrumentation lines:

- Submerged discharge pumps installed in dedicated casings (i.e. typically three cased pumps),
- The vacuum and pressure safety valves are connected to a central large diameter pipe, which is the boil-off line. It could be also used as a manhole if the storage was to be decommissioned,
- The other pipes are the usual filling lines (top and bottom lines), level gauge, densimeter, etc.
- The lines are either insulated (mainly process pipes) or not (mainly instrument pipes).

**Shaft and Cavity Dome**

All pipes pass down the operation shaft. The pipes are embedded in a concrete structure near the bottom of the shaft. This is the level where the penetrations are anchored and sealed against leakage. All the pipes must be free to contract from this point, requiring expansion loops at ground level. After the pipework is in place, with their insulation as applicable, the shaft is backfilled with sand. Inside the cavity the pipes are joined together by braces forming a pipe tower designed to permit differential pipe shrinkage.

## Comparison with Conventional LNG Storages

There is no need to stress the many advantages of underground storage in terms of safety (for instance regarding fire on plant or decreased potential damages in case of industrial accident nearby, multi-containment barrier with liner and ice ring, lower vulnerability regarding to earthquakes), security (especially relating to sabotage or terrorism acts) and environmental acceptability (no need of large reclaimed areas, less earthworks at ground level, better acceptability by people located nearby) compared with above-ground tanks and even in-ground tanks, since the dome of a buried tank is exposed to risk and of course, to view. Incidentally, it is contemplated to cover the future in-ground LNG tanks to be built in Japan with a flat roof below ground level in order to integrate completely the tanks in the landscape.

Nevertheless, some other advantages should be considered because of the minimisation of total space required for the LNG terminal. This is due to the fact that LNG storage is about 50 m underground: this can represent a huge cost saving especially in seashore areas where industries are already developed and free remaining areas are small and expensive. It is also the case in areas whose topography needs expensive reclaimed land.

However, underground and in-ground tank storage must be considered as mutually complementary since they are suitable for different types of host rock.

Small galleries should be avoided wherever possible by reason of their poorer capacity/area ratio. Underground storage in the form of a gallery of around 20m width by 30m height cross section, which is the most favourable studied geometry in cost terms versus rock behaviour, costs a little bit more than a surface tank with a prestressed concrete envelope and substantially less than an in-ground tank with diaphragm wall and reinforced concrete tank. This investment cost would be reduced in countries where labour cost, and thereafter rock excavation, is cheap.

Moreover, mining technologies and membrane containment system have such flexibility that unit storage capacity has "no limit" : LPG caverns are up to 500,000 m3 and LNG lined caverns can be increased till such value.

This comparative cost estimate between above-ground and cavern storage concerns only the storage itself and its equipment. It does not take into account the substantial cost saving which could be made, in the case of the cavern storage, for the safety equipment (impounding basin, peripheral retention wall, fire fighting systems, etc.) and possibly for the piping length and terminal surface area reduction. A global approach will be made soon for a complete terminal using this type of storage in order to determine the actual difference in price between both solutions. Moreover, operation costs, including maintenance are very attractive.

## DAEJON LNG PILOT CAVERN

Development of Daejon LNG pilot cavern is based on a partnership between SKEC (SK Engineering & Construction Co., Ltd.), GEOSTOCK and SN Technigaz.

SKEC partly participated in basic and detailed design of the underground cavern in co-operation with Géostock and SN Technigaz. SKEC is mainly responsible for the whole phase of construction and operation of the pilot cavern.

Geostock, in collaboration with SKEC, is in charge of basic and detail design of the underground cavity and providing assistance during construction and operation of the pilot cavern. The Geostock scope mainly focuses on the thermomechanical and hydrogeological aspects of the project.

SN Technigaz is in charge of basic and detail design of the containment system and associated process. Technical assistance during construction works on site and operation is also provided.

The three partners signed a Head of Agreement in April 2002 and after having issued basic design in the second half of 2002, decided to proceed to study in detail, to construct and to operate Daejon LNG pilot cavern.

### Description of the Daejon LNG pilot cavern

The LNG pilot cavern is located in Daejon, about 200 km south from Seoul, in an existing cavern implemented within the KIGAM (Korean Institute for Geology And Mineral resources) research centre area.

Access to the pilot cavern is provided through an existing horizontal tunnel and the experimental cavity roof lies at a depth of about 20 m below the ground. In order to have the containment system completed, a concrete wall closes the entrance of the cavern. Additionally, a platform above the entrance of the cavern is required to install instrumentation, manhole and piping. The internal dimensions of the completed pilot are a section of 3.5 by 3.5 m, a length of 10 m, inducing a working volume of 110 m3. (see figures 3 and 4).

### Containment system

The main differences between the usual SN Technigaz containment system and the one to be used for Daejon LNG pilot cavern are the following (see fig 1, page3):

- Smaller pitch between corrugations (identical to the one of LNG carriers); smaller panels and membrane sheets are used due to small area of pilot,
- The containment system is laid on all surfaces, even the vault. Top chamfers are designed to fit as much as possible with existing rock geometry of pilot cavern. Consequently, prefabricated dihedral and trihedral panels are considered with 90° and 135° angles,
- Insulating panels are made of reinforced P.U. foam due to the small thickness of panels and use of liquid nitrogen at –196°C inducing high temperature gradient.

### Process & instrumentation

For safety and practical reasons, liquid nitrogen is used instead of LNG. The cryogenic pilot is operated from a laboratory room located in the access tunnel. It houses the Data Control System (DCS) for the process of the cryogenic pilot and for the rock monitoring. An LN2 plant is installed outside on a fenced site, containing one LN2 storage tank, a vaporizer and associated control systems in order to inert, cool down the cryogenic pilot, fill it completely with LN2 and make up to compensate for the gaseous N2 boil-off (see figures 3 and 5).

These operations use two insulated pipes from this dedicated site to the cryogenic pilot through the access tunnel and the special local enlargement in the cavern roof to go over the plug and penetrate inside the cavern. The additional line is necessary to release overpressures if they happen.

Some instrumentation installed in the containment system and cavern permit to follow the main parameters (inner pressure and temperature, LN2 level, containment deformation) and to operate the pilot.

The pilot cavern and its surrounding rock mass is equipped with a full set of geotechnical instruments (see figure 6) which allows monitoring of:

- temperature profiles around the cavern,
- temperature induced displacements in the containment system, the concrete, some of the rock bolts and in the rock mass,
- concrete structure and rock mass deformations induced by thermomechanical effects,
- pore pressure distribution in fissures,
- water table level.

## Drainage and rock mass treatment

The pilot cavern is equipped with a drainage system made of drainage holes drilled around the cryogenic cavern. This drainage hole pattern is designed in order to:

- check all the time the pore pressure and water saturation in the surrounding fissured rock mass,
- reduce water intrusions in the cavern during installation of the containment system and,
- control the ice lenses formation in the fissured rock mass during the cooling phase.

Pre-grouting works of the main water bearing joints of the host rock mass were carried out improving the efficiency of the drainage system. Moreover, the contact between the concrete lining and the rock mass is grouted when deemed necessary.

## Objectives of the Daejon LNG pilot cavern

The objectives of such a cryogenic pilot are the following:

- demonstrate the feasibility of this technology,
- check adequacy between results of previous computations and laboratory tests regarding in-situ measurements,
- test the overall performance of the storage and
- have an efficient return of experience to improve construction and design of industrial scale projects.

## Schedule

The partners performed during the first half of 2003 the detail design of containment system, process, drainage system, thermal and geomechanical aspects. From site preparation to end of tightness test of the containment system, the construction phase took place from March to end of 2003. Operation started beginning of 2004 after purging air and drying internal volume of cavern using gaseous nitrogen, cooling down and filling of the pilot. Actually, cavern is maintained at nearly full level.

Numerous parameters are followed during operation such as level, temperatures, pressure and boil off rate for the containment system. The behaviour of rock and the containment system is recorded during three successive phases of operation :

- a first six months duration during which a full level is maintained by filling with LN2 in order to compensate for loss of boil-off,
- and a second one during which no more filling is performed and so the cavern empties naturally,
- the third one during which heat up till ambient temperature is followed and pilot cavern is decommissioned.

## CONCLUSION

The Daejon LNG pilot cavern represents a further important step in the validation of the technology of lined LNG underground storage. Even if the cavern is small, this project will permit to address, if any, the potential critical points for construction, start up and operation of such new technology. Besides, the study of the different parameters which will be recorded during operation will increase the knowledge of the partners in the behaviour of the containment system and its surrounding rock in order to be able to tackle an industrial project in the near future.

## List of Figures

# The Storage of Natural Gas in Underground Caverns

*Hugh D. Kaiser, P.E.*
*Principal Engineer*
*kaiser@pbworld.com*

*http://www.pbenergy.com*

*Prepared for Presentation at AIChE Spring National Meeting*
*4th Topical Conference on Natural Gas Utilization*
*New Orleans, LA, April 28, 2004*

## ABSTRACT

The production and importation of hydrocarbons used for fuel or plant feedstock occurs steadily and year-round, whereas consumption is seasonal. Seasonal consumption of hydrocarbons, specifically natural gas, depends on weather patterns that are difficult to predict. Unusually cold weather, which can sometimes last for long periods of time, can strain natural gas supplies, negatively impacting process plant operations, power generation and household budgets. Gas storage in underground caverns has lessened the impact of the unpredictable nature of weather-related demand. As the US population increases, demand for natural gas will increase along with the importation of greater amounts of LNG.

Caverns, constructed in salt domes, represent to a large degree, the underground volume available for hydrocarbon storage in the US. These cavern-types are found in the Southern US along the Gulf Coast. Bedded salt (non-salt dome type) and rock formations, located in other parts of the US, are used for hydrocarbons storage, but to a smaller degree than salt dome caverns. Salt caverns are ideally suited for storing natural gas, since the cavern depth may be easily selected to optimize stored gas volumes. With several LNG import terminals planned in the US, new caverns may be constructed and used for storing vaporized LNG prior to and in anticipation of seasonal demand.

A typical salt cavern is capable of storing the entire volume of vaporized LNG carried by one 125,000m$^3$ LNG tanker[1]. The cavern well design usually permits the total working gas volume to be withdrawn over a 10-day period or faster, when required. Caverns provide safer, more secure and more cost-effective storage than most other modes of hydrocarbon storage. Underground storage caverns have and will continue to play an important role in balancing natural gas supply with demand in the United States and in other parts of the world.

## INTRODUCTION

The underground storage of hydrocarbon plays an important role in the energy infrastructure of the United States and countries around the world. Various types of geologic structures allow for the efficient and economic development of large-scale underground storage facilities. Salt domes, bedded salt and hard rock deposits are three common types of underground geologic structures that are used for constructing hydrocarbon storage caverns. The construction of caverns in hard rock is relatively difficult when compared to salt and the

[1] Based on 22,000 standard ft$^3$ of natural gas per m$^3$ of LNG or 2.75 Billion Standard Cubic Feet (BCF)

relatively shallow depth of the rock limits maximum storage pressures to levels, which are too low for economical natural gas storage. Due to the ample size and number of salt domes found along the US coast of the Gulf of Mexico, construction of caverns of any size and depth in salt is relatively convenient and cost effective.

Cavern storage is now more desirable than ever due to anticipated increases in volumes of imported LNG. Caverns are ideally suited for receiving and securely storing the vaporized effluent gas from LNG receiving terminals. The proximity of available salt deposits is very close to a number of planned LNG import terminals as well as the existing transmission pipeline infrastructure. The high deliverability of cavern storage facilities makes this mode of gas storage suitable for satisfying high seasonal demand.

## OVERVIEW

### History of Underground Storage

Underground storage of natural gas was first attempted in salt deposits in the United States during the early 1900s, unsuccessfully. The first successful gas storage cavern was built in 1916 near Buffalo, New York. Other successful caverns were built in Kentucky and Pennsylvania, shortly, thereafter.

Cavern development reached a rapid pace during the 1950s. Development continued through the 1960s and began slowing during the 1970s and 80s. Cavern storage of natural gas peaked during the early 1980s and has changed little since then. During the 1970's, the Federal Government, in response to the Arab oil embargo and acting through the United States Department of Energy (DOE), funded the construction of the Strategic Petroleum Reserve (SPR) for crude oil storage.

### Geography – Where Have Caverns Been Built?

A large number of caverns in the US have been constructed in salt domes along the US Gulf of Mexico, in Texas, Louisiana, Mississippi and Alabama. Other salt caverns have been built in Arizona, Kansas, Michigan, New Mexico, New York, Oklahoma, Pennsylvania, and in parts of East and West Texas. Refer to Figure 1.

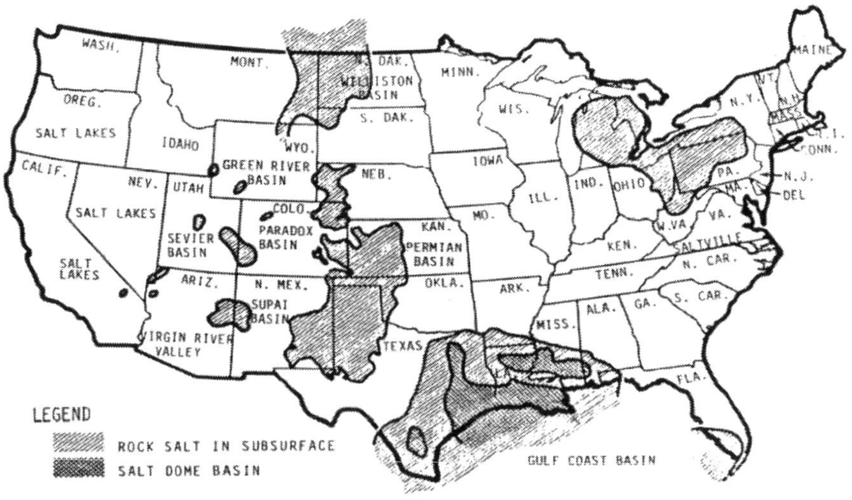


Figure 1. Principal Salt Deposits Within The United States

## Common Cavern Types and How They are Used for Storage

There are three common types of caverns used for storing hydrocarbons:

1. Hard Rock Cavern
2. Bedded-Salt Cavern
3. Salt Dome Cavern

The first cavern type – hard rock – is commonly used for storing liquid hydrocarbons with moderate vapor pressure, such as propane. The caverns are typically mined in shale or granite formations at depths of 300 to 400 feet below ground. At this depth range, the rock temperature is at a constant temperature of 55°F to 65°F, which allows propane to be stored at its vapor pressure without exceeding the maximum allowed operating pressure of the cavern. The maximum allowed pressure in a rock cavern is somewhat less than the hydrostatic gradient of water. Actual maximum operating pressures, however, may only be determined from a thorough geological evaluation of the region and the specific rock deposit. A typical hard rock cavern is depicted in Figure 2.

Figure 2. Hard Rock Cavern for Propane Storage

The second cavern type – bedded salt – is used for storing hydrocarbon liquids and gases. Bedded salt caverns consist of impure deposits of salt, found at various depths below ground. The feasibility of using bedded salt deposits depends on the thickness of the salt deposit and depth, since the cavern must be of a sufficient size to be economic and deep enough to contain pressure. A typical bedded salt cavern is shown below in Figure 3.

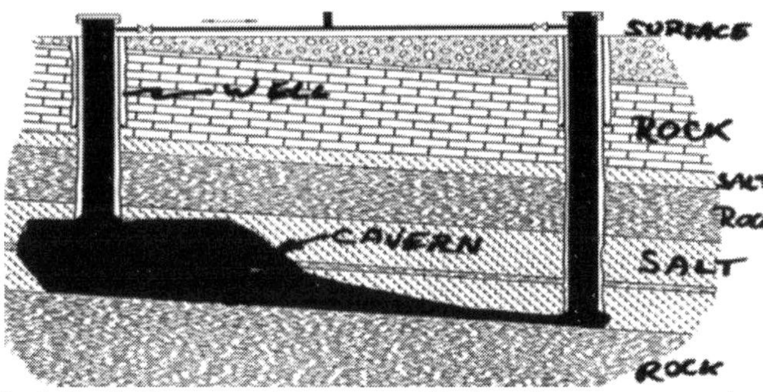


Figure 3. Bedded Salt Cavern (shown with fracture-connected cavern)

Salt dome caverns, or "salt caverns" are the most common and can be developed to virtually any desired size and placed at depths that are optimal for a specific storage application. Due to the size of a salt dome, groups of caverns may be placed within one storage facility and each cavern may be used for storing different types of hydrocarbons. A grouping of salt dome caverns is shown below in Figure 4.

Figure 4. Salt Dome Caverns

**Benefits of Cavern Storage**

The benefits of cavern storage include the following:

- Economical ($3.50 to $5 capital cost per US petroleum barrel of cavern volume)
- Safe - storage cavity is several thousand feet below ground
- High deliverability – storage wells have been sized to deliver 500 Million Standard Cubic Feet per Day (MMscfd) or more
- High storage volume capability in one facility – 6 BCF or more is not uncommon

## DISCUSSION

**Raw Water Supply and Brine Disposal Requirements**

Salt caverns are formed through solution mining or "leaching". Raw water, which is water obtained from underground aquifers or from rivers or lakes, is used for dissolving the underground salt and forming a cavity. When the water is injected into the salt dome, it produces salt-water, or brine. The brine is generally a nearly saturated solution of sodium chloride, which must be disposed in accordance with state and federal laws. In many cases, brine is disposed in wells that are drilled into porous underground formations that will easily accept the brine. The disposal wells are located about one mile from the salt dome.

If, for example, a leaching flow rate of 2,000 US gallons per minute (GPM) is specified per cavern and two caverns are developed simultaneously, a pumping capacity of 4,000 GPM would be required for both the raw water and brine.

## Development of a Cavern in a Salt Dome

Drilling the Storage Well

Once a salt dome is located and land and permits are acquired, a well is drilled into the salt structure. The innermost well (product) casing is cemented in place and its depth is one of the factors for determining the maximum operating pressure in the cavern. A maximum pressure gradient of 0.85 psi/ft is common (example: 3,400 psig maximum pressure for a 4,000 foot cemented product casing depth). The depth of the product casing string will depends on geologic factors as well as the required maximum operating pressure of the cavern. Shallower caverns are suited for storing liquid hydrocarbons, and cost less to drill than deeper caverns. Storing natural gas may benefit from a deeper cavern, due to its higher storage pressure capability and compressibility. A typical storage well is shown below in Figure 5.

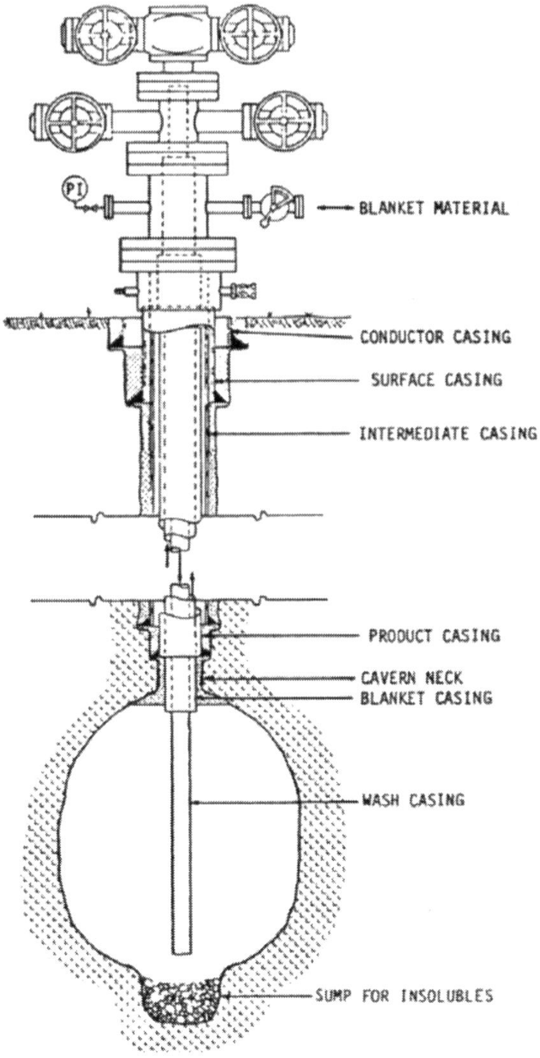


Figure 5. Typical Storage Well Configuration

## Cavern Operations Considerations

During normal cavern operations, gas is injected for storage up to the maximum cavern operating pressure. The maximum cavern operating pressure is determined from the cemented depth of the product casing or "casing seat depth." The maximum pressure is determined by multiplying the casing seat depth by the maximum allowed operating pressure gradient, which is commonly 0.85 psi/ft for gas storage in salt dome caverns. The gradient is determined after conducting a study of the salt characteristics in the vicinity of the planned cavern.

A minimum operating pressure gradient must also be observed and is also governed by subsurface geology considerations. This minimum gradient is commonly 0.3 psi/ft and, like the maximum operating pressure gradient, must also be determined after studying the specific salt deposit. A cavern with a 4,000-foot casing seat depth may require that a pressure of no less than 1,200 psig be maintained at the casing seat. Maintaining a minimum pressure in the cavern reduces "salt creep," which is a phenomenon where the salt dome is deformed by the pressure exerted on it by the earth. When under compression, salt undergoes plastic deformation, thereby moving or "creeping," resulting in cavern closure and loss of available storage volume. The rate of creep can be controlled through proper cavern operation and periodic leaching to regain lost volume.

The gas that may be withdrawn is referred to as "working gas," which is the total gas volume minus the cushion gas volume that must remain in the cavern. The cushion gas is recovered only when absolutely necessary by displacing it from the cavern with water or brine.

As the cavern (casing seat) depth is increased, the higher the maximum cavern operating pressure and the more natural gas stored per unit of cavern volume. For example, a 4 million barrel (net) capacity cavern with a casing seat depth of 4,000 feet (the bottom of the cavern may be 5,000 feet deep) may have the capability of storing as working gas the volume generated by vaporizing 125,000 $m^3$ of LNG (1 tanker) or about 2.75 BCF of natural gas. If the casing seat depth is 3,000 feet, the cavern size must be increased to 4.5 million barrels in order to compensate for the lower maximum operating pressure.

Although a shallower cavern must be larger for holding the same quantity of working gas, its minimum operating pressure is lower, and thus the amount of cushion gas required may be lower. Also, a lower cavern operating pressure requires less compression horsepower as well as piping and vessels that are less massive when compared with facilities having deeper caverns for the same stored working gas volume. The larger cavern size, however, requires more time to develop and, therefore, additional power consumption and cost.

## Cavern Development

After the storage well is completed and connected to the water and brine pumping systems, cavern development begins. Initially, a sump is developed at the bottom of the cavern, which serves as a collection area for insoluble materials contained in the salt. Since these materials will not dissolve in water and, having a higher density than brine, will settle in the sump. Development of the sump may require about several months.

Once the sump is completed, the primary cavern volume is developed. The leaching process produces 1 barrel of storage volume for every 6 to 10 barrels of water circulated. A 4 million barrel cavern may require about 14 to 24 months to develop assuming a 2,000 GPM circulation rate. The development time is based on 7,000 annual hours of leaching and a 10 percent allowance for unusable volume.

During the leaching process, a "blanket" of hydrocarbon liquid is placed in the developing cavern for the purpose of preventing the undermining of the well casing and undesired leaching of the cavern roof. The blanket will float on top of any water in the cavern and acts as a barrier between the water and the cavern roof, where the end of the cemented well casing is located. A schematic of the cavern leaching process is shown below in Figure 6.

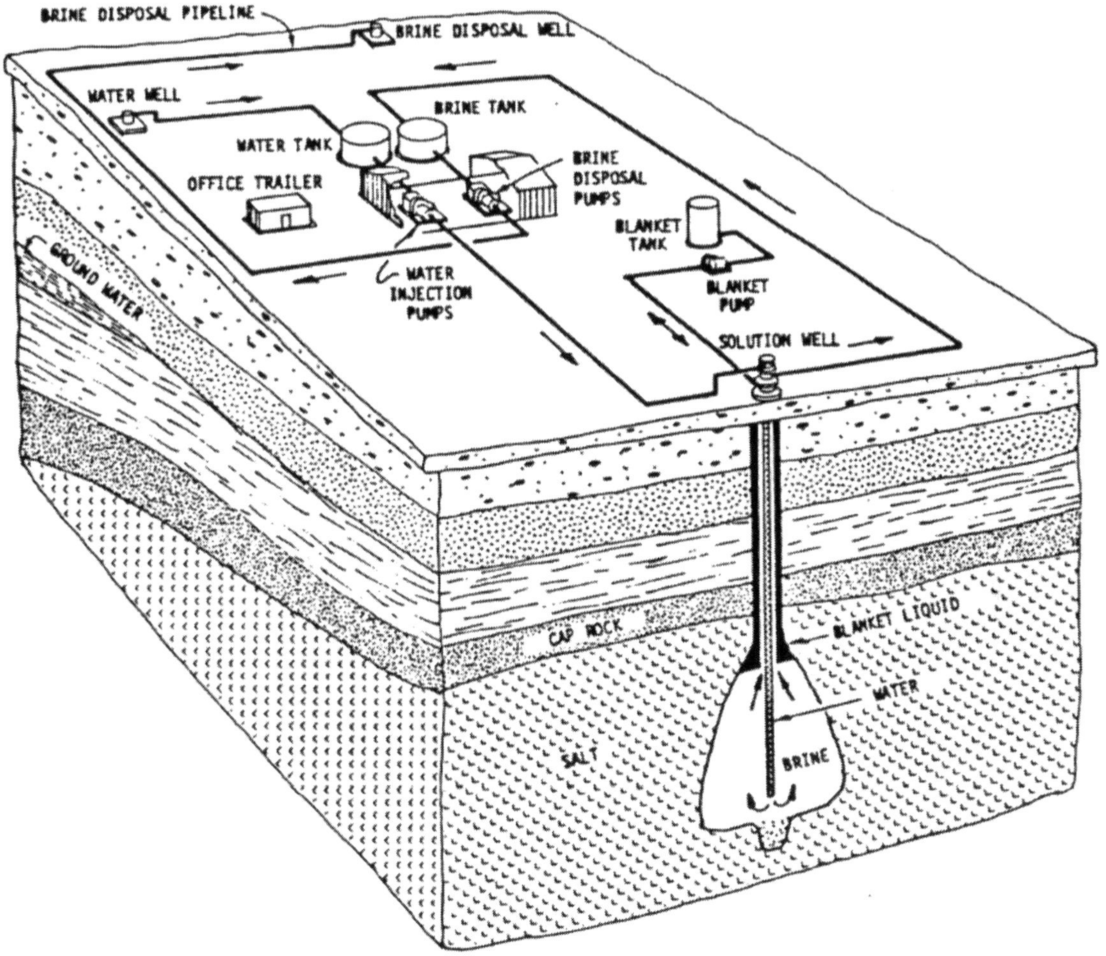


Figure 6. Cavern Leaching Process

"Dewatering" the Cavern

Once the leaching operation is complete, the cavern is emptied of brine by injecting the natural gas that will be stored in the cavern using the injection compressors (see below). The gas is injected through the annulus formed by the "dewatering" string and the product casing. Brine is displaced upward through the dewatering string and is recovered for eventual disposal. Injection of gas occurs relatively slowly to prevent over-pressuring the cavern; therefore, dewatering may require 2-1/2 to 3-months to complete.

## Gas Storage Operations

Injection

Once the cavern has been dewatered (although, some brine will always be present), normal storage operations begin. Generally, the gas compression equipment is sized to fill a cavern over a 20-day period. A cavern designed to hold 2.75 BCF of natural gas would receive the gas at a rate of 138 MMscfd. If gas is delivered to the storage facility at 800 psig and injected into the cavern at a maximum pressure of 3,200 psig, approximately 13,000 HP of compression would be required. Typically, because of their inherent operating flexibility, reciprocating compressors, driven by piston engines or electric motors, have been used. A simplified schematic of the gas handling facility is shown below in Figure 7.

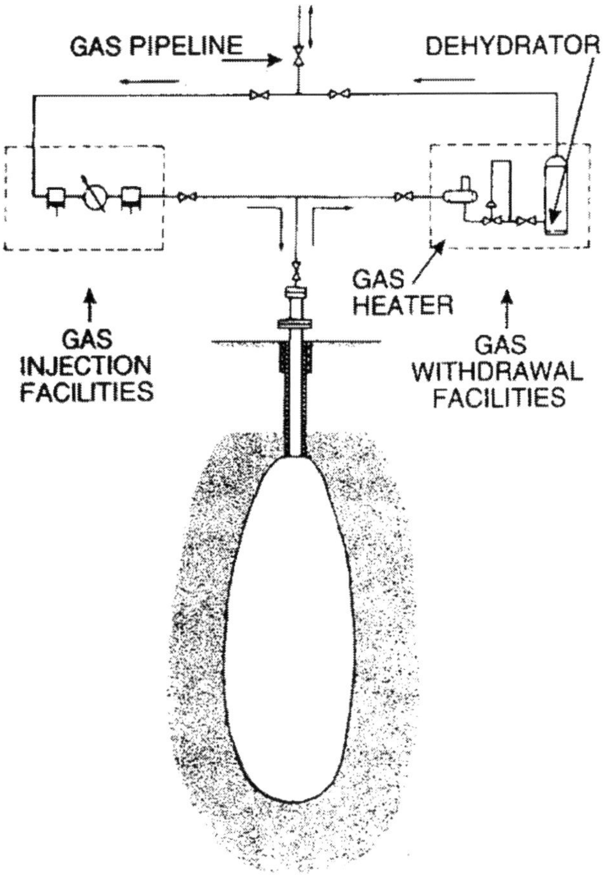


Figure 7. Gas Handling Facility for Injection and Withdrawal

During injection, gas is taken from a pipeline, then passes through a separator to remove entrained liquids and is then compressed to a pressure required to flow into the cavern. The compressed gas is cooled to prevent damage to the well from thermal stress.

### Withdrawal

Since the gas pressure in the cavern usually far exceeds the required delivery pressure of many pipeline systems, gas withdrawal often occurs without the need for compression. If, for example, the gas pressure at the cavern wellhead is at 3,000 psig and the receiving pipeline is at 800 psig, the gas could rapidly flow from the cavern and into the pipeline. During withdrawal, the gas flow is regulated, by holding backpressure on the cavern. The useable cavern inventory may, for example, be withdrawn over a 10-day period. Therefore, if 2.75 BCF of gas is stored in the cavern, it would be withdrawn at a rate of 275 MMscfd for 10 days.

Since the cavern may contain small quantities of residual brine, the withdrawn gas stream may contain some moisture as well as solid particles. Gas is then passed through a high-pressure separator downstream of the wellhead to knockout any condensed water and solids. The gas must then take a substantial pressure drop through a two-stage pressure reducing station. Before the pressure reduction may be taken, the gas may require preheating in order to prevent the formation of methane hydrates.

Once the gas pressure is reduced, it passes through a filter/coalescer in order to remove additional water that may condense after the gas pressure is reduced. The gas then enters a glycol contactor, where the gas flows countercurrent to and is dried by a triethylene glycol desiccant to pipeline specifications. The gas is then measured and transferred to the pipeline.

## Periodic Maintenance Requirements

### Mechanical Integrity Testing (MIT)

Most states require periodic cavern pressure testing in order to verify cavern integrity. The testing is carried out over several days and provides data that is used to determine if the cavern is able to hold pressure or is leaking, which would indicate serious problems and need of immediate attention.

### Sonar Survey

Sonar surveys are used for measuring the physical size of a cavern. This survey is conducted by, lowering a transducer into the cavern and bouncing sound waves off the cavern walls. The data collected from the transducer is used to generate a three-dimensional image of the cavern and to calculate an approximate cavern volume. Due to the stresses imparted on the cavern by the earth, the salt has a tendency to creep (discussed above), resulting in a gradual loss of cavern volume. A comparison among periodic sonar surveys can help in determining the degree and rate of salt creep and any need for corrective action.

## References


Halbouty, Michel T., Salt Domes, Gulf Region, United States & Mexico, Gulf Publishing Company, Houston, Texas (1967)

Various company reports and internal documents, PB Energy Storage Services, Inc.

Sodium Chloride: The Production and Properties of Salt and Brine, Dale W. Kaufmann, Editor, ACS Monograph 145, American Chemical Society, Washington, DC (1978)

Thoms, R. L. and R. M. Gehle, A Brief History of Salt Cavern Use, Solution Mining Research Institute, Encinitas, California (2000)

# THE BISHOP PROCESS™: ELIMINATING CRYOGENIC STORAGE OF LNG AT THE RECEIVING TERMINAL YIELDS IMPROVED SECURITY, ECONOMICS, AND SCALABILITY IN DEEPER WATER LOCATIONS

*James F. Davis, P.E.,*
*Paragon Engineering Services, Inc.*
*10777 Clay Road, Houston, Texas 77041-5497*

*L. Terry Boatman, P.E.*
*FMC SOFEC Floating Systems*
*14741 Yorktown Plaza Dr., Houston, Texas 77040-4063*

*Prepared for 2004 AIChE Spring National Meeting*
*LNG VI – Storage & Facilities*



## ABSTRACT

The Bishop Process™ is a novel method of unloading and re-gasifying liquefied natural gas (LNG) directly from ocean tankers for storage in underground salt caverns. Initial indications are that a salt cavern-based LNG terminal can be built at about half the cost and twice the storage capacity of a conventional cryogenic liquid tank terminal.

Bishop Process™ receiving terminals have the potential to significantly increase world LNG trade and provide a highly secure, economical and flexible way to expand LNG imports and augment the United State's energy supply. The Bishop Process™ entails receiving LNG directly from an offshore tanker, pressurizing and warming it to 40° F, and then injecting the natural gas into underground salt caverns for storage, thereby effectively eliminating the need to build expensive above-ground cryogenic storage tanks, while simultaneously increasing the storage capacity potential of LNG imports.

The patented Bishop Process™ will enable the development of LNG receiving and storage facilities that are faster to build, less expensive, and have higher storage and send out capabilities than today's tank terminals. The Bishop Process™ is currently undergoing operational field testing and evaluation of critical elements needed to reach operational success. The National Energy Technology Laboratory (NETL), Conversion Gas Imports, LP and industry partners have joined in these tests, which include the Bishop Process™ Heat Exchanger (BPE), high pressure LNG offloading pumps, and various methods for applying Bishop Process™ technology offshore.

For water depths greater than 40 m (131 ft), the entire Bishop Process™, with power generation, pumps, controls, and all support facilities can be installed on a floating and weathervaning terminal. The reduced motion characteristics of a column-stabilized floating platform offer increased availability for offloading when compared to a ship-hull type floating structure. Additional operational safety is gained with a novel platform concept for mooring the carrier to the platform.

## INTRODUCTION

The use of salt caverns for storage of hydrocarbon liquids and gases is a well established proven technology [1] [2]. The U.S. Strategic Petroleum Reserve relies heavily upon salt cavern storage for its 700+ million barrels of oil. A multi-participant study, sponsored in part by the National Energy Technology Laboratory of the U.S. Department of Energy, will establish the use of salt caverns as an integral part of LNG receiving terminals with the single-most important feature being to provide very large gas storage volumes (> 18 billion standard cubic feet at one existing site) at a very low cost – and the caverns are not visible to an observer. [3]

The technical and economic advantages of salt cavern storage apply whether the LNG is unloaded at dockside or in open water. The unique capability to handle high volume gas flow rates matches well with the need for LNG tankers to unload quickly at rates from 8,000 - 14,000 cubic meters of LNG per hour (an equivalent flow rate of 4 to 7 billion cubic feet per day of gas). Similarly, these caverns exhibit high-deliverability when customers of the pipelines connected to the cavern need the gas. The reaction time to turn from low flow rate to full flow is on the order of 30 minutes making caverns ideal for highly variable loads and for back-filling interruptible sources.

A self-contained floating terminal that weathervanes, offers improved maneuvering safety for LNG carrier mooring, requires no LNG cargo storage, and uses proven technology throughout, is an attractive concept for a Gulf of Mexico import terminal.

## BISHOP PROCESS™ EQUIPMENT

### Heat Exchanger

To take full advantage of the cavern's capability, a high volume flow rate heater is needed to warm the LNG as it is offloaded for immediate injection into the cavern or into a send-out pipeline. Warming the LNG from a nominal minus (–) 260°F up to a cavern gas injection temperature of 40°F requires a specially designed process unit, a vaporizer that can operate at send-out pressures varying from 1000 to 2200 psi to accommodate both pipeline and cavern requirements. The patented Bishop Process™ is the enabling vaporizer technology that has been documented by McCall et al [1] for application in Gulf of

Mexico shallow water locations such as Vermilion 179 (100 feet of water). The Bishop Process™ Exchanger (BPE) is a simple, modular pipe-in-pipe design comprised of an inner 6-inch 316L cryogenic pipe for the LNG and an outer 12-inch UV-resistant coated polymer pipe. Each BPE will warm up to 300 cubic meters of LNG per hour at pressures up to 2200 psi (ANSI 900) into about 6.5 million standard cubic feet per hour (155 million std $ft^3$/day equivalent flow rate). A key advantage of the Bishop Process™ is maintaining the LNG/gas within the dense phase (see below). Within the annulus, seawater or fresh water is pumped to warm the LNG. The total length of the single cryogenic pipe within a BPE is 2000 feet in contrast to the several hundred small bore tubes in conventional tube exchanger designs. These conventional designs may include shorter overall length but they are 5 to 6 times the diameter of the BPE and may require alloys such as AL6XN to achieve an ANSI 600 service rating. As another example the entire BPE can be easily configured into different physical lengths for a given application by bending it once into two runs (1000 feet), two times into three runs (675 feet), or three times into four runs (500 feet), or even fives runs (200 feet). The flexibility of the design is demonstrated by "folding" the 4-run configuration in half creating a 250-foot long unit for space saving on fixed platform topsides. This special configuration is termed a folded four pass. The term "pass" refers to the number of separate water injection points along the length of the inner pipe required to achieve the proper heat transfer.

For the FMC SOFEC floating platform facility, the BPE will be designed as a three-run/ three-pass configuration (**Figure 1**) taking advantage of the length of the supporting structure and accommodating the LNG entering via the unloading arms near the stern and the gas exiting near the mooring turret.

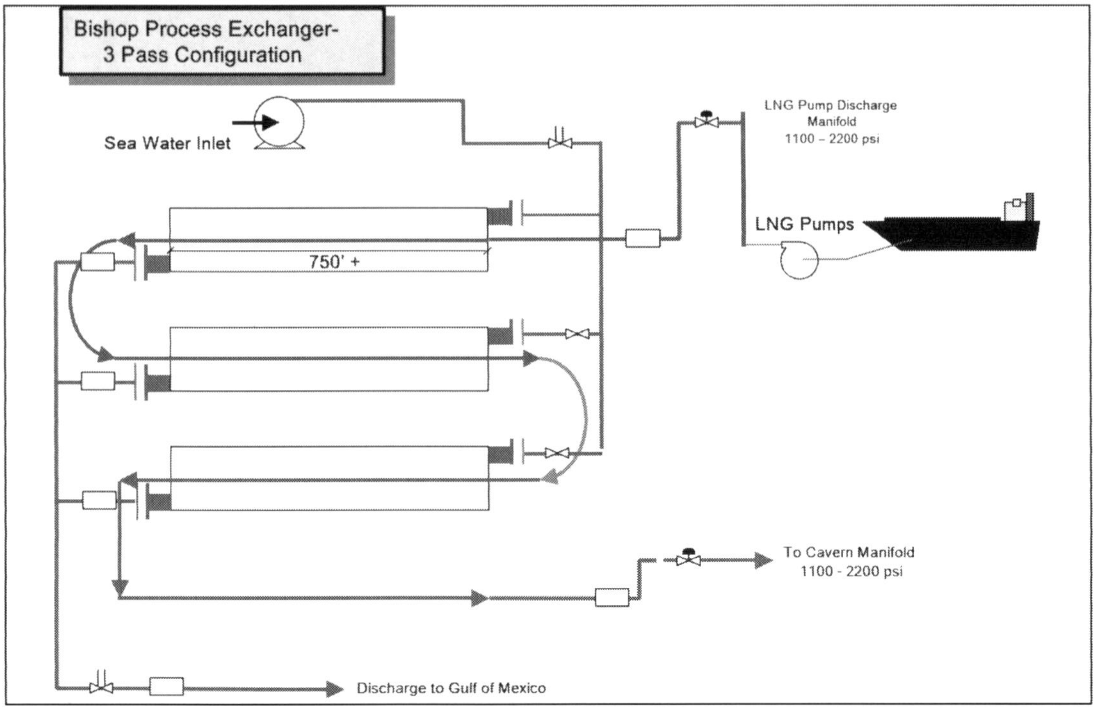


**Figure 1: BPE for Floating Platform**

This configuration will reduce the water pumping cost by 25% vs. the stationary platform configured BPE and will incorporate heat recovery from the gas turbine (GT) power island. Waste heat will offset up to 6% of the vaporizing load further reducing the pumping cost. Since this facility may be located in deeper water, there may be yet another advantage by using the greater temperature spread between inlet water near the surface and discharge near the bottom increasing the 20°F temperature drop across the BPE to perhaps as much as 25°F. At the water volumes required for warming the LNG at the rapid unloading rates, this small temperature change would translate into significant savings.

## LNG Pumps

All of the equipment including the BPE and high-pressure LNG pumps are being proven for platform and land-based implementation. There are no new BPE process components required for a floating terminal, and all components will function when being subjected to some platform motion. The LNG pumps have been proven "in kind" as part of other LNG transport and offloading concepts that utilize shipboard vaporizers and pumps. The BPE should be unaffected by platform motions that occur during operation since it operates in the dense phase and above the critical point of even a rich LNG mixture e.g. Nigeria LNG (**Figure 2**).

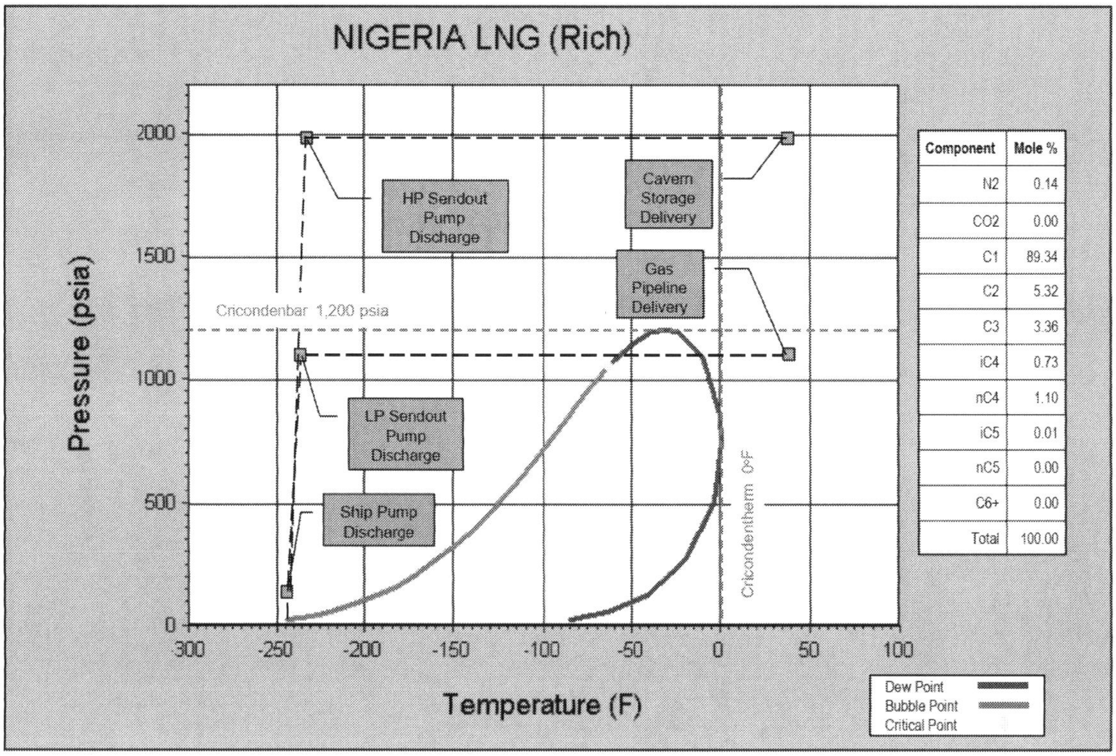


**Figure 2: Nigeria LNG**

The BPE complex can be tailored to match a specific location and operating requirement. It can be designed for the specific cavern pressure operating range and storage capacity as well as for the pipeline send-out flow rate and pressure. For example, if the pipeline pressures are below 1300 psi, a low pressure BPE can be designed with ANSI 600 (1480psi) specifications supplied by cryogenic pumps readily available from several manufacturers. For injection into un-compensated salt caverns at pressures varying from 1000 to 2000 psi (depending upon how much gas is currently stored), a high pressure BPE will be designed with ANSI 900 (2220 psi) ratings. Paragon Engineering has extensively modeled LNG pump configurations required to work against this variable pressure at the required flow-rates. By linking multiple high pressure BPEs into a common high-pressure header, currently available LNG pumps can satisfy the pressure range from 1100 psi to 2200 psi at the flow rates required to offload a LNG tanker in the allotted time frame using pumps as shown in **Figure 3**.

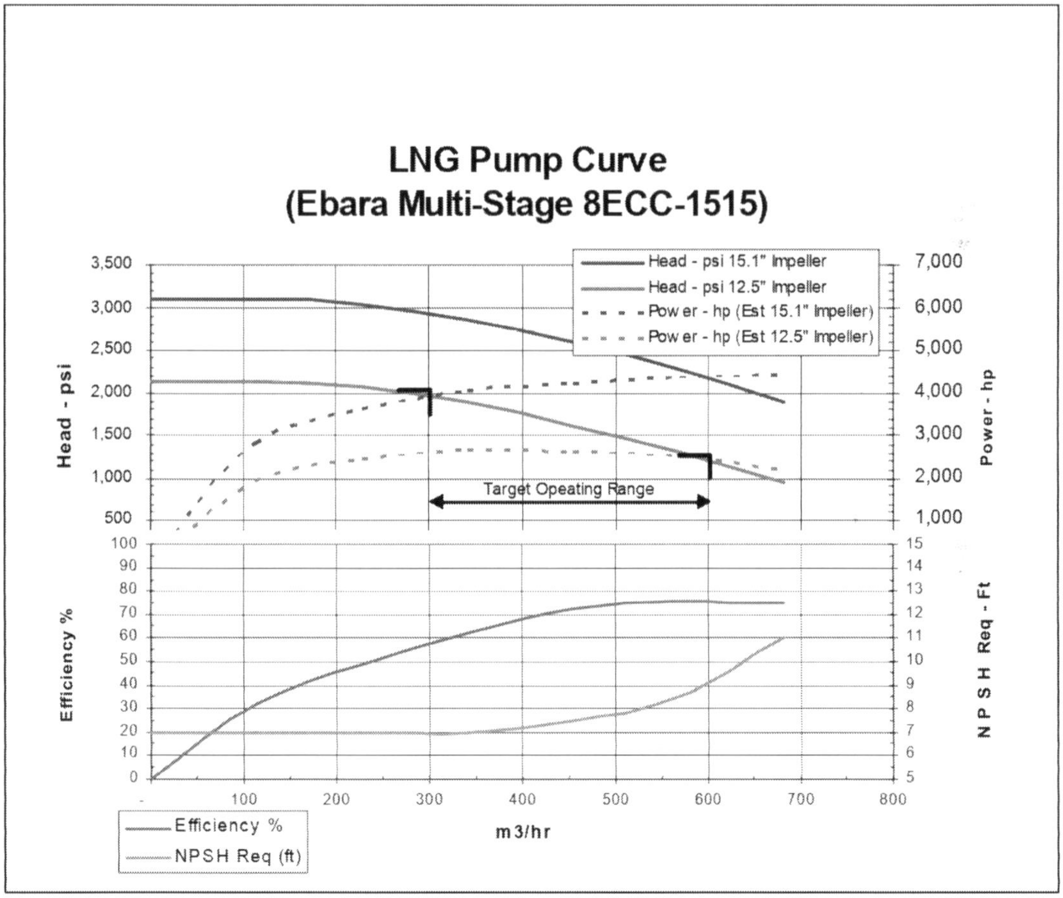


**Figure 3: LNG Pump Curve**

Multiple low and high-pressure BPE units can be combined into a Vaporizer System tailored for a particular location. As part of design activity on other projects utilizing the Bishop Process™ Exchanger, extensive finite element modeling of the stresses within the cryogenic pipe will be conducted. Each BPE can be made from simple, readily available and low cost 316L SS and polymer (GRP, HDPE, etc.) materials, fittings and valves. Other cryogenic pipe alloys that would reduce weight would be AL6XN or 25/6Mo but would tend to increase capital cost. The BPE units will be fabricated, inspected and hydro-tested in shop conditions to reduce costs and assure long service life. The entire process facility would be installed onto the platform while floating at dockside and tested before float-out.

## General Arrangement of Terminal Facility

**Figure 4** schematically shows one of many possible arrangements of the floating platform, pipelines and cavern storage. The flexible risers form the platform connect to a pipeline end manifold (PLEM) on the seafloor under the platform turret. One set of gas pipelines run from the PLEM to subsea wellheads at the caverns. A second set of gas pipelines connect directly to sales pipelines for send-out while the LNG carrier is offloading. Between carrier off-loadings, the gas manifold on the floating platform can provide capability for having gas from the caverns routed back to the platform and then send that gas directly to the sales pipeline.

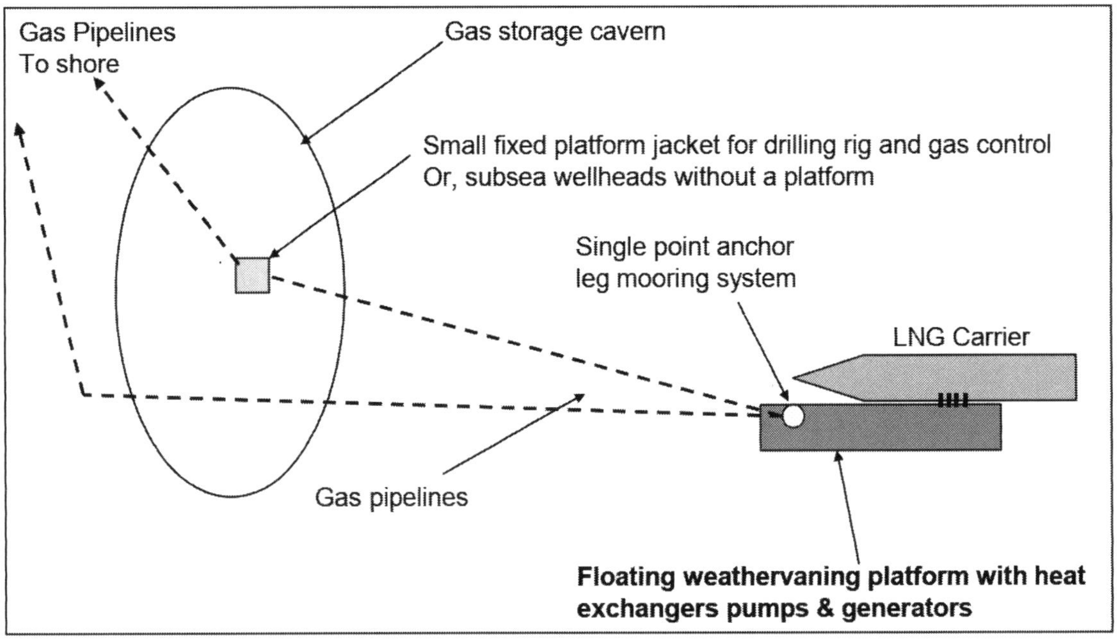


**Figure 4: General Arrangement of Receiving Terminal**

## FLOATING LNG RECEIVING TERMINAL

The LNG Floating Terminal proposed by FMC SOFEC (**Figures 5, 6,** and **7**) is a proprietary concept comprised of a column-stabilized platform, a mooring turret, and marine thrusters. The terminal can be located virtually anywhere close to salt caverns and gas markets and thus can serve many differing requirements for offload rate, storage capacity, and send-out capacity at a range of delivery pressures. The floating platform is readily moored in deep water using conventional catenary anchor legs. It can also be moored in shallow water using a soft yoke mooring connected to a stationary tower on the seafloor. A key aspect of this concept is the absence of cryogenic LNG submerged marine pipelines. This facility can be located at a distance from the cavern – limited only by the gas pipeline flow to the caverns. This could make it easier to gain access to shipping lanes for the LNG carriers. The greater stability of the FMC SOFEC column stabilized structure in a variety of sea states reduces relative motions with the moored LNG carrier, as compared to mooring side-by-side with a converted oil tanker hull.

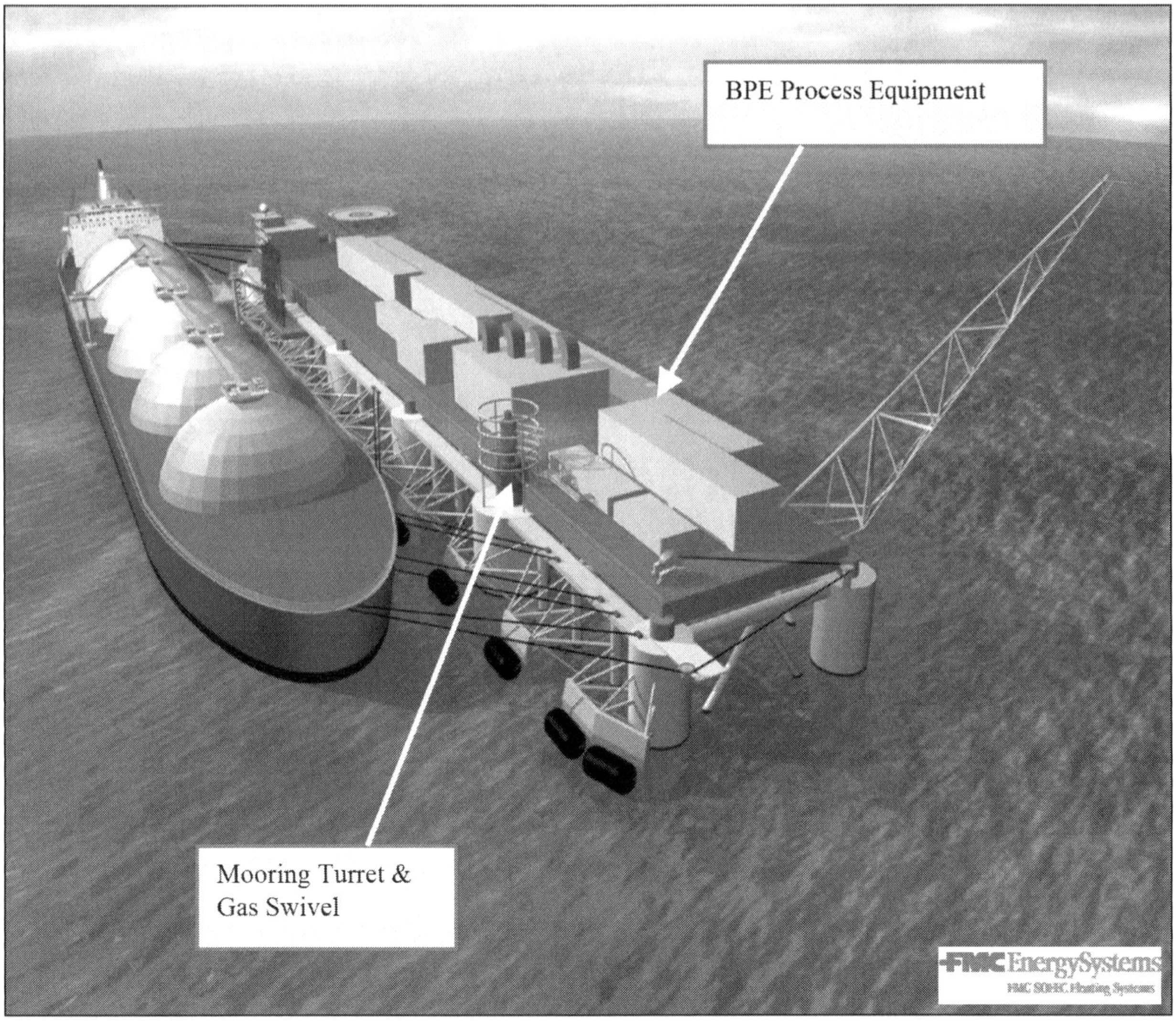


**Figure 5: Floating LNG Receiving Terminal**

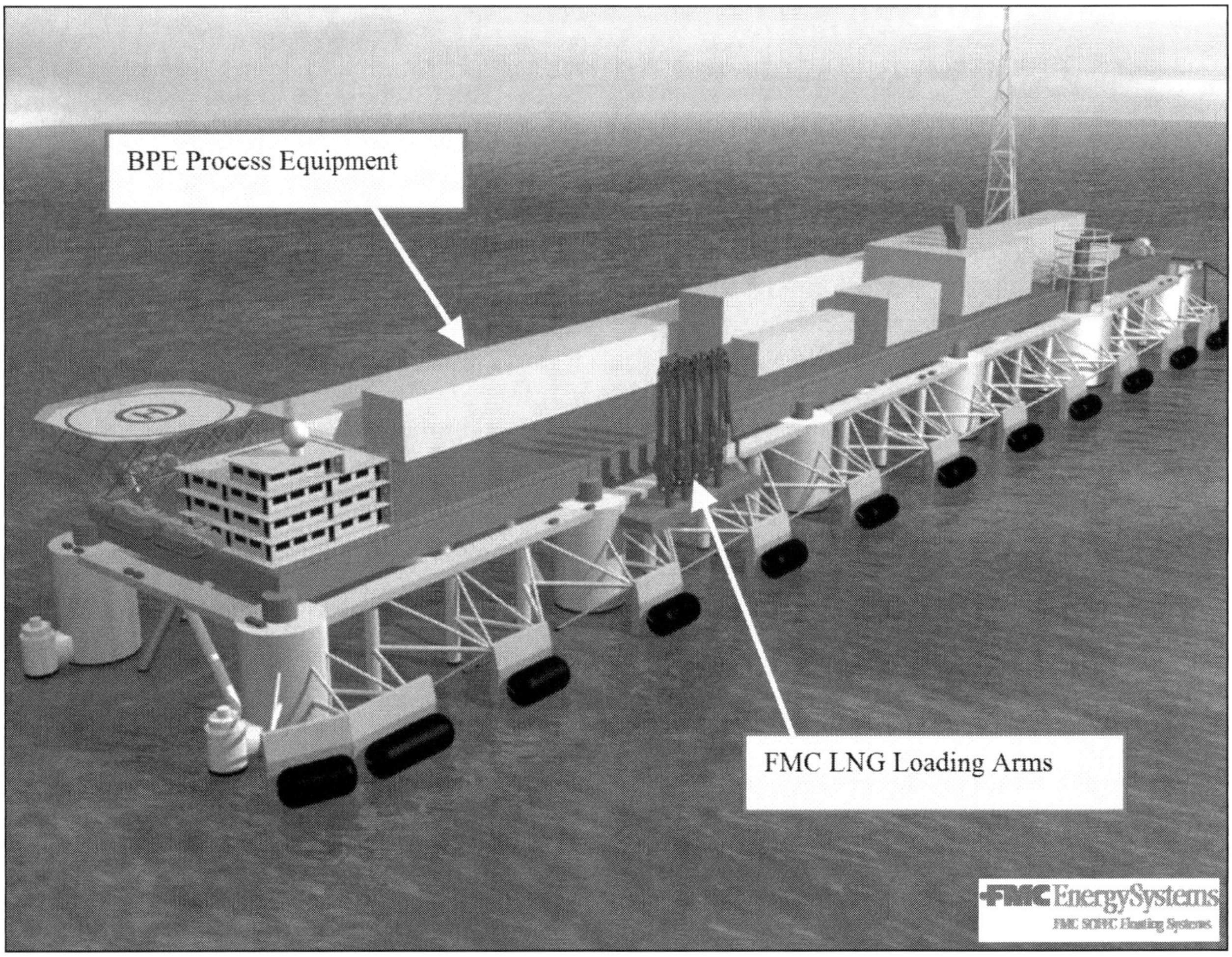


**Figure 6: Floating LNG Receiving Terminal**

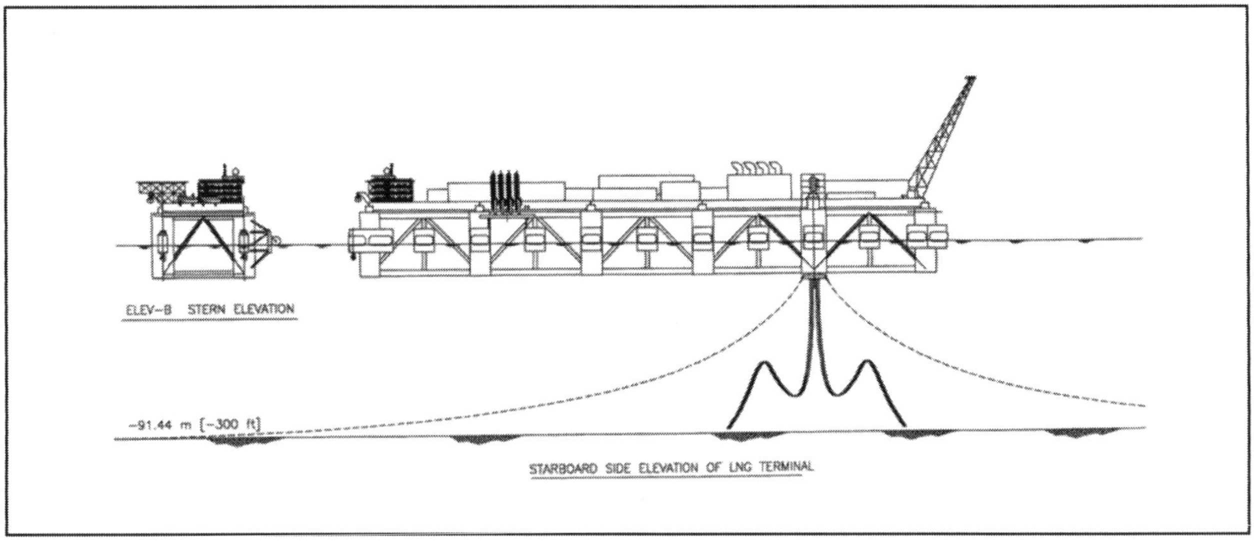


**Figure 7: General Arrangement of Floating Platform**

## Design Basis – Environment

The environmental design basis for the preliminary platform design is given in **Table 1**. These conditions are for a Gulf of Mexico location in a water depth of about 91m (300ft). The operating offloading environment for preliminary design has been increased from the generally accepted 2.5 m (8.2 ft) to 3.5 m (11.5 ft) significant wave height to take advantage of the reduced motion characteristic of the platform.

The availability for offloading LNG carriers is estimated to increase to 99% of the time, corresponding to the selected 3.5m significant wave height. If limited to operation in 2.5m waves, availability is about 95%, depending on the area of the Gulf chosen.

| | Environment | 100-Year Storm Event Seastates | | | 99.0% Availability | Units |
|---|---|---|---|---|---|---|
| | Directionality ---> | Collinear | Crossed 1 | Crossed 2 | Offloading | |
| **Wave** | | | | | | |
| | Wave Spectral Model | JONSWAP | JONSWAP | JONSWAP | P-M | |
| | Significant Wave Height | 11 | 11 | 7.5 | 3.5 | m |
| | Peak Period | 15.1 | 15.1 | 12.4 | 8.5 | sec |
| | $\gamma$ | 2.75 | 2.75 | 2.75 | | |
| | Direction | 180 | 180 | 180 | 180 | deg |
| **Wind** | Velocity | 80.0 | 80.0 | 48.6 | 32.0 | knots |
| | Wind Spectral Model | API | API | API | API | |
| | Direction | 180 | 210 | 260 | 225 | deg |
| **Current** | Velocity | 2.14 | 2.14 | 1.55 | 1.00 | knots |
| | Direction | 180 | 210 | 210 | 210 | deg |

**Table 1: Design Environment**

## Platform Motions

The platform is designed to remain anchored and survive the 100 year storm load conditions. Extreme motions of the platform are given in **Table 2** for the survival condition

| | |
|---|---|
| Water depth | 91m (300 ft) |
| Platform offset maximum | 31.5 m (300 ft) |
| Vertical motion | ± 9.4 m (30.8 ft) |
| Roll Motion | ± 0.3 deg |
| Pitch Motion | ± 3.72 deg |

**Table 2: Maximum Platform Motions, 100 Year Storm**

Relative motion between the LNG carrier and the platform is important for these reasons:

- At the loading arms, relative motion is limited by practical considerations of loading arm reach and arm dynamics, particularly when making the initial hook-up of the flanges to the carrier
- At the fenders, the motion causes higher fender loads.
- In the carrier-to-platform mooring lines, larger motion causes larger line loads

Another inherent advantage of this open structure design is reduction of motion coupling between the two floating bodies. The platform is considered somewhat "transparent" to the wave action. Two conventional hull shapes moored side-by-side are known to influence each other's motion, and the phenomenon can cause increased relative motions between the hulls. This coupling effect is also difficult to quantify with numerical analysis. However, the independent motion characteristic of this platform allows for better initial numerical analysis of relative motion than can be done between two conventional hulls.

Preliminary platform motion analysis in the offloading seastate (3.5m Hs) predicts the RAOs for relative motion between the LNG carrier and the platform, at the LNG loading arms. See **Figure 8**. At 3.5 m Hs, the preliminary estimate of relative motion at the loading arms is below ±1.5 m. A Moss type 165,000$m^3$ LNG carrier was used for the analysis.

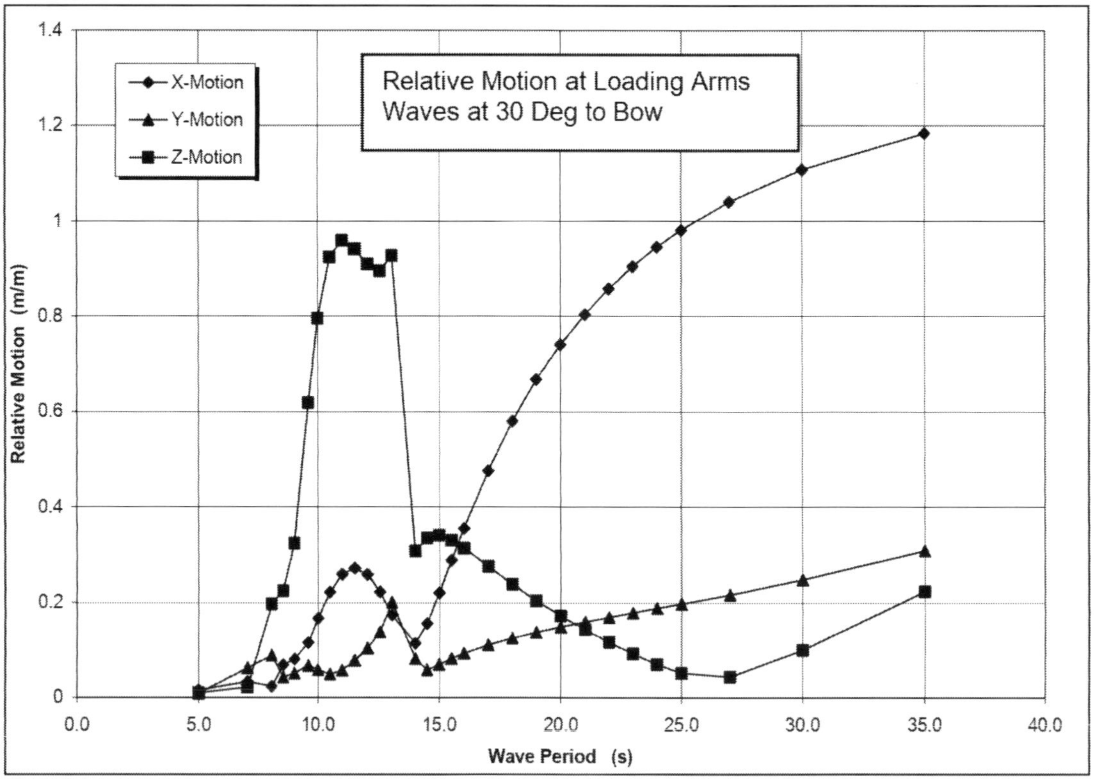


**Figure 8: Relative Motion RAOs Between Carrier and Platform**

### Floating Platform Description

General characteristics of the column stabilized floating terminal are given in **Table 3**. Notice the displacement (total weight or mass) of the platform is about 24,000 metric tons. This compares with a fully loaded displacement of about 140,000 tons for a 165,000 m3 LNG carrier, an important difference considering the interactive motions of the two floating bodies.

Fenders are placed sufficiently outboard to minimize contact by the carrier bow shape if collision occurs at included heading angles up to about 30 degrees to the platform.

| | |
|---|---|
| Deck Load, process equipment | 6,000 metric ton |
| Platform displacement | 24,000 metric ton |
| Structure dimensions | 290m Length, 50.6 m width, 35.5 m Depth |
| Draft | 15.2 m (49.9 ft) |
| Under deck air space above waterline | 15 m (19.2 ft) |
| Gross deck space | 280 m X 40 m, 11,200 $m^2$ |
| Anchor leg mooring system | Catenary chain and drag anchors<br>Eight leg equal spaced, or nine leg in 3 groups of 3 legs (3x3), depending on the location |
| Riser system | Six each 16" flexible risers<br>Two each control umbilical |

**Table 3: Platform Particulars**

### Process Deck Layout

**Figure 9** illustrates a preliminary (*but not optimized*) deck layout of equipment. The process flow layout characteristics are:

- LNG flows generally forward from the loading arms through the vaporizer
- Cooling water pumps are located forward, picking up water under the forward end of the platform. Water flows generally rearward through the vaporizer, and the cooler water is dispersed at the aft end of the platform in an underwater diffuser.
- Warmed gas flows forward into gas manifolds and metering units to the turret and swivel stack on the forward starboard side.

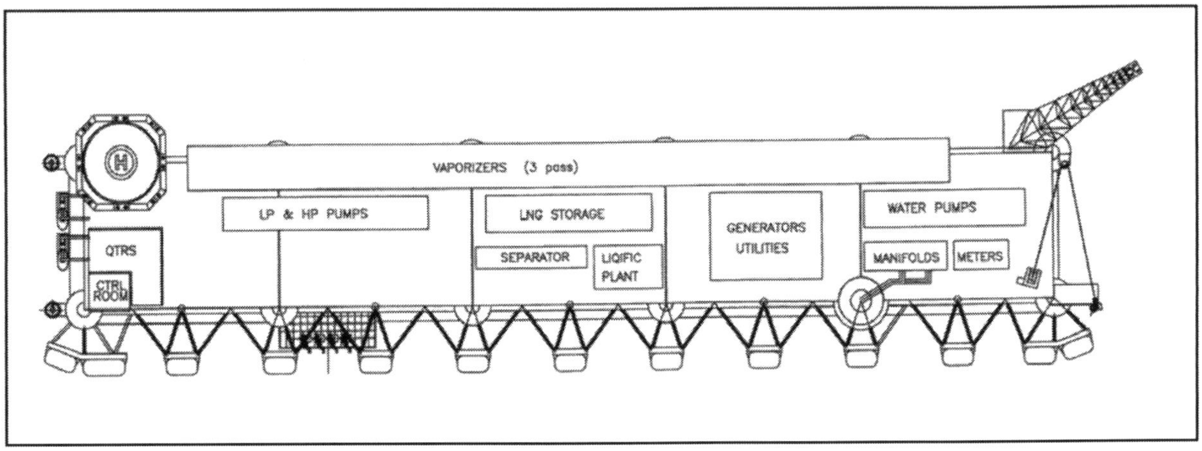


**Figure 9: Deck Layout**

## Safety Drivers for Equipment Layout

An LNG receiving terminal will have similar safety drivers as are known for floating LNG plants (FLNG, or LNG FPSO vessels).
This platform concept corresponds with many features described by Michael Persaud, et al, in Reference [4]. The significant characteristics of this deck layout are:

- Visiting LNG carriers are always required to moor with their accommodations near the platform stern, thereby keeping accommodations of both vessels at the stern
- Living quarters are on the aft end, shielded by blast walls, distant from higher risk areas (high pressure gas). The accommodations may be more optimally located on the port side under the helicopter deck, instead of as shown.
- Lifeboats are on the aft end to facilitate easier escape and rescue from the stern.
- Large flat deck areas provide space for large gaps between process modules to mitigate jet fires and explosions
- Control tower and accommodations are aft of a large open utility space for increased safety
- Gas turbines are forward, surrounded by ceiling and walls for containment of blade and rotor failure fragments
- Side mounted internal turret and gas swivel stack reduces congestion
- No hydrocarbon pump rooms in confined spaces
- Flare boom on forward end, outboard, and on opposite side from LNG carrier
- Supply boat unloading is only permitted at the port aft end of the platform, near the accommodations, as far away as possible from the risers, to minimize risk of dropped objects damaging the risers

- Helicopter pad is located on aft end for optimal flight path into the wind, and conveniently near the accommodations. Minimizes flight path over the process equipment.
- The open frame structure of the platform facilitates open stainless steel drip pan draining downward to the sea to carry away any LNG spills from the LNG offloading and LNG piping. Some of the seawater warmant can be used in water flood style LNG drip pan drains.

**LNG Carrier Maneuvering to Platform**

A particularly important aspect of this platform concept involves the improved safety of operations during the process of mooring the LNG carrier to the floating platform, as compared to mooring the carrier side-by-side with a conventional hull or to a stationary platform. Whether or not a tug is necessary and to what extent will be the subject of considerable study of carrier berthing procedures.

Refer to **Figures 10, 11, and 12**. The fundamental principles of the concept are:

- The platform is self-powered to swing away from the approaching LNG carrier, about 30 to 40 degrees into the prevailing environment. **Figure 10A**.
- The LNG carrier is brought to a stop near an optimal position. **Figure 10B**.
- The mooring hawser winch system and hawser fairlead are located considerably forward of the turret. This feature, in conjunction with the side location of the mooring turret, provides a unique ability to pull the carrier into the optimal position near the platform under resistance from a slow astern thrust by the ship. Upon stopping all headway, with heading maintained by the bow hawser and assisted by a tug, the platform will swing around the turret center with fenders tangent to the carrier side. This action provides added reliability and safety to mooring operations in the more severe environments. If the carrier is perceived to overrun, the platform can attempt to swing back around parallel to the ship and present its entire side fender array to the carrier. If the carrier continues forward, the hawser fairlead is thus moved out of the ship path, without collision damage such as can occur with some proposed outboard hawser boom arrangements.
- The platform is powered toward the LNG carrier until contact is made with the aft fenders of the platform, near midship, where the minimum relative motion occurs. **Figure 11A** and **Figure 11B**.
- The LNG carrier is moved forward to achieve alignment of the LNG manifold with the LNG loading arms. **Figure 12**.
- Mooring lines are run from the carrier to the platform and attached to quick release hooks on the platform. Mooring lines are placed in

accordance with OCIMF guidelines. The bow hawser provides additional longitudinal mooring load capacity. **Figure 13**.

- LNG carrier departure is performed generally in the reverse order; with the carrier backing away from the platform while the platform rotates away for additional clearance.

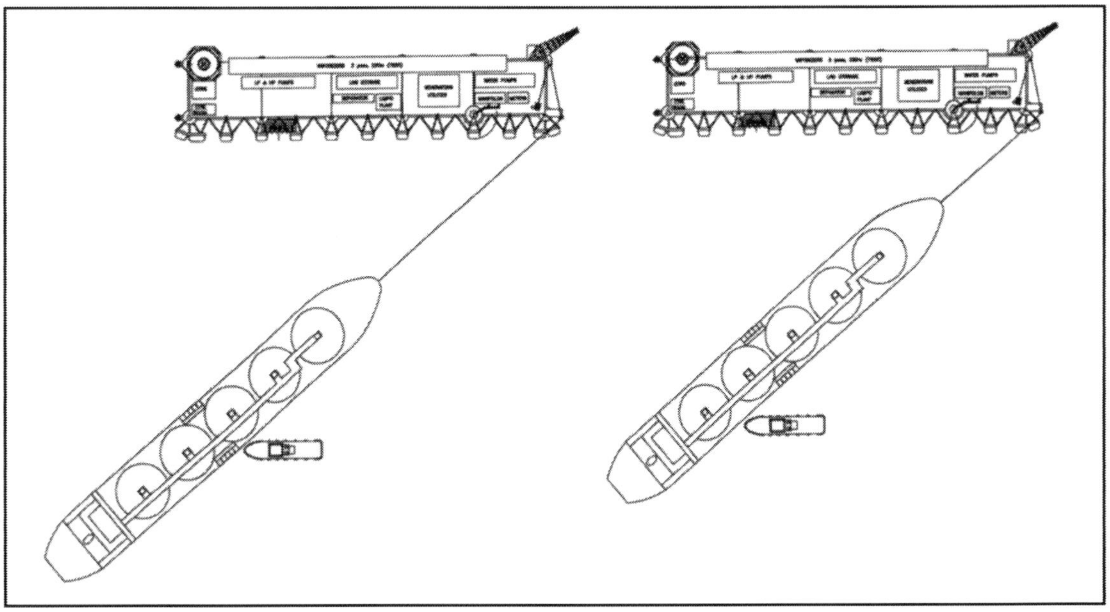

**Figure 10A: Carrier Approaches** **Figure 10B: Carrier Slows to Stop**

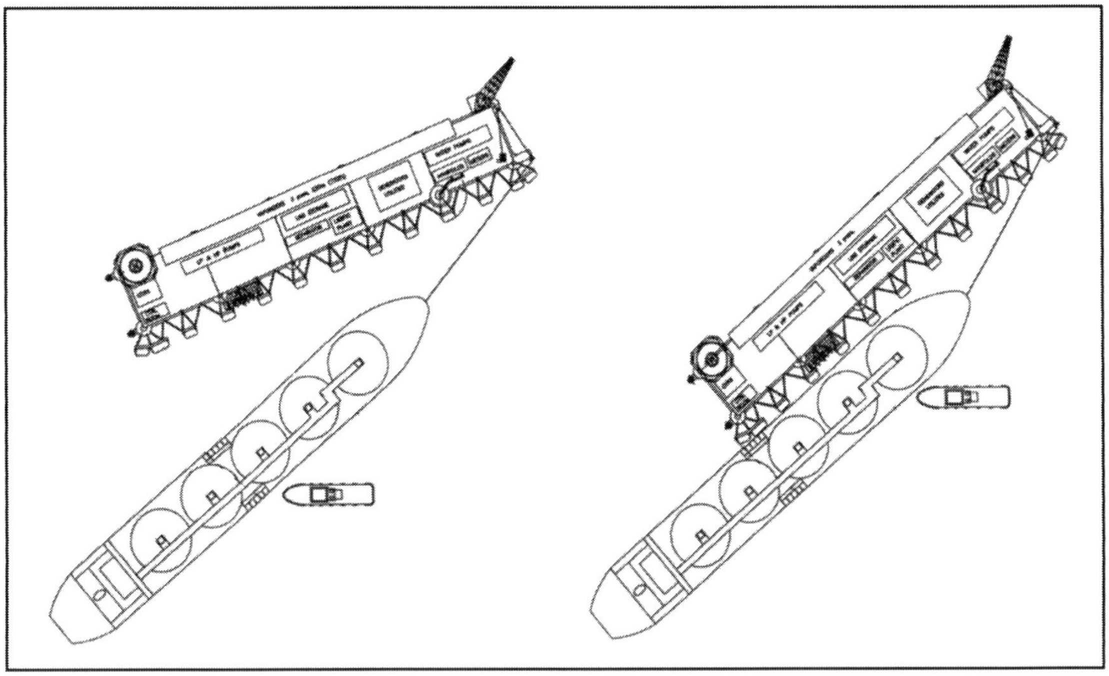

**Figure 11A: Platform Swings** **Figure 11B: Platform Nudges Ship**

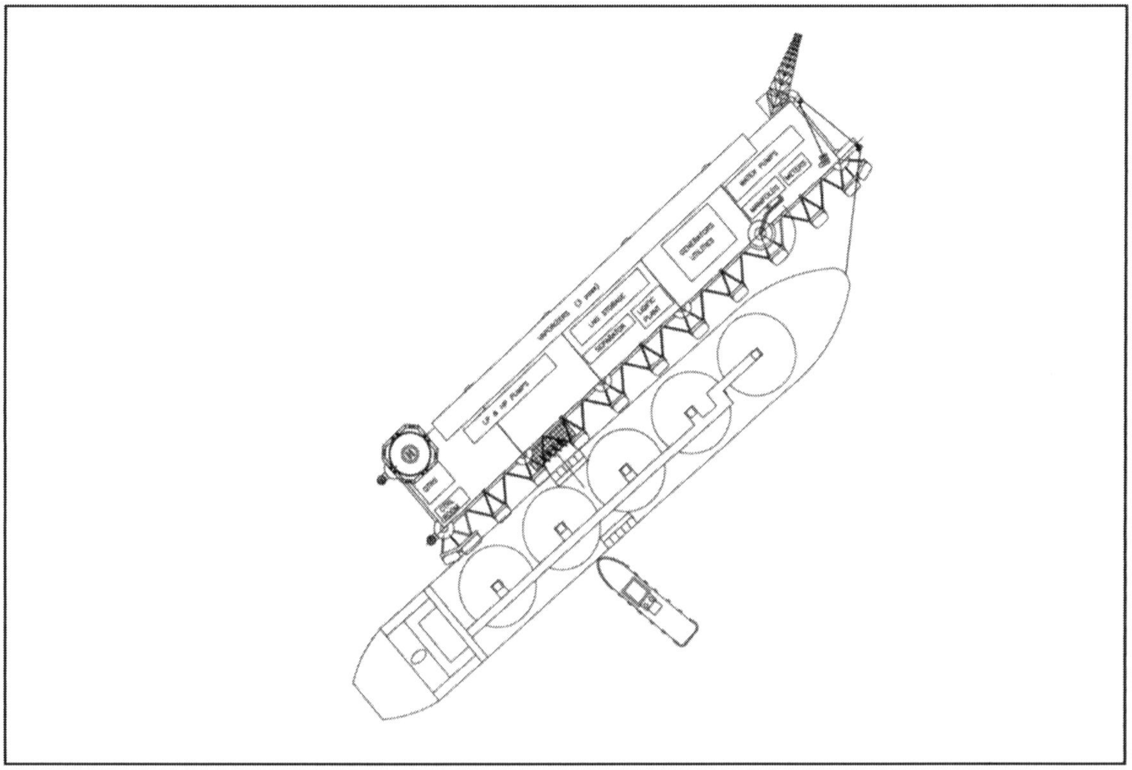

Figure 12: Carrier Moves to Align with Loading Arms

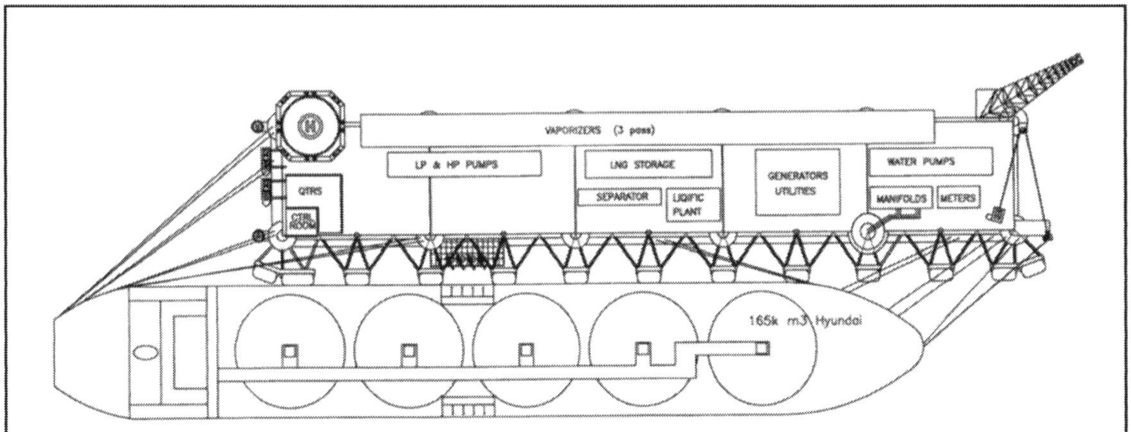


Figure 13: Mooring Lines in Place

**Summary of Advantages of the FMC SOFEC Floating Platform Concept**

- A weathervaning platform reduces undesirable carrier roll motions resulting from quartering and beam sea conditions associated with mooring to stationary structures
- Berthing an LNG carrier utilizing a maneuverable platform of much lower mass then the carrier is safer than berthing the carrier to a conventional hull regasification vessel or to a stationary structure
- Risk of serious collision damage to the LNG carrier is lessened by the relatively low mass of the platform
- Conventional LNG loading arms are used that do not require significant modification of the LNG carrier manifold. This feature eliminates risk of unproven LNG transfer technology, minimizes cost, and facilitates spot trading.
- No underwater cryogenic LNG pipelines are required
- Improved safety for maneuvering of LNG carrier to platform
    - Self powered platform moves out of way of carrier until carrier is in position
    - Platform moves to ship, rather than moving ship to the platform, or to a large tanker hull, or to a stationary structure
- Availability for LNG carrier off-loading is increased due to a more severe allowable seastate. Column stabilized construction, similar to a semi-submersible, facilitates the low motions
- Platform concept contributes to considerably lower overall cost of the complete import terminal facility
    - Simple tubular construction that can be fabricated in local areas utilizing techniques similar to Gulf Coast offshore platform jacket construction – easy to build in modules
    - Dockside installation and testing of all on-board process equipment and utility systems
    - Offshore construction spread cost is reduced. Installation of mooring system, riser system, and hook-up of floating platform is reduced to a few weeks time.
    - Control of salt cavern subsea wellheads can be done from the floating platform by a control umbilical. Alternatively if small fixed platforms are used with dry trees on the caverns, then these can also be controlled from the floating platform.

## CONCLUSION

The BISHOP PROCESS™ for natural gas storage in salt domes located in deeper water can be combined with a self-contained floating platform designed for minimum motions and maximum operational safety. This approach can provide a quickly built and cost efficient facility for expanding the import, storage, and distribution of natural gas.

## References

[1] McCall, M. M. et al: "Examine and Evaluate a Process to Use Salt Caverns to Receive Ship Borne Liquefied Natural Gas", Final Report DOE Award Number DE-FC26-02NT41653, April 29, 2003

[2] Baan, J. de et al: "Offshore Transfer, Re-Gasification and Salt Dome Storage of LNG", Paper OTC 15301, presented at the 2003 OTC Conference, Houston 5-8 May 2003

[3] Energy Information Administration: "Annual Energy Outlook 2004 with Projections to 2025", DOE/EIA-0383(2004), January 2004

[4] Persaud, M. et al, Shell Global Solutions, Intl.: "Safety Drivers in the Layout of Floating LNG Plants", Third Topical Conference on Natural Gas Utilization, AIChE Spring National Meeting, Mar. 30 to April 3, 2003

## Conversion of Peak Shaving LNG Tanks to Import Terminal Duty, and Revalidation for an Additional 25-Year Lifespan

*Ing. Paul Sullivan*
*Business Development Director*
*Skanska Whessoe*
*4139 E. Northampton Place*
*Houston, TX 77098*
*Paul.Sullivan@skanska.co.uk*

**ABSTRACT**

Skanska Whessoe is undertaking the structural revalidation and modifications of four 50,000 cubic metre (1 bcf) LNG storage tanks on the Isle of Grain in the United Kingdom. The work is being done as part of a contract to convert the terminal from a Peak Shaving facility to an LNG Import Facility.

The project scope includes:

- The warm up, inspection, repair/modification and return to service of the tanks.
- Assessment and design of modifications to the tank roof frames.
- Design and detailing of new pipework and support structures.
- Design and detailing of new tank valve and access structures.
- Revalidation of the 20+-year-old tanks for a further 25 years of modified service.

### INTRODUCTION

The Isle of Grain LNG peakshaving installation in the UK is being converted to an LNG importation terminal. Whessoe describe in this paper the planning and execution of a novel programme of LNG storage tank modification and revalidation.

In consequence, the four 50,000 $m^3$ double containment LNG tanks needed to be de-commissioned to install bigger fill lines, boil-off lines, relief systems, updated gauging and instrumentation and finally, inspected and re-certified so that they can receive the high filling rates from LNG carriers.

Two of the tanks were originally designed and built by Whessoe in 1977/80. The project is of particular interest to the LNG community, because it is the first of its type in Europe, where similarly to the USA, there are a considerable number of LNG tanks towards the end of their original design life.

### THE CLIENT REMIT AND EARLY DEVELOPMENT

Through the FEED contract in 2002 the client aided by Whessoe engineers, refined the project requirements to include the following:

- Extension of the tank life based on engineering calculation and physical testing/inspection

- The maintenance of two full tanks in operation for the duration of the contract
- The delivery of all four tanks in time to receive gas shipments for the 2004/5 winter season

In order to facilitate these requirements the client chose to enter a single source negotiation with Whessoe on a partial lumpsum and partial re-imbursement basis. The client-contractor team then developed a detailed risk based execution strategy with a view to minimising schedule risk without over investing in possible replacement materials, which in the end might not be required.

## EXECUTION STRATEGY

The Isle of Grain peak shave facility is the nearest such facility to London and the southeast of England, which is the UK's largest regional gas market. The commercial value of the facility to its owners is considerable; hence any disruption to the operation would carry heavy financial penalties. It was therefore decided that the most important issue was to determine the extent of work which would be required to bring the tanks back into service as part of an importation terminal regime and to extend their operational life by 25 years.

Under the previous peak shaving regime the tanks were filled using surplus pipeline gas during periods of lower demand. The peak shave filling season amounted to some 200 days per year. Under the new importation regime, the tanks could be emptied and filled as often as once weekly, given the send out capability.

The strategy developed was to take only one tank out of service in the first instance, in case more serious repairs than anticipated were required. This would have allowed the owner to predict more accurately the extent of repairs required on the other three tanks and to set the schedules accordingly.

This strategy was underwritten by an early engineering programme to determine from the original design records, any serious design deficiency when compared to new structural codes. Engineering preparation for revalidation was carried out in parallel with the decommissioning of the first tank.

In addition a program of materials testing was agreed to ensure sufficient information would be available to support the re-certification of the tanks for a further 25 years.

## ENGINEERING

### *Structural revalidation and modification*

The structural work scope included:

- Assessment and design of modifications to the tank roof frames
- Design and detailing of new pipe support structures
- Design and detailing of new tank roof valve access structures

The aim was to produce safe, economical designs for the modifications. Whessoe undertook all engineering, including desk studies and reviews of existing calculations and drawings, followed by site surveys and detailed modelling. Structural design and analysis was then completed prior to production of general arrangements and detailed fabrication drawings.

The objective of the structural analysis was to assess the long-term integrity of the roof structures under increased loading levels from the three 24 inch pipeline nozzles for the new 24 inch bottom fill, top fill and boil-off gas lines, ensuring compliance with the latest codes of practice (BS5950-2000), so that long term service could be ensured.

In order to construct a computerised 3-D model of the roof a search of the Client's archives was made to ascertain the details of the structure. A copy of the original Whessoe calculations was found and used for the model construction. The geometry of the tanks as envisaged in their design concept was modelled with a high degree of confidence. This provided the sizes and arrangement of the existing members, their methods of connections and the loading criteria used. Details were subsequently confirmed by inspection on site.

In brief, the roof structure consisted of a steel I-beam frame of main radial rafters set at equal spacing, curved to the roof radius. Steel angles were used as circumferential bracing between the rafters. At the top of the dome the rafters connect to a centre-drum fabricated from plate and at the outer circumference they are bracketed to the tank wall. Additional structural members were provided in order to reinforce the area local to the new nozzles. The plates of the roof were lap-welded but not connected to the rafters. The roof structure also supports a plywood suspended deck carrying insulation over the inner tank.

In order to check the adequacy of the roof, 3D analysis models were created using the STAAD.Pro 2002 (Structural Analysis And Design) package which has full in-built code checking facilities to BS5950 : Part 1 : 2000. Following analysis it was concluded that some purlin/wind bracing members may have required new gusset plates and required an increase in section size to cater for the increase in axial forces. It was agreed with the Client that the replacement of these members would be subject to a site survey based on a visual defects basis, i.e. if any of the purlins showed signs of yielding, buckling or cracking then these members would have been removed and replaced. The subsequent survey showed that all members were in a good condition; hence, they were not replaced.

The new members required to support the nozzles were designed to be cut short of the rafter flanges to allow ease of erection. Fitted fin plates with horizontal stiffeners were site welded to the roof rafters, which in turn were welded to the webs of the new beams. The top and bottom flanges on one side of the new beams were stripped back so that a sufficient fin plate depth can be achieved. The stiffened fin plate connections were required, as the members will be subject to significant loads from the nozzle anchors including biaxial bending moments.

### *Validation of the tanks for a further 25 years service*

An important aspect in determining the suitability of the Isle of Grain tanks for import duty was the validation of the existing 20+ -year old tanks, for a further 25 years service. To this end Whessoe employed both finite element analysis techniques and fracture mechanics assessments.

The finite element analysis of the inner tank shell was carried out in order to determine the stress levels in the tank for various historical and future load conditions. From this analysis the fatigue life of the tank, particularly at the shell / annular junction, was determined and permitted the calculation of the predicted future life. This was done using Appendix C of PD5500 with relevant safety factors on life.

Carrying out an Engineering Critical Assessment (ECA) of the tanks reinforced this work. This involved a fracture mechanics assessment of the structures in accordance with BS7910. The results of the stress analysis were used in the ECA, as part of the input data in order to determine critical defect sizes. The number of cycles taken to achieve these critical defect sizes was calculated, enabling the determination of the future design life of the tanks. Thus by combining both techniques the tanks were validated for a further 25 years service.

### *Control of tank pressure through relief valve design*

In peak shave operation, over pressure protection of the tanks is provided by two sets of relief valves:

- "Process" pressure relief valves (PRVs), set point 125 mbar(g)
- "Emergency" relief valves (ERVs), set point 140 mbar(g)

The process PRVs are 2-off (1 plus 1 spare) 10 inch pilot operated devices per tank, while each tank has 6-off (5 plus 1 spare) 10 inch weight-loaded pallet type units acting as ERVs.

On conversion from peak shave to importation duty, the required capacity of the process relief valves is greatly increased, while the existing emergency relief capacity is still adequate. It was therefore initially proposed that the pallet type ERVs would be retained; with the process PRVs replaced with 2-off new higher capacity 12 inch units. However, this would have necessitated the installation of new relief valve inlet pipework from beneath the suspended deck, through the tank roof via a heat break arrangement and an isolation valve to a new set of nozzles on each tank.

It was considered desirable to avoid this requirement for new roof penetrations, and to utilise existing nozzles if at all possible. A revised proposal was therefore developed, to replace both the existing process PRVs and the existing ERVs with 6-off new pilot operated PRVs mounted on the nozzles previously accommodating the pallet-type ERVs. The higher capacity of the pilot operated valves compared to the pallet type meant that only 3-off (2 plus 1 spare) new ERVs are required, with the remaining 3-off (2 plus 1 spare) new pilot operated valves providing the required capacity for the increased process PRV duty. This revised design has been adopted, allowing existing nozzles to be re-used and eliminating the requirement for new roof penetrations.

## SITE ACTIVITIES

### *Specialist techniques for schedule enhancement*

Many LNG tanks built in the 1970s/1980s and later, with a 25/30-year life, did not have means of emptying the last meter or so of liquid from the tank bottom once minimum pump-down with the in-tank pumps was achieved. This was the case with the four Isle of Grain LNG tanks.

Leaving the tank contents to boil off by atmospheric heat leak to the tank takes several months causing an extended period of lost operation and product loss of some hundreds of tonnes.

Whessoe, with LNG tank experience going back to the mid 1950s, were requested to propose a solution to empty the tank to absolute minimum level. This involved investigating, verifying, finalising and confirming the best, safest and simplest way to remove the last level of liquid in each tank (to say 50 mm). This reduced the time to fully de-commission the tanks from months to a few weeks. Tank de- commissioning was actually six steps: remove liquid to the minimum practical level, boil-off the remainder, warm up from - 162 °C to ambient, nitrogen purge, air purge, isolate and make all systems safe for entry handover certificate.

After reviewing all possible methods of liquid removal and the time it would take, it was apparent that the simplest solution would be to install through the domed roof a very small pump similar in type to the much larger existing in-tank pumps, provided that the minimum pump-down could be achieved. The problem was to fit a mini-pump through a suitable 10-inch dia. roof nozzle, lower it 45 m to the tank bottom and then be able to pump out the liquid to another tank. A suitable design and method were finalised with the selected pump vendor for the type and size of pump, the pipe and pump installation method in segments, and the gas sealing arrangement. The method statement and operating procedure were prepared and risk analysis carried out with the client.

The January 2003 weather was very bad with rain, high winds and cold. The lifting of equipment to the tank roof, installing the pump, installing the temporary discharge pipework, supports and insulation took several days. Following on from this the tank pump-down was achieved in 45 hours from 380 mm to 41 mm. The remaining five de-commissioning steps were successfully carried out ahead of programme, ready for tank modifications and re-certification.

### *Construction and modification*

In order to convert the four tanks without any disruption to the commercial operations of Grain LNG, the construction was split into 3 phases. The first phase was to decommission, modify and $N_2$ purge one tank in the first half of 2003, followed by the second tank in the latter half of 2003. Once both tanks were modified and $N_2$ purged, they were then cooled down consecutively and filled with LNG from the remaining two in-service tanks. This allowed the two other tanks to be emptied of LNG and available for warm up in the first quarter of 2004, ready to be decommissioned, modified and brought back into service in parallel, by August 2004.

Safety was paramount throughout the site works. There was little knowledge of what to expect inside tanks that were constructed over 20 years ago. As the facility had to continue in operation during construction, both Whessoe and Grain LNG had to be extremely vigilant. All work was carried out under permit to work systems. All personnel working on the site were required to be holders of safety passports, a recognised UK site safety qualification. All personnel that were to work inside the tank were trained for work in confined spaces as well as being trained in the use of Breathing Apparatus (BA), for use in emergency rescue. Two rescue teams were trained to ensure that work colleagues could be removed from the tanks in an emergency.

With two tanks still in service, a second means of emergency egress had to be in place prior to construction. Each tank had an emergency egress tower erected to the side of the tank, so that if a situation arose on either of the in-service tanks, construction personnel would be able to exit the tanks safely.

The condition of the inside of the tanks, in particular the condition of the suspended deck, was not known. Opening and entering the tank after air purging, was carried out under strict supervision, using rope access technicians wearing BA. This ensured that in the possible event of suspended deck collapse, or problems with the inner tank access ladder, there was a means to evacuate the personnel. Using BA was a necessity to ensure that rope access personnel could carry out gas checks on the tank floor to verify the there were no pockets of gas trapped within the tank after purging. Rope access systems were then set up to allow personnel working in the tank a safe means of access to the work fronts.

To ensure the tanks were acceptable to operate under the new importation conditions, Whessoe were required to carryout a number of visual inspections and NDE of the inside of the tanks to allow for them to be revalidated. Purlins and rafters were visually inspected to ensure there were no signs of deterioration or defects. The suspended decks were load tested, to allow personnel to walk on them during modifications. Inner tank floor plates were vacuum box tested and shell plates and weld tees were UT inspected. The roof compression ring weld was also UT inspected, as well as a number of thickness checks on both the inner and outer tank shell plates. Existing pump columns were visually inspected for in-service defects, both internally and externally.

The roof steelwork adjacent to the positions of the two new heat breaks for the Boil Off Gas (BOG) and the Liquid Top Fill (LTF) required modifications and strengthening. This required temporarily supporting three roof rafters to enable the centre rafter to be cut and additional steelwork installed to support the rafters and heat breaks.

Whessoe strategy throughout the construction phase was to minimise the site activities and to prefabricate as much as possible, at as large a size as practicable. Short schedule windows and consideration of the transport and cranage restrictions when working in a live gas plant drove this. The heat breaks and inner tank pipework were prefabricated as one unit to enable them to be installed into the tank roof openings. The Stilling wells and Thermo wells were prefabricated as assembled units and installed through the roof openings also.

The external pre-painted pipe support tower was assembled on the ground and erected as one unit to the side of the tanks, with BOG and LTF pipework spools erected in the structure. Both lines were pneumatically tested to 6 bar(g) and 11 bar(g) respectively from the double block and bleeds at the base of the tank to the thermal heat breaks on the roof of the tank. The pipework was then traditionally insulated and clad in-situ after pneumatic testing.

All redundant gauging systems were removed and replaced with new systems, together with relocating the existing LTD system. Existing in-tank pumps were removed to allow for testing and refurbishment. This enabled the client personnel to be trained in the in-tank pump removal and reinstatement procedure for future use on the plant.

To comply with API 620 appendix Q, the tanks required pneumatic testing, to a test pressure of 170 mbar(g), allowing all new welds to be soapy water tested. During the test the new pressure relief systems were tested individually to 125 mbar(g) and 140 mbar(g), ensuring the correct lift and seating pressures were obtained. Refurbished vacuum relief valves were also tested to –2 mbar(g) after the tank had been depressurised.

Operating on a live gas plant, Whessoe carried out the site works under a permit to work system, with continuous gas monitoring. The geographical position of the site on the South East coast of the UK at the mouth of the river Thames gave it an exposed position to high winds and heavy rains, which all had to be taken into consideration in the 16 week construction schedule.

To ensure that the tanks were kept dry whilst they were open and until commencement of $N_2$ purging, compressed dried air was fed into the tank outside of the working hours to keep a positive pressure within the tanks. This prevented the suspended deck insulation and interspace resilient blanket and Perlite from absorbing moisture during construction and ultimately quickened the $N_2$ purging process.

## CONCLUSION

The success of the tank refurbishment contract at the Isle of Grain is being achieved through the blending of all of the EPC capabilities at Whessoe combined with a fully engaged client with extensive operations capability. The close integration of the entire team in handling engineering and construction interfaces has been invaluable in maintaining schedule and safe and continuous operation of the plant. Whessoe were also successful in securing the contract for the new marine facilities and Regasification plant in competitive tender. First gas importation is expected in January 2005.

# LNG TANDEM OFFLOADING SYSTEM

*L. Terry Boatman, P.E.*
*FMC SOFEC Floating Systems, Inc*
*14741 Yorktown Plaza, Houston Texas 77040-4063*

*William B. Mitchell, P.E.*
*FMC Loading Systems*
*1803 Gears Road, Houston, TX 77067*

*Prepared for 2004 AIChE Spring National Meeting*
*LNG VII – Offshore & FPSOs*



## Abstract

Marine loading arms for handling LNG have been supplied by FMC Loading Systems™ for forty years. FMC SOFEC™ has provided offshore vessel mooring systems for over thirty years. Through this experience, an LNG offloading system using proven components from these two related technologies provides for LNG transfer rates up to 15,000 $m^3$ LNG/hr in seas up to 4.3 m significant wave heights. The new tandem mooring arrangement provides increased vessel safety and higher availability through increased vessel separation and by the methods of berthing the LNG carrier. Improved economics are achieved by the reduced need for large tug auxiliary vessels, and also by the reduced need for thrusters and DP controls on the LNG carrier.

Concepts for tandem offloading of LNG have been in development for over ten years by FMC Loading Systems. Model basin testing has been done in industry JIP studies. A large-scale working model has been tested using vessel motion data from the model basin tests. The new yoke system that utilizes a duplex mooring yoke has been model basin tested for operation in harsh offshore environments.

In 2003 FMC Energy Systems launched an improved LNG offloading concept that provides significant advantages to tandem moored LNG transfer. This new concept utilizes a new revolving crane arrangement to add separation distance between vessels, simplify access for maintenance and reduce the overall elevation height of the system. The revolving crane swings the pantograph back on board the FPSO for parking but does not take up any more deck space than the previous stationary boom concept. LNG transfer is accomplished with the crane boom securely locked in the outboard position. The duplex yoke maintains the LNG carrier motion within the operating envelope of the pantograph with the crane boom fixed. No movement of the crane boom is required while offloading. Another advantage of this concept is variable boom height capability that allows for greater relative draft change between the two vessels. If large draft changes of the vessels are not involved, then further simplification and lower cost will result from a non-adjustable height boom.

## Introduction

Offshore transfer of liquefied natural gas (LNG) from one vessel to another is a technical and operational challenge, but side-by-side (SBS) transfer of liquid hydrocarbons and minus 48C° LPG is commonplace and is a proven practice in mild environments. Also well known is tandem (one vessel directly behind the other) transfer of oil through floating hose systems for floating production systems to oil tankers. However, transferring LNG at minus 160°C between two vessels is not common, and has not been done at all in a production capacity in any serious weather locations. SBS transfer of LNG with conventional loading arms has been proposed and studied for many years; and model testing and full size loading arm testing have been performed by FMC Loading Systems to prove those systems are satisfactory for moderate environments. Vessel motion is the limiting factor in the severity of environment that these systems can successfully operate in, especially during the flange connection phase of the operation. One example of an exposed weather LNG terminal facility that experiences fairly constant wave frequency vessel motion is the Shell Bruni loading terminal. This is a facility about 3km offshore, located on a jetty without a breakwater, where LNG was originally transferred through a tandem loading system consisting of multiple degree-of-freedom flexing LNG piping arrangement connected with cryogenic swivel joints. The original loading system served for over twenty years. The facility now operates with conventional FMC loading arms that utilize constant motion cryogenic swivel joints to load LNG.

The technology for transferring LNG between two floating vessels is a crucial component in floating LNG facilities now in the planning stage. One of the primary applications for this technology is the floating hydrocarbon production unit for field developments where large quantities of gas are present. Unfortunately, many of the largest offshore gas fields are situated in areas of unfavorable weather, such as in the Timor Sea and offshore South Africa. There are some in the industry that consider offshore LNG transfer technology to be a “blocking” technology, or at least one more hindrance to the progress of some major LNG projects. Although the general weather conditions at some promising offshore West Africa locations are not severe, a persistent sea swell is expected to cause undesirable vessel roll motions during LNG loading. An LNG transfer system that provides maximum operational availability and safety is needed as a part of commercial justification of several LNG production projects.

Berthing an LNG carrier to another floating vessel, and remaining moored there for 12 to 24 hours will require perfected berthing procedures and robust dependable equipment. A series of developments by FMC Energy Systems in the evolution of tandem vessel berthing and LNG transfer equipment has led to the system described in this paper, a system ideally suited for environments such as West Africa and other locations where a tandem LNG transfer system is the preferred solution.

## Developing the Tandem Offloading System

### Early Developments – The First Boom to Tanker (BTT)

The Shell Bruni LNG facility was the first boom-to-tanker system, and operated successfully for over twenty years. Although this loading system was mounted at the end of a jetty, not on a floating vessel, the system was required to connect with a moving vessel and load LNG while the vessel underwent wave frequency motion in its spread mooring. See Figure 1.

**Figure 1: Shell Bruni LNG Loading Facility**

The Bruni facility is now operated with conventional FMC LNG loading arms that incorporate constant motion cryogenic swivel joints as illustrated in Figure 2.

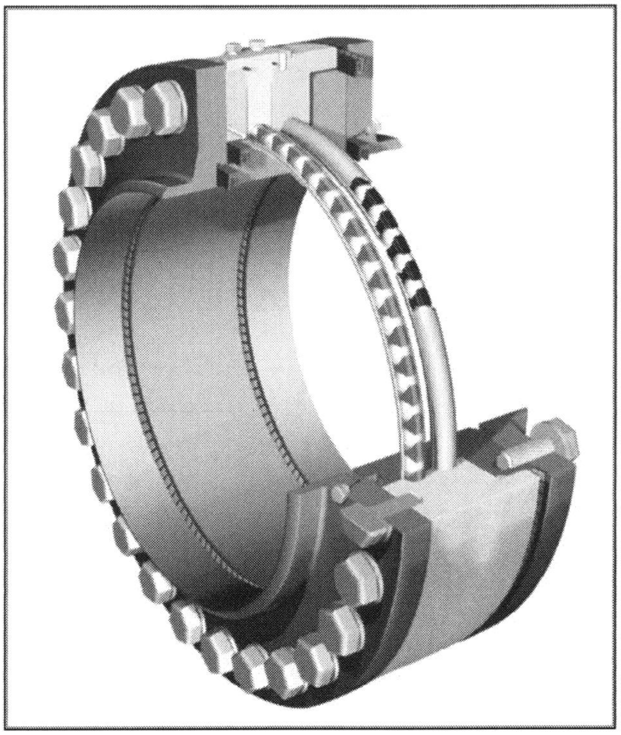

**Figure 2: Chiksan Cryogenic Constant Motion Swivel Joint**

Loading arms for a typical ship-to-ship transfer system arranged in a side-by-side berthing is shown in Figure 3.

**Figure 3:Offshore Ship-to-Ship Transfer Loading Arms**

## Boom to Tanker (BTT) for Tandem Ship to Ship LNG Transfer

All of the proposed BTT ship-to-ship LNG transfer systems utilize the piping pantograph as the flexible piping component that allows relative motion between the vessels. Constant motion cryogenic swivel joints and stainless steel pipe provide all required degrees of freedom of motion. Figure 4 illustrates the original BTT system concept. A carrier connection manifold system is required on the bow of each LNG carrier. Because the system used a nylon hawser to connect the vessels, large sway motions of the carrier required the crane boom to follow the motion of the carrier bow manifold. Otherwise the manifold travel could over-reach the physical limits of the pantograph, especially when combined with significant vertical vessel motion.

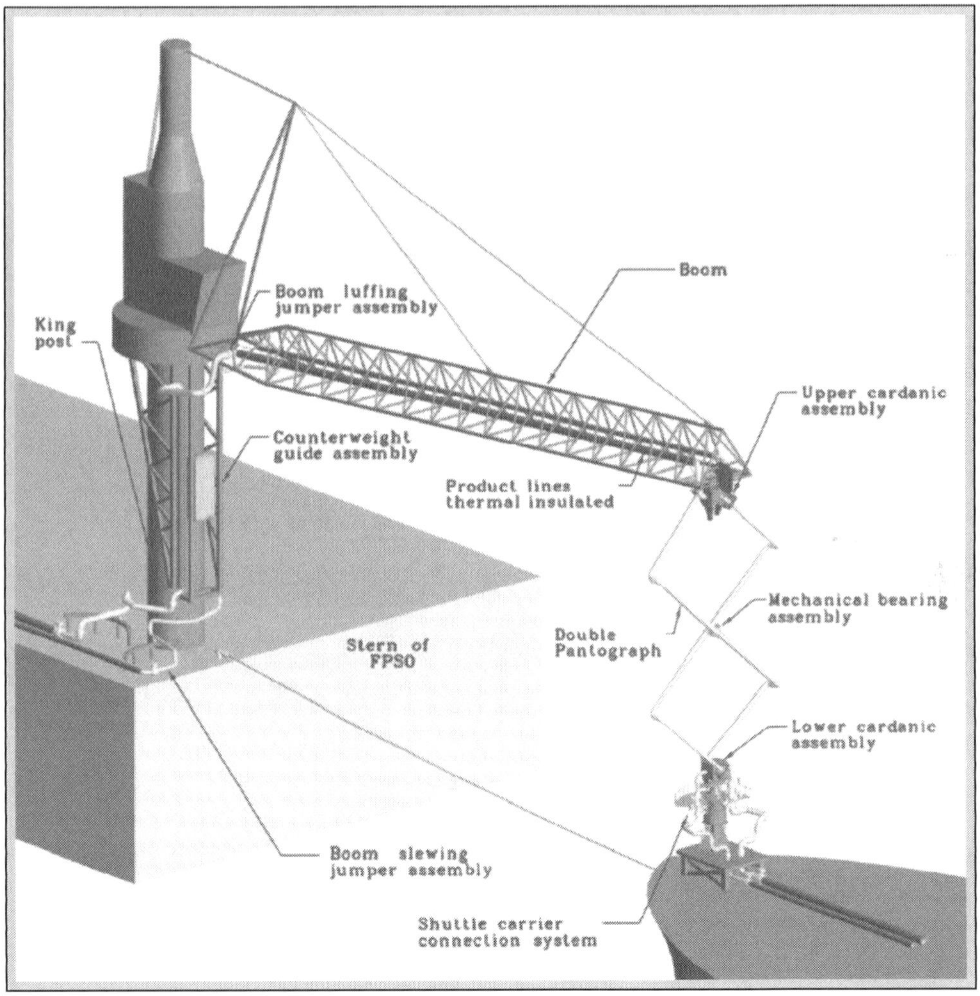


Figure 4: BTT for LNG FPSO

Figure 5 shows the one fifth of actual size working model of the BTT system that is on display and is operable at the FMC Loading Systems factory in Sens, France. This working model includes a computer controlled motion table that operates to simulate motion between the two vessels.

**Figure 5: One-Fifth Scale Working Model of BTT**

## Methods for Berthing the LNG Carrier

Three methods have been evaluated for berthing, or connecting, the LNG carrier to the LNG FPSO. The original BTT system concept used a nylon hawser to connect the vessels, but this method has little resistance for side-to-side motions, and was limited to moderate sea states. A conventional soft yoke system was evaluated and found to allow about 36% of the hawser moored vessel motions, still excessive unless the pantograph boom follows the LNG carrier motion. A new yoke concept, the "duplex yoke" was developed to further minimize the lateral relative motion between the two vessels. This made it possible to keep the pantograph support boom stationary with respect to the FPSO. For comparison, and with all other variables equal, the lateral motion allowed by the duplex yoke is only about ±10.5m relative to the FPSO. Refer to Figure 6 for an illustration of comparative motions between the three methods.

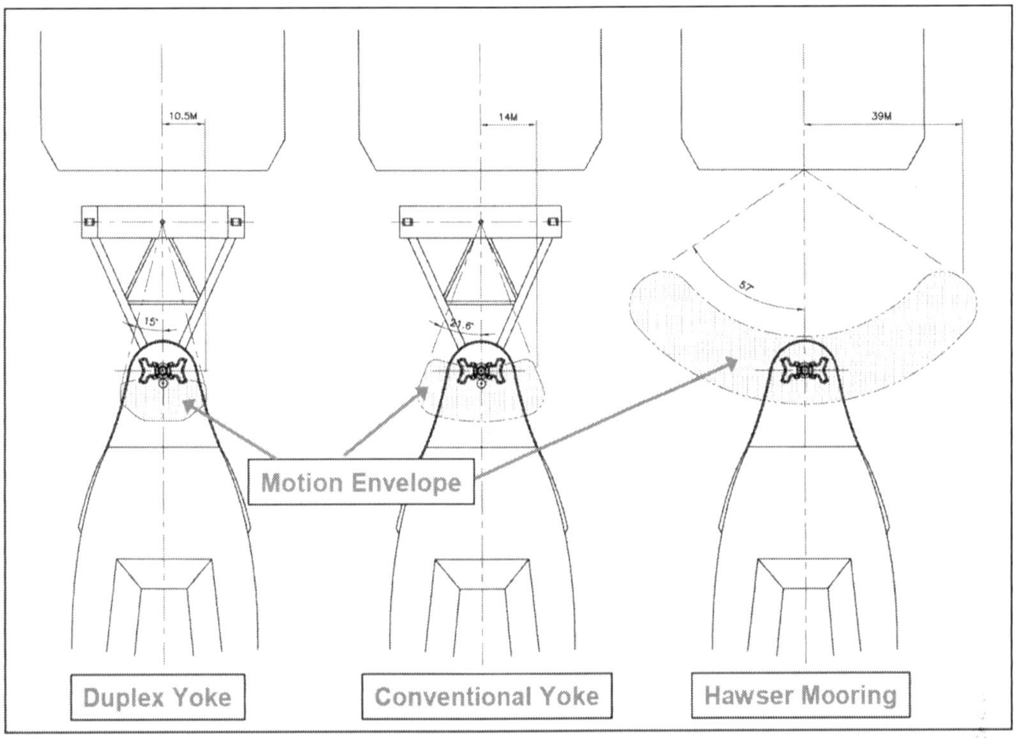


**Figure 6: Motion Envelopes at LNG Carrier Manifold**

Figure 7 shows the duplex yoke for connecting the LNG FPSO to the LNG carrier while the carrier is loading LNG.

**Figure 7: Duplex Yoke Mooring System**

**Stationary Boom Concept for Tandem Offloading**

The stationary boom concept is illustrated in Figures 8 and 9. In this system, the piping pantograph remains suspended at the end of a stationary boom. When the pantograph is not being used it is retracted upward and a service platform moves out to surround and secure the pantograph. Any service or testing of the pantograph would be done from this platform.

**Figure 8: LNG FPSO with Stationary Boom Tandem Loading System**

**Figure 9: Stationary Boom Tandem Offloading System in Operation**

### Revolving Boom Tandem Offloading Loading System

The latest evolution of the tandem offloading system is illustrated in Figures 10, 11, and Figure 12. This concept retains the advantages of the stationary boom when in operation, that advantage being the boom does not follow the motions of the LNG carrier. In this case, the boom is locked in the outboard position and does not rotate while loading LNG. The rugged box construction of the boom safely allows for all required roll motions and side loads applied to the piping pantograph. However, the boom can be raised or lowered ± 4.5 meters to allow for large variations in draft of the two vessels. When the system is not loading an LNG carrier, the boom can be revolved around 180° to secure the boom onto a boom-rest. Then all necessary inspection and maintenance work is readily done on board the FPSO, as illustrated in Figure 13.

**Figure 10: LNG FPSO with Revolving Boom Loading System**

**Figure 11: Revolving Boom Tandem Loading System**

Figure 12: Revolving Boom Tandem Offloading System

Figure 13: Tandem System in Parked Position

## Comparison of Stationary Boom and Revolving Boom Tandem Loading Systems

The improvement between system configurations is evident in Figure 14.

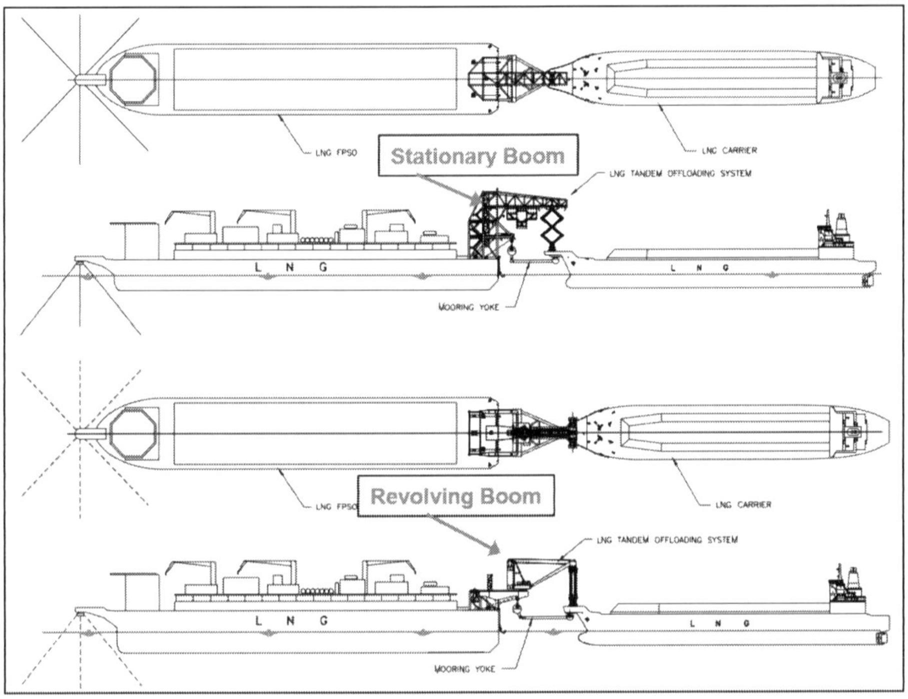


**Figure 14: Comparison of Stationary Boom and Revolving Boom Systems**

## Advantages of the New Revolving Boom Concept

Concept improvements over the stationary boom concept include the following advantages:

- There are 16m (23%) added clearance between the two vessels
- Between vessel perpendiculars measures 75 meters
- The overall height of the structure above waterline is reduced 10m (15%)
- The boom height is adjustable ± 4.5 m, to provide for large variable draft difference between the two vessels
- Boom swings 180° to park piping pantograph onto service platform for improved safety and service access
- Because of the outboard location of the boom swing bearing, no additional deck space is required over the space previously needed for the stationary boom concept.

## Conclusion

The FMC Energy Systems revolving boom concept for tandem LNG transfer between two vessels offers reliability and safety for operation in harsh sea conditions. The system is comprised of components based on service proven equipment and this feature allows for the shortest possible time for designing, classing, construction, and commissioning.

# Maximizing Value-Added LNG Regasification: Modular ORV for Offshore LNG Terminals

*Nobuya Himoto*
*General Manager, Heat Exchangers Engineering*

*Yoshihisa Nakamura*
*General Manager, Heat Exchangers Marketing*

*Yoshiro Tonoike*
*Manager, LNG Vaporizers Engineering*

*Sumitomo Precision Products, Co. Ltd.*

*Prepared for Presentation at AIChE Spring National Meeting*
*4$^{th}$ Topical Conference on Natural Gas Utilization*
*New Orleans, LA, April 2004*

## ABSTRACT

Time consuming environmental and permitting issues force project developers to alter the paradigms of re-gasification terminal locations from onshore sites to offshore locations. Since an Open Rack Vaporizer (ORV) uses seawater as a heat source, it is much more economical and environmentally friendly than any other vaporization method. The ORV upholds low operating costs and minimal maintenance requirements and does not emit harmful greenhouse gases. This fulfills the energy project developers' obligation and promise to protect the environment.

Modular engineering and construction technology have been developed and applied to the ORV design for offshore applications. The modularized vaporizer is pre-fabricated at the manufacturer's shop and the modules are transported to the Gravity-Based Structure (GBS) construction yard. This modular approach minimizes site fabrication and increases constructability, resulting in a reduction in the total cost.

This paper presents the development of the modular ORV as it specifically applies to offshore regasification terminals. Several important concerns, such as the motion effect on even seawater distribution, the developed methodology of shop fabrication and its integration at the GBS construction yard, the weight of light equipment, and the equipment's lifespan in relation to anti-corrosion and anti-erosion are also discussed. The paper also presents the optimum design of an ORV using the technically innovative large star-fin tubes and improvements in operation and maintenance.

## INTRODUCTION

Import LNG is a critical supplement to the US natural gas supply. As immediately available resources are depleted, LNG will play an increasingly larger role in the country's energy supply. An LNG receiving terminal must be in close proximity to the markets and easily accessible to LNG carriers. However, the receiving end of an LNG chain is often located in highly populated areas where land is scarce. Moreover, due to the lengthy environmental regulatory processes and the strict permitting issues, developers have shifted their focuses towards offshore regasification terminals.

The process of offshore LNG Terminals constructed on a gravity-based structure (GBS) consisting of a couple of large concrete caissons, which will be floated to the site and lowered to rest on the seafloor in about 50 - 100 ft of water, is similar to the land base LNG terminals. LNG carriers coming from the natural gas liquefaction plants will offload cargoes into the GBS storage tanks (using concrete or steel hulls). The LNG will then be regasified by the Open Rack Vaporizers (ORVs) and the regasified LNG will then be transported by pipelines to existing trunklines for final delivery to consumers. Since an ORV is a relatively simple device which uses seawater as a heat source, the operation and maintenance costs remain low and the avoidance of the emission of the greenhouse gases is achieved.

The concept of modular engineering and construction is not new. The modular concept has been used in petroleum and chemical industries for decades, namely to build modules for use in gas and crude oil process plants on the North Slope and on barge-mounted process plants destined for use in remote areas of the world. The modularized vaporizer is shop fabricated and the pre-assembled modules are transported to the GBS construction yard for installation. The modular approach is especially attractive in congested or remote areas, such as GBS, where an adequate labor force is not readily available, or where severe weather makes working conditions difficult. Consequently, much of the work can take place under controlled conditions within the module assembly area.

This paper addresses the development of the modular ORV design specifically applicable to offshore regasification terminals. The important issues pertaining to ORV modularization, such as modularization screening factors, modular engineering, size, transport, fabrication, and schedule effect are also discussed. This study presents the cost analysis and benefits of modularization. Technical features of the advanced heat exchanger tube to be used for the modular ORV are also presented. In summation, this paper will expound on the fit-for-purpose and cost effective modularization of the ORV applicable to the GBS LNG terminal.

## ADVANCED DESIGN OF THE MODULAR ORV

### Conventional ORV Structure

ORV's are constructed of coated aluminum alloy tubes configured in a panel arrangement. Inlet LNG and outlet natural gas manifold piping connects the panels. Ancillary equipment includes seawater piping, seawater distribution equipment, structural supports and a vaporizer basin. The heat exchanger tubes generally have extended surfaces on the outside, and some form of a heat transfer enhancement device (such as a turbulator) on the inside.

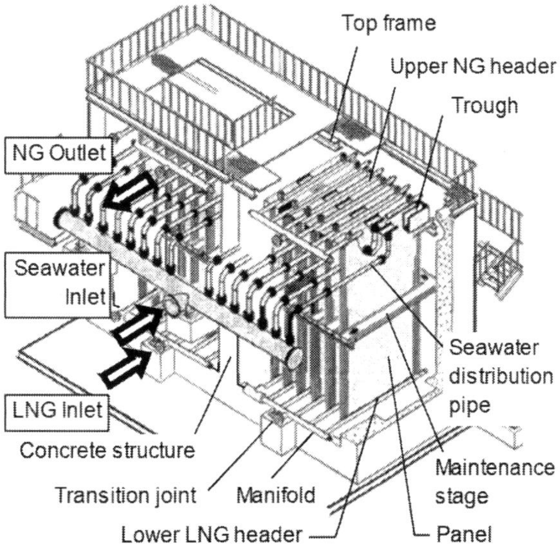


**Fig. 1 – Conventional ORV Schematic**

The ORV utilizes seawater as a heat source. The seawater is fed from an overhead distributor, which allows it to fall over aluminum panels. The water is then collected at the bottom from where it is routed and discharged back into the sea. In the ORV, a panel consists of the vertical exchange tubes with fins, and heat is exchanged between the LNG flowing inside the tubes and seawater falling on the outside surfaces of the tubes. Figures 1 and 2 illustrate the schematic structure of the conventional ORV and seawater distribution on each tube panel, respectively.

### Advanced Heat Exchange Tubes

The vaporization capacity is governed by many parameters. These include heat exchange surface areas, heat transfer coefficients, temperature differences between inlet and outlet seawater, and the amount of seawater. For an LNG vaporizer of the seawater falling film type, a limit in the reduction of the seawater flow rate from the standpoint of thermal balance exists because the usable heat of seawater is restricted. Increasing the outside wet area, which also means to increase the LNG vaporization rate per unit tube, is essential.

The HiPerV is a technologically advanced ORV, developed by Sumitomo Precision Products, Co., Ltd. along with the cooperation of Tokyo Gas Co., Ltd. The advanced exchanger tube used for HiPerV is supported and exemplified through LNG tests carried out with the cooperation of the Tokyo Gas Co., Ltd.

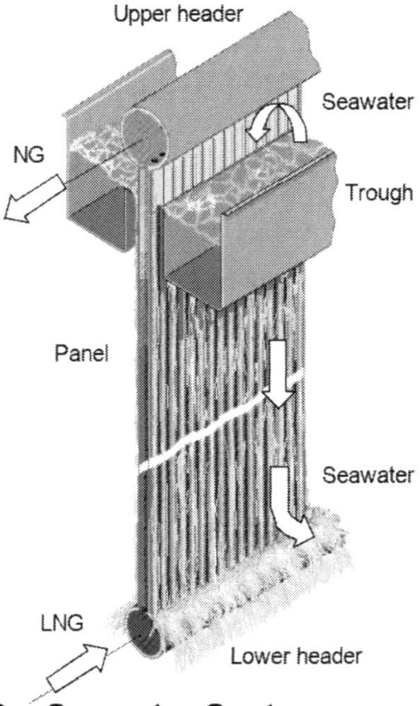


**Fig. 2 – Seawater System on ORV Heat Exchange**

The newly developed HiPerV tube increases surface area per unit length and improves thermal transfer rates. Table 1 presents the technical features of the HiPerV in comparison with the conventional star-fin tubes. This improvement of heating surface on the seawater side has expanded the range of variation in the seawater flow rates, and has enabled the selection of an economical seawater flow rate that corresponds with the seawater inlet temperature and the LNG flow rate. Within the parameters of the design, if the HiPerV has the same plot area as that of a conventional ORV, seawater flow rates can decrease by about 20%, or if the HiPerV flows the same quantity of seawater, then panel areas can reduce by about 30%.

**Table 1 – Technical Features and Comparison of Star-Finned Tubes**

| Description | **HiPerV** | **Conventional Type** |
|---|---|---|
| Heat exchanger tube | Large size 64 tubes / panel | Normal size 90 tubes / panel |
| LNG and seawater heat exchange capacity | Extraordinarily High<br>• Reduction in icing height by optimization of tube shape and heat transfer accelerator combination<br>• Improvement of seawater holding capacity by increasing the outside wet area | High |
| Tube strength | Very Good<br>• A5052 aluminum alloy<br>• Larger cross-sectional area and section modulus | Good<br>• A3003 aluminum alloy |
| Plot area (180 t/h basis) | 6 (W) x 11 (L) = 66 $m^2$<br>(10 panels / unit) | 6 (W) x 15 (L) = 90 $m^2$<br>(18 panels / unit) |

Since 1969, over 30+ years, Sumitomo Precision Products has continued to develop ORV technology. Roughly speaking, there are four (4) generations of heat exchanger tubes used for the ORV. The first is a Pair-finned tube, the second is an 8-Star-finned tube, the third is a 12-Star-finned tube (we refer to this tube as the conventional one), and the last one involves the large 12-Star-finned tube for HiPerV. The large 12-Star-finned tube was developed as a result of the new design method to effectively determine the optimal shape of heat exchanger tubes.

## Modular ORV Specification

The modular ORV specifications, based on HiPerV tubes, stem from fit-for-purpose design and depend on project conditions. The following are typical modular ORV specifications:

- Unit LNG vaporization capacity: 90 – 270 t/h (100 – 310 MMscfd)
- Design pressure: up to 150 barg (2,175 psig)
- Seawater temperature: 5 – 30°C
- Typical seawater delta Temp.: 3 – 7°C

## Modular ORV Structure

Each ORV unit is divided into several pre-fabricated ORV modules. Each module contains several panels, consisting of 64 exchange tubes, all of which are connected by means of manifolds at the upper and lower ends, thus creating a panel type heat exchange unit. Figure 3 shows the schematic drawings of the modular ORV.

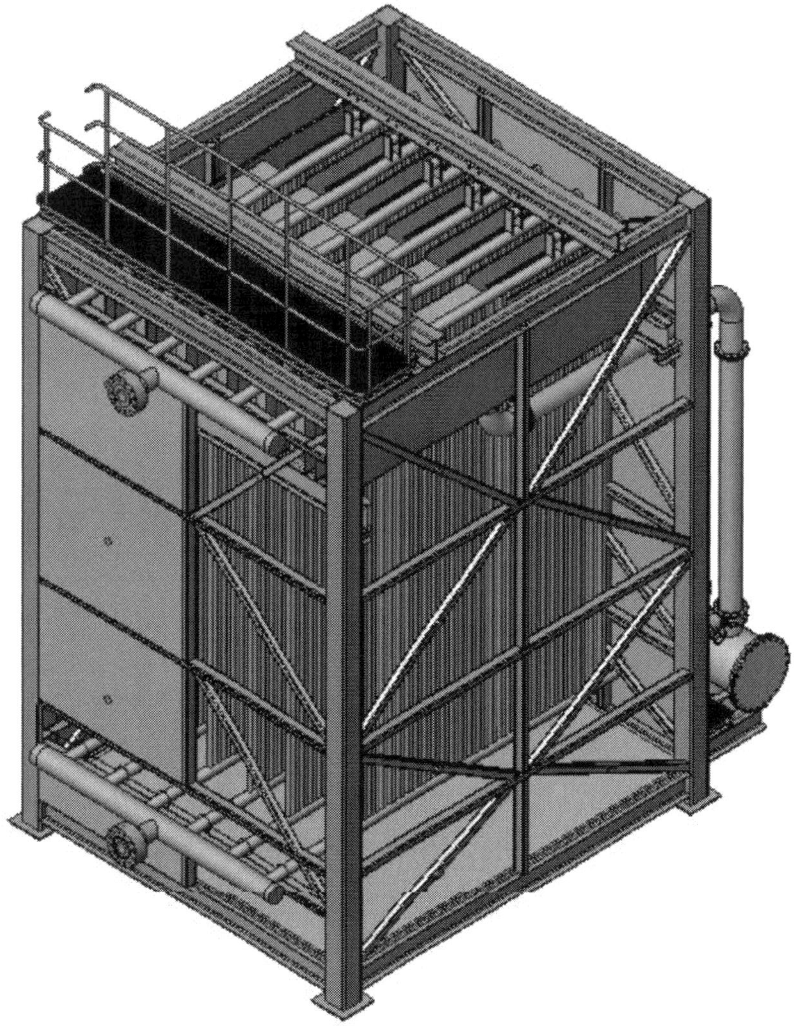

**Fig. 3 – Modular ORV**

A sample configuration of the developed modular ORV with HiPerV tubes is as follows:

Design Conditions:

| | |
|---|---|
| • LNG regasification rate: | 180 t/h |
| • Seawater inlet temperature: | 10° C |
| • Seawater outlet temperature: | 4° C |
| • ORV type: | HiPerV-6m tubes |

Modular ORV Configuration: (Typical)

| | |
|---|---|
| • Total number of modules per unit: | 2 modules |
| • Total number of panels per module: | 5 panels per module |
| • Total number of tubes per panel: | 64 tubes per panel |
| • Total number of tubes per unit: | 640 tubes |

- Required seawater flow rate: 5,600 $m^3$/h
- Overall dimensions of a module: 6 x 5.5 x 8 m (W x L x H)
- Overall dimensions of a unit: 6 x 11 x 8 m (W x L x H)

### Motion Effect on Seawater Distribution

Seawater distribution is very important for successful ORV operation. Even the slightest level of tilting can arouse serious effects on ORV operation. Required leveling accuracy for one seawater trough is 0.04%. The modular ORV is designed to allow for a total of 0.3% level of tilting during long-term operation. A 0.3% level of tilting can be adjusted by using thin shim plates that re-adjust the trough within a 0.04% level of tilting every year or couple of years.

### Anti-Corrosion of the Structure

The standard material of the structure is carbon steel on which heavy duty anti-corrosion painting is applied. Stainless steel, hot dipped galvanized carbon steel, or aluminum coated carbon steel are also applicable as an option.

## MODULARIZATION SCREENING

### Drivers of Modularization

Five 'Factors of Influence' are consolidated into the following categorized drivers. In order of importance, these are: Site Location, Labor Availability, Construction Characteristics, Environmental Factors, and Project Risks. Each of these drivers has a series of attributes that help define its importance.

#### Site Location

Site location is one of the most significant factors affecting the decision to modularize. At offshore LNG Terminals, offsite construction may be beneficial because of the shortage of skilled craftsmen within the local workforce. Since the offshore LNG terminal possesses limited site accessibility during construction, modularization may relieve the need to mobilize craftsmen to remote areas.

#### Labor Availability

Substantial differences between effective labor costs at the module fabricator and the jobsite must exist because adequate skilled labor or support infrastructure is not available. The cost differential of the labor performed at the jobsite and the same work done in a shop environment is a powerful driving force which aids in the promotion of the benefits of modular construction methods. The reduced jobsite labor force and more stable labor relations contribute to the completion of scheduled goals and may result in higher field efficiency. However, in certain industrial areas, the political influences upon labor and the location upon which the modules are fabricated may affect the extent of modularization [1].

#### Construction Characteristics

Construction characteristics must be carefully evaluated in order to achieve the benefits of offsite prefabrication. Since the offshore GBS LNG terminal encompasses limited deck space, and because both the plot space and working areas are limited, congested working conditions result. This can be alleviated by offsite prefabrication and module construction at GBS construction yards. Reduced equipment weight is an essential parameter in the design of the GBS system. The modular ORV implements the use of steel frames instead of concrete structures. As a result, the design provides flexibility in equipment arrangement and operation and provides greater maintenance space.

Environmental Factors

Another factor to consider in evaluating the modular approach is the effect of weather conditions. Severe weather conditions may reduce productivity and may require more time to complete construction, compared to the onshore LNG terminals. Weather guarded working environment increases the productivity and the quality of workmanship by skilled craftsmen. As a result, the modular concept, which requires more engineering than the conventional ORV, can bring benefits in job quality and compensate for the high engineering costs.

Project Risks

Although additional front-end planning efforts are necessary with modularization, these efforts improve the overall GBS regasification terminal design and construction schedule. An integrated planning effort might have a positive effect on both cost and schedule risks [1]. Prefabrication at offsite locations is not affected by the pre-work process, which could include establishing an equipment foundation as required for conventional construction. Construction at one or more offsite locations permits flexibility to accommodate for unforeseen events. In some cases, this flexibility might contribute to a reduction in project risks.

The summarized screening factors to be considered in the evaluation of modularization are presented in Table 2.

**Table 2- Modularization Screening Factors**

| |
|---|
| Site location and accessibility |
| Job site weather |
| Availability of skilled construction labor |
| Cost and productivity of local labor |
| Plant layout/process configuration |
| Construction/installation approach |
| Size and weight constraints |
| Schedule |
| Transportation |
| Impact on plant operation/maintenance |

## Advantages and Disadvantages

Modularization has been successful at offshore platforms, arctic locations, and plants in remote locations. Restricted site space, potential availability of skilled labor, and the likely availability of sea access prompted the initial feasibility. The needs for additional costs are often cited as reasons against modularization. Although the need for extra materials, monitoring, and increased transportation is inevitable, a number of other important benefits which offset the cost argument also exist. In addition, a significant investment of time devoted before the start of the project to establish the detailed plans of the engineering, supervision, materials usage, and transportation areas is vital to ensure the success of the project [2].

Since the modular ORV has been developed with steel structured frames instead of with concrete structures, as used for the conventional ORV, the weight of the ORV is significantly reduced. The reduced weight is one of the great advantages of the modular designed ORV applicable to GBS LNG regasification terminals. Improved quality standards are a key benefit of modularization. The off-site pre-fabrication of modules under controlled conditions, in purpose-designed surroundings readily accessible to special equipment, leads to more efficient and effective workmanship. This also provides higher acceptance rates and ultimately, lower overall construction costs.

Modularization transfers labor hours from the field to a controlled area, such as a GBS construction yard, or another construction site that can be set up and organized with multi-craft skills specifically selected for the type of modules to be built. The modular design limits the possibility of a decrease in productivity in remote areas where existing infrastructures create the inability to fabricate the elaborate ORV module because of the harsh climates.

If the start of construction is delayed due to the late receipt of permits or the limited availability of access ways or land, then modularization may allow much of the plant to be built elsewhere. The constructed parts of the plant can then be shipped to the job site when physical work can start, thus saving schedule time [3]. Modules can be designed without undue congestion and safety hazards. Adequate maintenance access can be provided.

Pre-fabrication under controlled conditions optimizes productivity, safety and quality. QA/QC requirements can be better applied and better controlled. Safety conditions are also improved as parent sites become less congested and as the requirements for in-site fabrication in confined or high-level situations are minimized.

The total installed cost and/or schedule of some projects is improved where site labor costs are exceptionally high, where labor productivity is exceptionally low, or where sufficient amount of skilled labor is not available locally.

## Modularization disadvantages

Realistically, for any given project, the advantages offered by modularization must be weighed against certain disadvantages. Modules will take longer to construct and entail higher costs. Designing the modules requires more engineering time because a more detailed design must be completed before module fabrication can begin.

A modularization decision must be made up front in order to allow for an engineering team to create a design for the modules, to identify the fabrication/assembly areas, and to justify the economics of transportation.

Another potential problem with prefabrication/modularization involves the precise definition of work left for the field to do. Packaged units or modules are not assembled as often desired. Therefore, detailed specification and good quality control is required to ensure that those responsible for prefabrication and modularization successfully complete all work required.

The modules must be self-supporting and transportable. In some cases, temporary bracing and supports can be used for transportation. But in the modular ORV designs, much of the additional steel is incorporated into the module. Moreover, shipping involves more time.

## MODULAR ENGINEERING OF ORV

### Modular Engineering and Construction

Modular Construction is generally defined as the use of prefabricated and packaged plant assemblies or subassemblies that can be transported and installed at remote locations. In modular construction, the emphasis is on the packaging and transport [4].

On the other hand, Modular Engineering can yield significant improvements in plant quality and costs effectiveness regardless of the type of plant, its location, or the construction method used. Benefits are achieved by improved process and equipment designs, and more efficient plant arrangements [4]. The application of modular engineering concepts in an ORV design will almost always facilitate modular construction of the GBS regasification terminal.

A modular system costs more to design because of the coordinating efforts required by the design team, and because of the premium placed on the compact design. A module typically contains 60-80% more structural weight. However, steel is a relatively low-cost component pound for pound, and structural steel is one of the cheapest bulk items in a plant. Higher capacity cranes are needed to install modules. On the other hand, structural erection costs are usually reduced, and foundation work is simplified [5].

The key to the success of such projects is the integration of engineering and construction planning. Both planning and construction engineers must meet regularly to review plans and drawings to ensure that the design is constructible. It cannot be stressed enough how important it is to have a constructible design from the outset.

### Modular Size

The size and weight of modules should be as large as feasibly practical to minimize the cost of handling, transportation and field labor. However, sizes and weights are influenced by factors such as: the availability of land and sea transportation systems, handling equipment (cranes and jacks), plant layout, module transportation restrictions, and structural integrity [6].

In order to generate standard dimensions of the modular ORV, the following modular concept parameters of design and fabrication must be considered:

- Module size and weight limits
- Acceleration loads imposed by shipping
- Foundation design parameters
- Means of marine and land transport

Each module size restriction is determined to be maximum 6.5 x 6 x 8 (W x L x H) because this dimension affords safe and easy handling by land/marine transport. The weight of the module base frame and framing steel per ton of supported equipment increases in direct proportion with the increase in module size and weight. As weight does not normally present a problem, the maximum weight expected for a typical modular ORV is less than 50 tons. Modules of this size are not costly to transport and no limitation on the type of contract on which the technique can be used exists.

### Transport

Generally, ocean shipment of modules is done by barge or heavy-lift ships. Barges are either self-propelled or towed. Heavy-lift ships and certain self-propelled barges have higher speeds and therefore can more readily avoid storms, resulting in safer voyages. Barges tend to have higher horizontal, but lower vertical acceleration forces than do heavy-lift ships [6].

The vessel size will determine the number and size of the modules that can be shipped on each voyage, thus influencing transportation costs. The shipyard's barge crane can routinely lift single modules weighing up to 630 tons. Transporters in the shipyard also transport modules weighing up to 500 tons [7]. This system of handling permits the modules to rest directly on the ship's lower and upper decks, eliminating the need for support trestles.

Module transportation was studied in detail and an overall plan was developed which included ocean transport, loading and unloading methods, and the setting of modules. It was determined that equipment was available or could be fabricated to handle modules up to 500 tons. This information was then used to set module size and weight criteria.

### Fabrication

Assembly of modules can be done in a large GBS construction yard. With the advent of offshore construction technology, many facilities exist worldwide for fabrication and assembly of large and heavy modules.

Dimensional accuracy of the module is critical in modular construction. Modules must precisely fit with each other at the plant construction site. The basic engineering philosophy for the allowable fabrication tolerance for piping at the matching point of each module must be no more than plus/minus three (3) millimeters. In order to adhere to this tolerance, great care must be taken in the fabrication of panel blocks and structures. If adjustments are required based on the final measurement, a special shop adjustment will be performed to maintain the allowable tolerance.

Assembly work functions normally done at the jobsite such as receiving, warehousing, fabricating, and erection are done by the same classifications of personnel at the assembly facility. Early in the project, the area which will allow for simultaneous work on several modules is determined. Supervision is handled by people qualified to do similar work at the jobsite.

### Schedule

Significant differences in the engineering schedules for modular and conventional plants exist. For a conventional project, engineering work can not begin without having complete information on the location of equipment and piping. In a modularized plant, the detailed design of the modules can start without knowledge of equipment and piping locations, their weights, and other support related information. Consequently, these differences play a role in the procurement strategy to be adopted for the modularization project.

A successful modularized project requires more careful planning, more disciplined execution, and earlier sequencing of a number of activities. Timely issuance of the drawings, accuracy of drawings, coordination of equipment deliveries, rapid solution of day-to-day problems, and coordination of fabricators are all key factors in meeting the project schedule requirements. With proper planning and execution, modular projects consume less time from kick-off to commissioning than do field erected projects [8]. Site preparation and preparation of the foundation occurs in parallel with equipment manufacturing and shop fabrication activities. Weather related delays are non-existent in a fabrication shop. Earlier plant start-up translates into an early market entry and early cash flow [8].

## COST ANALYSIS AND BENEFITS

ORV cost compositions for the conventional and Modular types are shown in Fig. 4. The compositions are estimated based on the following conditions:

- ORV capacity: 180 t/h (205 MMscfd)
- Seawater delta Temp.: 6°C
- Seawater requirement: 5,600 $m^3/h$

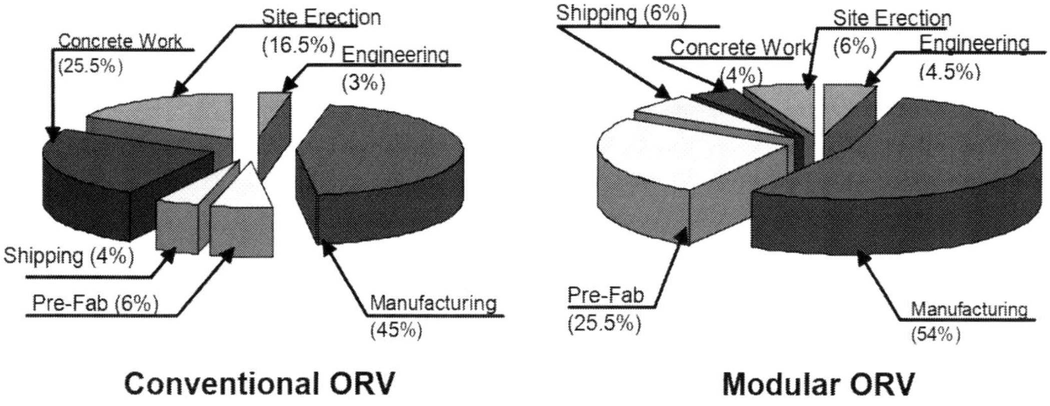


**Fig. 4 – ORV Cost Comparison**

Engineering and planning for modular construction requires more man hours even when the calendar time allotted to engineering does not change. Engineering man hours, which represent about 3% of the total project cost, have increased to 4.5%. Most of this effort is in "closure-type" engineering, formerly done during construction. Construction drawings are provided with much more detailed information now, and the design is essentially complete before construction begins.

Foundation for the conventional ORV frequently requires concrete piles, which are quite expensive, and involve long construction periods. This study does not include the piling cost because its cost varies depending on the soil conditions. Since the concrete structure is replaced by the shop fabricated module steel structure, foundation work is simplified with a modular design. The cost share decreases to 4.0% from 25.5%. The concrete water basin is the only concrete work needed for the modular ORV.

The modular ORV reduces field erection cost composition to 6% from 16.5%. Larger modules are typically preferred since this reduces the overall number of modules and field labor. However, larger modules are naturally heavier, which causes an increase in transportation costs and limits the transportation options.

An increase in shipping cost is due to an increase in module weight. The module size and weight for a typical modular ORV are 6m x 5.5m x 8m (W x L x H) and 50 tons, respectively. The maximum shipping size and weight for a block of the conventional ORV is 6.5m x 6m x 7.5m (W x L x H) and 25 tons. As the module weight increases, module handling becomes more difficult, resulting in an increase in the handling and shipping costs.

Figure 5 illustrates the relative cost analysis of the modular ORV compared to the conventional ORV. The cost evaluation of the modular ORV includes a series of trade-offs between shop fabrication and transportation costs, field concrete work, and installation costs. A modular unit requires cranes and other lifting equipment with heavier lifting capacity; however, the duration for which this equipment is required on-site is limited. The overall total installed cost of the modular ORV can save 7% compared to that of the conventional ORV.

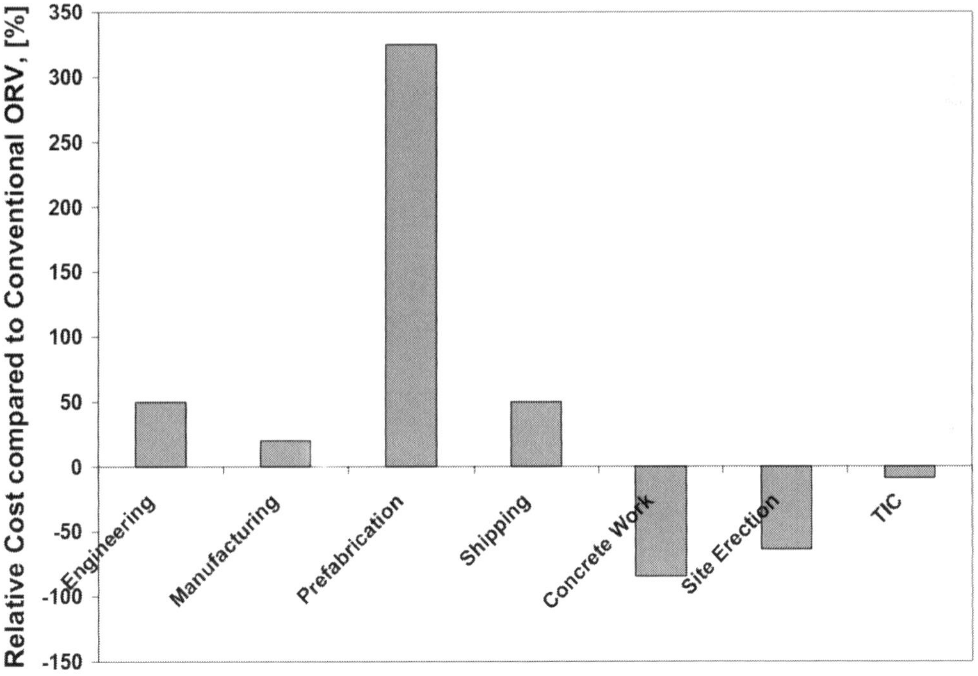


**Fig. 5 – Relative Costs Analysis of Modular ORV Compared to a Conventional ORV (based on 205 MMscfd, dT = 6° C)**

Preassembly at the manufacturer's shop is expected to have beneficial effects on the cost and timing of the project. The cost savings increase as productive use of labor in a more stable environment, with better working conditions, also improves. The weather will have less of an influence and safety measures will be enhanced. Wage costs tend to be lower where continuity of employment is greater and the setting up and overhead charges should also decrease.

The modularization also offers the potential for reducing the time period of the project since preassembly can overlap with GBS body construction. The manufacturer's facility can operate simultaneously and the remaining work onsite is freed of the limitations congestion can cause.

## CONCLUSIONS

The newly developed HiPerV tube has enabled the selection of an economical seawater flow rate that corresponds with the seawater inlet temperature and the LNG flow rate. Within the parameters of the design, if the HiPerV has the same plot area as that of the conventional ORV, seawater flow rates can decrease by about 20%, or if the HiPerV flows the same quantity of seawater, then panel areas can reduce by about 30%.

The modular ORV has been developed to be applicable to the offshore LNG regasification terminals and it implements the use of steel frames instead of concrete structures. The modular ORV reduces its weight by one fifth, compared to that of the conventional ORV with a concrete structure. As a result, the design provides flexibility in the upper deck arrangement of GBS and provides operation maintenance space.

The modular ORV at GBS LNG regasification terminals requires low capital costs in seawater lift structure. The vaporization by use of the ORV is much more economical and environmentally friendly than any other vaporization method because the ORV upholds low operating costs and minimal maintenance requirements and does not emit harmful greenhouse gases.

Though modular ORV increases the engineering, prefabrication, and shipping costs, it eventually reduces fabrication work, resulting in a reduction of the total installed cost by about 7% based on a typical reference case as specified in the Section of Cost Analysis and Benefits. Additional intangible benefits also exist, which have not been quantified in this study, such as: the advantages of being light weight, the simplicity in the use of the upper deck arrangement, the flexibility of schedule management, the benefits of labor control, and the achievement of high quality.

## ACKNOWLEDGEMENT

The authors wish to thank Sumitomo Precision Products Co., Ltd. (SPP) for permission to prepare and present this paper. Sincere thanks are offered to Tokyo Gas Co., Ltd. for their cooperation in developing HiPerV. Special thanks are also extended particularly to Shozo Hotta at the SPP New York Office for his kind assistance and valuable discussions.

## REFERENCE CITED

1. D. L. Stubbs and P. D. Emes: "Prefabricating a Process Plant", Mechanical engineering, p. 63-65, Nov. 1990.

2. D. Pope: "Offsite Modularization Proves Cost Effective at Mobil Refinery", HTI Autumn Quarterly, p.55-59, 1994.

3. A.E. Kerridge: "Plant for Constructability", Hydrocarbon Processing, p. 135-145, Jan. 1993.

4. R.R. Tarakad, C.A. Durr, and R. Hunt: "Modular Engineering – Application in Liquefaction Plant Design", presented at the GasTech 1986, Nov. 1986.

5. R. Clement: "Deciding When to Use Modular Construction", Chemical Engineering, p. 169-170, Aug. 1989.

6. M.A. Tan, R.P. Kumar, and G. Kuilanoff: "Modular Design Construction", Chemical Engineering, p. 89-96, 28 May 1984.

7. C.P. Boudreaux, W.W. Lewis, E.E. Smith: "The Application of Shipyard Modular Construction Techniques to Other Industrial Projects", presented at the AIChE Spring National Meeting, New Orleans, 07 April 1986,

8. A.I. Malik, R.D. Key, and H.M. Hudson: "Modularization: A Practical and Economical Construction Technique for Process Systems in the Gas Processing, Petrochemical, Chemical, and Oil Refining Industry", presented at the AIChE Spring National Meeting, April 1990.

# $CO_2$ as Precooling Refrigerant in Floating LNG Production Plants

*Jostein Pettersen*
*(corresponding author)*
*Statoil ASA*
*Research and Technology*
*NO-7005 Trondheim*
*Norway*
*jospet@statoil.com*

*Heinz C. Bauer*
*Linde AG*
*Engineering Division*
*DE-82049*
*Hoellriegelskreuth*
*Germany*
*heinz.bauer@linde-le.com*

*Pål Rushfeldt*
*Aker Kværner*
*Engineering &*
*Technology*
*NO-1326 Lysaker*
*Norway*
*pal.rushfeldt@akerkvaerner.com*

*Prepared for Presentation at AIChE Spring National Meeting*
*4th Topical Conference on Natural Gas Utilization*
*New Orleans, LA, April 2004*

## ABSTRACT

Floating LNG production is a promising concept for exploitation of offshore gas fields in remote areas. When locating the LNG process plant on a floating structure, the safety issues with flammable refrigerants become more critical. One way of reducing this problem is to replace the hydrocarbon precooling cycle refrigerant with carbon dioxide. This may also give a more compact liquefaction module with reduced weight, due to the high vapor pressure of $CO_2$. Owing to the low critical temperature of carbon dioxide, the cycle efficiency is an issue. By taking in cold cooling water from large depth, this disadvantage may be reduced. The paper provides results from a pre-study on $CO_2$ precooling for floating LNG plants, using the load-balanced MFC®s3 process developed by the Statoil-Linde Technology Alliance. Key features of the process and its performance are outlined, and the practical potential of $CO_2$ precooling is evaluated.

## INTRODUCTION

Floating LNG production, storage and offloading facilities (LNG FPSOs) are now considered as a realistic option for remote offshore gas fields that cannot be economically exploited with conventional onshore technology. A floating concept may therefore become the preferred solution for production of the deep-water gas reserves in the NnwaDoro blocks off the coast of Nigeria. The key features of a deep-water LNG FPSO for tropical conditions were outlined by Berger, Sparby and Bölt (2003), with a concept using cold seawater from a depth of 1000 m for cooling the high-efficiency Statoil-Linde MFC®s3 (Mixed Fluid Cascade) liquefaction process and the gas turbine intake air. Further aspects of large-scale LNG FPSOs were outlined by Graff (2003).

The FPSO consists of a permanently moored floating hull, with integrated LNG and LPG storage tanks. Topsides facilities include gas receiving and treatment, liquefaction, and utilities. One area of concern is the total safety of the installation. Here, a number of factors are important, including safety in connection/disconnection and loading situations, storage tank integrity, and fire/explosion in the various areas. Regarding fire and explosion hazards, the design and location of process equipment plays an important role.

Use of non-flammable refrigerant in the liquefaction process may reduce the hazard in this part of the system. Earlier studies have focused on nitrogen expander cycles for offshore LNG, but one key issue here is the high power consumption of such a process. Another non-flammable and inert refrigerant option is carbon dioxide, which may operate in a vapor compression cycle giving reasonable efficiency. With a freezing point of –56.6°C, this is of course an option for the precooling process only (Fredheim and Paurola, 2000). Since most of the hydrocarbon refrigerant inventory is in the precooling cycle, a change over to $CO_2$ may still improve the safety of the liquefaction process significantly. On this background it was decided to conduct a study on $CO_2$ precooling, considering factors like process efficiency and capacity, as well as impacts on liquefaction module size, weight and cost, in addition to safety. The present paper gives a summary of the findings from this joint study.

## REFRIGERANT CHARACTERISTICS OF CARBON DIOXIDE

Apart from being non-flammable, carbon dioxide is distinguished from the common hydrocarbon refrigerants for natural gas liquefaction by its high triple point (freezing) temperature of -56.6°C. This restricts the minimum evaporating temperature in precooling due to risk of dry ice formation. For a refrigerant that is to be operated up to ambient temperature, $CO_2$ also has a quite low critical temperature (31.1°C), which is comparable to that of ethane (32.3°C).

As may be observed from Figure 1, pressures in a $CO_2$ precooling circuit will be relatively high compared to $C_3$ and even compared to $C_3/C_2$ mixtures. This may give reductions in suction piping diameter, suction drum size, and compressor dimensions due to the smaller vapor volume flow rates. Compared to $C_3/C_2$ mixed refrigerant, the pressure level is roughly doubled, thus reducing the volume flow by about 50%, giving a piping diameter reduction of 30%.

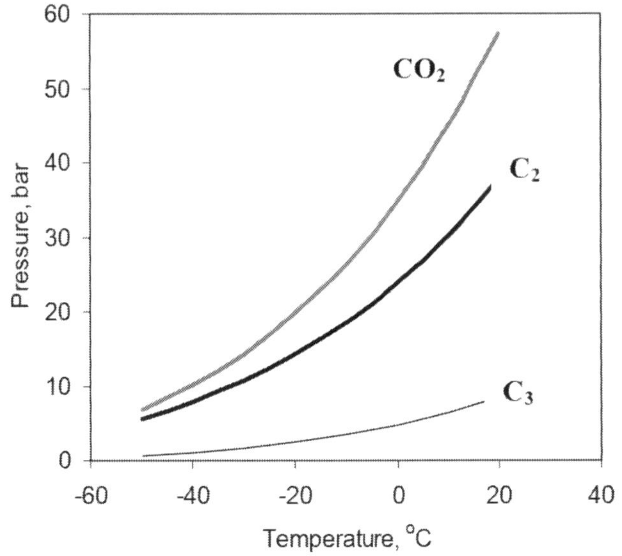


***Figure 1** – Saturation pressure curves of refrigerants*

The low critical temperature of $CO_2$ is a disadvantage since the throttling loss and heat rejection loss in the refrigerating cycle will be larger than for $C_3$ and $C_3/C_2$ mixtures. In addition, the heat transfer loss will be larger than with mixed refrigerant due to constant-temperature evaporation.

## DESIGN CONDITIONS

Table 1 gives the key design basis data for the liquefaction process. The plant will produce LNG with the specified gross calorific value, in addition to LPG and condensate.

***Table 1** – Design data*

| | | |
|---|---|---|
| LNG production capacity | 5.8 | mtpa |
| Gross calorific value of LNG | 40 | $MJ/Sm^3$ |
| Feed gas pressure (liquefaction inlet) | 69 | bar |
| Sea cooling water temperature | 5 | °C |

## LIQUEFACTION PROCESS CONCEPT

Initially, we studied a process concept where a $C_3/C_2$ precooling process with 3 compressor suction pressures was replaced by an equivalent $CO_2$ process, all other parameters left unchanged. Some key comments and findings from this initial screening are:

- Evaporating temperature in the first stage of the $CO_2$ precooling cycle was only a few degrees higher than the $CO_2$ triple point. This could lead to operational problems.
- The total power consumption for liquefaction increased by 9%, which was considered too much for acceptance.
- Recent developments of the MFC® process (Bauer et al., 2004) gives a potential for larger train capacity, and also fits well for $CO_2$ precooling since the compressor load in this part of the process is reduced. In addition, a new concept for NGL extraction leads to lower specific energy requirement (Bauer et al., 2004).

Thus, an updated load-balanced MFC®s3 process was chosen as a baseline; using three mixed hydrocarbon refrigerants and two precooling pressure levels. A corresponding $CO_2$-precooled process was designed, also having equal driver capacity at each stage, but using three precooling pressure levels. The load-balanced process gave a higher minimum $CO_2$ evaporation temperature, and reduced the impact of reduced thermodynamic efficiency of $CO_2$ compared to $C_3/C_2$ in precooling. Figure 2 shows simplified process flow diagrams of the $CO_2$ precooled case (left) and the mixed-refrigerant case (right).

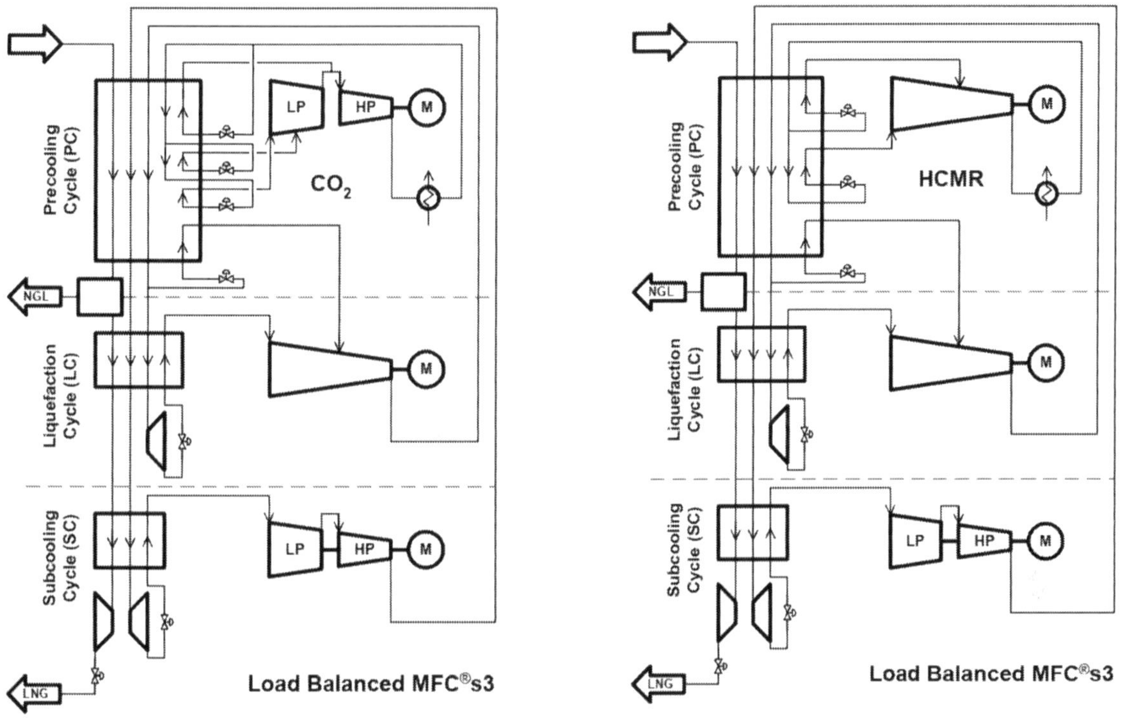


***Figure 2** – Simplified PFD for symmetrical MFC process with $CO_2$ precooling (left) and hydrocarbon precooling (right)*

In both cases, the total compressor power is equally distributed on the three drivers. This "load balancing" is achieved by allowing the liquefaction compressor to take over some of the precooling cycle load (Bauer et al., 2004), leading to a "symmetrical" process.

Owing to the higher operating pressure, the $CO_2$ precooling compressor was split into two casings, with a barrel type casing for the high-pressure stage. Not shown in the above figure is an additional subcooling heat exchanger on the $CO_2$ circuit, using seawater to subcool the liquid refrigerant after the receiver, in order to improve process efficiency.

## POWER REQUIREMENTS AND PLANT CAPACITY

Process simulations of the above two cases gave power requirement data as shown in Table 2, using the design data as shown in Table 1. As a result of the load-balanced process, the power input to the $CO_2$-precooled case was only 4.4% higher than the baseline. For a given maximum available power as defined by the hydrocarbon process case, this would correspond to a LNG capacity of 95.6% with $CO_2$ precooling unless more driver capacity is installed.

*Table 2 – Power requirements of the $CO_2$-precooled and baseline process*

| | $CO_2$ precooling | $C_2/C_3$ precooling | |
|---|---|---|---|
| Total shaft power | 162.7 (104%) | 155.8 (100%) | MW |
| Precooling compressor | 49.6 | 47.6 | MW |
| Liquefaction compressor | 50.5 | 47.7 | MW |
| Subcooling compressor | 50.5 | 48.5 | MW |
| Other power consumers | 12.1 | 12.0 | MW |

Temperature profiles in the form of hot/cold composite curves for the two cases are shown in Figure 3. The three $CO_2$ precooling temperature levels are easily observed in the left diagram. The highest pressure level to the liquefaction compressor is also considered part of precooling. Changes in the subcooling process are minimal between the two cases.

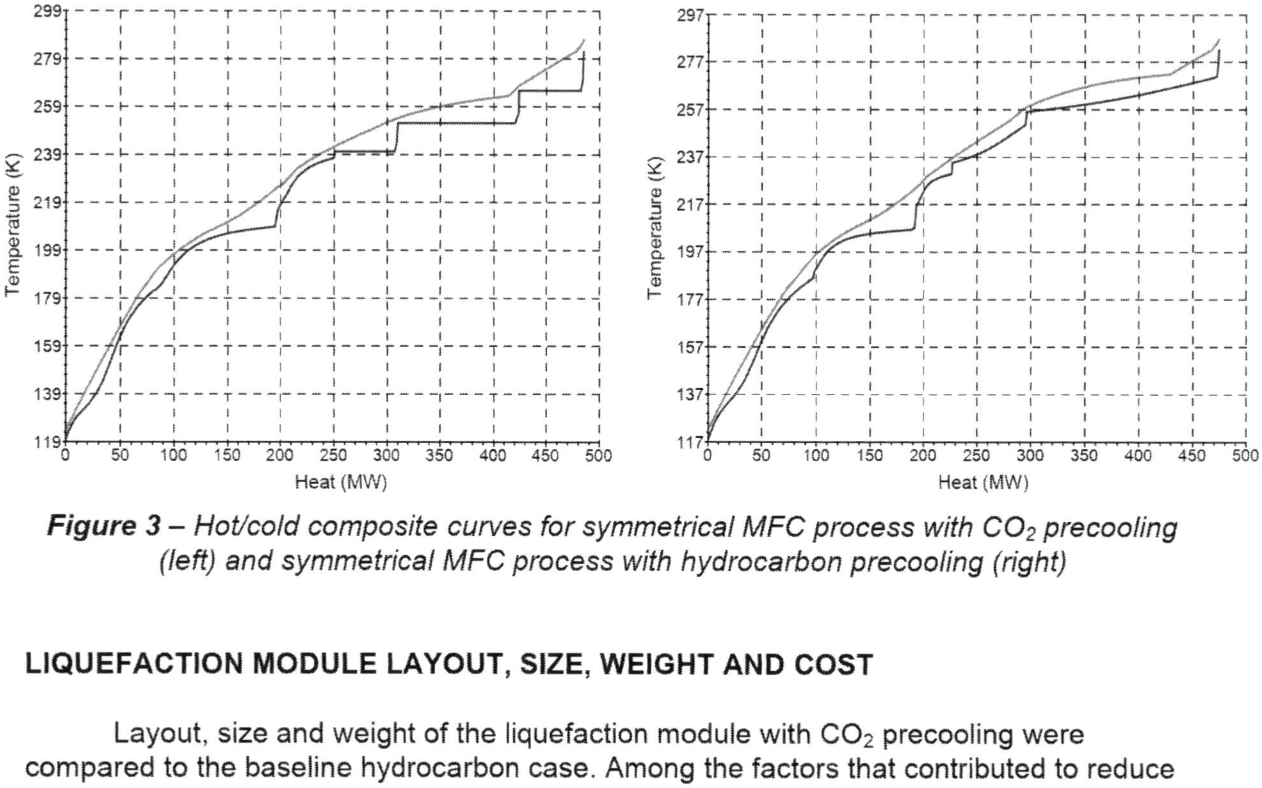


***Figure 3*** *– Hot/cold composite curves for symmetrical MFC process with $CO_2$ precooling (left) and symmetrical MFC process with hydrocarbon precooling (right)*

## LIQUEFACTION MODULE LAYOUT, SIZE, WEIGHT AND COST

Layout, size and weight of the liquefaction module with $CO_2$ precooling were compared to the baseline hydrocarbon case. Among the factors that contributed to reduce the equipment footprints and give smaller dimensions with $CO_2$ were reduced precooling compressor suction drum sizes and smaller precooling piping dimensions. Additional equipment caused by the third precooling pressure level/drum and the installation of a refrigerant subcooler made the net reduction in footprint area marginal, however. The plate-fin heat exchangers were reduced in size due to larger LMTD and less internal duty, but it was decided to maintain the sizes of the cold boxes. Some of the major pipe sizes in the liquefaction and subcooling circuit did not change much, and it is these pipes that to a large degree set out the deck heights, so no changes were envisaged relative to deck elevations. In total, it was concluded that the liquefaction module size would be the same for both cases, even though a few square meter reduction could have been possible for the $CO_2$ case. The weight of the module dropped by 100 tons, which is insignificant compared to the total weight of the module, which was estimated to 11 000 tons.

A rough estimate of the impact of $CO_2$ precooling on the equipment cost for the natural gas liquefaction and fractionation process units showed a very small cost increase in the order of magnitude of 10 million USD. This is well below the accuracy of the estimates, and insignificant in relation to the total cost of the modules. Changes in bulk material and further cost for changes in make up system, storage etc was not considered.

## SAFETY

A major safety concern of the LNG process with hydrocarbon precooling, especially when applied offshore, is the possible formation of a flammable and explosive hydrocarbon/air mixture in case of a major leakage in one of the refrigerant circuits. Thus, the minimization of hydrocarbon refrigerant inventory is quite important in terms of safety.

Figure 4 shows the refrigerant inventories in the $C_2/C_3$-precooled and $CO_2$-precooled processes, respectively.

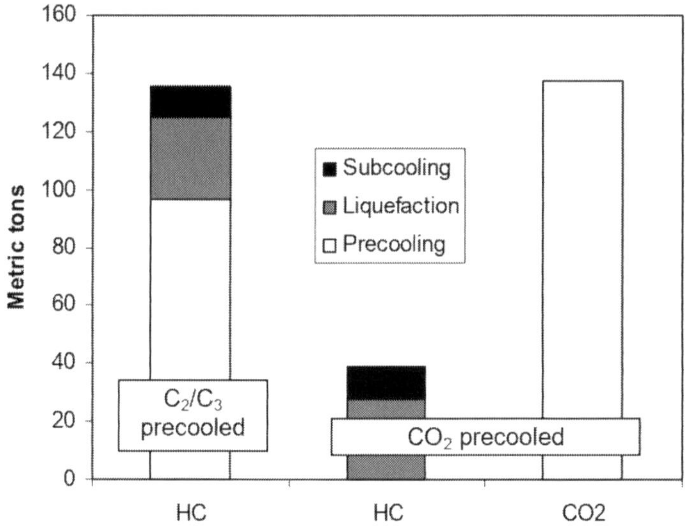


***Figure 4*** *– Refrigerant inventories in the process with $C_2/C_3$-precooling (left bar) and $CO_2$ precooling (two right bars)*

As may be observed from the diagrams, the HC refrigerant inventory is reduced by about 70% in the $CO_2$-precooled process. The reduced hydrocarbon charge is positive in relation to loss prevention and to the availability of the three main safety functions of the LNG barge, which are (i) main structural strength, (ii) main escape routes, and (iii) means of evacuation.

If the molecular weight of the hydrocarbon refrigerant is higher than that of air, a flammable cloud can accumulate inside or between the modules, and on the deck surfaces. Thus, in addition to minimizing the total hydrocarbon inventory it is of special importance to eliminate the heavier components, especially propane (52% heavier than air), but also ethane (4% heavier than air). By replacing the hydrocarbon precooling with $CO_2$, all propane is eliminated from the liquefaction module, and even though ethane is present in the liquefaction and subcooling refrigerants, both these mixtures have a molar mass that is lower than air. With $CO_2$ as precooling refrigerant, suffocation may be a safety issue (as it would for un-ignited propane which has the same molar mass). Especially with the low wind speeds that prevail off the West African coast this may be an issue with both $CO_2$ and propane.

## DISCUSSION AND CONCLUSION

From the above results we have found that the introduction of $CO_2$ precooling in a load-balanced MFC®s3 process does not give a significant increase in specific power requirement, or equipment size/weight/cost, while the safety of the process can be improved. The actual economic potential of the concept remains to be investigated and this will depend on the "value" assigned to safety improvements. It should be kept in mind, however, that the above results are based on cold seawater cooling. Cold water is of course a benefit for all LNG processes, but due to the low critical temperature of $CO_2$, this process is especially sensitive to increased cooling water inlet temperature. The $CO_2$ process efficiency would drop more than a corresponding HC process if the coolant (sea water or air) temperature were increased.

In relation to the above safety discussion it is interesting to note the guidance notes given by Lloyd's (2002) for offshore LNG production and storage installations, where a basic principle is that the design of the installation should adopt where possible aspects of inherent safety in order to prevent hazards occurring. Minimizing the size of hydrocarbon inventories to prevent escalation is one of the specific measures that are mentioned.

A quantitative risk assessment (QRA) is needed in order to provide numerical estimates of the risk (combinations of expected frequency and consequences of potential accidents) associated with liquefaction processes with $CO_2$ and $C_2/C_3$ precooling, respectively. On this basis the safety characteristics can be quantified and compared. Effects on overall risk of FPSO operation should also be evaluated.

A number of additional factors also need to be investigated, including blow-down situations, maintenance, and reliability/availability factors. The amount and type of equipment in a $CO_2$ precooling circuit is more or less the same as for hydrocarbon-based precooling, and it is not likely that the availability will be much different. A $CO_2$ circuit may be allowed to remain pressurized in some shutdown situations, and/or may be allowed to depressurize more rapidly. The risk of dry ice formation in piping and valves during depressurization of $CO_2$ is an issue here. Start-up situations also need to be considered, where the use of a single-component refrigerant like $CO_2$ may give a quicker procedure. The liquefaction refrigerant of the $CO_2$-precooled process only consists of two components, making start-up and operation simpler than with the baseline MFC®s3 process, where the liquefaction refrigerant has three hydrocarbon components.

Finally, we should note that a load-balanced process opens up for increased capacity of the process train. Calculations show that the installation of additional driver capacity can raise the LNG capacity under the above conditions by about 20%, to 7.0 mtpa, while still remaining within proven sizes of compressors, heat exchangers, and drivers. Owing to the lower specific power input, the all-hydrocarbon process capacity can be increased even further, to about 7.3 mtpa.

Conclusions from the study are as follows:

- Weight, size and cost of the liquefaction module is not affected significantly by a change over to $CO_2$ as precooling refrigerant.
- As long as the cooling water temperature is low, the power consumption for liquefaction is only a few percent higher, or LNG capacity is reduced a few percent for given power production.
- Hydrocarbon refrigerant inventory is reduced by 70%, which improves safety
- The effects of reduced hydrocarbon inventory on safety have not been quantified, and a QRA is needed in order to fully evaluate this.
- In total, the symmetrical MFC®s3 process with $CO_2$ precooling provides an interesting "improved safety" option for offshore LNG, where cold seawater cooling is available.
- The concept could even be considered for onshore plants in cold climates.

## REFERENCES


Bauer, H. C, Franke, H. E., Schier, M. W, Owren, G. A., 2004. MFC®s3 – the Formula of High Capacity LNG Base Load Plants. 2004 AIChE Spring National Meeting – 4th Natural Gas Utilization Conference, New Orleans, Louisiana, April 25-29

Berger, E., Sparby, B., and Bölt, M., 2003. The Benefits of Cold Deep Sea Water for LNG FPSO in Tropical Seas. 2003 AIChE Spring National Meeting, New Orleans, March 30 – 3 April 2003

Fredheim, A. O., and Paurola, P., 2000. Natural Gas Liquefaction Process. International Patent Application WO01/69149, filed 15 March 2000

Graff, O. Fr., 2003. Offshore LNG – Recent Developments. IBC's 18th Annual Conference & Exhibition - Floating Production Systems, London, 1 & 2 December 2003

Lloyd's, 2002. Classification of Offshore LNG production and storage installations: Guidance notes. DRAFT. Report No GN02015, Rev 1, October 2002.

# Qualification of a Flexible Pipe for use in a Transfer System for LNG Between Two Floating Vessels in Harsh Environment – an Update

Jorgen Eide, Harald Holden, Knut Erik Steen, Peter Jenkins – The OCL Group

## Abstract OCL System

This report describes the status of the qualification of the flexible pipe for LNG transfer as of 2004.

Many oil companies presently work on plans for production of Liquefied Natural Gas on floating vessels and offshore LNG storage. All these schemes require suitable systems for offshore loading of the LNG between floating production vessels or the floating storage terminals and transport tankers.

The OCL project has focused on developing a offloading system for turret moored FLPO's (Floating LNG Production and Offloading) to LNG tankers. The OCL system is a stern to bow, crane and flexible pipe solution, ref figure 1. The OCL system is based on extensive experience from tandem offloading operations of crude oil in the North Sea in harsh weather conditions. The system is well suited for LNG transfer where high regularity is required and a large number of cargoes are shipped every year.

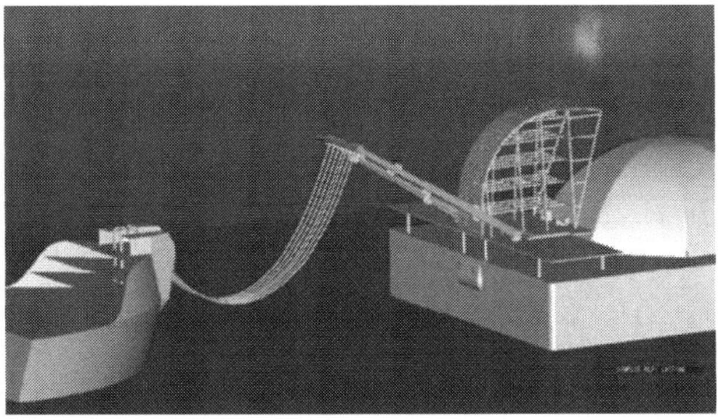

**Figure 1 The OCL LNG transfer system**

The LNG tanker is a standard dedicated LNG transport vessel with modifications in the bow to suit purpose built hawser and loading facilities.

During offshore loading, the FLPO and the LNG tanker will be moored together in a "crowfoot" hawser configuration reducing the relative movements between the FLPO and the bow of the tanker. The LNG tanker will be operating on constant stern thrust to ensure the required stability of the offloading system in both calm and harsh weather conditions.

Components and sub-systems have been selected based on functional requirements. Critical components have been qualified in a step-by-step process. The qualification included development of mathematical tools that were verified through material testing, model testing and full scale testing. System tests included manufacture and testing of the complete pull-in and connection system.

Vessel model tests were carried out for verification and calibration of the mathematical models and to ensure accurate calculation of the two vessel's motions in the loading area and thus good prediction of the loads on the transfer system.

The flexible pipe is the most critical part of the transfer system. The Nexans Kabelmetal's "Cryodyn™" super insulated cryogenic pipe has been the basis for the development of the flexible pipe, Cryodyn™. The Cryodyn™ pipe is based on a longitudinally welded, thin-wall stainless steel tube corrugated to provide the required bending flexibility. The finished flexible transfer pipe is a lightweight, double-wall construction with vacuum in-between the pipes as thermal insulation.
The Cryodyn™ pipe has been fully qualified for offshore transfer of LNG between two floating vessels in harsh environment. The qualification process has included:

- full scale flow tests with water and LNG
- full scale fatigue testing
- small scale fatigue testing
- torsion testing
- etc.

All tests and calculations showed that the OCL system accommodates safe and economic transfer of LNG from a floating production vessel to a tanker, thus allowing exploitation of stranded gas reserves.

## Summary of the Cryodyn™ pipe design features

The pipe is designed for 10 bar internal pressure and –164°C design temperature. The structural integrity have been checked for a typical FLPO and LNG tanker with maximum loading operating wave height of $H_S$=5.5 m.

Fatigue life is specified to 5 years operation including 100 connections and disconnection's per year, each with 24 hours loading duration.

**Table 1 Summary of Cryodyn™ pipe properties**

| Item | Data |
|---|---|
| Outside diameter | 358.0[mm] |
| Internal volume | 0.0619 [$m^3$/m] |
| Dry weight | 36.5[kg/m] |
| Weight LNG filled | 64.3[kg/m] |
| Axial stiffness, AE | 5200 [kN] |
| Bending stiffness, EI | 41.0 [$kNm^2$] |
| Torsion stiffness, GJ | 3487 [$kNm^2$] |
| Minimum bending radius for storage | 3.2 [m] |

The thermal exchange coefficient of the Cryodyn™ pipe is less than 0.1W/m°C. Outside surface metal temperature for a pipe filled with liquid nitrogen is in the range of 5-10°C below ambient temperature.

## Design of the Flexible LNG Pipe

The table below summarizes the build-up of the cross section of the Cryodyn™ pipe. Each part is also described in more detail in the following.

| | | |
|---|---|---|
| 1 | Helically corrugated and longitudinal welded stainless steel inner pipe | Provides internal fluid integrity and vacuum leak tightness of the structure and bending flexibility. The inner pipe also sustains the net internal pressure (hoop stress) and partial axial loading due to content, self-weight, net internal pressure and pressure drops. |
| 2,4 | Superinsulation consisting of reflecting foils and vacuum space. | The foils reflect heat radiation from outside (control radiation) while the vacuum space provides thermal insulation (control convection). |
| 3 | Spacer with low thermal loss | The spacer is an element to prevent direct contact of the inner corrugated pipe to the outer corrugated pipe without being a thermal conductor. |
| 5 | Helically corrugated and longitudinally welded stainless steel outer pipe | Ensure vacuum leak tightness and bending flexibility. The outer pipe sustains partial axial loading due to self-weight, net internal pressure (vacuum) and load transfer between steel pipes and reinforced rubber hose. |
| 6 | Tensile reinforcement | Provide axial stiffness and axial capacity of the flexible pipe, and in addition to protect the corrugated stainless steel pipe against mechanical damage. |

**Figure 2 Cross section of the Cryodyn™ pipe**

Corrugated steel Pipe

The main parts of the Cryodyn™ flexible pipe are the inner and outer corrugated steel pipes. These provide the pressure resistance, the main part of the axial and bending capacity during normal operation and are essential for the thermal insulation. The material used in the pipes is EN 1.4404 (or ASTM 316L), which is a molybdenum-alloyed austenitic chromium-nickel stainless steel with extra low carbon content. This material has been chosen for the corrugated pipes due to the following properties:

- High strength when cold worked
- High strength, high toughness and ductile behavior at LNG temperature
- Good fatigue performance both at room temperature and LNG temperature
- Good weldability
- Good corrosion resistance
- Good formability during the corrugation process.
- Combination of cost and availability

The corrugated steel pipes are thin walled, single ply pipes where the corrugation profiles have been optimized to provide adequate strength and flexibility for this application.

Dimensions, corrugation profiles and materials have been selected within the existing capabilities of the pipe manufacturer Nexans AG.

Spacer & Vacuum Insulation

The thermal insulation is provided by vacuum between the corrugated pipes. Spacers are used to avoid contact between the pipes. The spacers must be compatible with the temperature, have acceptable thermal conduction, be reliable such that contact between the corrugated pipes is avoided during the entire lifetime and not release molecules, which will influence the vacuum.

The spacers are mounted on the outside of the inner pipe before manufacturing of the outer pipe takes place. The spacers are not free to move.

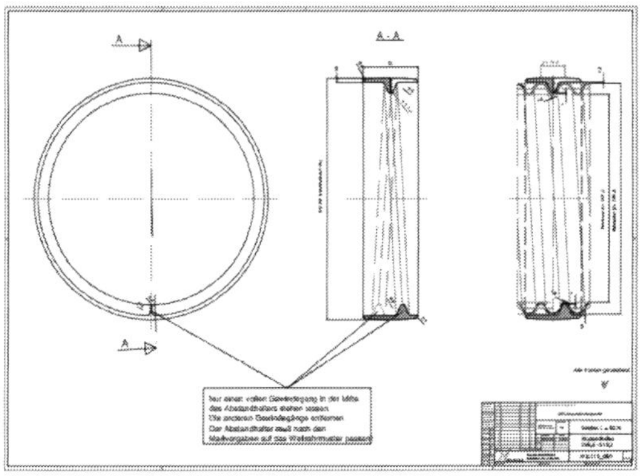

**Figure 3 Spacer Design**

Suppression of flow induced vibrations

Super insulation consisting of polyester fabric and reflecting aluminum tape is used to to improve the thermal insulation. The spacer and the super insulation are also designed to provide structural damping.

Extensive testing has proved that vibrations are effectively damped with this pipe construction.

End terminations and bend stiffeners

The corrugated pipes are terminated in rigid, stainless steel nuts welded to the pipe as shown in Figure 4. The end termination nuts are designed to take up axial loads and torsion moments, as well as providing a pressure tight transition between the two corrugated pipes in order to maintain the vacuum between the pipes. The end termination is fitted with a vacuum-measuring device that enables online monitoring of the vacuum. This device may be used for an automatic shut down of a line in case the pipe is damaged and vacuum is lost.

A slip-on bend stiffener takes up all bending moments in the pipe for the end termination. The bend stiffener is designed to give literally no bending in the pipe for a length of approximately one OD from the end termination. Outside this, the bending is allowed to increase gradually, with a smooth variation of bending moment within design limitations.

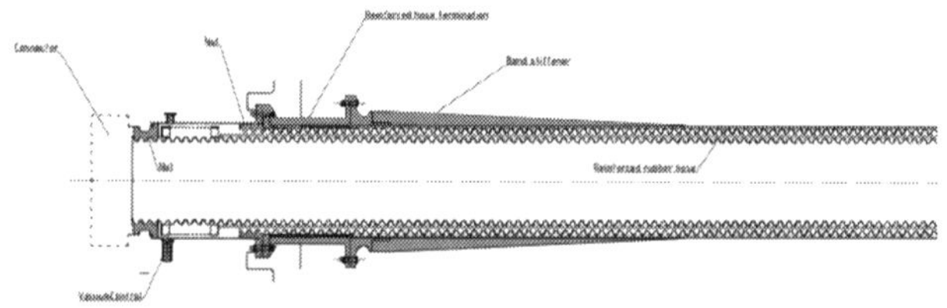

**Figure 4 End termination and bend stiffener design**

The bend stiffener is made from polyurethane and is is designed according to standards used for flexible lines used for oil production. The design life of the bend stiffener far exceeds the desired design life of LNG loading lines. The design temperature for the polyurethane bend stiffener is –40°C to +70°C, which is well within the operating surface temperature for the Cryodyn™ pipe, which is within 10°C of ambient temperature.

<u>Rubber Hose</u>

The protective rubber hose provides the flexible pipe with impact protection and additional tensile reinforcement.

For accidental axial loads the pipe capacity will go through three-stages:

1. The elastic stress in the corrugated pipes will provide the initial part of the capacity. For normal operation this capacity will never be exceeded.

2. At a certain load the corrugated pipes will start to yield in bending and the corrugations will be extended in the axial direction. The corrugated steel pipe always maintains a positive stiffness but during the initial part of the yielding this stiffness is less than the stiffness of the protective rubber hose. The loads will then be gradually transferred to the rubber hose. The rubber hose will be designed to withstand all accidental loadcases.

3. If the elongation exceeds the breaking strain of the rubber hose, the rubber hose will fail and the corrugated steel pipes will take the full load. The corrugations will then be fully extended. However, the tensile capacity of the steel pipes with extended corrugations is many times greater than the accidental loads, thus no leak will occur. After an accidental event where the corrugation are elongated the pipe will however loose its flexibility and it will no longer be fit for dynamic operation.

Design temperature for the rubber hose is related to the outside temperature of the cryogenic line. The cryogenic line will never expose the rubber hose to temperatures less than 10 degrees Celsius below the ambient temperature. If the design ambient temperature is set to –20°C, the design temperature for the rubber hose is minus 30°C.

## Analysis of the Cryodyn™ pipe

<u>Local stress in the corrugated pipes</u>

When a corrugated pipe is subjected to axial deformations, bending occurs in the corrugation root and crest. The axial stress will hence be dominated by bending stress (ref. Fig 1). In the hoop direction, the corrugation root will try to move inwards and the crest will try to move outwards during axial deformations, setting up hoop membrane stresses.

During the development of a fatigue SN curve based on coupon tests it is important to have equations for the local stress to ensure that the transition from the test coupons to a pipe is done correctly. Equations for pipe stiffness and local bending and membrane stress in corrugated bellow are presented in prEN 13445. The corrugation profile of the corrugated steel pipes manufactured by Nexans is outside the limitations for the stress equations presented in prEN 13445. FE analyses have shown that the performance of the prEN equations is poor for the Nexans produced pipe.

A new set of equations for pipe stiffness, bending and membrane stress in the root and crest of the corrugations have been developed based on Finite Element (FE) analyses of 19 different corrugation profiles. The equations include pipe stiffness and stress for a pipe subjected to internal pressure, axial strain, bending and torsion.

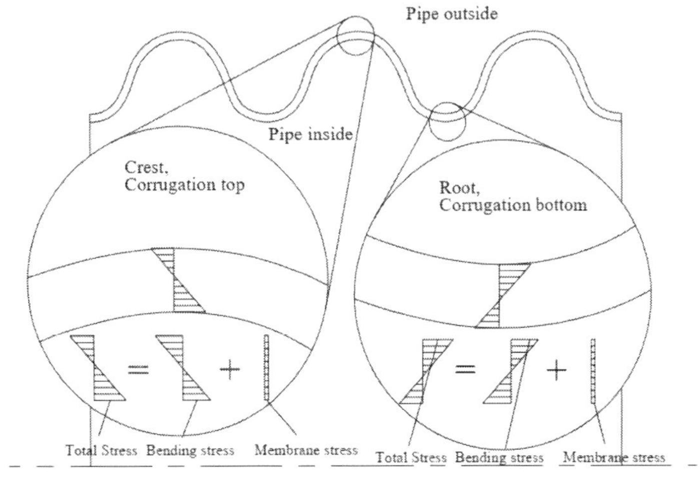


**Figure 5 Stress in a corrugated pipe due to axial deformation.**

These design equations are used in a von Mises stress check where the equivalent stress should satisfy the following requirements:

$$(\sigma_e)_{Pm} \leq \eta \cdot f$$
$$(\sigma_e)_P \leq 1{,}5 \cdot \eta \cdot f$$
$$(\Delta\sigma_e)_{PQ} \leq 3{,}0 \cdot \eta \cdot f$$

where

| | | |
|---|---|---|
| $(\sigma_e)_{Pm}$ | is von Mises primary membrane stress | N/mm$^2$ |
| $(\sigma_e)_P$ | is von Mises primary membrane plus bending stresses | N/mm$^2$ |
| $(\Delta\sigma_e)_{PQ}$ | is von Mises primary plus secondary stress range for normal operating | N/mm$^2$ |
| $\eta$ | is the design case usage factor | |
| $f$ | is the material strength | N/mm$^2$ |

The design equations have been verified by both strain gauge measurements of coupons from a corrugated pipe tested with axial deformations and test data from Nexans' pressure tests of corrugated steel pipes. The torsion capacity was verified by torsion tests of short pipe samples.

Load model for integrated pipe

The Cryodyn™ consists of two concentric corrugated steel pipes with an external tensile reinforcement. To be able to calculate the load in each component a model for load transition between the three pipes has been developed. The load model includes loads due to pipe weight, internal pressure, internal fluid flow, bending, torsion and tension. The load transition model was verified by FE analyses and full-scale pressure testing of an 8" ID double walled corrugated pipe.

Global pipe behavior

Global FE analyses of the OCL system consisting of a turret moored FLPO, the OCL system with ship to ship mooring and an LNG tanker were performed. The analyses results were calibrated against model tests.

Motion time series of the two vessels in 6 degrees of freedom were used as input to the local FE model of the Cryodyn™ pipe. The extreme response during 3-hour simulations was used in a design check of the corrugated steel pipes and the end fittings.

Fatigue analysis of the OCL system was also performed, and the fatigue damage was found to be acceptable for 5 years of operation in a harsh North Sea environment. The fatigue calculations included both fatigue failure due to axial (bending, internal pressure and axial tension) and torsion fatigue failure. The fatigue analysis was based on the project developed SN curve for the corrugated pipes. The fatigue life of the Cryodyn™ pipe has also been verified by a full scale bending test of a 7.5m pipe sample.

## Testing

The test program carried out for qualification of the Cryodyn™ pipe is summarized in the table below. More details of the tests are given in the following.

**Table 2 Summary of Cryodyn™ qualification program**

| Property | Test description |
|---|---|
| Fatigue testing | Small-scale tests of coupons cut from a corrugated steel pipe. The tests were performed both at cryogenic and room temperature. |
| | Full scale rotary bending test of single corrugated pipe |
| | Four point bending test of full scale vacuum insulated pipe at cryogenic temperature simulating 5 years of operation in harsh environment. |
| | Torsion fatigue |
| | Full-scale fatigue test of assembled pipe at cryogenic temperature. Tested at Marintek simulating entire lifecycle of pipe in harsh environment. |
| Structural capacity | Small scale tests of coupons cut out from a corrugated steel pipe |
| | Pressure test of full-scale vacuum insulated pipe. The test were performed both with curved and straight pipe. |
| | Tension capacity of terminated rubber hose used as protection against impact and accidental tensile loads |
| | Axial elongation and compression |
| | Torsion |
| Material properties | Small-scale tests of coupons cut from a corrugated steel pipe |
| LNG transfer rate | Full-scale test of flow through the inner corrugated steel pipe. |
| | Full scale test of LNG flow through Cryodyn™ pipe (with bend stiffener) at Ebara test facilities in Nevada. |

Axial direction (bending and tension)

Fatigue design curves, S-N curves, were established from testing of manufactured corrugated pipes for normal stresses (axial/bending deflection). The tests were performed on small-scale test specimens at room temperature and liquid nitrogen temperature.

Further, two full-scale verification fatigue tests have been performed with a vacuum insulated pipe with liquid nitrogen inside. See Figure 9, which shows the test at Marintek.

Torsion

Torsion fatigue tests were performed on 10,5" ID pipes with two end nuts welded on the pipe. Based on this test, a conservative fatigue resistance S-N curves for shear stress were proposed based on published design curves for welded structural steel. Ref figure 14.

7.5m Nitrogen insulation test (Nexans)

After trial manufacturing of the 10.5" ID Cryodyn™, the pipe was filled with Nitrogen to check the insulation capacity. The resulting temperature was well within the expected range and not below 10deg.

Water flow test & LNG flow test

During a flow test with water through the inner corrugated pipe, flow induced vibrations in the corrugated pipe was observed. The same test was repeated with the super insulating material wrapped around the pipe to introduce the correct damping properties of the pipe. The pipe was instrumented with several strain gauges measuring with a sampling frequency in the range of ~2400Hz.

The typical response included response below 10Hz and a high frequency response peak near 230Hz. The measured response at both frequencies is shown together with the SN curve in fig 9.

A test was also performed on a fully assembled pipe, both with water and LNG. The results from the flow test with LNG were better than with water, hence using the results from the water test is conservative. The LNG flow test arrangement is shown in figure 15.

The tests proved that fatigue damage due to flow-induced vibration is not a problem for the Cryodyn™ pipe.

Reference is made to figures 7 & 8 & 10

Full scale fatigue test at cryogenic temperature

A full-scale bending test of the 10.5" ID Cryodyn™ pipe was carried out at Marintek in Trondheim, Norway.

The purpose of the test was to simulate five years of operation in harsh environments. The test was designed to simulate five years of operation on the Fylla field in extreme harsh environment. 1 000 000 cycles were run, divided into 6 blocks.

The following instrumentation was used during the test :

- Tension – The input tension was continuously monitored.
- Angle – The input bending angle was continuously monitored.
- Internal temperature – The pipe contained liquid nitrogen throughout the entire test. The internal temperature was hence cryogenic throughout the test.
- External temperature – The outside temperature at the bend stiffener end fitting and at the pipe below the tip of the bend stiffener was measured throughout the test.

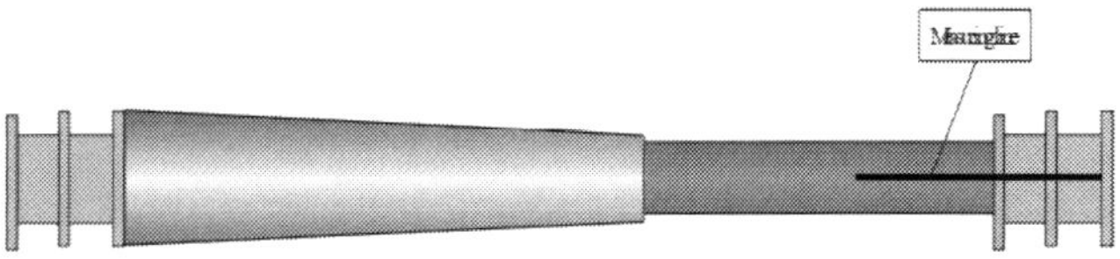

**Figure 6 Full scale test pipe**

After having completed the fatigue test the pipe was dissected and all components were inspected and found to be in good condition. See Figure 15 and Figure 16

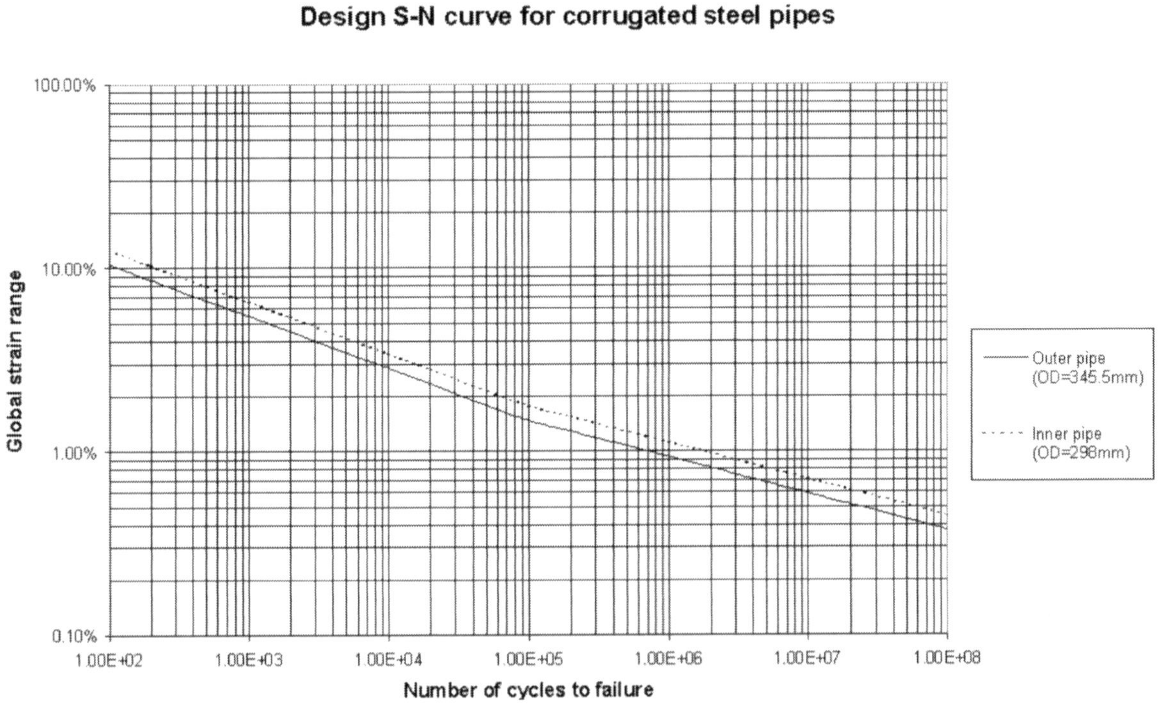


**Figure 7 Design SN-curve for corrugated stainless steel pipes.**

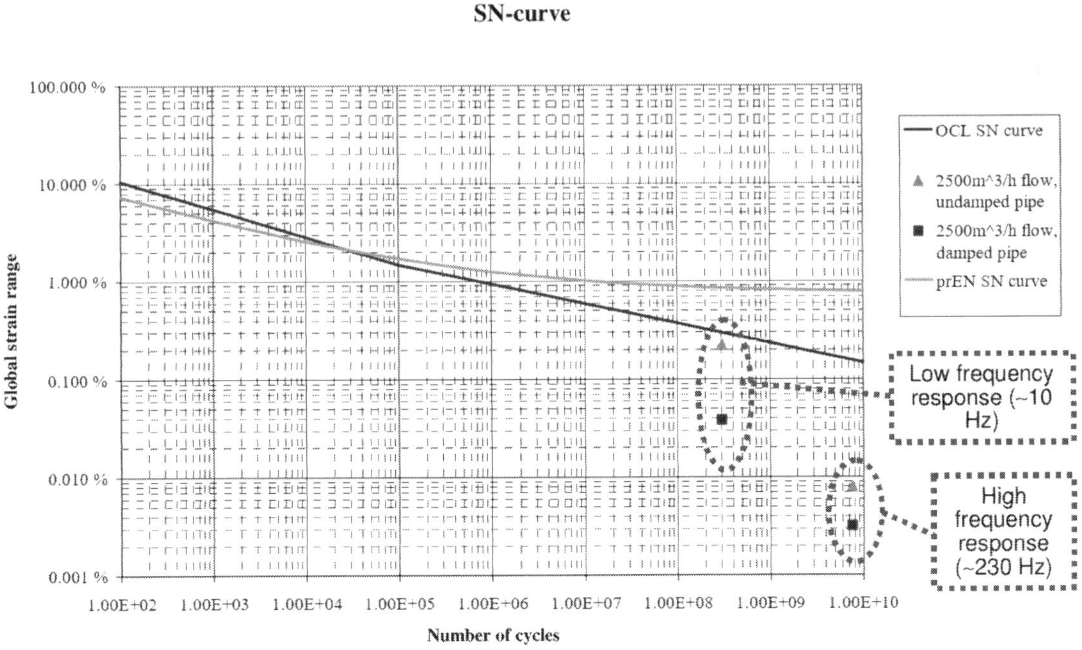


**Figure 8 Flow induced vibration test results with water flow.**

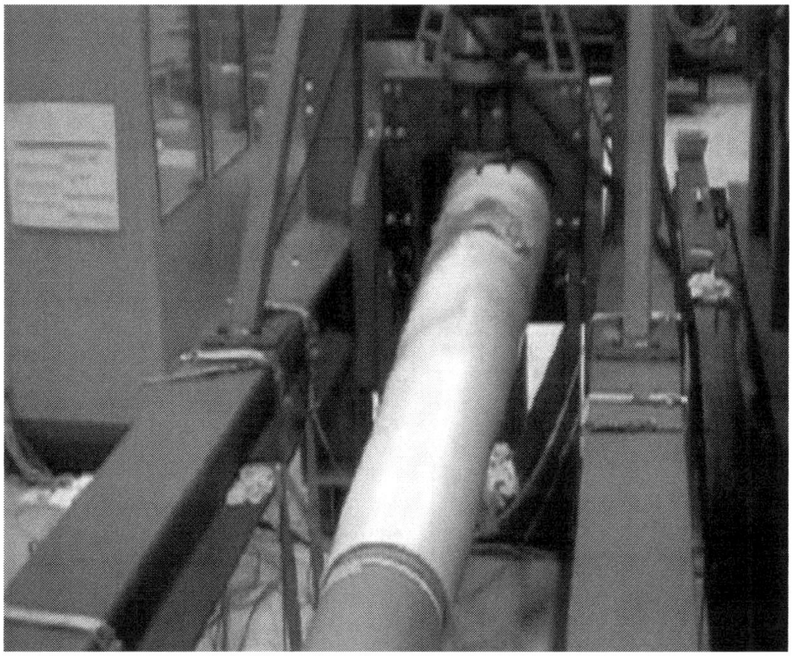

**Figure 9 Full scale fatigue test of Cryodyn™ pipe with liquid nitrogen inside at Marintek.**

**Figure 10 Flow test with water through bent pipe. Tested up to 2500m$^3$/hour. The sag was intentially included to simulate suspended and bent pipe.**

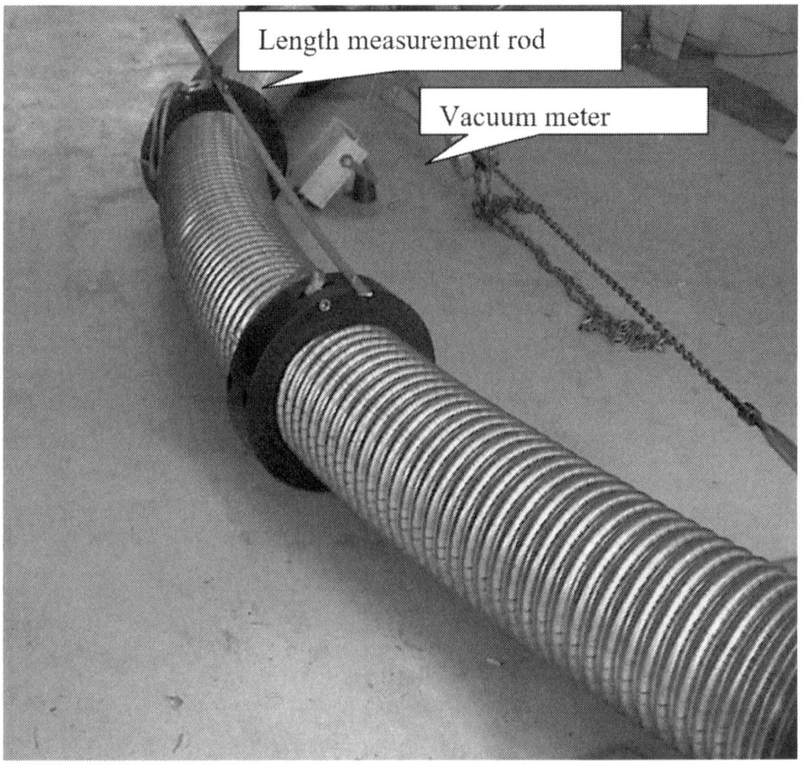


**Figure 11 Pressure test of bent pipe**

**Figure 12 Tension and compression test of single wall, corrugated pipes. Un-tested sample shown in the middle.**

Figure 13 Dynamic torsion test of single wall corrugated pipe.

**Figure 14 LNG flow test of Cryodyn™ with bend stiffener at Ebara, USA**

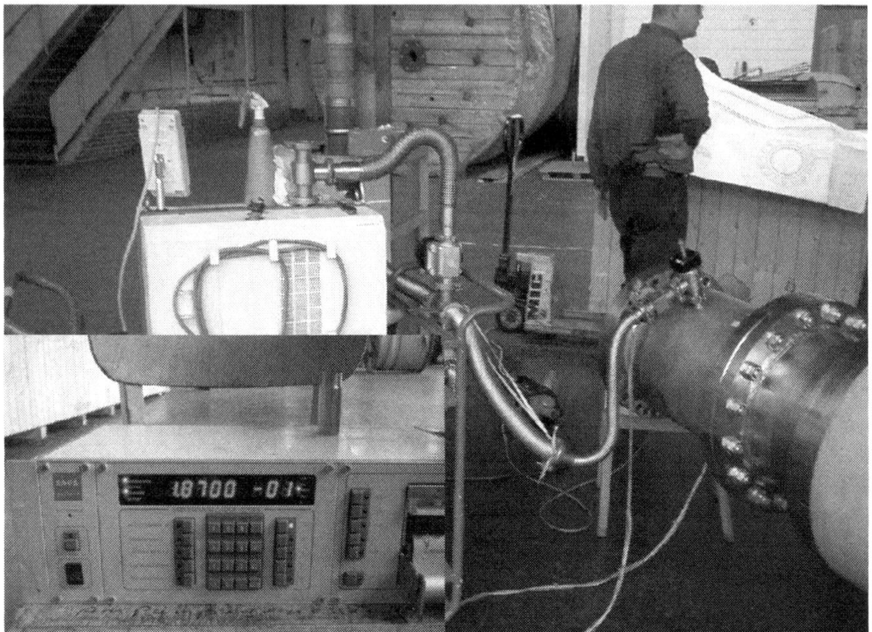

**Figure 15 Vacuum test after fatigue testing. Vacuum found to be $1.9*10^{-1}$** bar.

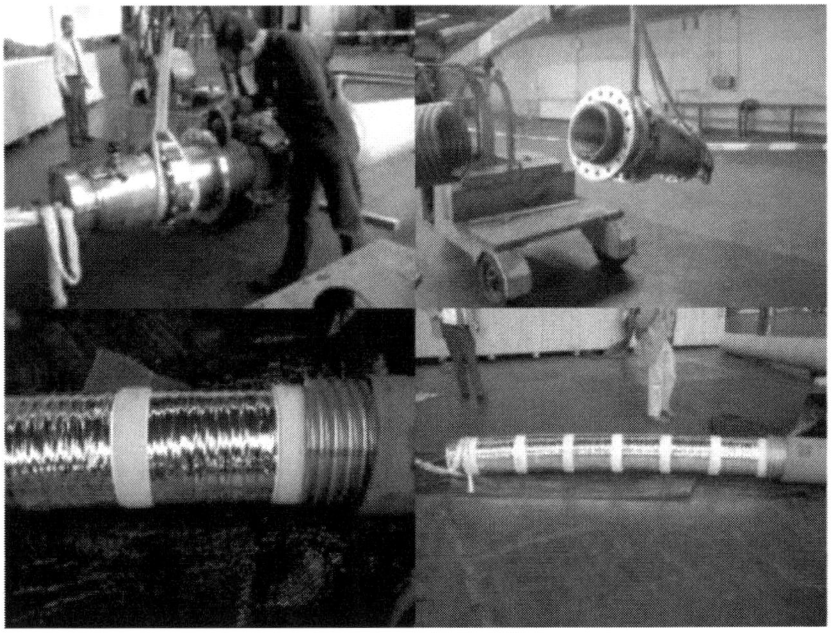

**Figure 16 Pipe dissection after fatigue test. No damage found on any components.**

# Microchannel Heat Exchanger for Liquefying Natural Gas

Ravi Arora, Anna Lee Tonkovich, Laura Silva
Velocys, Incorporated, 7950 Corporate Boulevard, Plain City, OH 43064
Prepared for Presentation at 2004 AIChE Spring National Meeting, April 25-29
Fourth Natural Gas Utilization Topical Conference


## Abstract

Velocys has developed a new approach to cryogenic heat exchange with the potential to significantly reduce heat exchanger size and refrigerant compression requirements. Current liquefied natural gas (LNG) production processes include the steps of compressing a refrigerant and flowing it through a spiral-wound or brazed aluminum heat exchanger. The capacity of a single-train LNG liquefaction cycle is generally limited by the capacity of compression equipment, the maximum size of the liquefaction heat exchanger, or both. Velocys is applying microchannel process technology to develop multi-stream heat exchangers, using massively parallel channels to achieve the desired capacity. The heat duty is accomplished in channels of short length, which decreases the refrigerant pressure drop and therefore reduces compression requirements. Additionally, the overall size of the heat exchanger is significantly reduced, leading to opportunities for offshore LNG development. This paper will present the results of a rigorous thermal design and flow sheet analysis for a single, mixed-refrigerant LNG loop.

## Introduction

As liquefied natural gas (LNG) becomes an increasingly attractive option for transporting natural gas, LNG producers desire to increase the capacity and reduce the costs of LNG production trains. LNG production is governed by the liquefaction train size which is usually limited by the maximum capacity of the available compressor. LNG facilities are generally optimized to yield the maximum production from a single, large, high reliability compressor (e.g., GE Frame 7).

In some cases, the size of the liquefaction (cryogenic) heat exchanger is the limiting constraint. The commonly used liquefaction heat exchangers are spiral wound heat exchangers and brazed plate fin aluminum heat exchangers. Spiral wound heat exchangers are the most commonly used type but the size of a spiral wound heat exchanger is restricted by transportation constraints for some facility locations. Brazed plate fin heat exchangers are much smaller but their size could be a limitation for offshore LNG plants or other highly space constrained locations.

In summary, compressor capacity and/or heat exchanger size generally limit the production capacity of a natural gas liquefaction train. This paper discusses how microchannel heat exchanger technology improves the production of LNG by greatly reducing the size of the heat exchanger while reducing overall compression requirements.

**Microtechnology**

Microtechnology is one of the fastest growing technologies. The origin of microsystems can be traced back to the first silicon sensor and actuator developed almost half century ago. With the development of microfabrication techniques, the field of microelectro-mechanical systems emerged. Now, novel microengineered structures can be made that perform many of the standard operations of interests to chemical systems. The chemistry involved in the chemical reaction is scale independent but the transport phenomenon is not. Microchannel process technology intensifies heat and mass transfer rates by reducing heat and mass transfer distances and thus decreasing transfer resistances between process fluids and channel walls. The review paper by Gavrilidis [1] reviews new advancement and current status of microengineered reactors.

Velocys has designed and built more than 50 microchannel devices for chemical reaction and heat exchanger applications. Velocys' intellectual property portfolio for microchannel process technology is unmatched and includes 25 issued patents, 50 submitted patent applications, and numerous paper publications [2], [3], [4]. Recently Velocys was issued a patent on novel microchannel heat exchanger design for liquefying natural gas [Patent #6,622,519][5].

**Microchannel Heat Exchanger**

A microchannel heat exchanger employs circular or polygonal microchannels as flow passages. A microchannel is a channel with smallest dimension less than 2 mm. Figure 1 shows the comparison between a microchannel heat exchanger and conventional heat exchanger, which typically have channels with minimum flow path dimensions greater than 10 mm.

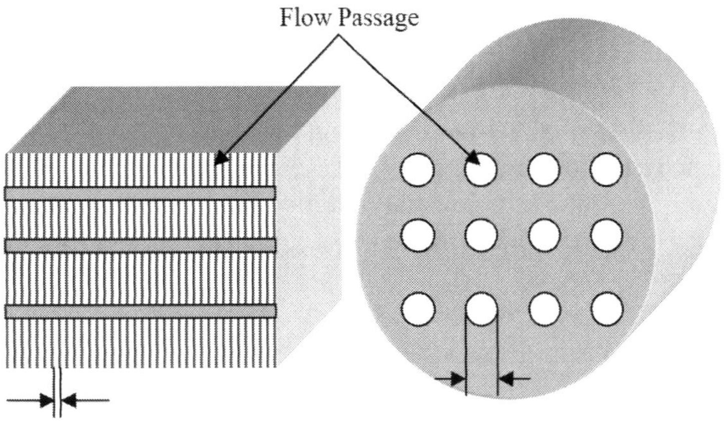


A microchannel heat exchanger offers excellent convective heat transfer performance. The dense channel packing provides high surface area per unit volume. Additionally, the small dimension of microchannels enable high convective heat transfer coefficients, because the coefficients are inversely proportional to hydraulic diameter, for a constant Nusselt number (Eq-1).

Microchannel heat exchanger can also offer low pressure drops. High heat transfer properties results in shorter channel lengths for a given heat transfer. Also, in a microchannel heat exchanger, the flow is distributed over a large number of microchannels. This distribution reduces the flow velocity and pressure drop in each channel (Eq-2). Therefore, the shorter length combined with distributed flow gives low pressure drop in microchannel heat exchangers. Figure 2 illustrates factors that contribute to lower pressure drop in microchannel heat exchanger than conventional technology.

$$h = \frac{Nu \times k}{d} \qquad \textbf{Eq-1}$$

$$\Delta P = 4f \frac{L}{d} \frac{\rho v^2}{2} \qquad \textbf{Eq-2}$$

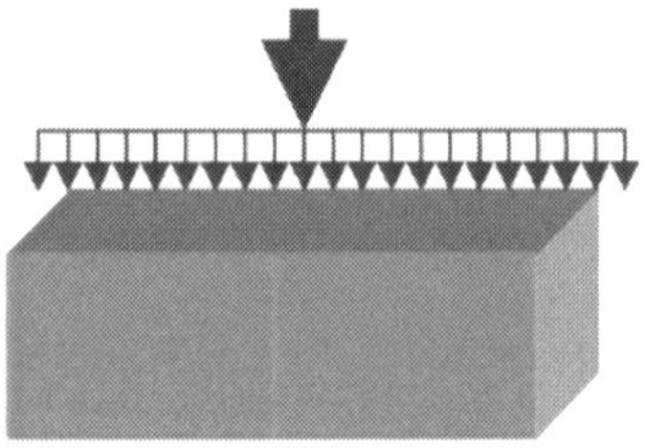

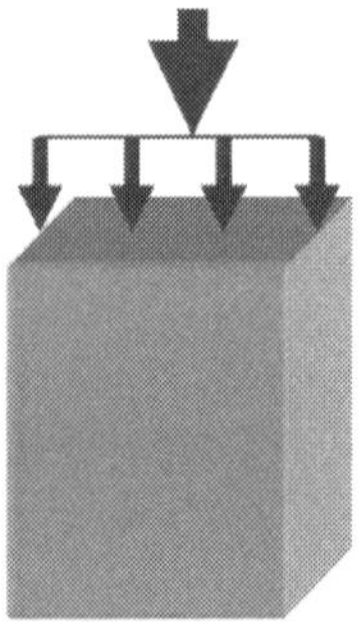

| **Microchannel Heat Exchanger** | **Conventional Heat Exchanger** |
|---|---|
| • **High heat transfer**<br>○ Small hydraulic diameter<br>○ High surface area/volume ratio<br>○ High heat transfer per unit volume<br>• **Low Pressure drop**<br>○ Distributed flow<br>○ Short length | • **Low heat transfer**<br>○ Large hydraulic diameter<br>○ Low surface area/volume ratio<br>○ Low heat transfer per unit volume<br>• **High Pressure drop**<br>○ Bulk flow<br>○ Long length |

Figure 2: Microchannel heat exchanger advantages

**Technology Status**

Commonly used liquefaction heat exchangers are brazed plate fin aluminum heat exchangers and spiral wound heat exchangers. The length of brazed plate fin heat exchangers can range from 6 meters to 15 meters while the spiral wound heat exchanger can be up to 50 meters long. For these exchangers, the minimum dimension of the flow path is greater than 10 millimeters and the available surface area per unit heat exchanger volume ranges from 500 to 1500 $m^2/m^3$.

Aluminum is the commonly used construction material for LNG heat exchangers because of its high thermal conductivity. One shortcoming of using highly conductive material, like aluminum, for heat exchanger material is the loss of thermal efficiency due to axial conduction. Axial conduction is the transfer of heat from hot end of the heat exchanger to its cold end. Typical temperature difference between hot and cold end of a liquefaction heat exchanger in LNG plants is between 100 to 200°C. Axial conduction can be quantified by a dimensionless parameter λ (Eq-3). The axial conduction is proportional to thermal conductivity of the metal and axial cross-sectional area and inversely proportional to length of heat exchanger. Therefore, as process channel sizes in conventional LNG heat exchangers are reduced, axial conduction increases. Offsetting this axial conduction requires lengthening heat exchangers which increases pressure drops and compressor power requirements.

$$\lambda = \frac{kA_c}{LC_{\min}} \qquad \text{Eq-3}$$

Velocys has developed microchannel based heat exchangers fabricated by a thermal diffusion bonding method. The heat exchanger is built by stacking thin metal sheets with embossed channel geometries. The stack is then subjected to high pressure and high temperature to diffuse metal from one sheet to adjacent sheets. The result is a continuous solid stack with internal microchannels of specific dimensions, regularity and connectivity. Figure 3 shows a microchannel device being stacked at Velocys. Figure 4 shows a thermal diffusion bonded device developed by Velocys.

Figure 3: Stacked Velocys device

Figure 4: Thermal diffusion bonded device by Velocys

These microchannel heat exchangers have high heat transfer surface area per unit volume generally, greater than 1500 $m^2/m^3$, and have high convective heat transfer coefficients. The excellent heat transfer performance of microchannel heat exchangers enables use of less conductive metal which reduces axial conduction and shortens heat exchanger length. Typically, microchannel heat exchangers are 5 to 20 times smaller than conventional brazed aluminum heat exchangers.

Other companies have developed micro/milli scale heat exchangers formed by diffusion bonding plates. In contrast to Velocys' heat exchanger design, these exchangers are typically constructed of aluminum and have longer flow paths. Thus, the micro/milli heat exchangers offer significant size reduction but do not provide a pressure drop advantage.

## Case Study

A case study was performed to compare the size and performance of a conventional brazed aluminum heat exchanger to a Velocys microchannel heat exchanger for a single mixed refrigerant LNG process using multi-stream heat exchangers. A simple schematic of the heat exchanger is shown in Figure 5. The high pressure refrigerant stream (HPR) and the natural gas stream are cooled by counter flowing low pressure refrigerant stream (LPR). The inlet temperature and pressure of different streams is shown in Figure 5. The required outlet temperature of liquefied natural gas is -153.9°C and the capacity is 7.7 MMSCFD of natural gas.

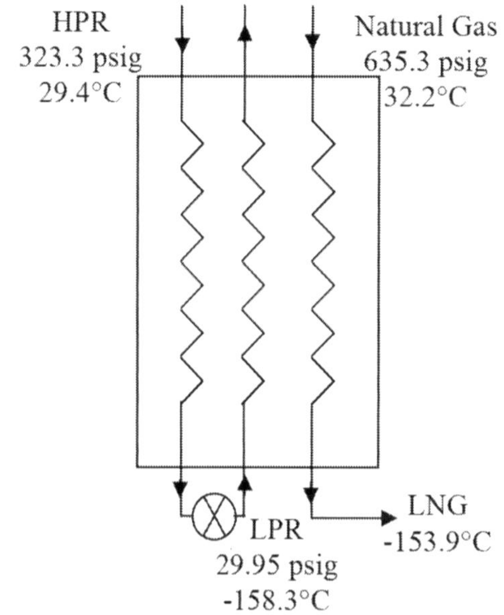

Figure 5: Schematic of single mixed refrigerant heat exchanger

In the case study, four conventional brazed aluminum cores were required to cool and liquefy natural gas to the required temperature. The total volume of four brazed aluminum heat exchanger cores was 21.7 $m^3$. A microchannel heat exchanger with 4.85 $m^3$ core volume was sufficient to replace all four brazed aluminum heat exchanger cores. Figure 6 shows the comparison between brazed aluminum heat exchanger cores and microchannel heat exchanger core.

The Velocys microchannel heat exchanger design was based on using stainless steel 304 as the construction material. Using stainless steel reduces the axial conduction effect. For the same cooling capacity, the ineffectiveness caused by axial conduction is <1% if stainless steel is used compared to 5% if aluminum is used as construction material for this microchannel heat exchanger design.

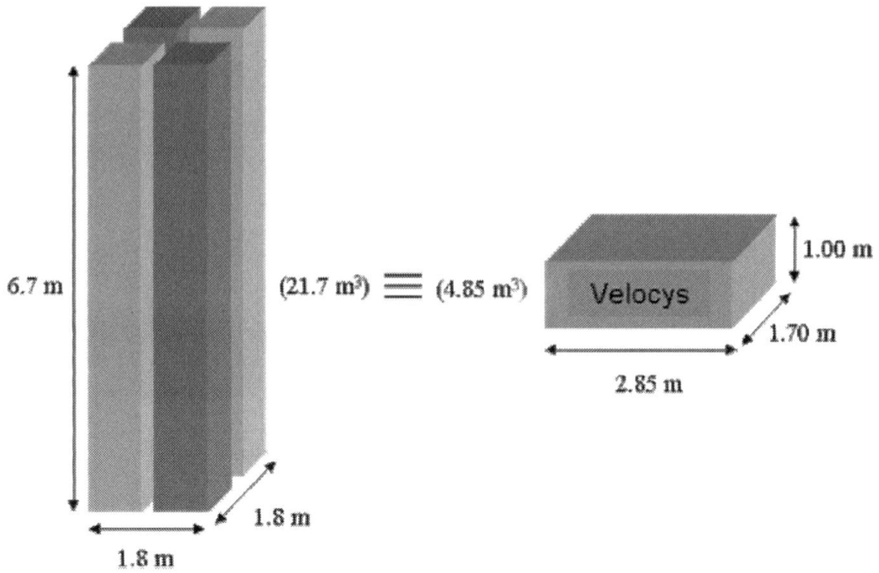


Figure 6: Size comparison between brazed aluminum and microchannel heat exchanger cores

Another significant advantage offered by microchannel heat exchangers is reduced pressure drop. The pressure drop in low pressure refrigerant stream was compared between brazed aluminum and microchannel heat exchanger. Table 1 shows the pressure drop comparison.

**Table 1: Comparison of Pressure Drop in LPR Stream**

| | Brazed Aluminum Heat Exchanger | Microchannel Heat Exchanger |
|---|---|---|
| LPR Stream Pressure Drop | 10.0 psi | 2.1 psi |
| Compression Ratio | 10.0 | 8.2 |

Compressor operating power is directly proportional to the pressure drop in LPR stream. In comparison to a single mixed refrigerant LNG process using brazed aluminum heat exchanger, the system using microchannel heat exchanger can reduce compressor power by an estimated 18-22%.

## Summary

Microchannel heat exchanger technology offers significant benefits over conventional heat exchange technology. Microchannel technology provides high heat transfer properties and low pressure drops which reduce overall system size and power consumption. The high heat transfer properties enables use of low conductivity material to reduce axial conduction effect in short heat exchangers.

## Nomenclature

$A_c$ = axial cross-sectional area, $m^2$
$C_{min}$ = minimum heat capacity, W/K
d = hydraulic diameter, m
f = fanning friction factor
h = heat transfer coefficient, $W/m^2/K$
k = thermal condutivity of fluid, W/m/K
L = length of heat exchanger, m
Nu = Nusselt number
v = velocity of fluid, m/s
$\rho$ = density of fluid, $kg/m^3$
$\Delta P$ = pressure drop in fluid, Pa
$\lambda$ = dimensionless axial conduction parameter

## References


1. Gravrilidis A. etal, "Technology and Applications of Microengineered Reactors", Institution of Chemical Engineers, Trans. ChemE, Vol 80, Part A, January 2002.
2. Tonkovich AY, JL Zilka, MJ Lamont, Y Wang, RS Wegeng. 1998. "Microchannel Reactors for Fuel Processing Applications. I. Water Gas Shift Reactor." *Chemical Engineering Science.*
3. Tonkovich AY, CJ Call, DM Jimenez, and MK Drost. 1996. "Microchannel Heat Exchangers for Chemical Reactors." *Proceedings of the 1996 National Heat Transfer Conference*, Houston.
4. Tonkovich AY, JL Zilka, DM Jimenez, MJ Lamont, and RS Wegeng. March 1998, "Microchannel Chemical Reactors for Fuel Processing",AIChE 1998 Spring National Meeting, New Orleans. PNNL-SA-29747
5. Patent # 6,662,519, "Process for Cooling a Product in a Heat Exchanger Emplying Microchannel for the Flow of Refrigerant and Product"

# HiLoad LNG Regasification Facility

## AIChE Spring National Meeting
## April 25 - 29, 2004, Hyatt Regency, New Orleans, Louisiana

**Author: Claes W. Olsen, VP Business Development, Remora Technology, Houston**

## Summary

This document gives a status and an overview of the work carried out with respect to using the HiLoad technology for an LNG import and regasification facility.

The HiLoad LNG Regas facility will accommodate any LNG carrier, without modifications or special arrangements, with regasification rates of as high as 2 BCFD. It will be possible to operate safely in Hs=4,5 m.

The HiLoad will provide a very cost competitive solution, with a regasification cost of <10 ¢/MMBTU.

## Introduction

The HiLoad® Technology development was started in 2000, and has been carried out by HitecVision in co-operation with ConocoPhillips. HitecVision subsidiary Remora Technology was established in 2002 in order to further develop and commercialize the HiLoad® Technology.

The **LNG Regas** version of HiLoad has been based on the detailed engineering work carried out for an oil loading version. The "chassis" (hull) and most of the systems onboard HiLoad will be very similar to the oil version. The HiLoad Technology has been properly presented at previous occasions, and reference is made to a vast amount of back-up information that can be made available upon request. This paper will mainly focus on how an LNG regasification plant has been incorporated in the HiLoad design.

The HiLoad LNG Regas Facility has been designed with particular focus on:

- Use of non-dedicated ships
- Safety and reliability
- Use of proven technology
- Cost efficiency
- Flexibility

Import of LNG into the US has been used as an example, with a regas rate as high as 2 BCFD. Through a Department of Energy funded study, Remora is currently working on a solution where the HiLoad LNG Regas is used in combination with salt cavern storage for continuous supply.

The feasibility study and conceptual design carried out to date has verified that the solution is technically feasible with no identified showstoppers. DNV has been engaged as a 3rd party verifier, and Approval In Principle is expected to be issued shortly.

## System Description

The proposed layout for the system is shown below.

The HiLoad can be docked onto any LNG carrier, where the HiLoad is equipped with regasification equipment of up to 2 BCFD capacity.

The system comprises the following main elements:

- HiLoad Unit
- SALM Buoy
- PLEM and Swivel
- Flexible Risers
- Remote Power and Control System

The layout shown below is the base case with the use of a SALM buoy and bow mooring of the LNG carrier since this solution can be utilized for all the LNG carriers operating in the world.

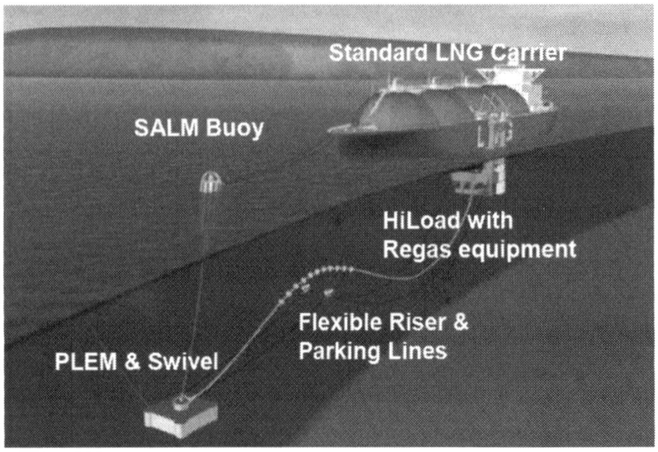


**Figure 1 - Layout of HiLoad and SALM buoy**

### *HiLoad Unit*

The HiLoad Technology is built up around a semi-floating L-shaped loading terminal that can dock onto any ship in a similar way as a forklift picks up a pallet. The HiLoad is equipped with thrusters, and will be easy to maneuver into position on a slow-moving or moored ship. The HiLoad will attach to the ship using buoyancy as well as the hydrostatic pressure present at the draft of the vessel's bottom. It has been properly documented through analysis, model testing and Class review that the HiLoad will not cause any damage to the ships.

The main components of the LNG Regas System on the HiLoad are shown on Figure 2. For a description of each of these components, reference is made to separate sections of this report.

Most of the LNG Regas equipment is located outside the HiLoad chassis as shown in the figure above. Compared to the basic oil version, additional buoyancy has been added as required due to the weight of the LNG regas equipment.

The hull is divided into three main sections: pontoon, towers and keel. It is constructed of steel plates and standard profiles for simple fabrication at any yard. The room layout is designed according to the DNV Rules for Mobile Offshore Units. The General Arrangement is designed with symmetry about the centerline of the unit, which gives redundancy and simpler fabrication.

The unit is equipped with a high capacity ballast system in order to keep the time needed to connect at a minimum (about 2-3 min).

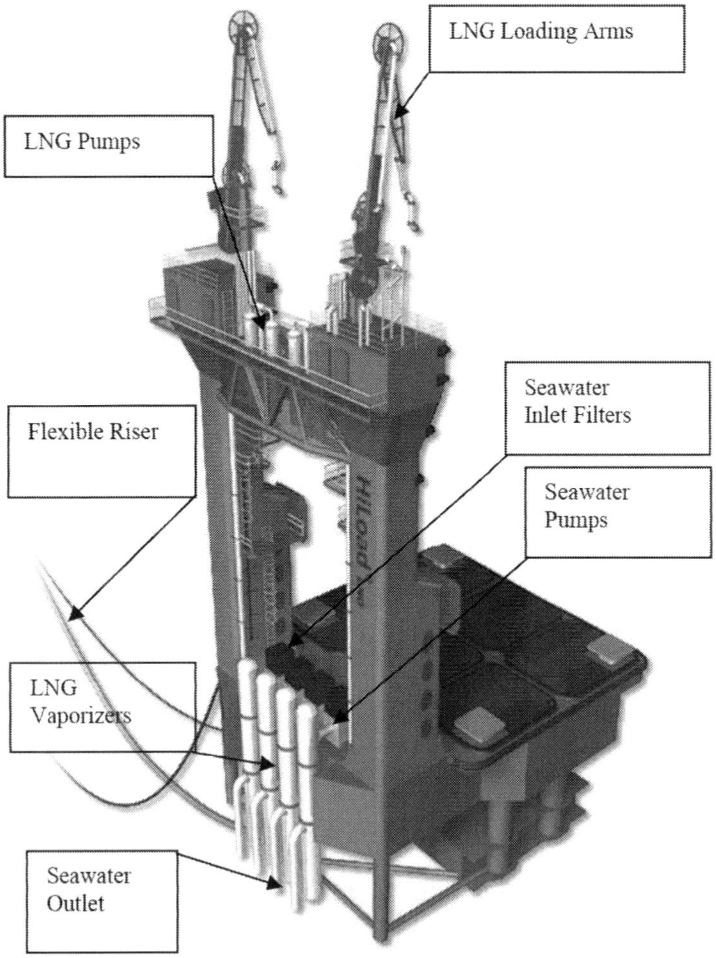


**Figure 2 - HiLoad Unit, Main LNG Regas Systems**

The top of the pontoon is equipped with a patented friction attachment system, which generates a strong grip onto the carrier resulting in a safety factor of 10-15 depending on the draft of the carrier. The system consists of standard and well proven equipment which has been full-scale tested on a ship. A considerable amount of documentation is available.

The pontoon and towers on the HiLoad are equipped with soft shock absorbing fenders to limit the contact forces between HiLoad and the carrier during connection. The tanker interface has been addressed in significant detail during the development, including assessment of:

- Optimum waterline area for towers
- Cushion effect between pontoon and ship bottom
- Keel design to increase damping and reduce motions
- Thrusters used for motion damping
- Fenders designed according to international standards

Docking and undocking onto tankers have been model tested in wave heights of Hs=4,5m with very satisfying results.

## Position Keeping System

It has been selected to use a traditional SALM (Single Anchor Leg Mooring) buoy for mooring of the LNG carrier during offloading. In addition the azimuth thrusters on the HiLoad can be used for reducing any vessel movements, and the snap loads acting on the SALM mooring system.

The SALM buoy is a proven single point mooring system used worldwide for 30 years for mooring of crude oil tankers. The distance between the LNG carrier and the SALM buoy will be approximately 70-80 m. One or two mooring hawsers will be connected to the OCIMF strong point(s) in the bow of the LNG carrier.

The SALM system will take most of the environmental loads acting on the LNG carrier like a normal single point mooring system. However, the thrusters on HiLoad can be utilized for reduction of vessel movements and to reduce the snap load acting on the SALM mooring system, i.e. the system can act as "thruster assisted mooring system".

The SALM alternative will make use of the OCIMF "Emergency Towing Bracket" on the LNG carrier for connection of the bow mooring hawser. The classification society Det Norske Veritas (DNV) has carried out an evaluation regarding use of this system for connection of the mooring hawser to the SALM, concluding that it is an acceptable solution.

A SIMO simulation of Position Keeping System for HiLoad and an LNG carrier has been carried out by Marintek with very promising results.

## LNG Vaporizers

The HiLoad LNG Regas Facility will be equipped with direct seawater heated shell and tube vaporizers. This type of vaporizers is available from a number of suppliers.

Photos of a direct seawater heated shell and tube LNG vaporizer are shown in Figure 3.

**Figure 3 - LNG Vaporizer (Chicago Power & Process)**

This type of LNG vaporizer has a unique and robust design with no movable parts. A minimum of maintenance is required which makes them very suitable for the HiLoad system.

The LNG vaporizers will be installed vertically (as recommended and approved by the manufacturers) and located below seawater level during normal operation. Advantages with this location are:

- Power demand for the seawater pumps will be reduced with at least 50 % due to no "lifting height"
- Increased safety, due to submergence and not close to deck on LNG carrier. The vaporizers are located on the outboard side of HiLoad, facing away from the LNG carrier.

The LNG vaporizers will be made of material AL-6XN or equivalent. This material has excellent corrosion resistant properties and has a very long track record in warm seawater (20 years, 50 million feet of tubes).

Limiting the marine growth inside the vaporizers can be done through 2-way flushing, and by introducing a hypochlorite solution to the seawater. Further evaluation of this subject will have to be carried out in each specific case based on input data such as seawater quality, temperature, etc.

In general, the performance of seawater heated LNG vaporizers will depend on the inlet seawater temperature, the temperature drop trough the vaporizer and the seawater flow rate.

The LNG vaporizers will be able to regas 6 x 0.333 BCFD with an inlet seawater temperature of 20 °C, representing the minimum seawater temperature for a typical US Gulf Of Mexico location.

The seawater flow required for vaporizing 0.333 BCFD at this temperature will depend on the minimum seawater outlet temperature as recommended by the different suppliers, in order to avoid internal ice formation.

However, US environmental authorities have set a limit of 20°F for the maximum sea water temperature drop.

The 2 BCFD regasification capacity requires a seawater flow of approximately 5 000 m³/h for each vaporizer or 30 000 m³/h in total, based on a delta T = 20 °F.

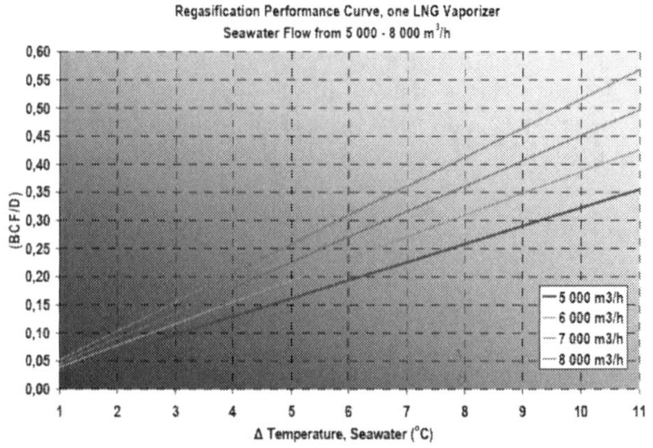


**Figure 4 – Performance Curve, LNG Vaporizer**

Use of seawater as heating source for vaporizing of LNG is a very economical and environmental friendly solution because it will avoid burning gas for the heating process as is normal for many of the LNG import terminals in the United States. By using gas for heating, approx 1.8 % of the imported gas is used for heating. There is hence a considerable cost saving to use seawater.

Local cool-down of the sea needs to be studied for the given LNG terminal location. Water depth and current are two important parameters in this respect. In areas with strong current, the local cooldown will be less of an issue as "new warm seawater" will continuously flow to the terminal. With low current and shallow water (10-20 m, typical for a GBS-terminal), the local cool-down may be an issue. The HiLoad will typically require a minimum water depth of 60 m for acceptable working conditions for the flexible riser. At this water depth it is expected that the effect of local cool-down will not be an issue, but this will naturally be addressed in each specific case.

## LNG Pumps

The HiLoad will be equipped with six vertical, vessel contained, submerged-motor high-pressure pumps located above the splash zone, protected on a deck between the two towers.

The pumps are cooled and lubricated by the product flow and are therefore vulnerable to loss of flow. The pumps are therefore equipped with protection devices to avoid dry running of the pump.

The total regas rate of 2 BCFD gives a required LNG flow of **3 550-3 720 m$^3$**/h based on LNG with specific gravity of 0,48 – 0,45 respectively.

## Sea Water System for Heating of Vaporizers

Seawater is used to heat the Liquefied Natural Gas from -163°C to 5°C. This requires a considerable large amount of seawater for the heating process through the LNG vaporizers. It is vital for the process that the seawater flow trough the vaporizers is maintained, to avoid freezing.

Four submerged seawater pumps will be located on top of the HiLoad pontoon. They are located near the LNG vaporizers in order to keep the amount of large diameter piping to a minimum. During the regas operation the pumps and strainers will be submerged. This will give significant power saving (> 50 %) compared to a ship or platform deck solution due to no "lifting height" for the pump.

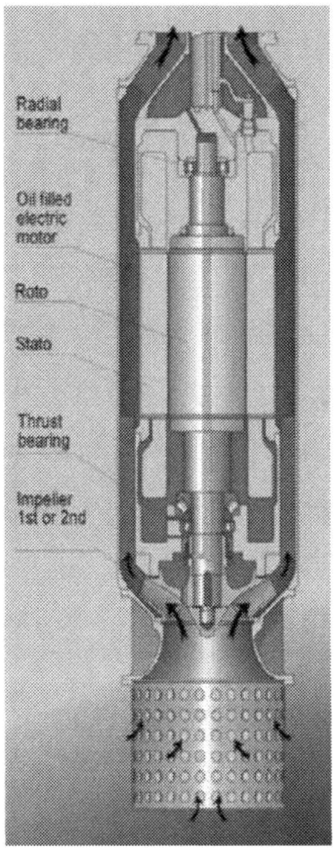


**Figure 5 - Sea Water Pump and El. motor (Framo type)**

The seawater pumps are designed with capacity to give full regas rate with a seawater temperature of 68 °F with a total flow rate of **30 000 m$^3$/h**.

In order to avoid any risk of freezing in the LNG vaporizers, it is important to design a robust and redundant seawater system. It has therefore been concluded that the four pumps shall be installed with three pumps operating and one spare, where 3 pumps shall be able to supply sufficient seawater to ensure full regas capacity at the defined seawater temperature. A solution is that each pump shall be designed for **9 700 m$^3$/h** with a total maximum flow rate of 38 800 m$^3$/h when all 4 pumps are running. The seawater flow from all the pumps will be pumped into a common manifold before the flow enters into the LNG vaporizers. Monitoring of the seawater flow and outlet temperature will also be designed with full redundancy.

The pumps are dimensioned for a pressure drop through the vaporizer and pipes of 1.5 bar. The pressure drop through the vaporizer varies for the different suppliers, but the pumps are designed for the vaporizer that has the highest pressure drop.

The electric motors for the seawater pumps are suitable for subsea use; ref. Figure 5. The pumps and strainers can be accessed for inspection and maintenance when the HiLoad is deballasted to service draft (1m freeboard for the pontoon).

Each seawater pump will be equipped with an automatic self cleaning strainer. The strainer is suitable for subsea operation.

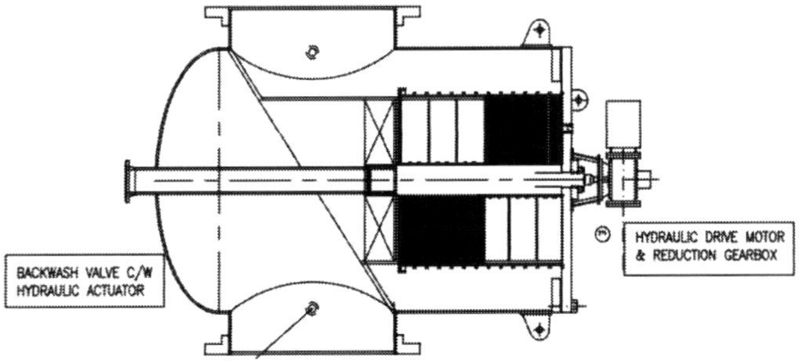


**Figure 6 – Automatic self cleaning strainer**

The strainers can be made for the required capacity and are designed for continuous operation. Due to high filtration requirement for filter screen (3000μnm, normally 6000-8000μnm), an automatic self cleaning strainer is used. A hydraulic drive motor initiates the backwashing of the filter. This can be done in controlled manner at pre-decided intervals or when the pressure differential transmitter indicates high differential pressure over the filter. Normal pressure drop through a clean filter is 0,16 Bar.

## LNG Loading Arms

Two standard LNG transfer arms will be installed on top of the towers on HiLoad, where each arm has sufficient flow capacity to give a full regas rate of 2 BCFD (LNG discharge rate of 3 720 m$^3$/h), giving 100% redundancy. It is planned to always connect both loading arms and run 50% flow through each of them.

Due to **no** relative motion between HiLoad and the LNG carrier after connection, the operation envelope for the LNG transfer arms on HiLoad will be considerably reduced compared to conventional Loading Arms for terminals. The deciding factor is the longitudinal docking accuracy of HiLoad onto the LNG carrier and the different manifold elevations for the carriers.

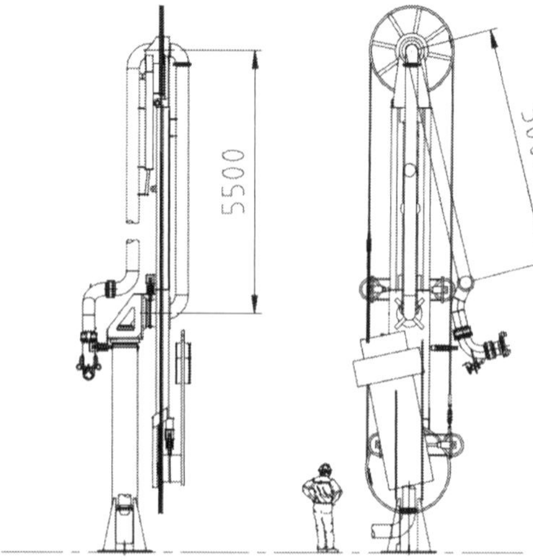


**Figure 7 – Proposed LNG Loading Arms for HiLoad**

The 3D illustration below is showing the LNG transfer arms connected to the midship manifold on the carrier.

**Figure 8 – HiLoad Loading Arms connected to the LNG Carrier**

The LNG Loading Arms will be equipped with emergency quick release couplers that allow release of the loading arms without LNG spill.

**Figure 9 – Typical Quick Release Coupling**

The system can be designed with a vapor return system. However, as all LNG carriers are equipped with LNG vaporizers, it is foreseen that the ship's vaporizer is put into operation to generate sufficient amount of gas required to maintained positive tank pressure during discharging and supply gas fuel to the boilers while being docked to HiLoad. For LNG discharge rates below 1,2-1,4 BCFD, the natural boil off will be sufficient to maintain the required over-pressure in the cargo tanks.

## Gas Export System and Swivel

A flexible gas riser system with 2 x 18 " risers will be fitted from HiLoad and down to the PLEM on the seabed. The approximate length is 200 m, but will naturally depend on the actual water depth. This type of riser is a standard product and can be manufactured by three different suppliers.

A typical construction of the flexible riser is shown on the figure below.

**Figure 10 – Typical Flexible Riser construction (NKT Design)**

A subsea swivel will be located on top of the PLEM, consisting of 2 paths for the high pressure gas, power & signal swivels.

The electrical part of the subsea swivel will be oil filled and pressure compensated.

The PLEM is the base for the SALM, subsea swivel and pipeline connection. The pipeline will be manifolded and the PLEM will be equipped with a 20" isolation valve on each of the gas pipeline connections to the swivel.

The soil conditions for the selected site will determine whether the PLEM shall be of pile or suction type. Exact design of the PLEM is site specific and detailed engineering will have to be carried out when the actual site is selected and the soil conditions are known.

## Power and control system

As base case, electric power for operation of HiLoad will be supplied from a nearby platform, about 2 km away from HiLoad. Power requirement for HiLoad will be 18 – 20 MW.

Due to the high power demand, 22 kV/60 Hz will be used for line voltage to HiLoad. The power system is based on a 3-phase system, insulated 3-wire system for the distribution systems of 6,6 kV, 690 V and 230 V.

The electrical plant onboard the HiLoad will be powered through a subsea cable from the platform to the PLEM and by two independent umbilicals connecting the PLEM and the HiLoad.

For power supply, control and optical signals, a static sea cable will be used from the platform to the PLEM and dynamic umbilicals clamped to the flexible riser will be used from PLEM and up to HiLoad.

The dynamic umbilicals from the PLEM to HiLoad are proposed with a redundant configuration using 2 umbilicals clamped to the flexible riser.

**Figure 11 - HiLoad Operator Control Console**

**American Institute of Chemical Engineers**
3 Park Avenue
New York, New York 10016-5991

**ISBN: 978-1-60423-477-0**